Recrystallization and Grain Growth

Volume One

Springer

Berlin
Heidelberg
New York
Barcelona
Hong Kong
London
Milan
Paris
Tokyo

Recrystallization and Grain Growth

Proceedings of the
First Joint International Conference
August 27-31, 2001
RWTH Aachen, Germany

Edited by
Günter Gottstein and Dmitri A. Molodov

Springer

Prof. Dr. Günter Gottstein
Institut für Metallkunde und Metallphysik
RWTH Aachen
52056 Aachen

PD Dr. Dmitri A. Molodov
Institut für Metallkunde und Metallphysik
RWTH Aachen
52056 Aachen

ISBN 3-540-41837-7 Springer-Verlag Berlin Heidelberg New York

Springer-Verlag Berlin Heidelberg New York
a member of BertelsmannSpringer Science+Business Media GmbH

Typesetting: Camera-ready copy from editor
Cover-Design: de´blik, Berlin
Printing and Binding: Druck- und Verlag Mainz, 52072 Aachen
Printed on acid-free paper SPIN: 10833756 62/3020 CU -5 4 3 2 1 0-

The First Joint International Conference on Recrystallization and Grain Growth
(ReX&GG)

Conference Chairman:

Günter Gottstein

International Advisory Committees:

ReX

G. Abbruzzese (Italy)
T. Chandra (Australia)
R. Doherty (USA)
M. Fuentes (Spain)
G. Gottstein (Germany)
F. J. Humphreys (UK)
N. Hansen (Denmark)
M. Hillert (Sweden)
J. J. Jonas (Canada)
O. A. Kaibyshev (Russia)
T. R. McNelley (USA)
F. Montheillet (France)
T. Sakai (Japan)
C. M. Sellars (UK)
D. J. Srolovitz (USA)
T. Watanabe (Japan)

GG

G. Abruzzese (Italy)
J. Driver (France)
G. Gottstein (Germany)
T. Y. Hsu (Xu Zuyao) (China)
F. J. Humphreys (UK)
W. B. Hutchinson (Sweden)
J. J. Jonas (Canada)
D. Juul-Jensen (Denmark)
K. Lücke
N. Ryum (Norway)
L. S. Shvindlerman (Russia)
D. J. Srolovitz (USA)
R. Vandermeer (USA)
T. Watanabe (Japan)
D. Y. Yoon (Korea)

Organizing Commitee:

D. A. Molodov
M. Goerdeler
J. Hirsch
U. Lotter
D. Mattissen
D. Raabe
M. Winning
S. Lakrache
Chr. Dieckelmann
H. Linckens
M. Wulfes
I. Zeferer

ReX&GG Sponsors:

Deutsche Forschungsgemeinschaft
Deutsche Gesellschaft für Materialkunde
Verein Deutscher Eisenhüttenleute
Alcan International Limited
Thyssen-Krupp Stahl AG
VAW aluminium AG

Preface

During heat treatment of crystalline solids recrystallization (ReX) and grain growth (GG) are the major processes that modify the granular microstructure besides phase transformations. Owing to their strong impact on material properties and subsequent processing these phenomena have been extensively investigated in the past with the first publication on ReX dating back to 1881, i.e. more than 100 years ago and at the same time Lord Kelvin (then Sir William Thompson) studies soap froth to determine curved interfaces and optimum position of space.

A huge amount of data has been collected in particular on *ReX&GG* of commercial alloys, and the fundamental underlying phenomena have been identified. Nevertheless, despite of all these efforts, there remains a substantial lack of understanding of the essential mechanisms of ReX and GG. This is mainly due to two reasons:

a) *ReX&GG* are not equilibrium phase transformations and, therefore, do not comply with equilibrium thermodynamics. Rather they are microstructural instability phenomena and, therefore, subject of reversible thermodynamics in terms of entropy production.

b) Instability phenomena proceed from locations which are untypical for the average microstructure of the material. In fact homogeneous microstructures strongly resist recrystallization and markedly slow down grain growth. The crucial microstructural features for nucleation of recrystallization have, in principle, been identified but their mechanisms are not understood. Equally the growth of recrystallized or coarsening grains are commonly perceived as grain boundary motion in a homogeneous environment, which may be far from the truth. Only very recently it was demonstrated that in connected grain boundary systems there are also other microstructural elements that may play an essential role, like grain boundary junctions which are comparable in number to the frequency of grain boundaries. Another issue is the entire negligence of the role of internal stresses on nucleation of recrystallization and grain boundary migration, and until recently, high angle grain boundaries were tacitly assumed not to couple with mechanical stresses at all and, therefore, to remain unaffected by internal stresses except for the stored dislocation energy acting as thermodynamic driving force for recrystallization.

Owing to an insufficiently deep understanding of *ReX&GG* and, therefore, the inability of making reliable predictions on microstructure evolution, in particular crystallographic texture, during heat treatment of deformed materials, this research field was essentially dormant for decades, and the collection of data bases rather than the development of predictive power was the guideline of materials producers.

In the beginning of the 1990's interest in this field was suddenly rekindled in particular due to the development of novel research tools, like high resolution

TEM, orientation mapping by EBSD, and computer simulation owing to the availability of high performance computers and new theoretical concepts of microstructure modeling.

Besides workshops and topical conferences like the RISØ International Symposium, International conferences on ReX and GG were initiated separately in Wollongong, Australia (1990) and Rome (1991), respectively. Conferences on ReX followed in San Sebastian, Spain (1992), Monterey, USA (1996), and Tsukuba, Japan (1999). The Rome conference on GG was succeeded by ICGG-2 in Kitabyushu, Japan (1995) and ICGG-3 in Pittsburgh, USA (1998).

While the ReX conferences were more strongly oriented towards applications, the GG conferences focussed on the scientific aspects of the evolution and behavior of granular microstructures. The complementarity of both approaches attracted essentially the same core audience so that both conference series decided not to merge but to have a joint conference with common sessions of general interest and specific topical sessions on the core issues of both conference series. The contributions to these proceedings are the manuscripts presented in both conferences and are designed to provide both, a general but comprehensive review and detailed information on current research in *ReX&GG.*

We hope the reader will find valuable information to guide his own work or to acquaint him with new concepts and modern issues in *ReX&GG.*

G. Gottstein and D.A. Molodov

Aachen, August 2001

CONTENTS

VOLUME ONE

Opening Address

Keynote Papers

Grain Growth

Invited Papers

Fundamentals, Modeling and Simulation of Grain Growth

Stagnation Effects in Grain Growth

Grain Boundary and Triple Junction Properties

Steels

Ultra-Fine Grain Structures

Advanced Experimental Methods

VOLUME TWO

Recrystallization

Invited Papers

Mechanisms and Kinetics of Recrystallization

Dynamic Recrystallization

Recrystallization Modeling

Steels

Aluminium Alloys

Non-Ferrous Alloys

Non-Metallic Materials

Author Index

Keyword Index

Opening Address

Main Steps in the Development of the Statistical Theory of Grain Growth

Kurt Lücke

Institut für Metallkunde und Metallphysik, RWTH Aachen, Germany

There is no question that the clarification of lattice defects – namely points defects, dislocations, and grain boundaries (GB) – has strongly contributed to the progress in metal physics. Strangely enough, the defects which are known for the longest time – namely the grain boundaries which are known longer than hundred years – are the ones which took longest to be understood. For this reason, at present time there are very few conferences on the fundamentals of point defects and dislocations, but frequently conferences on grain boundaries, like the present one.

Physical research on lattice defects started mainly after the last world war. The foundation of it for GBs was essentially laid by Cyril Stanley Smith, who worked first in the copper industry, than in the Manhattan Project, later became director of the Institute for the Studies of Metals in Chicago and after his emeritation institute professor at MIT. I am proud to have been one of his personal friends.

Already Smith divided the field of grain boundaries in two main subfields, the field of recrystallization and the field of grain boundary motion. Accordingly, also the following international conferences were dedicated mainly to one or to the other of these two subjects. Since there are considerable overlappings, in this conference both topics will be treated together.

But as the time for this introductory talk is too short to cover both areas, I will limit myself to the field of grain growth where I myself worked most in the last years. In this field there exist two main approaches: direct computer simulations and the statistical theory of grain growth, which was more of analytical nature. In the following a short survey about the main development of the statistical theory in 2D homogeneous materials shall be given (see Table 1).

The early statistical treatment of grain boundary kinetics were generally found to be in poor agreement with the experimental results. The agreement, however, improved with the number of parameters the values of which have been used to characterize the various grains.

According to Smith (1952) for an individual grain boundary one can describe the velocity of GB motion $v = - m\gamma K = -m\gamma/\rho$, where m is the mobility, (the specific energy, K the curvature and Δ the radius of curvature of an individual grain boundary. Burke (1952) and Beck (1954), in a very rough approach, treated a microstructure which they considered to consist of grains of all the same size given by the average grain size Ř. Assuming Ř to be also proportional to the radius of curvature Δ one obtains by introducing this into the equation for v for dŘ/dt a differential equation (Table 1) which gives as solution normal grain growth with the $t^{0.5}$-law. That law however has never been observed experimentally.

Hillert (1965), in his pioneering work, developed the first true theory by introducing the grain radius R for characterizing the different grains and thus obtained a grain size distribution function (SDF) φ(R). With uniform GBs he assumed that grains larger than a critical grain size are growing, and the smaller ones are shrinking (Fig. 1).

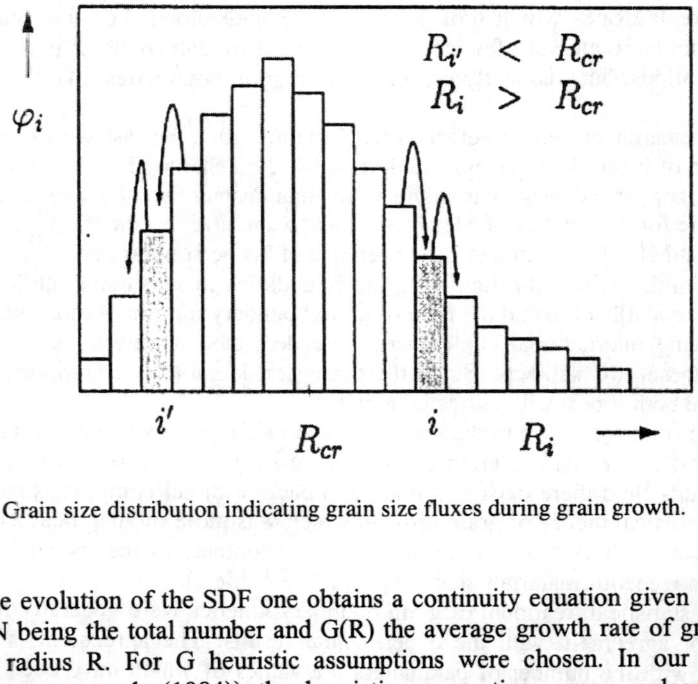

Fig. 1. Grain size distribution indicating grain size fluxes during grain growth.

For the evolution of the SDF one obtains a continuity equation given in Table 1, with N being the total number and G(R) the average growth rate of grains of the given radius R. For G heuristic assumptions were chosen. In our own work (Abbruzzese et al. (1984)) the heuristic assumptions were replaced by first principle topological relationships. The solution of the continuity equation yields normal grain growth leading again to the $t^{0.5}$-law and for long time an asymptotic self similar SDF (Hillert distribution) which also, however, never was observed experimentally.

Table 1. Main Steps in the Development of the Statistical Theory of GG

Characterization of Polycrystalline Samples			**Authors**	**Suppositions**	**Main Achievements**
Parameters: Number	Parameters: Nature	Distribution function			
0			Smith 1952	**Single GB** velocity v, mobility m, energy γ	$v = -m\gamma K$ $= -m\gamma/\rho$
		Mean radius: Ř	Burke 1952 Beck 1954	**Rough approach:** $\frac{d\check{R}}{dt} \propto \frac{1}{\check{R}}$	**Normal GG** $\check{R}^2(t) - \check{R}^2 \propto t$ ($t^{0.5}$ - law)
1	Radius: R	Size distribution function SDF: $\phi(R)$	Hillert 1965 Abbruzzese Lücke 1984	**Uniform GBs** Continuity equation: $\frac{\partial[N\phi(R)]}{\partial t} = -\frac{\partial[N\phi(R)G(R)]}{\partial R}$ For G(R) heuristic assumptions	**Normal GG** $t^{0.5}$ – law Self similar (Hillert) distribution $\phi_H(R/\check{R})$
2			v. Neumann 1952 Mullins 1956	**Single grain** 120° angles at vertices, grain area A	$\frac{dA}{dt} = \frac{m\gamma\pi}{3}(n-6)$
	Radius: R Number of sides: n	Size coordination distribution function SCDF: $\phi(R,n)$	Abbruzzese Lücke Heckelmann Brandt 1984-1999	**Uniform GBs** 120° angles at vertices Continuity equation: $\frac{\partial[N\phi(R)]}{\partial t} = -\frac{\partial[N\phi(R)G(R,n_i)]}{\partial R}$ For $G(R,n_i)$ first principles: Topological relationship $n = n(R)$	**Normal and abnormal GG** Time evolution of SDF: $\phi(R,t)$, $\check{R}(t)$ $t^{0.5}$ – law General self similar distribution
3	Radius: R Number of sides: n Orientation: g	Size coordination orientation distribution function SCODF $\phi^g(R,n)$	Lücke Abbruzzese Eichelkraut Brandt 1986-1999	**Non uniform GBs** Equilibrium angles β at vertices Set of continuity equations: $\frac{\partial[N\phi^g(R)]}{\partial t} = -\frac{\partial[N\phi^g(R)G^g(R,n_i)]}{\partial R}$ For $G^g(R)$ first principles: Use of topological functions	**Abnormal GG** Time evolution of SDF and ODF: $\phi^g(R,t)$, $\check{R}^g(t)$ No $t^{0.5}$ - law No self similar distribution

A further important progress was achieved by v. Neumann (1952) and Mullins (1956) who recognized that G not only depends on R but also on the number n of sides of grains. I.e. one obtains G(R,n) and the size coordination distribution function (SCDF) $\phi(R,n)$. With 120° angles at the vertices one could than calculate the average curvature of the grain boundary pieces and could thus calculate the value of G(R,n). In our own work (Abbruzzese et al. (1992)) we introduced these relations G(R,n) and $\phi(R,n)$ instead of G(R) and $\phi(R)$ into the continuity equation (see Table 1), calculated the topological relationship n(R) and averaged finally over all n. This led to the result that one could obtain normal and abnormal grain growth and the time evolutions of the SDF and G(R,t). Moreover, as asymptotic long time distribution one gets a general SDF, not anymore the Hillert-distribution. It shall be emphazised that also this latter derivation is based on first principles without heuristic assumptions.

A final approach in treating curvature driven grain growth is obtained for the case of non uniform grain boundaries which was also achieved by our own work (Lücke et al. (1999)). By giving different orientations g to the grains one obtains a size coordination orientation distribution function (SCDOF) ϕ^g (R,n) by which also differently natured grain boundaries are achieved with different mobilities and energies. Moreover, the dihedral angles at the vertices are then determined by mechanical equilibrium of the different grain boundary tensions. For the growth rate G^g (R,n) various topological relationships also based on first principles must be used. Furthermore, instead of a single continuity equation, a whole set of such equations, one for each orientation class, must be applied. Introducing the quantities ϕ^g (R) and G^g (R) leads to solutions which are very different from those obtained for uniform grain boundaries. One obtains only abnormal grain growth and generally SDFs, CDFs, and ODFs are changing with time. Neither the $t^{0.5}$ – law nor self similar distributions do occur at all.

In the foregoing the development of the statistical theory of grain growth in 2D-trivalent mosaics has been outlined going from the very rough approaches up to the most modern form including non uniform grain boundaries. The development was due to the use of more realistic models and by replacing heuristic assumptions by rigorous topological relationships. Also by increasing the number of parameters applied for characterization of the grains (R,n,g) more accurate derivations are obtained. It appears now to be possible to get a better agreement with experiment than for older versions.
But not all possible improvements have been made already. For instance, the motion and the drag of the triplepoints have been neglected. However, this problem has been approximately (use of regular grains) solved in our work (Lücke and Brandt (2000)). It results in the introduction of a triplepoint mobility parameter $\psi = \pi\, m^{TP}\, \breve{R}/m$. Fig. 2 for example shows the rate G of grain growth as function of the radius R of a grain for various ψ in comparison to the curve for $\psi = 0$, i.e. with infinite drag. Finally, of really true interest is the 3D-case instead of the 2D-case which could be solved in good approximation by simple transposing the 2D-derivation into 3D-form (Brandt and Lücke (2001)).

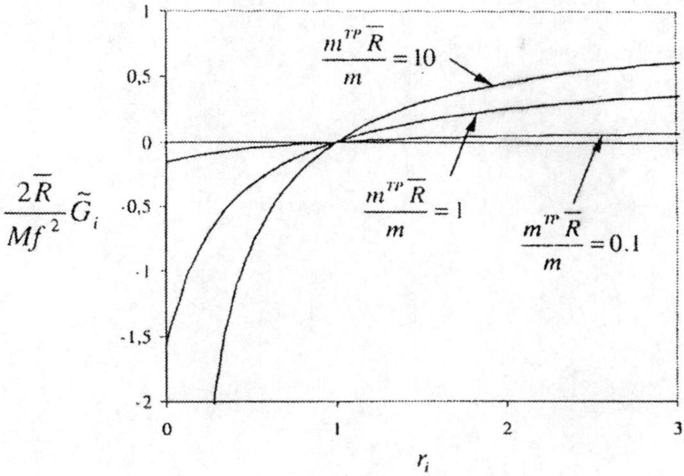

Fig. 2. Rate of grain growth G(R) as function of the radius $r = R/\check{R}$ for different values of ψ.

However, theoretical or experimental results for the parameters m, γ, ψ, and their orientation dependences are needed in order to get true improvements in this field. In order to get the necessary information I recommend to visit the recrystallization section of this conference.

The author likes to thank Dr. H. Schuster for his engaged assistance in the preparation of the manuscript.

Keynote Papers

External Field Applied Grain Boundary Engineering for High Performance Materials

Tadao Watanabe

Laboratory of Materials Design and Interface Engineering, Department of Machine Intelligence and Systems Engineering, Graduate School of Engineering, Tohoku University, Sendai, Japan

Keywords: grain boundary engineering, grain boundary microstructure, magnetic annealing, magnetic sintering, magnetic crystallization.

Abstract. In order to find a new approach to the control of grain boundary microstructures, an overview of recent works has been made on the effect of the application of external field on grain boundary related phenomena. Magnetic annealing and sintering have been found to be useful for the grain boundary engineering. The observed restrictive effect of magnetic annealing and sintering on abnormal grain growth is useful to produce uniform microstructure. Magnetic sintering can enhance the densification of iron powder compacts and improve magnetic and mechanical properties of the material. The usefulness of the application of external field has been discussed in connection with future development of the grain boundary engineering for high performance materials

1 Introduction

The generation of new grains and their subsequent growth or shrinkage are elementary processes of the evolution and modification of microstructure in a polycrystalline material system. Most of engineering materials are polycrystalline so that the control of the microstructure associated with grain boundaries ("grain boundary microstructure") is one of key issues in designing and development of high performance structural or functional materials. The presence of grain boundaries can affect both beneficial or detrimental effects on the polycrystalline material system. The concept of "Grain Boundary Design and Control" [1,2] is based on how to optimize the grain boundary microstructure by maximizing beneficial effects and minimizing detrimental effects of grain boundaries through appropriate fabrication processing, in order to endow the material with high performance or a new or desirable function. The present paper will introduce a new approach to the grain boundary engineering which we have recently developed, that is, the application of external field during material processing. It is likely that the application of some external field during fabrication processing may affect either beneficial or detrimental effects on the bulk properties of the material. McLean has pointed out that the presence of externally applied potential gradients may cause a microstructural change in a material because the stability of microstructure can be affected by the presence of potential gradient associated with an external field [3]. We discuss the potential of different types of external field which can be applied to the control of grain boundary microstructure and the feature of those microstructures which are produced by recrystallization and grain growth.

2 The Application of External Field to Microstructural Control

2.1 Types of external field and material

It has been widely recognized that the feature of the microstructure in polycrystalline material depends on processing method [4]. For instance, fine-grained and randomly oriented polycrystals of metal and alloy can be produced by rapid solidification from the melt, while grain-elongated and textured polycrystals can be produced by unidirectional solidification at a slow solidification speed. This is because some directional flow of atoms can be induced due to the presence of temperature gradient. Thus any type of directional flow of atoms under external field may induce some modification of the microstructure produced by conventional processing method without any external field. The following external fields are considered practically applicable: (1) temperature gradient (heat energy), (2) magnetic field (magnetic energy), (3) electric field (electric energy), (4) stress field (mechanical energy), (5) chemical field (chemical potential associated with compositional gradient). Such external fields can generate the gradient of the energy in a material system, resulting in the driving force for metallurgical phenomena. The level of the energy and the driving force may be restricted depending on the type of external field concerned. It is also reasonable to expect that a combined application of different external fields may result in some additional effect due to the contribution of the second or third external fields. Furthermore it is likely that the type and level of applicable external field depends on the type and the state of the material. For instance the application of a magnetic field can result in a great difference between magnetic and non-magnetic material, and also between ferromagnetic and paramagnetic state. Stuwe has estimated the level of the driving force for boundary migration in consideration of different types of internal or external potential [5].

2.2 Importance of grain boundary microstructure to bulk properties

A polycrystalline material is an aggregate consisting of a huge numbers of grains connecting to each other and grain boundaries between adjoining grains. The nucleation of grains and their subsequent grain growth strongly depends on the initial microstructure controlled by processing method and condition. The processing method and condition can be chosen to produce only useful influence which endow a polycrystal with beneficial and useful function and performance. For optimizing microstructure, the grain size, grain shape, grain orientation have been generally used as important parameters. Moreover such parameters associated with grain boundaries as the grain boundary character, boundary inclination, and the grain boundary character distribution and grain boundary connectivity, grain boundary chemistry have been used [6]. Among those parameters, the character and structure of grain boundary are key parameters which affect the grain boundary microstructure. Recent development of a new computer-aided and automated on-line orientation analyzing technique, so called "orientation imaging microscopy (OIM)" has enabled us to characterize a large number of grains and grain boundaries in a polycrystal with a wide range of grain size ranging from a few tens of nm to mm without much difficulty [7]. As mentioned earlier, the nucleation and migration of grain boundaries are inevitably involved in the evolution and modification of grain boundary microstructure. Therefore it is very likely that the application of some external field can affect the nucleation and migration of individual grain boundaries, resulting in the formation of an unique grain structure and grain boundary microstructures in the three-dimensional (3D) space in bulky specimen and also in 1D or 2D space in thin wire and thin film or ribbon, respectively. So far several investigators have studied the effect of the application of external fields, e.g.,magnetic field on recrystallization and grain growth in metallic materials [8-10].

3 The Application of External Field to Grain Boundary Engineering

According to McLean [3], the reduction of free energy that causes any microstructural changes arises from either change in chemical energy associated with phase transformation, or defect energy from dislocations, surface and internal interfaces like grain boundaries. It is suggested that the presence of an external field of varying potential gradient can modify the driving force for diffusional processes playing important roles in microstructural evolution and modification. Here we are confined ourselves to grain boundary microstructure in connection with the types of external field and material.

3.1 Magnetic annealing and grain growth

The effect of a magnetic field on recrystallization and grain growth was studied very early at the end of 1940's in iron-cobalt alloy [11] and more recently in iron [8] and iron-cobalt [9] and iron-silicon [10] alloys. The driving force originating from a magnetic field is suggested to be sufficient for metallurgical phenomena to occur. For iron, the maximum driving force by the application of a magnetic field was estimated to be the order of $0.1 N/cm^2$ in consideration of the difference of magnetic energy due to the magnetic anisotropy between <100> and <111>. The above mentioned value is almost comparable to the driving force originating from the anisotropy of surface energy which controls grain boundary migration and grain growth at the tertiary stage of recrystallization. It is interesting to know how and to what extent the application of a magnetic field can affect the grain boundary microstructure controlling bulk properties in polycrystalline materials.

Martikainnen and Lindroos found that the retardation of recrystallization and the texture evolution characterized by an increasing fraction of {100} component occurred during annealing at 973K in deformed iron by the application of a magnetic field of 1.5T [8]. They suggested that the application of a magnetic field during annealing generated grain boundaries with low mobility between magnetically oriented nuclei and deformed matrix in the ferromagnetic temperature region, resulting in the retardation of recrystallization in which grain boundary migration plays a key role. Unfortunately they could not provide any direct evidence for the presence of the grain boundaries with low mobility.

Watanabe et al. [9] also found that the application of a magnetic field could retard recrystallization in rolled iron-9mol%cobalt alloy at 1073K below Curie temperature. They quantitatively analyzed the type and the frequency of grain boundaries, termed the grain boundary character distribution (GBCD) in magnetically annealed specimens as a function of the strength of magnetic field, by using the SEM-ECP technique. The result is shown in Fig. 1 which clearly demonstrates that the frequency of low-angle boundaries with the misorientation angles smaller than 15° increased by almost 4 times with increasing magnetic field strength up to 5 kOe. The frequency of coincidence boundaries with Σ values smaller than 29 was about 12% without magnetic field. It once decreased and then increased with increasing the magnetic strength. We know that low-angle boundaries have low mobility and that grains bordered by low-angle boundaries are difficult to grow. Therefore it is reasonable to think that the origin of retardation of recrystallization in magnetic annealing is connected to the presence of a higher frequency of low-angle boundaries. It is likely that the application of a magnetic field may restrain grain boundaries which can have higher mobility than the low-angle regime.

More recently we have studied the effect of a magnetic field on grain growth in rapidly solidified iron-50mol%cobalt alloy ribbons with thickness of 50μm and 4mm wide. Several important finding were obtained. As shown in Fig. 2a,b, the application of a magnetic field to the normal direction to the ribbon surface restricted of abnormal grain growth during annealing, forming more uniform grain size distribution indicated by Fig. 2b. On the other hand non-magnetic annealing shows a more wide spread of grain size distribution than magnetic annealing, as shown in Fig. 2a. Another important finding is that both retarding effect and enhancement of grain growth can be observed depending on whether a magnetic field is applied to the normal or parallel direction to the ribbon surface. When a magnetic field is applied in the normal direction, the retarding effect can occur, while an enhancement of grain growth was observed in the same material when a magnetic field is applied in the direction parallel to the surface. Moreover the direction of a magnetic field affects the frequency of coincidence boundaries.

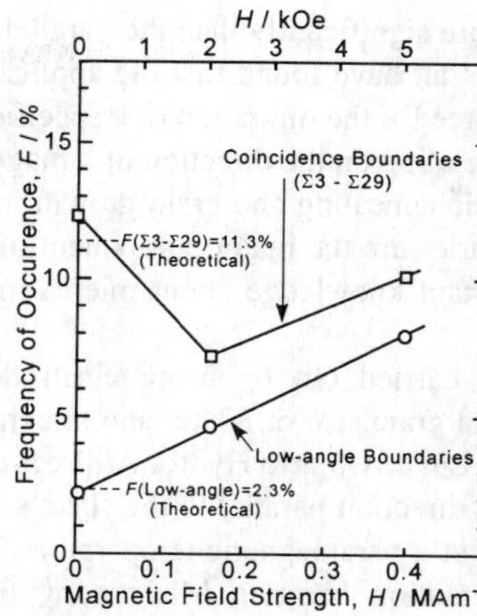

Fig. 1. Magnetic field strength dependence of the frequency of low-angle boundaries and coincidence boundaries for magnetically annealed iron-9mol% cobalt alloy polycrystals (T=1073K, t=18ks).

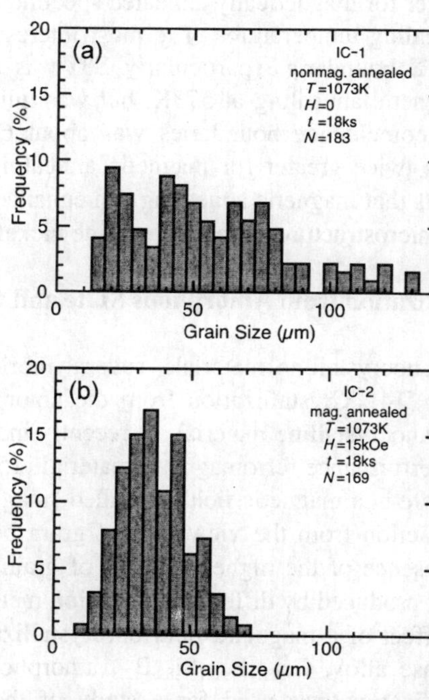

Fig. 2. The distributions of the grain size for iron-50mol% cobalt alloy ribbons produced by rapid solidification and annealed at 1073K (a) without a magnetic field and (b) with a magnetic field (H=15kOe).

We know that low-angle boundaries (including so called "subboundaries") have low mobility and that grains bordered by low-angle normal direction appeared to

increase the frequency more significantly than the parallel direction to the surface. Quite recently Molodov et al. have found that the application of a magnetic field can provide the driving force for the migration of a specific type of grain boundary in bismuth bicrystals, depending on the direction of a magnetic field [12,13].
It is expected that magnetic annealing and grain growth in nanocrystalline materials in which grain boundaries are the major component of the microstructure may provide us another important knowledge about microstructural homogeneity and thermal stability [14].
Magnetic annealing was carried out by using electrodeposited polycrystalline nickel sheet with the initial grain size of 30 nm and thickness of 0.5 mm, provided by Dr.G. Palumbo and P. Lin of Ontario Hydro. A direct-current magnetic field of 15kOe was applied in the direction parallel to the sheet surface at a ferromagnetic temperature of 573K and at a paramagnetic temperature 773K for 1.8ks. The micrographs taken by Orientation Imaging Microscopy (OIM) with a FE-SEM-EBSP/OIM for non-magnetically or magnetically annealed specimens are shown in Fig. 3. It appears that magnetically annealed specimen (right) has more uniform grain structure than non-magnetically annealed one (left). The average grain size tends to be slightly larger for magnetically annealed specimens both ferromagnetic and paramagnetic annealing temperatures. The most interesting finding was that the frequency of low Σ boundaries (particularly $\Sigma 3$) was almost the same for magnetic and non-magnetic annealing at 573K, but was quite different at 773K. The frequency of $\Sigma 3$ coincidence boundaries was about 6% for non-magnetic annealing while it was twice greater for magnetic annealing than non-magnetic annealing. This suggests that magnetic annealing can enhance the thermal stability of the grain boundary microstructure in nanocrystalline nickel.

3.2 Magnetic Crystallization from Amorphous State and Grain Growth

In order to produce nanocrystalline materials, several fabrication methods have been developed so far [14]. Crystallization from the amorphous state is a new method to produce a nanocrystalline material, as recently increasingly applied for development of high performance ferromagnetic material [15]. The crystallization from the amorphous state in a magnetic field, so called "magnetic crystallization " has an unrevealed attraction from the viewpoint of grain boundary engineering. We can expect the presence of the highest density of grain boundaries in nanocrystalline polycrystals produced by different fabrication method. For this purpose we have studied the effect of a magnetic field on crystallization and subsequent grain growth in Co-base alloy. $Co_{83.1}Fe_{6.1}Si_{8.5}B_{2.3}$ amorphous alloy ribbon produced by rapid solidification was used for a study of the effect of magnetic annealing on crystallization and grain growth resulting in the evolution of the grain boundary microstructure in nanocrystalline specimen. There are a large variety of annealing processes based on combination crystallization and subsequent grain growth, with or without a magnetic field. Moreover annealing

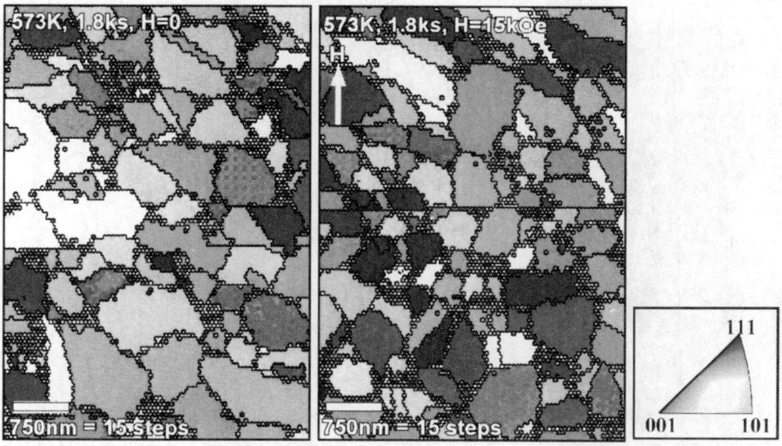

Fig. 3. OIM analyses for nanocrystalline nickel sheet specimens annealed at 573K without (left) and with a magnetic field of 15kOe (right).

temperature is another important factor which may affect crystallization and grain growth, whether it is below or above Curie temperature (1388K for crystalline Co) and crystallization temperature (around 800K). Fig. 4 shows SEM micrograph (left) and OIM micrograph of only fcc cobalt phase (right) for the specimen received two step annealing, crystallization at 773K without a magnetic field and then grain growth at 973K with a magnetic field of 15kOe. For OIM analysis 50 nm step size was taken and the area with minimum four steps in cluster was recognized as a grain. The SEM micrograph indicates complete crystallization from the amorphous state and the presence of submicron grain size. It is evident from the OIM micrograph that the size and orientation of cobalt grains have a large variety. However it was observed magnetic crystallization and grain growth produced more uniform grain size. Magnetic properties (saturated magnetization, coercive force) and mechanical property (hardness) of crystallized specimens of the alloy appear to related to the average grain size of fcc cobalt phase. A large variety of magnetic annealing processes combining magnetic crystallization and grain growth will provide us with a higher potential as a tool of engineering the grain boundary microstructure.

Quite recently we have found that the application of a magnetic field can enhance the densification during sintering of carbonyl iron powder compacts [16,17]. Magnetic sintering was conducted in a DC magnetic field with field strength up to 15kOe in vacuum of 3~5x10^{-3}Pa. Fig. 5 shows measurements of the relative density of the powder compacts after magnetic sintering and non-magnetic sintering at 973K in the ferromagnetic temperature region below the Curie temperature (T_c=1043K). It is evident that the application of a magnetic field of 15kOe could shorten the sintering time for 95% relative density by almost 1/6 from 60hrs to

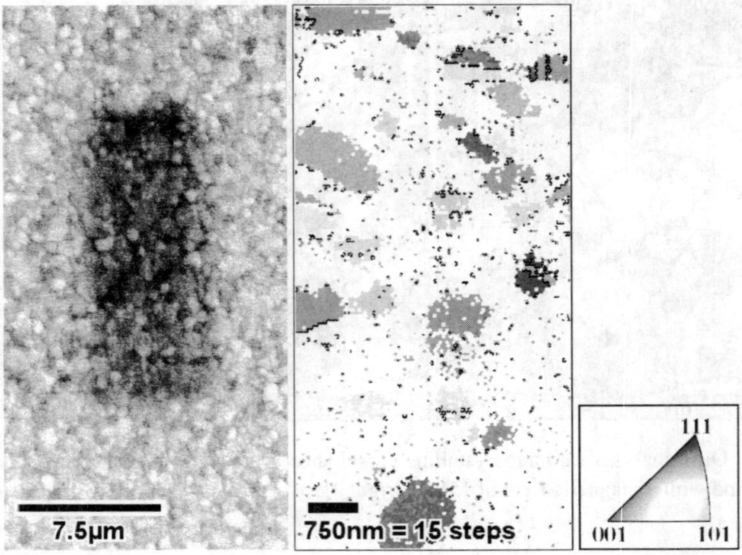

Fig. 4. SEM and OIM images observed on cobalt-based ribbon crystallized at 773K without a magnetic field from the amorphous state and then further annealed at 973K with a magnetic field of 15kOe.

10hrs. In particular, the application of a magnetic field was found to enhance more effectively the densification at early stage and intermediate stages of sintering. This is probably because sintering process requires the formation and migration of grain boundaries. This involves the relative rotation between contacting particles and the elimination of pores from triple junctions by grain boundary migration. The origin of enhancement of the densification has not been fully understood yet, but an increased boundary mobility likely promotes the elimination of pores existing near grain boundaries since the migration of grain boundary in iron silicon bicrystal has been found to be enhanced by applying a magnetic field. The driving force associated with the application of a magnetic field may be added to that for ordinary sintering without a magnetic field. It was also found that uniform microstructure could be formed by magnetic sintering, restricting abnormal grain growth and that magnetic properties (saturated magnetization, susceptibility) and mechanical property (hardness) of magnetically sintered specimens were improved in comparison with non-magnetically sintered ones. Thus we have found that magnetic sintering has a great technological importance to materials processing associated with grain boundary engineering and also scientific curiosity about possible mechanism of enhancement of densification by magnetic sintering. Much remains to be done in the future.

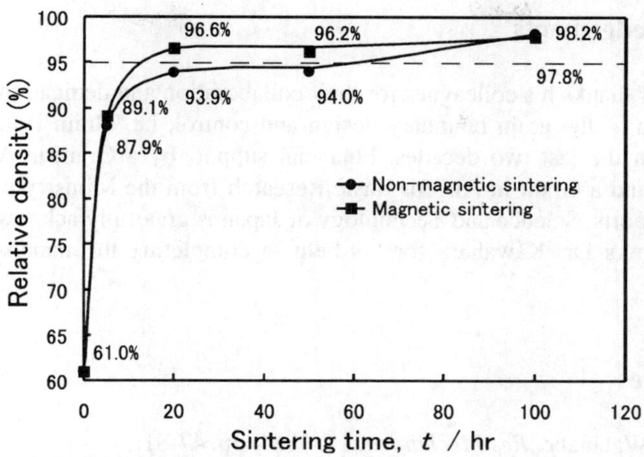

Fig. 5. Comparison of the densification for iron powder compacts non-magnetically sintered and magnetically (H=15kOe) sintered at 973K.

4 Prospect of External Field Applied Grain Boundary Engineering

In order to find a new approach to the grain boundary engineering for development of high performance structural or functional materials, we have overviewed recent studies of the effects of the application of external field on the evolution and modification of the grain boundary microstructure. The application of external field can provide us with a presently unrevealed potential for future materials development, particularly as seen from recent observations of magnetic annealing, involving crystallization and grain growth, magnetic sintering. These new methods for material processing have their own potential and a large variety which are not replaceable by other. There is much need for further work in this new area in order to establish a new discipline of the grain boundary engineering or interface engineering.

5 Conclusion

The application of external field has a great potential and a large variety which may be utilized for further development of grain boundary engineering for high performance structural and functional materials. In particular, the potential of magnetic annealing and sintering should be recognized which inevitably involve the formation of new grains and grain growth through the formation and migration of grain boundaries. The findings of enhancement of the densification during sintering by magnetic annealing and the restrictive effect of magnetic annealing on abnormal grain growth are of engineering importance.

Acknowledgements

The author thanks his colleagues for their collaboration and dedication to establish a new area of the grain boundary design and control, i.e."Grain Boundary Engineering" in the last two decades. Financial support by a Grant-in-Aid for COE Research and a Grant-in-Aid for Basic Research from the Ministry of Education, Culture, Sports, Science and Technology of Japan is gratefully acknowledged. The author thanks Dr. Kawahara for his help in completing the manuscript of this paper.

References

[1] T. Watanabe, *Res Mechanica*, 11, 1984, pp. 47-81.
[2] T. Watanabe and S. Tsurekawa, *Acta Mat.*, 47, 1999, pp. 4171-4185.
[3] M. McLean, *Metal Science*, 16, 1982, pp. 31-36.
[4] T. Watanabe, *Textures and Microstructures*,20, 1993, pp. 195-216.
[5] H.P. Stüwe, in "Recrystallization of Metallic Materials", ed. F.Haessner, Dr. Riederer-Verlag, 1978, pp. 11-21.
[6] T. Watanabe, *Mater.Sci.Eng.*, A176, 1994, pp. 39-49.
[7] B. Adams, S. Wright and K. Kunze, *Met.Trans.*, A24, 1993, pp. 819-831.
[8] H.M. Martikainen and V.K. Lindroos, *Scan.J.Metallurgy*, 10, 1981, pp. 3-8.
[9] T. Watanabe, Y. Suzuki, S. Tanii and H. Oikawa, *Phil.Mag.Letters*, 62, 1990, pp. 9-17.
[10] N. Masahashi, M. Matuo and K. Watanabe, *J.Mater.Res.*,13, 1998, pp. 457-461.
[11] R. Smolukowski and R.W. Turner, *J.Appl.Phys.*, 20, 1949, pp. 745-747.
[12] D.A. Molodov, G. Gottstein, F. Heringhaus and L.S. Shvindlerman, *Scripta Mater.*,37, 1997, pp. 1207-1213.
[13] D.A. Molodov, G. Gottstein, F. Heringhaus and L.S. Shvindlerman, *Acta Mater.*, 46, 1998, pp. 5627-5632.
[14] H. Gleiter, *Progress in Mat. Sci.*, 33, 1989, pp. 223-315.
[15] C.K. Kim and R.C. O'Handley, *J.Microscopy*, 180, 1995, pp. 70-79.
[16] T. Matsuzaki, T. Sasaki, S. Tsurekawa and T. Watanabe, in "Proc.4th Intern.Conf. on Recrystallization and Related Phenomena", Japan Inst. Metals, 1999, pp. 529-534.
[17] S. Tsurekawa, K. Harada, T. Sasaki, T. Matsuzaki, and T. Watanabe, *Mater. Trans. JIM.*, 41, 2000, pp. 1843-1851.

Grain Boundary Character – a Key Factor for Grain Boundary Control

Dmitri A. Molodov

Institut für Metallkunde und Metallphysik, RWTH Aachen, D-52056 Aachen, Germany

Keywords: grain boundary, junctions, mobility, driving force, segregation.

Abstract. The current research on grain boundary dynamics in metals is reviewed. The boundary dynamics, i.e. the boundary reaction to applied forces, is strongly dependent on the grain boundary character, which is commonly reduced to the orientation relationship between adjacent grains and impurity segregation on the boundary. The misorientation dependence of the motion of specific capillary driven grain boundaries, the effect of inclination on the mobility of planar boundaries and the motion of connected grain boundary systems with triple junctions are considered. The role of segregation behaviour and boundary crystallography for the orientation dependence of boundary mobility is addressed. The compensation effect for the motion of structurally different grain boundaries is discussed.

1 Introduction

Although the history of metallurgy, of the metal producing, began more than 5000 years ago, it has been comparatively recently, only in the last, 20th century, recognized that the properties of metallic materials are less determined by the overall chemical composition of the material rather than by its microstructure, that means - the spatial distribution of elements, phases, defects, and orientations. Therefore a control of microstructure promises both an improvement of material performence and a development of advanced materials with taylored properties.

Important microstructural changes in metallic materials are introduced by heat treatment subsequent to plastic deformation. By recrystallization and grain growth the defect density and orientation distribution of grains are markedly affected which in turn drastically changes the material properties. The fundamental process of recrystallization and grain growth is the motion of grain boundaries, by which the crystalline solid is reconstructed atom by atom. Polycrystalline solids possess a wide variety of structurally different boundaries, which quite differently react to exerted forces. This provides a powerful tool for microstructure control and an elegant method for analysis of microstructure evolution. The effect of grain boundary character, determined by both orientation relationship and segregation, on grain boundary dynamics will be addressed in this paper.

2 Charakterization of Grain Boundary Dynamics

2.1 Force for grain boundary motion

A driving force p for grain boundary migration occurs, if the displacement $d\ell$ of boundary area s leads to decrease of the total free energy dG of the system :

$$p = -\frac{dG}{s \cdot d\ell} = -\frac{dG}{dV}, \tag{1}$$

where V is the volume swept by the boundary area during its displacement. The gain of free energy can result from a change of the grain boundary energy due to a reduction of the grain boundary area in case of curved capillary driven boundaries, or from a volume free energy difference across the boundary e.g. by the anisotropy of free energy density in a magnetic, electric or elastic field.

2.2 Measurements of grain boundary motion

Since there is no average representative grain boundary in a crystalline solid, grain boundary properties are most appropriately measured on specific individual grain boundaries, i.e. in specially prepared bicrystals. Equivalently, measurement of connected grain boundary systems with junctions requires tricrystal experiments, etc.

There are two different ways to determine the grain boundary velocity. In the discontinuous method the location of the boundary is determined at discrete time intervals by the position of a boundary groove. The advantage of this stepwise annealing method is its simplicity, but its main shortcoming is that the measured boundary velocity is averaged over the large interval of time between consecutive observations. In contrast, the continuous method requires to determine the boundary position at any moment of time without forcing the boundary to stop. This is achieved by utilizing the discontinuity of crystal orientation at the grain boundary. There are various techniques to distinguish different crystal orientations. Most measurements so far were conducted with the XICTD (X-ray Interface Continuous Tracking Device) [1]. The method employs X-ray diffraction to determine the GB position and, therefore, does not interfere with the boundary migration process itself. The device can measure a boundary velocity in a wide range between 1 μm/s to 1000 μm/s. Its inaccuracy amounts to less than 2% [1]. For the investigation of connected grain boundary systems (junctions, networks) the orientation dependence of backscattered electron intensity in an SEM can be used [2]. In materials with non cubic crystal symmetry, the anisotropy of reflectivity of visible light can be utilized for orientation contrast to determine the boundary location.

2.2 Grain boundary mobility

Grain boundary motion consists of the transfer of lattice sites across the grain boundary, which results in the physical displacement of the grain boundary with regard to an external reference frame. Atoms crossing the boundary will change their energy state (Fig. 1) which results in an asymmetry of the atomic migration energy and thus an imbalance of the jump rate across the boundary in both directions.

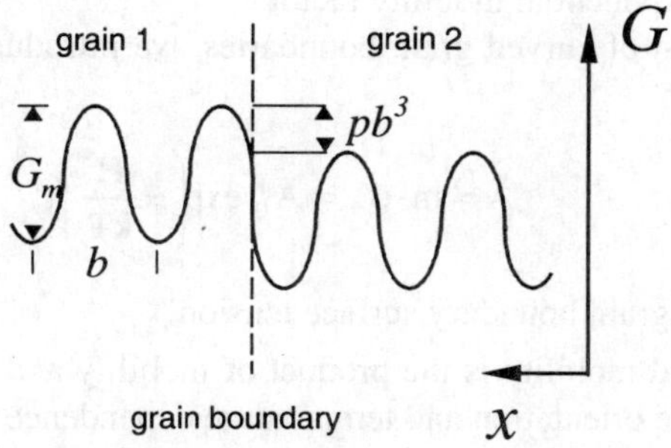

Fig. 1. The free energy of a moving atom changes by the driving force pb^3 when it crosses the boundary. G_m is the free energy barrier for atom tranfer.

Each atom of volume $\Omega_a \approx b^3$ will gain the free energy pb^3 when becoming attached to the growing grain but has to expend this free energy when moving in the opposite direction. Correspondingly, with a boundary displacement b per atom the boundary velocity reads

$$v = b\nu_D \left(e^{-\frac{G_m}{kT}} - e^{-\frac{G_m + pb^3}{kT}}\right) \tag{2}$$

For all practical cases $pb^3 << kT$ at temperatures where boundaries are observed to move ($T \geq 0.3T_m$) and, therefore

$$v = \frac{b^4 \nu_D}{kT} e^{-\frac{G_m}{kT}} \cdot p \equiv m \cdot p, \tag{3}$$

where ν_D is Debye-frequency and m is referred to as grain boundary mobility.
Grain boundary migration is a thermally activated process. Thus, its kinetics follow an Arrhenius type temperature dependence

$$v = v_0 \exp\left(-\frac{H}{kT}\right) \tag{4}$$

Since the driving force is essentially independent of temperature, the temperature dependence of v is the temperature dependence of the grain boundary mobility

$$m = v / P = m_0 \exp\left(-\frac{H}{kT}\right) \tag{5}$$

where H is the activation enthalpy of grain boundary migration and m_0 the corresponding preexponential mobility factor.

For the migration of curved grain boundaries, we introduce for simplicity the reduced mobility

$$A = m \cdot \sigma_b = A_0 \exp\left(-\frac{H}{kT}\right), \tag{6}$$

where σ_b is the grain boundary surface tension.

Since the reduced mobility is the product of mobility and grain boundary energy, it reflects also the orientation and temperature dependence of grain boundary energy. However, according to computer simulations and measurements, the grain boundary energy may change by 20% or less but the reduced mobility may change by orders of magnitude. Therefore changes of the grain boundary energy can be neglected, if not absolute mobility values but activation parameters and mechanisms of grain boundary motion are considered, as in the case of investigations referred to in the following.

3 Orientation Dependence of Grain Boundary Mobility

3.1 Misorientation of tilt boundaries

It was already shown in the past in bicrystals experiments of Aust and Rutter [3] and Shvindlerman with coworkers [4-6] that for high angle grain boundaries the mobility depends on axis $<hkl>$ and angle φ of misorientation. Studies of the mobility of tilt grain boundaries in Al bicrystals [4] have shown that the mobility of low Σ coincidence boundaries (special boundaries) exceeds the mobility of random (non-special) boundaries. Among all tilt boundaries those with <111> rotation axis and rotation angle of about 40° were found to have the highest mobility, which is associated with the special Σ7 tilt boundary.

However, from growth selection experiment [7-9] it was known that the rotation angle of the fastest boundary was invariably larger than 38.2° even consistently larger than 40°.

Owing to the importance of maximum growth rate boundaries for texture formation during recrystallization and grain growth we addressed this obvious discrepancy, and investigated the misorientation dependence of grain boundary mobility on a fine scale in the angular interval 37°-43°<111> with angular spacing 0.3°-0.6° [10]. The experiments revealed that both the activation enthalpy and the preexponential factor were at maximum for a misorientation angle φ=40.5° and at minimum for the exact Σ7 orientation (Fig. 2). Therefore, one is tempted to conclude that the Σ7 boundary has the highest mobility. However, the mobility of boundaries with different misorientation angles do have a different temperature dependence, and there is a temperature, the so-called compensation temperature T_c, where the mobilities of all investigated boundaries of differently misoriented grains are the same. As a result, for $T > T_c$, the mobility is higher for grain boundaries with higher activation energy, in particular it is at maximum for $\varphi = 40.5°$, while for $T < T_c$ the exact Σ7 boundary moves fastest (Fig. 2b).

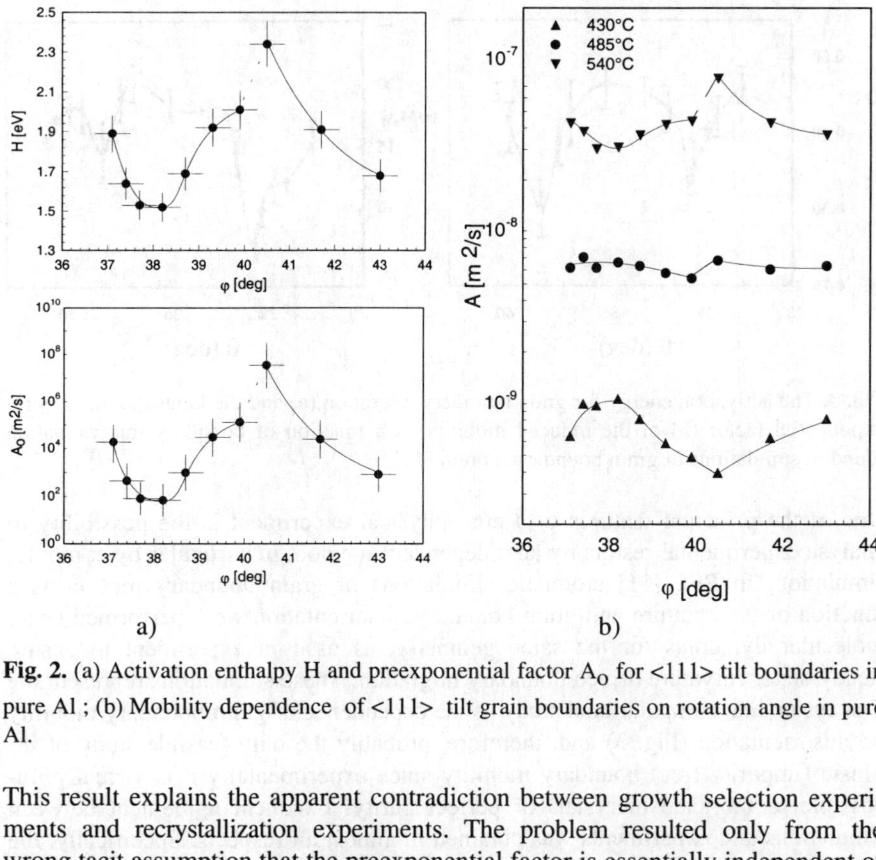

Fig. 2. (a) Activation enthalpy H and preexponential factor A_O for <111> tilt boundaries in pure Al ; (b) Mobility dependence of <111> tilt grain boundaries on rotation angle in pure Al.

This result explains the apparent contradiction between growth selection experiments and recrystallization experiments. The problem resulted only from the wrong tacit assumption that the preexponential factor is essentially independent of misorientation so that only the activation enthalpy controls mobility. Growth selection experiments have to be conducted at very high temperatures (above 600°C), i.e in the temperature regime, where the mobility of the 40.5°<111> is the highest due to its high preexponential factor.

At high temperatures the grain boundary mobility in this angular interval is obviously not dominated by the segregation behaviour of low Σ boundaries. The reason for the changing maximum mobility orientation in different temperature regimes is obviously the orientation dependence of both, the activation enthalpy and the preexponential factor. In fact, both are related to each other in a linear fashion (Fig. 6), i.e.

$$H = \alpha \ln A_0 + \beta \quad , \tag{7}$$

where α and β are constants. This correlation is referred to as the compensation effect and will be discussed in Section 4.

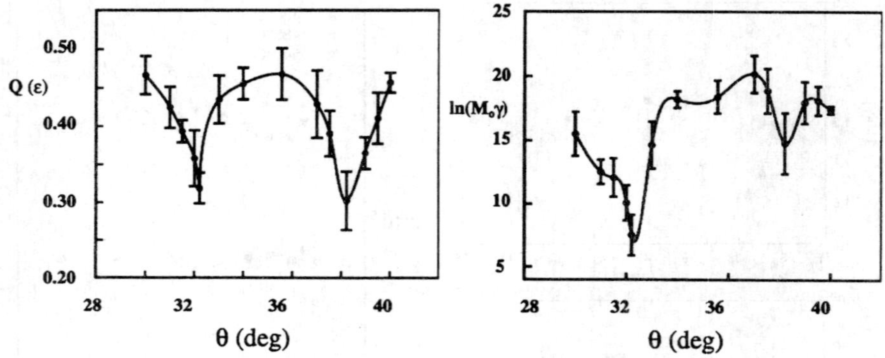

Fig. 3. The activation energy for grain boundary migration (a) and the logarithm of the pre-exponential factor (b) of the reduced mobility as a function of boundary misorientation found in simulations of grain boundary motion [11].

One of the principal features of a pure physical experiment is the possibility to analyse experimental results by an independent method, in particular by computer simulation. In Ref. [11] atomictic simulations of grain boundary motion as a function of temperature and grain boundary misorientation were performed using molecular dynamics for the same geometry, as used in experiment to ensure steady-state, curvature driven boundary migration. These simulations represent the first systematic computational study of the dependence of grain boundary mobility on misorientation (Fig. 3) and, therefore, probably the only feasible study of intrinsic (impurity free) boundary mobility since experimentally it is virtually impossible to fabricate bicrystals of perfect purity. Excellent agreement between simulations and experiments was obtained in almost all respects. Specifically, the activation energy and the logarithm of the pre-exponential factor of boundary mobility exhibited very similar variations with misorientation, including the presence of distinct cusps at low Σ misorientations. It is important to keep in mind that the results of these computer simulation pertain to "absolutely pure material" when discussing the effect of impurity on the orientation dependence of boundary mobility (Section 5).

3.2 Transition from low to high angle boundary

It is common to assume the border between low and high angle boundaries to be in the region of 15° for structural reasons, expressed by Brandon [12]. Recent measurements on stress driven boundaries can be considered as the first experimental determination or this border afforded by a difference in grain boundary property, namely boundary mobility [13]. In Ref. [13] a method to activate and investigate the migration of planar symmetrical and asymmetrical <111> and <112> tilt boundaries in aluminium bicrystals under the action of an external shear stress was introduced . It was shown that low angle as well as high angle tilt boundaries can be moved by such shear stress. The transition from low angle to high angle boundaries was found to manifest itself as a sharp step of the activation energy at a misorientation angle of 13.6±0.5° which, in the limited framework of experimentally obtained data, seems to be independent of tilt axis, impurity content and tilt boundary plane.

3.3 Effect of boundary plane

Grain boundary mobility is known to depend not only on misorientation, but also on the orientation of the grain boundary plane. This is particularly evident for coherent twin boundaries, which are much less mobile than incoherent twin boundaries despite of identical misorientation across the boundary. But anisotropy of grain boundary mobility can also be observed for misorientations other than twin relationships, in particular grain boundaries of a misorientation with <111> rotation axis. For such orientation relationships tilt boundaries can move orders of magnitude faster than pure twist boundaries [14].
However the influence of inclination on the mobility of pure tilt grain boundaries has remained unclear. In experiments on curved tilt grain boundaries [4-6,13], only the average mobility of the moving curved part of the boundary is measured and it is tacitly assumed that there is a uniform tilt boundary mobility. While this may be reasonable, although to be shown, in cubic crystals, it is questionable in materials with lower crystal symmetry. If there is an inclination dependence of tilt grain boundary mobility, it will affect the velocity of steady-state motion of a curved boundary [15].
It is impossible to directly study the effect of grain boundary orientation on its mobility by utilizing grain boundary curvature as a driving force, since the moving curved part of the boundary consists of different boundary planes. Rather, the motion of flat boundaries has to be investigated. The motion of planar grain boundaries under the action of a magnetic field in bicrystals of a material with anisotropic magnetic susceptibility was investigated in [16,17]. The experiments were carried out on bicrystals of high-purity bismuth. Symmetrical and asymmetrical ($\psi = 45°$) pure tilt grain boundaries with 90° <112> misorientation were examined. It was found that the motion of asymmetrical boundaries can be very different from symmetrical boundaries. Firstly, the migration activation enthalpy for asym-

metrical boundaries is nearly one order of magnitude higher than for a symmetrical one: 3.4-3.8 and 0.51 eV, respectively (Fig. 4a) with a compensation temperature close to the melting temperature of Bi.

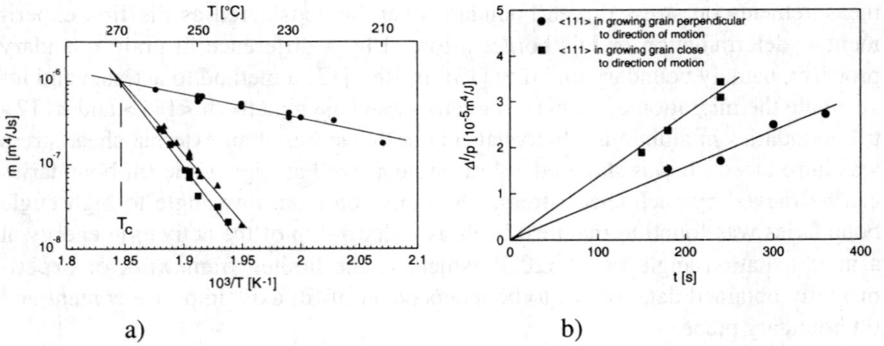

Fig. 4. (a) Temperature dependence of mobility of 90°<112> symmetrical (●) and asymmetrical (▲, ■) boundaries in Bi-bicrystals, moving in opposite directions [16,17]. Trigonal axis in the growing grain parallel (▲) or perpendicular (■) to the growth direction; (b) Normalized displacement vs. annealing time for asymmetrical tilt boundaries moving in opposite directions.

The most surprising feature is that in contrast to the symmetrical boundary, for an asymmetrical tilt boundary the measured mobility was found to be distinctly different for the motion in opposite directions (Fig. 4b). There are several potential reasons for this anisotropy. First, there is an essential difference in the distance between the crystallographic planes on each side of the boundary. An estimation shows that this factor may change the velocity of grain boundary motion, however this difference is unlikely to affect the velocity of boundary motion by more than 20%, which is distinctly less than the observed effect [16,17]. Second, because boundary motion in Bi-bicrystals may be influenced by impurity drag, the difference in the diffusivity of impurities in two opposite directions in the anisotropic structure of Bi should be taken into account. Finally, as shown recently, the motion of a grain boundary in a magnetic field can be considered as a motion of a conductor in a magnetic field, or more strictly, as the motion of a region with a conductivity different from that of the surrounding matrix in a magnetic field. Such a motion causes an electromotive force and as a consequence, an additional dissipation of energy in a magnetic field [18]. This dissipation should, on the one hand, be different for symmetric and asymmetric boundaries and, on the other hand, be different for boundary motion in opposite directions for asymmetric boundaries [18].
In any event, if this asymmetry of grain boundary mobility holds also for other metals, it will have a serious impact on our understanding of grain boundary motion, since the mobility of a grain boundary is commonly conceived as not dependent on its direction of motion.

3.4 Non-tilt boundaries

The orientation dependence of the boundary mobility depends sensitively on grain boundary character, in particular when approaching pure twist and tilt boundaries. This was reported for pure <111> twist boundaries in terms of the loss of anisotropic growth rate upon departure from pure twist boundaries [11].

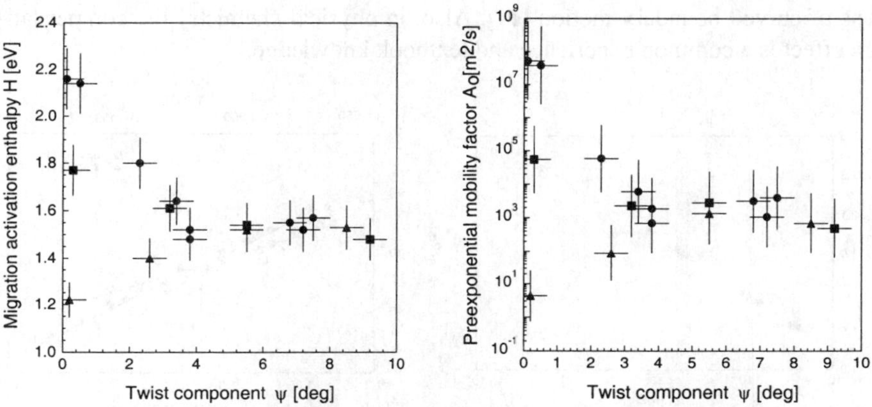

Fig. 5. Activation enthalpy and preexponential factor for migration of 40.8° (●), 37.9° (▲) and 38.6° (■) tilt grain boundaries in pure Al vs. an additional twist component of grains misorientation .

The effect of an additional twist component on the mobility of tilt grain boundaries was investigated on Σ7 and 41°<111> (mostly) tilt boundaries. In the respective bicrystals the tilt boundary was superimposed by a rotation around the axis perpendicular to the grain boundary plane by an angle ψ. The dependence of the activation enthalpy on the twist angle ψ reveals that the twist component degrades and eventually levels out the initial difference between the activation enthalpy for these two tilt boundaries (Fig. 5). The preexponential factor behaves in the same manner. The difference of the activation parameters disappears within $\psi = 4°$.

4 Compensation Effect in Grain Boundary Motion

It is well known that the temperature dependence of GB mobility follows an Arrhenius relation (Eq. 5). Commonly, for evalution of experimental data the activation enthalpy H is determined from the slope H/k of the Arrhenius plot $\ln m$ vs. $1/T$, and much less attention is paid to the preexponential factor m_0. However, there is a large body of experimental evidence that the preexponential factor is strongly related to the activation enthalpy: m_0 increases or decreases, if H increases or decreases according to relation (7). This correlation is referred to as the compensation effect, since it strongly moderates the effect of a variation of H on the value of the mobility, and therefore allows one to use annealing temperature as a selective tool for grain boundary control.

The compensation effect was repeatedly observed for the mobility of pure tilt boundaries in materials of equal purity [6,17,19]. Fig. 6 shows the compensation effect for <111> tilt boundary migration in the vicinity of the special misorientation Σ7. The compensation effect was also observed in molecular dynamics simulations of curved boundary motion [11]. Also, in physical chemistry the compensation effect is a common experience and textbook knowledge.

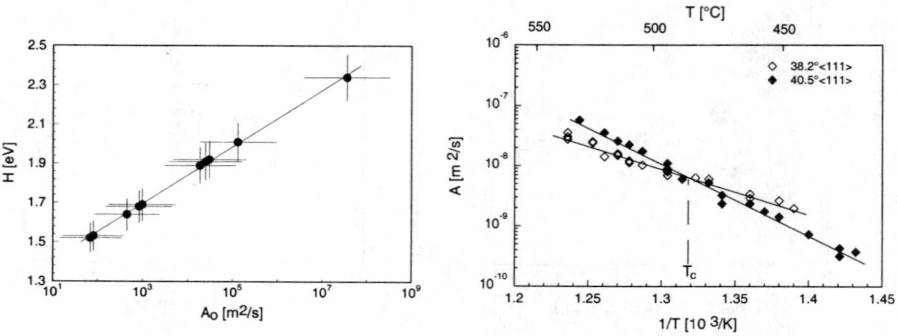

a) b)

Fig. 6. (a) Dependence of migration activation enthalpy on mobility pre-exponential factor for <111> tilt grain boundaries in pure Al [10]; (b) Temperature dependence of the reduced mobility A for 38.2° and 40.5° <111> tilt grain boundaries in pure Al [10].

A consequence of the specific linear dependence between the activation enthalpy and the logarithm of the preexponential factor is the existence of the so-called compensation temperature $T_c = \alpha/k$, at which the mobilities are equal and at which the kinetic lines in Arrhenius co-ordinates intersect at one point (Fig. 6b). In this context it is worthy to note that the compensation effect was also observed for ^{71}Ge diffusion along <111> tilt grain boundaries with different angles in the vicinity of the Σ7 misorientation in pure Al [20].

The compensation effect can be associated with the fact that the activated state is not a random energy fluctuation in space and time but a defined and thus reproducible although unstable state, which is described by its respective thermodynamic functions [21]. The linear compensation relation and the expression for the compensation temperature can be derived under these conditions. In particular, the compensation temperature can be expressed as $T_c = dH/dS|_{\lambda=\lambda_0}$, where the parameter λ denotes some intensive structural or chemical specification, like angle of misorientation, composition, etc. In this approach T_c is the equilibrium temperature between the ground state and an activated state.

5 Segregation Effects on Orientation Dependence of Boundary Mobility

It is well known that the purity of a material has a great influence on grain boundary motion. Impurity atoms almost always reduce the rate of grain boundary migration. However, very little is known on how solute atoms affect grain boundary motion besides the exerted drag, in particular, how impurities influence and thus interact with the grain boundary structure.
The strong interaction of impurities and grain boundary structure is particularly obvious in <100> tilt boundaries in Al (Fig. 7a [24]). For ultrapure and very impure material the mobility of <100> tilt boundaries was found to be independent of rotation angle, irrespective whether special or non-special boundary. For intermediate (although high) purity material, the mobility strongly depends on rotation angle, distinguishing special and non-special boundaries. Such behavior was however never reported for tilt boundaries in Al with axis other than <100>.
The common understanding of the orientation dependence of grain boundary mobility and the effect of solutes on this dependence is mainly founded on results mentioned above [24] as well as on the classical work of Aust and Rutter (Fig. 7b)[25]. According to this understanding the orientation dependence of grain boundary mobility is a segregation effect: strongly ordered boundaries, i.e. low Σ coincidence boundaries segregate less and, therefore, move faster than random boundaries.

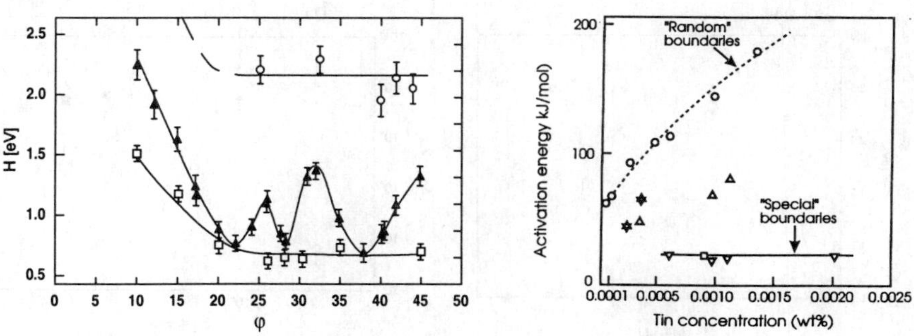

Fig. 7. (a) Dependence of the activation enthalpy of migration for <100> tilt grain boundaries in Al of different purity: □ - 99.99995 at%; ▲ - 99.9992 at%; ○ - 99.98 at%. [24]; (b) The rate of grain boundary migration vs. the concentration of tin in zone-refined lead at 300°C [25].

This classical concept, however, can not be confirmed by results of recent experiments [23,26] and computer simulations [11]. The experimental results reveal that the migration activation enthalpy is strongly affected by both, the boundary crystallography and material purity (Fig. 8). However, in the former case the preexponential factor A_0 rises with increasing H by several orders of magnitude, while in the latter case A_0 remains at the same level. Therefore, the preexponential factor A_0 in the investigated impurity concentration interval was found to be much less sensitive to the material purity than to a change of the misorientation angle. This result allows to conclude that the observed orientation dependence of mobility (Fig. 2b), determined by both H and A_0, does not simply reflect the different segregation behavior of coincidence and random boundaries, as frequently proposed [25], rather it provides evidence for an intrinsic dependence of grain boundary mobility on grain boundary structure. Also molecular dynamics simulations of boundary motion in absolutely pure material reveal strong evidence for the intrinsic structural character of the orientation dependence of boundary mobility [11].

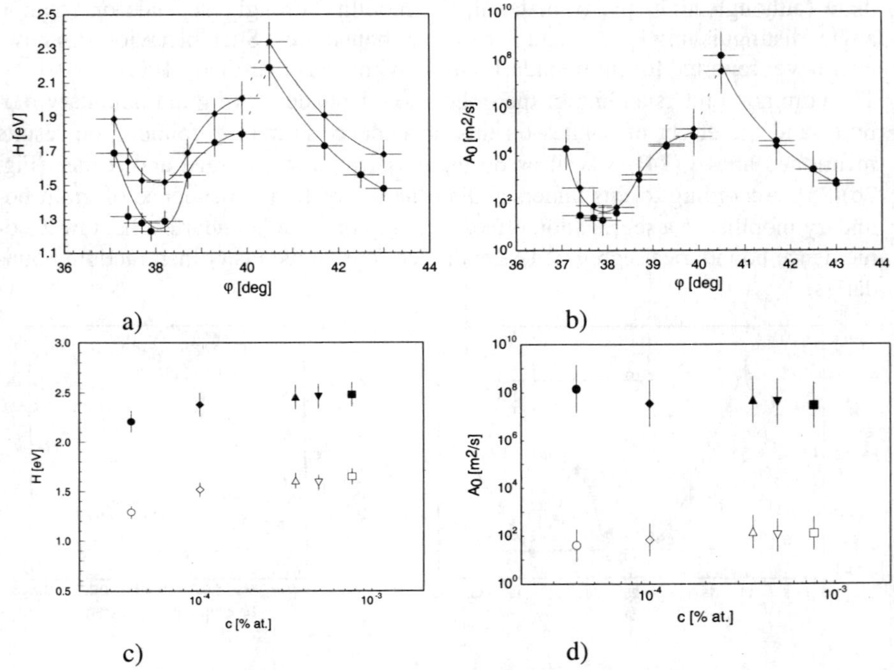

Fig. 8. (a) Activation enthalpy and (b) mobility preexponential factor for migration of <111> tilt grain boundaries as a function of misorientation angle for two different aluminium charges: ● - total impurity concentration c = 0.4 ppm; ◆ - c =1 ppm; (c) the dependence of activation enthalpy and (d) mobility preexponential factor for migration of 38.2° (open symbols) and 40.5° (filled symbols) <111> tilt grain boundaries on impurity concentration in differently pure aluminium.

Although the activation enthalpies of both the special and non-special boundary depend on impurity concentration (Fig. 8c) they remain vastly different in magnitude with H (random) > H (special). This may be associated with different mechanisms of grain boundary migration irrespective of material purity [26].

Aust and Rutter attributed the abnormally high values of activation enthalpy for grain boundary motion in lead to impurity effects on this process and interpreted their results in terms of diffusional mechanisms of boundary motion [25]. However, such interpretation requires to attribute the difference of activation enthalpy measured for different boundaries to the adsorption energy of impurity atoms in the pure boundary. This energy normally does not exceed 0.4-0.6 eV [27]. The difference of the activation enthalpy of motion for different boundaries in the experiment of Aust and Rutter amounts to 42.8-5.2=37.6 kcal/g-atom (1.63 eV) and thus, is too large to be interpreted as an adsorption energy of impurities.

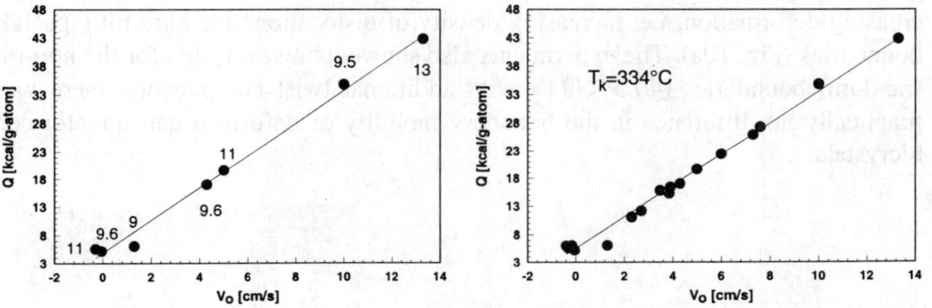

a) b)

Fig. 9. Relationship between the migration activation enthalpy and preexponential factor of the mobility equation for migration of grain boundaries in Pb as obtained from an analysis of results of Aust and Rutter [25] (a) for different boundaries in Pb of equal purity (concentration of Sn in Pb in ppm is noted in the Figure), and (b) for all 17 measured boundaries in Pb with different crystallography and concentration of Sn from 1 to 13 ppm.

Moreover, all 17 investigated boundaries in Ref. [25] were crystallographically different and the compensation effect, which obviously reflects the thermodynamic fundamentals of the migration mechanisms for boundaries with different grain misorientation, holds for the migration parameters in the work of Aust and Rutter as well (Fig. 9). The compensation effect for migration of crystallographically different boundaries in lead of almost equal purity (9-13 ppm) in Aust and Rutter's experiment (Fig. 9a) provides unambiguous evidence that the activation parameters do not increase due to an increase of tin concentration in lead, rather than due to different boundary crystallography. The compensation effect is actually observed for migration of all 17 boundaries investigated by Aust and Rutter, irrespective of tin concentration, with a compensation temperature T_c =334°C which by practical means is identical with the melting point of Pb (327°C) (Fig. 9b).

6 Dislocation Absorption by Moving Boundaries

An interesting example of the effect of the boundary character on grain boundary control is the influence of dislocation absorption on the migration rate of specific grain boundaries in Al-bicrystals. The grain boundary motion in slightly rolled (up to 1.3%) bicrystals of pure Al (99.999%) was investigated in [28]. In particular, the migration of <111> pure tilt grain boundaries and a boundary having an additional twist component (ranging between 5.9° and 9.2°) was measured. The boundaries moved under the action of a constant capillary driving force provided by the grain boundary surface tension of the curved boundary. For the small strain imposed, the grain boundary mobility was found to decrease substantially with increasing deformation, i.e. increasing density of dislocations for pure tilt (special) boundaries (Fig. 10a). The experiments also showed, however, that for the non-tilt (random) boundaries (40.5°<111> with additional twist components) there was practically no difference in the boundary mobility of deformed and undeformed bicrystals.

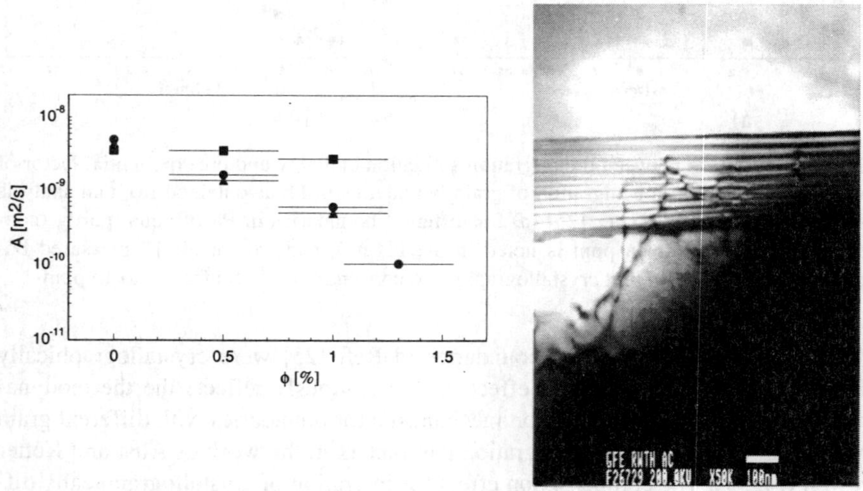

a) b)

Fig. 10. (a) Grain boundary mobility versus degree of rolling ϕ: ● - 40.5<111>; ▲ - 37.7<111>; ■ - 40.5°<111> with twist component 5.9°< ψ < 9.2°; (b) TEM image of moved grain boundary areas in rolled (1%) bicrystal containing a 38° <111> pure tilt grain boundary.

Microstructural TEM investigations of rolled bicrystals containing a 38°<111> boundary after its migration in a deformed sample reveal the reason for this different behaviour. The pure tilt boundary was always observed to contain extrinsic dislocations (Fig. 10b), apparently swept by the boundary during its migration and not yet absorbed. The presence of these dislocations in the boundary appears be the reason for a retardation of the boundary migration rate. If the dislocations are only adsorbed in the boundary without being decomposed or annihilated, their stress field - although certainly partly relaxed in the boundary - still contributes to the internal energy and is not released as driving force. In such case the boundary will not experience the full gain of stored dislocation energy, rather it has to drag along the swept dislocations, which degrades its mobility. Accordingly, random boundaries are obviously more capable of absorbing (annihilating) the swept dislocations than special boundaries. In fact, TEM investigations have shown substantially fewer extrinsic dislocations in the random grain boundaries than in special boundaries [28].

7 Motion of Grain Boundary Systems with Junction

Recent experiments carried out on Al-tricrystals revealed that the grain boundary geometry may play an important role in the motion of a grain boundary system with triple junction [29]. These experiments confirmed the results obtained on Zn-tricrystals and substantiated that the motion of grain boundary systems with triple junctions can be controlled by slowly moving triple junctions. The influence of triple junctions depends on temperature (Fig. 11). It is particularly strong at low temperatures and small grain size. In the high temperature regime the motion of a grain boundary system is less affected by the triple junction, and, therefore, effectively controlled by the grain boundary mobility.

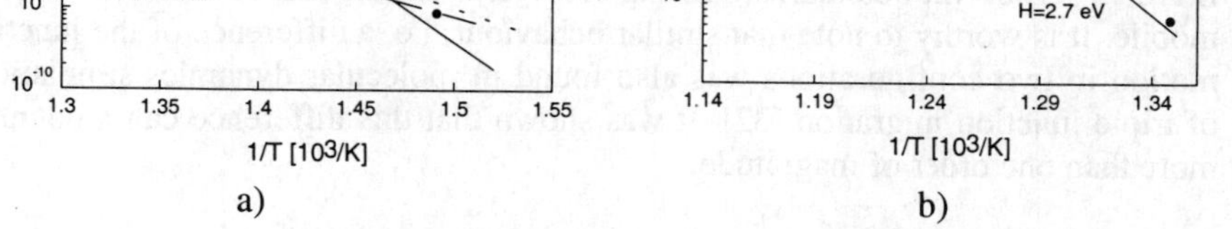

Fig. 11. Temperature dependence of triple junction (●) and grain boundary mobility (▲) in samples with (a) <111> and (b) <110> tilt grain boundaries [29]. Dotted line represents the mobility of a 21°<111> tilt boundary in Al as reconstructed from literature data [30] ($H = 1.6$ eV; $A_0 = 10^3$ m^2/s).

A drastic difference between activation enthalpy of grain boundary and triple junction motion was revealed by experiments [29] (Fig. 11). Therefore, there is a

temperature, below which triple junctions govern the motion of the connected boundary system and grain boundaries simply follow the triple junction.
It has been shown, that the transition range between grain boundary and triple junction control of migration kinetics may depend on grain boundary and correspondingly triple junction geometry (<111> and <110> tilt boundaries in Ref. [29]). Hence, particular grain boundary systems consisting of structurally different grain boundaries need to be adressed to probe the influence of triple junctions and their geometry on texture development during grain growth.

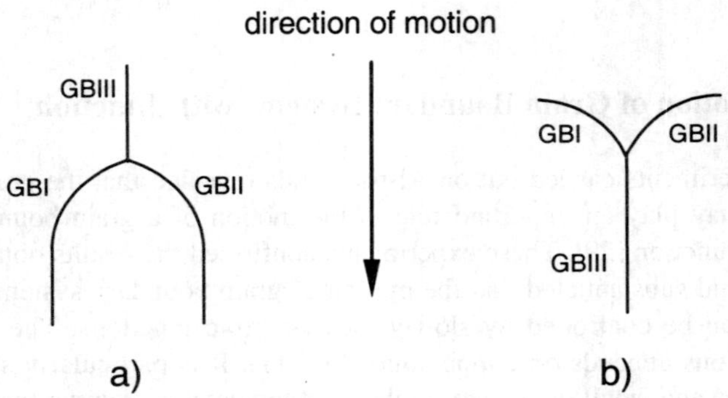

Fig. 12. Different configurations of the grain boundary system with triple junction.

Surprisingly, not only the grain boundary geometry but also the configuration of connected grain boundary system with junction is important for its motion (Fig. 12). Quite recent experiments on Al-tricrystals [31] reveal that the junction in configuration (b) is extremely stable. In experiments with commonly used driving forces (about 0.5 kJ/m^3, assuming grain boundary energy $\sigma \approx 0.5\ J/m^3$) such junction was found to be practically immobile in the entire investigated temperature range up to 600 °C, although the system with the same boundaries (GB I and GB II are 45°<100> tilt boundaries) having configuration (a), was observed to be quite mobile. It is worthy to note that similar behaviour, i.e. a difference of the junction motion in two configurations was also found in molecular dynamics simulations of triple junction migration [32]. It was shown that this difference can amount to more than one order of magnitude.

Acknowledgdement

The author expresses his gratitude to the Deutsche Forschungsgemeinschaft for financial support.

References

[1] U. Czubayko, D.A. Molodov, B.-C. Petersen, G. Gottstein, and L.S. Shvindlerman, *Meas. Sci. Technol.* 6, 1995, pp. 947-952.

[2] Y. Huang, F.J. Humphreys, *Acta Mater.,* 47, 1999, pp. 2259-2271.

[3] J. W. Rutter and K. T. Aust, *Acta Metall.*, 13, 1965, pp. 181-186.

[4] V. Yu. Aristov, V. L. Mirochnik, and L. S. Shvindlerman, *Sov. Phys. Solid State*, 18, 1976, pp. 137-142.

[5] V. G. Sursaeva, A. V. Andreeva, Ch. V. Kopezkii, and L. S. Shvindlerman, *Phys. met. metall.*, 41, 1976, pp. 98-103.

[6] D. A. Molodov, B. B. Straumal, and L. S. Shvindlerman, *Scripta Metall.*, 18, 1984, pp. 207-211.

[7] G. Ibe and K. Lücke, in „Recrystallization, Grain Growth and Textures“, ed. H. Margolin, ASM, Metals Park, Ohio, 1966, pp. 434-447.

[8] G. Ibe, W. Dietz, and A.-C. Fraker and K. Lücke, *Z. Metallkde*. 6, 1970, pp. 498-507.

[9] G. Ibe and K. Lücke *Texture,* 1, 1972, pp. 87-95.

[10] D. A. Molodov, U. Czubayko, G. Gottstein, and L. S. Shvindlerman, *Scripta Metall. Mater*. 32, 1995, pp. 529-534.

[11] M. Upmanyu, D. J. Srolovitz, L. S. Shvindlerman, G. Gottstein, *Acta Mater.*, 47, 1999, pp. 3901-3914 .

[12] D. G. Brandon, *Acta Metall*, 14, 1966, pp. 1479-1484.

[13] M. Winning, G. Gottstein, L. S. Shvindlerman, *Acta Mater.*, 49, 2001, pp. 211-219.

[14] G. Gottstein, H. C. Murmann, G. Renner, C. Simpson, and K. Lücke, in „Textures of Materials“, eds. G. Gottstein and K. Lücke, Springer Verlag, Berlin, 1978, pp. 521-531.

[15] V. E. Fradkov, and L. S. Shvindlerman, *Phys. Chem. Mech. Surfaces*, 1, 1982, pp. 180-191.

[16] D. A. Molodov, G. Gottstein, F. Heringhaus, and L. S. Shvindlerman, *Scripta Mater.*, 37, 1997, pp. 1207-1213.

[17] D. A. Molodov, G. Gottstein, F. Heringhaus, and L. S. Shvindlerman, *Acta Mater.*, 46, 1998, pp. 5627-5632.

[18] G. Gottstein, D. A. Molodov, E. Rabkin, L. S. Shvindlerman, I. Snapiro, to be published.

[19] M. Furtkamp, G. Gottstein, D. A. Molodov, V. N. Semenov, and L.S. Shvindlerman, *Acta Mater.*, 46, 1998, pp. 4103-4110.

[20] T. Surholt, D. A. Molodov, and Chr. Herzig, *Acta Mater.*, 46, 1998, pp. 5345-5355.

[21] G. Gottstein, L. S. Shvindlerman, *Interface Science*, 6, 1998, pp. 265-276.

[22] N. F. Mott, *Proc. Phys. Soc.*, 60, 1948, pp. 1-17.

[23] D. A. Molodov, U. Czubayko, G. Gottstein, and L.S. Shvindlerman, *Acta Mater.*, 46, 1998, pp. 553-564.

[24] E. M. Fridman, Ch.V. Kopetskii, L.S. Shvindlerman, *Z. Metallk.* 66, 1975, pp. 533-539.
[25] K. T. Aust and J. W. Rutter, *Trans. AIME*, 215, 1959, pp. 823-831.
[26] D. A. Molodov "Migration of High Angle Grain Boundaries in Metals", Shaker Verlag, Aachen, 1999.
[27] S. Hoffmann and P. Lejcek, *Interface Science*, 3, 1996, pp. 241-267.
[28] D. A. Molodov, P. Konijnenberg, G. Gottstein, W. Hu, and L. S. Shvindlerman, in „Recrystallization – Fundamental Aspects and Relation to Deformation Microstructure“, eds. N. Hansen et al., Risø National Laboratory, Roskilde, Denmark, 2000, pp. 451-457.
[29] S. G. Protasova, G. Gottstein, D. A. Molodov, V. G. Sursaeva, and L. S. Shvindlerman, *Acta Mater.*, 2001, in print.
[30] G. Gottstein, L. S. Shvindlerman, "Grain Boundary Migration in Metals: Thermodynamics, Kinetics, Applications", CRC Press, 1999.
[31] D. A. Molodov, S. I. Prokofjev, G. Gottstein, and L. S. Shvindlerman, will be published.
[32] M. Upmanyu "Grain Boundary Migration: Atomistic Simulation Studies", PhD Thesis, University of Michigan, Ann Arbor, 2001.

Recrystallization and Grain Growth
Proceedings of the First Joint International Conference
Eds. G. Gottstein and D.A. Molodov, Springer-Verlag, 2001

Modelling Microstructure Development in Annealing and Deformation

Pete Bate

Manchester Materials Science Centre, Manchester M1 7HS, England

Keywords: modelling, deformation structure, finite element method

Abstract. There have been great advances in the modelling- at the microscale- of microstructural development during annealing. Although many problems remain, the main difficulty with predicting recrystallisation microstructures lies in predicting the initial structure: that developed during deformation, dynamic recovery and the early stages of static recovery. The annealing process is highly sensitive to that initial structure and it is a major determinant of recrystallised grain size and texture. Finite element simulation can go some way to predicting deformation microstructures, but is subject to severe limitations. The prediction of recovered deformation microstructures is, then, likely to be the major challenge in modelling annealed microstructures in the future.

1 Introduction

The expansion of digital computing hardware and software has led to an explosion in numerical modelling. This is as true in the field of microstructural development as elsewhere, and there is now significant effort devoted to this general subject. Much work has been carried out on the modelling of microstructural changes during annealing and plastic deformation. It is convenient to classify models and, in order to make this discussion manageable, only those dealing explicitly with the spatial inhomogeneity of microstructure will be considered here. This will not be exhaustive, but will concentrate on the most common approaches currently being developed. Models which only use parameters averaged over macroscopic length scales or which deal only with statistical descriptions of microstructure are perhaps the most practically useful at present, but will not be dealt with.

Microstructural inhomogeneity is key to both plastic deformation and annealing and, in fact, very similar microstructural features would need to be incorporated into models of both those phenomena for them to be truly successful. The spatial distribution of crystal orientation is clearly important, and in much of annealing this becomes quite simple, because orientation changes occur at essentially two dimensional discontinuities: grain and sub-grain boundaries. If one had a model of plastic deformation which predicted orientation distributions correctly, then it

should be possible- in principle- to discretise such distributions and so produce an initial structure for models of annealing which can then deal with development of structures driven by boundary energies and controlled by their mobilities.
However- even if one accepts the questionable basis of dividing the problem in the way outlined above- there are problems with all aspects of that procedure. To demonstrate some of these, it is convenient to deal with it in reverse order and consider first the main methods of simulating the evolution during annealing of structures defined simply by boundaries.

2 Evolution in Annealing with Condensed Boundaries

As far as simulating the development of microstructure during annealing is concerned, there are currently two main approaches. These are the Monte Carlo – Potts (MCP) models and the 'Network' models. Usually the scope of both models is restricted to simulative models of the evolution of a condensed boundary cellular microstructure i.e. essentially grain and subgrain growth, though this is not always the case.
The MCP model has come to dominate annealing simulations at the micro-scale. It was first used for grain growth simulations [1,2], but has been extended to recrystallisation [3,4]. In this model, the structure is represented by a grid (2- or 3-d array) of sites, each of which is ascribed a state. The structure is updated by sequentially considering sites (at random), and at each site a potential change of state is examined and a choice made about whether that change occurs or not. The simplest case is 'curvature driven' grain growth. The state is then just some representation of crystal orientation. The potential changes are for the current site to adopt the state of one of its neighbours, chosen at random. To make the decision, the change in interaction energy of the system which would occur is calculated. Only local, neighbour, interactions are considered. To achieve the correct reduction in global energy, the configuration of the system needs to be subject to random perturbation, allowing a finite probability to increased energy changes. Generally, a Boltzmann- type expression is used to give the probability of such changes, although other forms are possible [5].
Network models deal with boundaries as entities. There are two variants, in one the boundaries are straight and their tensions (energies) are used to give driving forces at junctions [6,7] and in the other boundary curvature- derived from the equilibrium of tensions at boundary junctions- is used [8,9]. The introduction of additional 'vertices' to boundaries appears very promising [10], as it allows greater freedom of boundary configuration.
Both models can be extended to include energy terms internal to the 'grains'. This has been largely restricted to the MCP approach, but it is actually straightforward to include it in network models as well. The use of site-specific energies in MCP models has particularly been used for recrystallisation simulations, where it is taken to represent some dislocation density term [3,5].

The MCP method is easy to program, even for three dimensional problems, and this has undoubtedly contributed to its popularity. In contrast, network models are relatively difficult to code in two dimensions, and the scarcity of three dimensional models [11] testifies to the difficulty in producing the computer programs. Network models are, however, much quicker to run and have some other advantages as well.

It is straightforward to ascribe misorientation dependant boundary energies and mobilities in both models [5,7]. However, the character of a boundary, as well as the crystallographic misorientation, is known to affect its energy and mobility. Twins (Σ3) in FCC metals are a good example; these have significantly lower energy when they are spatially aligned with the common (111) than when they are not. Several workers have noted the high mobility of 40° <111> boundaries [12,13], also in FCC: this high mobility is only true for tilt boundaries [14,15]. It is difficult to imagine how the important effect of boundary character could be included in an MCP model, but clearly can be included in a network model, although at some computational cost and subject to an approximation depending on the assumed shape of the boundary.

It is informative to see the effect of including boundary character on model predictions. To do this, a linear boundary vertex model was used. The boundary specific energy, γ, and the baseline boundary mobility, μ, were set as simple functions of misorientation angle:

$$\gamma = 2\,(\omega/\omega_0)\,(1 - \tfrac{1}{2}\omega/\omega_0), \qquad \mu = (\omega/\omega_0)^2\,(3 - 2\omega/\omega_0) \qquad \omega < \omega_0$$
$$\gamma = 1, \qquad \mu = 1 \qquad \omega \geq \omega_0\,, \tag{1}$$

where ω_0, the transition angle above which the boundary is 'high angle', is 15°. To demonstrate the effect of higher mobility for boundaries near 40°<111>, the following addition was made to the mobility:

$$\mu^+ = A\,(\,3^{1.5} d_1 d_2 d_3\,)^{2n} \exp(\,-(\alpha\omega - \beta)^2) \tag{2}$$

-where ω and d are the misorientation angle and common crystal axis (using the minimum ω cubic symmetry variant). The other terms are parameters: n = 2, a = 5 and b = 3.5 (*radians*) gives a reasonable representation. If that addition is restricted to boundaries with tilt character, it is modified to:

$$\mu^{+*} = \mu^+ \cos(\,|\eta - \pi/2|\,) \tag{3}$$

-where η is the angle between the common axis and the boundary normal (0 for twist, π/2 for tilt). The results of Huang and Humphreys [15] indicate quite a large effect, and A=10 was used here.

Results showing the effect of enhanced near- 40° <111> mobility, without and with the dependence on character, on otherwise equivalent vertex model simulations are given in Fig. 1. The starting structure was derived from a finite element model with the assumption that a fully condensed subgrain structure was developed i.e. all the structural energy is represented by boundaries, though that is inconsequential here. What is significant is the difference in predicted structure. This naturally is associated with changes in predicted textures.

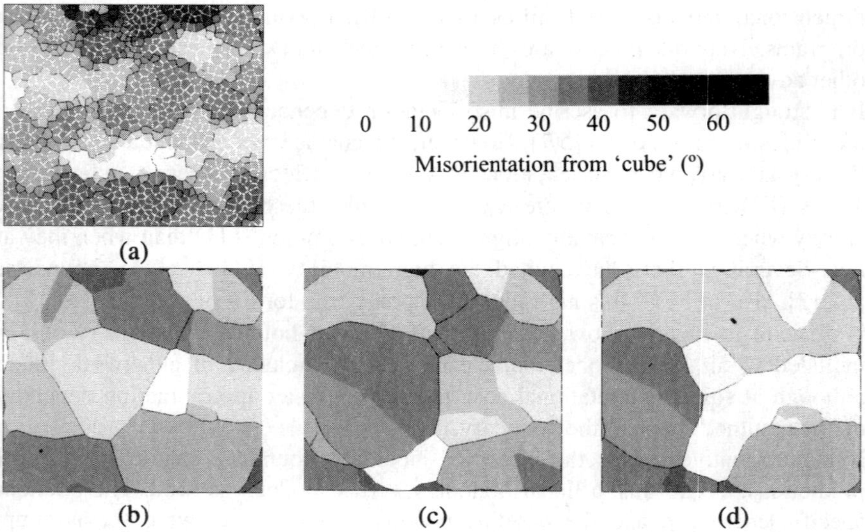

Fig. 1. Examples of the predicted effect of boundary mobility characteristics on the development of structure in a 2-d vertex model of annealing. The initial structure is shown in (a), with all boundaries shown: those in black have $\omega > 5°$. Using a simple dependance of mobility on ω, constant for all boundaries with $\omega > 15°$, is shown in (b). In (c), boundaries near 40° <111> have much enhanced mobility, and in (d) that enhancement is resticted to near- tilt boundaries. Only $\omega > 5°$ boundaries are shown in (b-d).

This leads to a problem. As noted above, it is difficult to see how boundary character could be dealt with in MCP models, especially if the relationships between character and boundary properties involve large first derivatives. Network models can deal with character, but three-dimensional versions are very difficult to produce and therefore rare, and assumptions about boundary shape are a limitation: discretising the surface in flat segments [10] is probably the best simple approach.

The requirement for an annealing model which can deal with boundary properties correctly is not the greatest difficulty. Although the structures and textures predicted are sensitive to those boundary properties, they are more sensitive to the initial structure used. The examples shown in Fig. 1 were derived from a finite element model. A simple relationship between the final, von Mises, flow stress and the 'sub-grain' density was used: the density was proportional to the square of the stress. This is an arbitrary translation, and if a fourth-power relationship is used, so that low flow stress regions are given a larger 'sub-grain' size, then a large difference in predicted annealed structure results, as shown in Fig. 2.

Irrespective of how the transformation between the output of a deformation model and the initial state of an annealing model is made, it is clear that any changes in the spatial distribution of orientation and other aspects of material state would affect the outcome of annealing simulations. The deformation model must be able to make a good approximation to the real deformed microstructure.

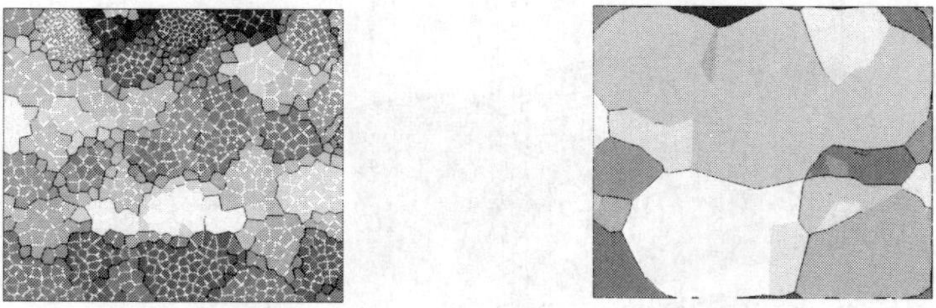

(a) (b)

Fig. 2. An example showing the effect of start structure on the development during annealing predicted by a 2-d vertex model. The initial structure shown in (a), as with Fig.1, was derived from CPFEM predictions of the deformed state. In the case of Fig. 1, the 'subgrain' density depended on the square of the von Mises final flow stress; here a fourth-power dependance was used. The simple boundary mobility function was used. The annealed structure, (b), can then be compared with Fig. 1b.

Irrespective of how the transformation between the output of a deformation model and the initial state of an annealing model is made, it is clear that any changes in the spatial distribution of orientation and other aspects of material state would affect the outcome of annealing simulations. The deformation model must be able to make a good approximation to the real deformed microstructure.

3 Predicting the Deformed Structure

The development of microstructure in models- as in real annealing- depends, then, critically on the initial conditions. This is largely determined by the output from the deformation model. The finite element method, using constitutive equations based on crystal plasticity (CPFEM) has become the dominant modelling tool. It promises, in particular, to be able to deal with the intergranular and intragranular inhomogeneity which is so important in the present context. The method has been developed and applied to a number of problems at the scale of the grain structure [16-20], and has yielded useful results. Output from CPFEM modelling has been used to give start structures for both MCP [21, 22] and network [20] annealing simulations.

An example of inhomogeneity predicted using CPFEM is shown in Fig. 3. In this, a repeating bicrystalline unit representing a layered structure of initially 'cube' and 'brass' orientations has been deformed in plane strain. Strain localisation has occurred, leading to a co-operative internal 'necking'. Although simple Taylor modelling predicts both orientations would be stable, the mechanical interaction means that this is not the case. Of interest is the crystal rotation about the compression direction which has led to material close to the 'P' orientation, {011}<111>. There is no other method available which could deal with this type of effect.

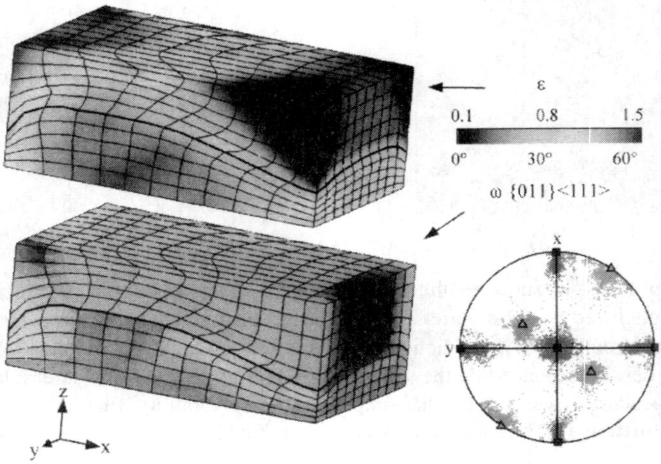

Fig. 3. CPFEM predictions of the deformation of a bicrystalline repeating unit, with the top orientation initially 'brass' and the bottom on initially 'cube', following plane strain (ε_{yy}=0) to ε_{xx}= 0.5 with octahedral glide and m=0.02. The von Mises accumulated strain distribution is shown at the top, and the orientations (as a 001 pole figure with initial orientations indicated) show a spread in the top 'grain' towards 'P' orientation: the lower Figure shows misorientation from that orientation.

Although CPFEM can help show the effects of grain arrangement and mechanical behaviour on the development of deformed microstructure, this class of deformation model cannot, at present, give a full simulation of the deformation microstructure for a variety of reasons [20], an important aspect that will be discussed below.

To demonstrate the sensitivity of annealing simulation to the results of CPFEM results, an example showing the effect of changing one factor in the constitutive equations on a subsequent annealing simulation is given in Fig. 4. An increase in the slip rate sensitivity index, m , from 0.02 used in the examples of Figs. 1 and 2, to 0.2 has changed the predicted deformation structure. Those changes are magnified in annealing simulation and give an obviously different structure- and texture- despite the texture after deformation being little different.

The examples given above all used a relationship between substructural scale and the final von Mises flow stress. As the Voce- type work hardening had 'saturated', this correlates closely with the local Taylor factor. A relationship between Taylor

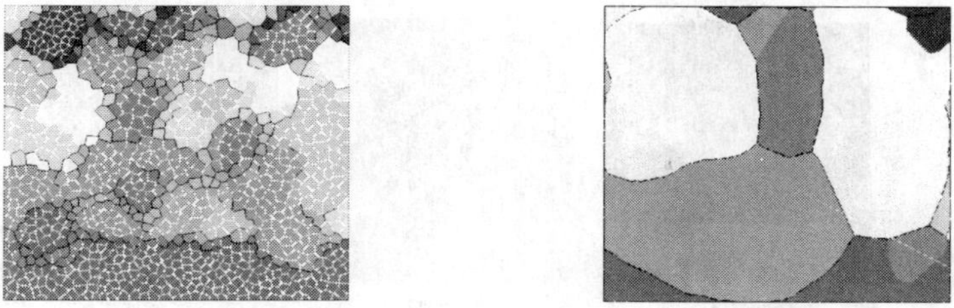

(a) (b)

Fig. 4. Example showing the effect of predicted deformation structure on the results from a 2-d vertex model. The initial structure shown in (a), as with Fig.1, was derived from CPFEM predictions of the deformed state but in this case a higher slip rate sensitivity was used: this has changed the overall pattern of deformation. As with Fig. 1, the 'subgrain' density depended on the square of the von Mises final flow stress. The annealed structure, (b), can then be compared with Fig. 1b.

factor and structure and stored energy has been observed, particularly in ferrite [23,24], and predicted in models [25]. It has been suggested [26] that high Taylor factor regions have a greater incentive to inhomogeneous deformation. This local inhomogeneity is, however, something that the CPFEM might be expected to predict explicitly.

Any expectation for CPFEM to fully predict inhomogeneity at sufficiently small scales for the direct determination of start structures for annealing models are over-optimistic. This is certainly the case where dynamic recovery is not sufficient to generate a condensed cell structure during deformation. Even if advances in 'strain gradient' plasticity [27] were to allow the inclusion of length scale and the direct effect of inhomogeneity on flow stress, the difference in these required for dislocation structures on one hand, and polycrystals on the other, is huge and would lead to very large models. A further problem is that structures generated in plastic deformation frequently show sharp discontinuities, and finite element models are best suited to continuous fields. Adaptive meshing would be needed- at considerable computational cost- to accommodate features such as shear bands and deformation bands correctly.

It seems then, that CPFEM must be used with the realization that much of the microstructural detail required for annealing models will remain described, in 'homogenized' form, by state variables in the constitutive equations. That information then has to be translated into variables suitable for use in annealing models. The von Mises flow stress involved in the examples above is a very simple example of this. To demonstrate further the significance of this, examples of an MCP simulation starting with data taken from the CPFEM results shown in Fig. 3, are given in Fig. 5. There is a vast difference between simulations using only orientation differences to give system energies, and those which include an additional site-specific energy determined (linearly) by the final von Mises flow stress.

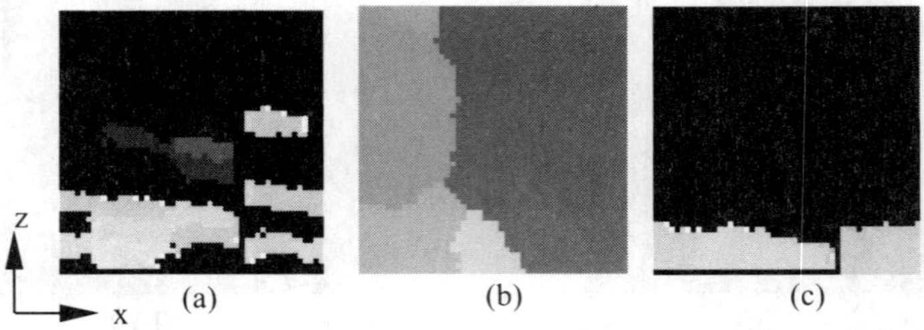

Fig. 5. Mid-plane sections from MCP simulations of annealing. The start structure (a) was derived from the right hand portion of the finite element simulation shown in Fig. 3. In (b), the MCP model has been run using inter-cell energies depending on misorientation, and the 'P' orientation becomes dominant. In (c), cell energies depending on the final von Mises flow stress were included as well, and a near-'cube' orientation grows. The 'cellular' appearance in (a) is simply due to the disposition of integration points in the finite element mesh.

This difference arises because the deformation simulation predicts greater orientation gradients in the 'cube' material, which has deformed more, but a higher von Mises stress in the near 'P' regions. The higher energy assumed with higher stress is notionally associated with greater inhomogeneity, at a scale below that explicitly dealt with in the CPFEM. The von Mises stress is only likely to be a very approximate indicator of substructural energy density. However, its use does demonstrate the need for some variable of state which adequately represents the real energy distribution and which can be deduced from CPFEM models. A very great problem arises if the deformation model cannot predict the distribution of crystal orientations in the deformed state, at least to the order of a small spread.
There has been work using dislocation dynamics (DD) which may allow rules for such a variable to be developed. Several statistical or stochastic approaches have been reported [28,29] with promising results. Coupled FE/DD is possible [30], but for the present application would need exceptionally large amounts of computation.

4 Summary

Some of the difficulties involved in modelling recrystallisation microstructures have been outlined above. In many important cases the separation into deformation and annealing models, as used here, would not be applicable: dynamic recrystallisation and grain growth are obvious cases, but even the ubiquitous phenomenon of dynamic recovery might need a coupled model to be truly effective. However, despite the attractiveness of coupled modelling and the fact that common aspects of microstructural description are involved, the separation is likely to remain for some time.

are limited because boundary character is important, and network/ surface models are difficult to code. A fuller knowledge of crystal boundary properties is certainly required. Experimental work on this is making significant progress [31], as is the use of atomic scale modelling [32,33]. The effects of solute interaction and particles, perhaps co-precipitating or dissolving during annealing, will only make this more difficult.

CPFEM is the most promising way of modelling inter- and intragranular inhomogeneity- and that due to particles- but it is a continuum method attempting to model a discontinuous phenomenon. Many microstructural details will remain hidden as averaged state variables. Where that averaging includes significant orientation spread, there will be great problems in formulation and implementation. Those state variables need to be translated into adequate field variables for annealing models. The static recovery process also has to be considered in this context as well, as do possible complications such as twinning during annealing.

The size of problems required to give a reasonable representation of microstructure is large, but given the rate at which computer capability is increasing, this is the least of the problems in the growing field of microstructural modelling.

Acknowledgement

The author thanks the EPSRC (GR/L65857) and Alcan International for their support. Thanks also to F. J. Humphreys, B. Hutchinson and G. Gottstein for their suggestions and comments.

References

[1] M.P. Anderson, D.J. Srolovitz, G.S. Grest, and P.S. Sahni, *Acta Metall.*, 32, 1984, pp. 783-791 and 793-802.

[2] M.P. Anderson, G.S. Grest, and D.J. Srolovitz, *Phil. Mag. B.*, 59, 1989, pp. 293-329.

[3] D.J. Srolovitz, G.S. Grest and M.P. Anderson, *Acta Metall.*, 34, 1986, pp. 1833-1845.

[4] P. Tavernier and J.A. Szpunar, *Acta Metall. Mater.*, 39, 1991, pp. 549-556 *and* 557-567.

[5] D. Raabe, *Acta Mater.*, 48, 2000, pp. 1617-1628.

[6] K. Kawasaki, T. Nagai and K. Nakashima, *Philos. Mag. B*, 60, 1989, pp. 399-421.

[7] F.J. Humphreys, *Scripta Metall. Mater.*, 27, 1992, pp. 1557-1562.

[8] H.J. Frost, C.V. Thompson, C.L. Howe and J.H. Wang, *Scripta Metall.*, 22, 1988, pp. 65-70.

[9] T.O. Saetre, O. Hunderi and N. Ryum, *Acta Metall. Mater.*, 37, 1989, pp. 1381-1387.
[10] D. Weygand, Y. Bréchet and J. Lépinoux, *Acta Mater.*, 47, 1999, pp. 961-970.
[11] Cl. Maurice, in „Proc. 21st Risø International Symp. on Materials Science", eds. N. Hansen *et al.*, Risø Nat. Lab., Denmark, 2000, pp. 431-437.
[12] B. Liebmann, K. Lücke and G. Masing, *Z. Metallk.*, 47, 1956, pp. 57-63.
[13] D. A. Molodov, U. Czubayko, G. Gottstein and L.S. Shvindlerman, *Scripta Metall. Mater.*, 32, 1995, pp. 529-534.
[14] M.G. Ardakani and F.J. Humphreys, *Acta Metall. Mater.*, 42, 1994, pp. 763-780.
[15] Y. Huang and F.J. Humphreys, *Acta Mater.*, 48, 2000, pp. 2017-2030.
[16] D. Peirce, R.J. Asaro and A. Needleman, *Acta Metall*, 31, 1983, pp. 1951-1976.
[17] A.J. Beaudoin, K.K. Mathur, P.R. Dawson and G.C. Johnson, *Int. J. Plasticity*, 9, 1993, pp. 833-860.
[18] R. Becker and S. Panchanadeeswaran, *Acta Metall. Mater.*, 43, 1995, pp. 2701-2719.
[19] G.B. Sarma, B. Radhakrishnan and T. Zacharia, *Comp. Mater. Sci.*, 12, 1998, pp. 105-123.
[20] P. Bate, *Philos. Trans. Roy. Soc. A*, 357, 1999, pp. 1589-1610.
[21] G. Sarma, B. Radhakrishnan and T. Zacharia, in „Proc. ReX'96", ed. T. R. McNelley, MIAS, California, 1996, pp. 555-562.
[22] B. Radhakrishnan, G.B. Sarma and T. Zacharia, *Acta Mater.*, 46, 1998, pp. 4415-4433.
[23] I.L. Dillamore, C.J.E. Smith and T.W. Watson, *Metal Sci. J.*, 1, 1967, pp. 49-54.
[24] R.L. Every and M. Hatherly, *Texture*, 4, 1974, pp. 183-194.
[25] A. Miroux, L. Badea, D. Arizmendi, B. Bacroix and L.J. Raphanel, *Archives of Metallurgy*, 44, 1999, pp. 241-252.
[26] B. Hutchinson, *Philos. Trans. Roy. Soc. A*, 357, 1999, pp. 1471-1485.
[27] J.W. Hutchinson, *Int. J. Solids and Structures*, 37, 2000, pp. 225-238.
[28] W. Pantleon, *Acta Mater.*, 46, 1998, pp. 451-456.
[29] I. Groma and B. Bakó, *Phys. Review Letter*, 84, 2000, pp. 1487-1490.
[30] M. Verdier, M. Fivel and I. Groma, *Mod. Sim. Mat. Sci. Eng.*, 6, 1998, pp. 755-770.
[31] B.L. Adams, D. Kinderlehrer, W.W. Mullins, A.D. Rollett and Shlomo Ta'asan, *Scripta Mater.*, 38, 1998, pp. 531-536.
[32] B. Schönfelder, D. Wolf, S.R. Phillpot and M. Furtkamp, *Interface Science*, 5, 1997, pp. 245-262.
[33] M. Upmanyu, D.J. Srolovitz, L.S. Shvindlerman and G. Gottstein, *Acta Mater.*, 1999, pp. 3901-3914.

Recrystallization and Grain Growth Textures in Low Carbon Steels

J.J. Jonas[1] and L. Kestens[2]

[1] McGill University, Department of Metallurgical Engineering, 3610 University Street, Montreal, Quebec, Canada, H3A 2B2
[2] Ghent University, Department of Metallurgy and Materials Science, Technologiepark 9, B-9052 Ghent, Belgium

Keywords: steel, recrystallization, textures, orientation imaging microscopy.

Abstract. The various nucleation and growth mechanisms that play a role during the formation of annealing textures in cold rolled low carbon steels are reviewed. The occurrence of oriented nucleation is discussed, both in terms of the stored plastic deformation energy and of the heterogeneities of the deformation substructure and microstructure. It is now generally accepted that, after medium to high rolling reductions, the first recrystallization nuclei appear in the microstructure at sites with an increased deformation energy, such as at the strain localizations produced by in-grain shear bands. In general, such high stored energy nucleation gives rise to a {111} nucleation texture without marked preference for any particular component of the {111} fibre.
During the subsequent growth stage, the growth velocity will be affected by the specific misorientation relationship that links the growing nucleus to the surrounding matrix material. Clear evidence of selective growth on the basis of an increased mobility associated with 30° <110> misorientations is observed in the case of very strong matrix textures, such as produced by severe rolling reductions. This mechanism accounts for the appearance of the {554}<225> and other non-{111} components of the annealing texture in the final stages of the recystallization process.

1 Introduction

The study of recrystallization textures in low carbon, single-phase ferritic steel sheets is a well established subject in the literature. Most of the investigations have been of a phenomenological nature; they are therefore not suitable for the quantitative prediction of annealing textures in alloys of a given chemistry subjected to a well-defined thermomechanical treatment. That is, our understanding of the subject has not yet led to a recrystallization model that can be employed to

predict texture-dependent properties such as deep drawability or magnetic properties with any accuracy.
The reason for this lack of precision must be found in the complexity of the recrystallization process itself. As will be reviewed below, there are several recrystallization mechanisms that can operate concurrently in a real material. A quantitative description is required for each of these mechanisms, each of which depends critically in turn on accurate knowledge of the deformed substructure. Although such single mechanism theories are not yet fully developed, they cannot readily provide a general theory of recrystallization because the various mechanisms operate on a competitive base and complex highly non-linear interactions can occur. In contrast to the theory of deformation textures, local heterogeneities (e.g. precipitates, shear bands and grain boundaries) cannot be treated as short range second order effects but, on the contrary, provoke long range first order effects during recrystallization.
Although a comprehensive theory has not yet been developed, a reasonable consensus has nevertheless been achieved at the level of a phenomenological description of textures and microstructures in cold rolled and annealed materials. This has been extensively reviewed by Hutchinson [1] as well as by Ray et al. [2]. The present paper will survey some of the more recent publications that have furthered our understanding of particular issues regarding both oriented nucleation and selective growth. In the present context, only those theories will be discussed that encompass orientation selection during recrystallization. Analyses that describe solely the microstructural or kinetic aspects (e.g. the JMAK analysis [3]) will not be treated. With regard to orientation selection, two fundamentally different types of theories can be discerned: those of *oriented nucleation* and of *selective growth*. The former are based on mechanisms that involve orientation selection during the nucleation stage of recrystallization, whereas the latter theories assume that orientation change takes place during the growth of nuclei into the deformed matrix. If site saturated oriented nucleation prevails, the recrystallization texture is fully determined at the onset of recrystallization. Conversely, when selective growth is taking place, a random nucleation texture gradually develops into the annealing texture during the growth stage following nucleation. These two simple hypothetical examples show that both oriented nucleation and selective growth can occur simultaneously.

2 Oriented Nucleation Theories

The nucleation stage of recrystallization is one of the least understood, probably because it is the most difficult to observe experimentally. It is well established that true nucleation in the thermodynamic sense does not occur during the recrystallization of cold deformed metals [4]. Nuclei can be defined as dislocation free crystallites present in the deformed structure with a critical volume of ~1 μm^3 and a sufficiently mobile external boundary. These nuclei are generated during the recovery stage either by a mechanism of subgrain coalescence [5] or by subgrain growth [6]. Alternatively, recrystallization nuclei can simply be embedded in the deformed substructure and not require the operation of a thermally activated process of sub- boundary displacement for their formation [7].

It is clear that, irrespective of the precise nature of the nucleation mechanism, the deformed structure is highly heterogeneous and some locations will be more prone to provide nuclei than others. These sites of highly potent nucleation seem to be well correlated with the deformation energy that is locally stored in the form of the dislocation substructure. Most important for nucleation in the steels destined for high formability applications are the sites where the deformation energy is higher than average. These are the *hard* orientations according to plasticity theory (i.e. the high Taylor factor orientations), as well as crystallites near grain boundaries, shear bands and second phase particles. It has frequently been observed that nucleation is favoured in these regions of high stored energy [8,9].

In order to model this type of high stored energy nucleation, Kestens and Jonas [10] proposed a nucleation theory in which each orientation g of the deformed matrix is assigned a probability $P_N(g)$ of generating a nucleus of the same orientation. This nucleation probability $P_N(g)$ was assumed to depend on the Taylor factor M according to the following criterion:

$$M < M_0 : P_N(g) = 0$$

$$M \geq M_0 : P_N(g) \propto (M(g)-M_0)^m \qquad (1)$$

According to Eq. 1, nucleation is prohibited when the Taylor factor of a deformed orientation g does not exceed an arbitrary critical threshold value M_0. In the above nucleation law, an exponent m is introduced that allows the dependence of the nucleation mechanism on the Taylor factor distribution M(g) to be varied.

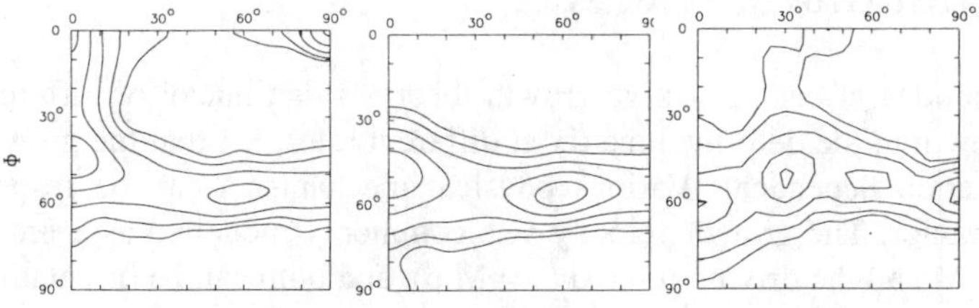

(a) (b) (c)

Fig. 1. $\varphi_2 = 45°$ sections of (a) an experimental texture of a Ti-Nb IF steel (0.0018mass%C, 0.079 mass%Ti and 0.010 mass%Nb) rolled to 85% reduction, (b) nucleation texture modelled by applying Eqs. 1 and 2 to (a) with m = 2 and M_0 = 3.0. (c) nucleation texture measured on a similar material with a recrystallized volume fraction of 5% [12]. Levels: 1-2-4-7-10.

The essential feature of this model is the assumption that the Taylor factor is an appropriate parameter for the quantification of the dislocation density of individual orientations following a large amount of plastic deformation and hence for the

probability of finding these orientations in the nucleation texture. Although this is not valid for arbitrary orientation paths, it seems to apply reasonably well to the simple case of rolling at room temperature, a conclusion that was recently confirmed by Miroux et al. [11].
Employing the nucleation probability of Eq. 1, the nucleation texture $f^N(g)$ can be calculated using the following formula, which states that the nucleation texture results from the operation of the nucleation mechanism on the deformation ODF:

$$f^N(g) \propto P_N(g)\, f^D(g) \tag{2}$$

where $f^D(g)$ represents the deformation texture. In Eqs. 1 and 2, the proportionality signs can be replaced by equality signs when the right-hand side expressions are normalized.
Given the cold rolling texture of a Ti-Nb stabilized interstitial-free (IF) steel (cf. Fig. 1a) and employing the nucleation potential of Eq. 1 (with $M_0 = 3.0$ and $m = 2$), the nucleation texture of Fig. 1b was calculated. This texture can be compared with the experimental nucleation ODF of Fig. 1c reported by Vanderschueren et al. [12], obtained by measuring more than 500 individual nucleus orientations employing the electron backscattering diffraction (EBSD) technique. Although the experimental and simulated ODFs display remarkable similarities, some differences remain, most noticeably with regard to the intensity distribution along the ND fibre and the exact position of the fibre skeleton. Compared to the modelled texture, the experimental texture exhibits a more marked curvature of the fibre skeleton with respect to the perfect ND fibre skeleton, located exactly at $\Phi = 54.7°$ and $\varphi_2 = 45°$. Another publication in this volume deals more specifically with the origin of the curvature in the fibre skeleton [13].

3 Selective Growth Theories

The basic idea of each selective growth theory is that nuclei of different orientations grow into the deformed matrix at different rates, i.e. that the growth velocity is orientation dependent. Various physical mechanisms can be responsible for such behavior. The growth velocity v is commonly modelled as a product of the mobility M and the driving force p ($v = M\,p$), and both can be orientation dependent. Hence, selective growth theories can be divided into two categories according to whether they rely on an orientation dependent driving force or an orientation dependent mobility.

3.1 Orientation Dependence of the Driving Force

The main driving force for growth is provided by the difference in dislocation density across the interface surrounding a growing nucleus. It was noted above that this dislocation density is highly orientation dependent. Apart from its effects on nucleation, these local differences in dislocation density can also generate local growth effects. On a statistical scale, however, the orientation dependence of the stored energy cannot be responsible for long range selective growth because the former does not depend on the local misorientation between the nucleus and the deformed matrix. In a mean field approach, each nucleus encounters the same *average* deformed matrix and so the average driving force encountered by each nucleus does not vary.

These remarks do not apply, however, to the theories in which the driving force is considered to depend on the local misorientation between the nucleus and the deformed grain. One such theory was proposed by Lee [14] and Park et al. [15]. They assume that the fastest growing nuclei are the ones in which the direction of minimum Young's modulus is aligned with the direction of maximum (absolute) principal stress in the deformed grain. The basic idea of this theory is that expansion along the softest elastic direction provides the fastest mode of relaxing the stress fields associated with the dislocation configurations within the deformed grains. The theory seems to be particularly successful in explaining the recrystallization behaviour of thin films on metallic substrates [16], but has also been applied to bulk BCC materials (e.g. IF steels) submitted to various deformation modes prior to annealing [15,17].

Another theory of selective growth that employs the orientation dependence of the driving force was proposed by Hayakawa and Szpunar [18]. They do not consider the driving force produced by the gradient of dislocation density across the interface but that associated with the energy of the interface itself, i.e. the grain boundary energy. As a result, their theory applies equally to normal and abnormal grain growth. They rely on data showing that the grain boundary energy is highest for boundaries displaying misorientations of near 30°, without regard to the misorientation axis. Because these high energy boundaries will be removed preferentially from the microstructure, this leads to a pronounced selective growth effect. According to Hayakawa and Szpunar [19], this effect is responsible for the development of the Goss component during the secondary recrystallization of electrical steels.

3.2 Orientation Dependence of the Mobility

It is generally accepted that boundaries with misorientations of less than 15 ° will exhibit reduced mobility. According to Juul Jensen [20], this property is responsible for a pinning effect affecting the orientations that display the smallest average misorientations with respect to their local environments. Employing a technique of growth rate determination combined with local orientation measurements, Juul Jensen [21] has shown that orientation pinning is primarily responsible for the development of the cube texture component during the recrystallization of Cu and Al.

According to a more controversial theory, specific crystallographic misorientations are associated with increased mobility of the interface. Although grain boundary mobility has been extensively studied for FCC materials, there are virtually no basic experimental data available for BCC structures [22-24]. Even for FCC materials, the reported mobility rates are circumstantial and depend critically on temperature, purity and type of interface (i.e. twist or tilt boundary). Nevertheless, the data show that the grain boundary mobility of FCC metals can be strongly dependent on misorientation and in general the <111> 40° misorientation is associated with increased mobility.

Because of the scarcity of such basic data, they have been supplemented with the results of annealing experiments in which texture components that are observed to grow preferentially are interpreted as resulting from their favourable misorientation relations with respect to the disappearing components. From such experiments, it has been concluded that ≈<110>30° is a preferential fast growth misorientation for BCC materials. Perhaps the best known experiment of this type was carried out by Ibe and Lücke [25] on single crystals of Fe-3% Si that were slightly deformed and in which a random nucleation texture was artificially introduced. The exact misorientation of rapid growth is commonly quoted to be <110>26.5°, because this corresponds to the misorientation associated with a Σ19 coincident site lattice (CSL) boundary.

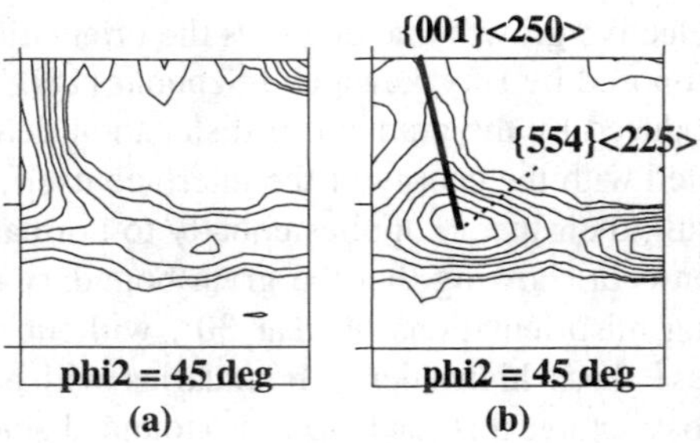

Fig. 2. (a) Deformation texture of the 95% cold rolled ULC steel and (b) recrystallization texture of the same steel after annealing at 750°C for 60 s. Levels: 1-2-4-6-8-11-14.

3.3 Experimental Verification of Selective Growth by Orientation Imaging Microscopy

It is now possible to list some necessary conditions that must be met by each selective growth theory, irrespective of the precise nature of the theory itself:

(i) Grains that are favored by growth selection must exhibit a larger than average size after the growth stage.

(ii) There must be a close correlation between the increasing intensities of the favoured components and the decreasing intensities of the deformation components that are misoriented according to the rapid growth relation.

(iii) The growth advantage of a selected component must be attributed to a selection mechanism and not to a size advantage at the start of the growth stage;

i.e. early nucleation cannot be responsible for growth selection in the later stages of recrystallization.

(iv) The growth selectivity must be reversible. This argument is based on the assumption that if component A is selected to grow into component B, the opposite will occur when the driving force is reversed.

Employing the technique of orientation imaging microscopy (OIM), Kestens et al. [26,27] have recently described experiments in which all these conditions were met in the case of a heavily deformed ultra low carbon (ULC) steel containing 24 ppm C. The deformation texture obtained after a rolling reduction of 95% is displayed in Fig. 2a, whereas Fig. 2b shows the recrystallization ODF produced by annealing at 750°C for 60 s. It is evident that a typical ND fibre texture developed during recrystallization, with a strong maximum at the {554}<225> component. Furthermore, a second, much weaker fibre is present running from the {554}<225> to the {001}<250> component. In the $\varphi_2 = 45°$ section, this fibre is almost parallel to the RD fibre.

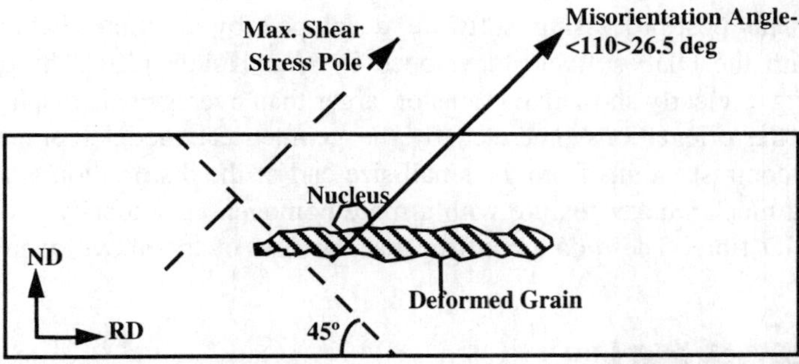

Fig. 3. Schematic representation of the rotation axis associated with higher grain boundary mobility.

According to the model of Kestens and Jonas [10], given a particular deformation texture, the growth probability $P_N(g)$ can be calculated for any arbitrary nucleus orientation g on the basis of a particular geometric relationship for preferred growth, e.g. <110>26.5°. The model also implies a variant selection rule according to which only those <110> rotation axes are selected that are closest to the poles of maximum shear stress. This variant selection rule was derived empirically from observations on both FCC and BCC materials and for different deformation modes (torsion and rolling) [28,29]. This geometrical condition for rapid growth of a nucleus into the deformed matrix is represented schematically in Fig. 3.

When this model is applied to the rolling texture of the 95% cold rolled ULC steel (cf. Fig. 2a), the growth probability distribution of Fig. 4 is obtained, which clearly displays a maximum at the {554}<225> component and a minor vertical fibre, parallel to the RD fibre. The model predicts a local maximum on this vertical fibre at $(\varphi_1, \Phi, \varphi_2) = (20, 20, 45°)$. Although such a local maximum is not evident in the macroscopic texture of Fig. 2b, it can be readily distinguished in the local orientation data recorded by OIM.

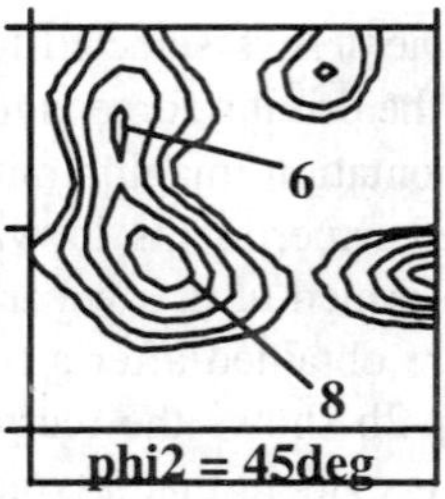

Fig. 4. Growth probability distribution calculated by applying the model of Kestens and Jonas [10] to the rolling texture of Fig. 2a and assuming the growth law of Fig. 3.

When the fully recrystallized sample were mapped by OIM, the micro-structure, grain size distribution and local ODFs of Fig. 5 were obtained; this was done by employing the post-processing software developed by TexSem Laboratories® together with the ODF software developed by Van Houtte [30]. The data displayed in Fig. 5 clearly show that grains of larger than average size display strong maxima at the orientations predicted by the geometrical model (compare with Fig. 4). By contrast, grains from the small size end of the distribution are characterized by a much weaker texture with a more homogenous intensity distribution along the ND fibre. These data confirm that the first of the above conditions is fulfilled.

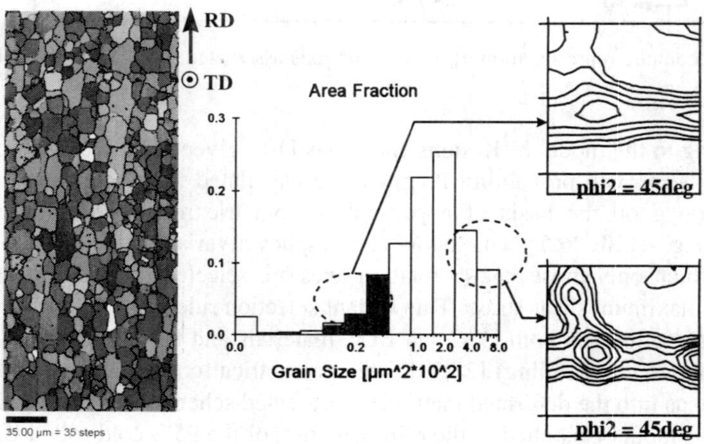

Fig. 5. OIM scan of sample quenched after heating to 720°C. The textures of the coarse and fine grains are shown separately, together with the grain size distribution. Levels: 1-2-4-6-8-11-14.

A partial recrystallization experiment carried out on the same material permitted the second condition to be verified. As shown in Fig. 6, the intensity along the RD fibre only disappears during the late stages of recrystallization when the recrystallized volume fraction gradually reaches 100%. This figure also reveals that the disappearance of the RD fibre components is closely correlated with the increase in intensity of the {111}<211> (or near {554}<225>) component in the recrystallization texture. This indicates that the {111}<211> orientations grow more quickly into the remaining RD fibre of the deformed matrix than the {111}<110> grains. This orientation change, which occurs during the final stages of recrystallization, is unlikely to be controlled by nucleation because all the high stored energy sites of the deformed substructure have already been consumed by nucleation in the earlier stages.

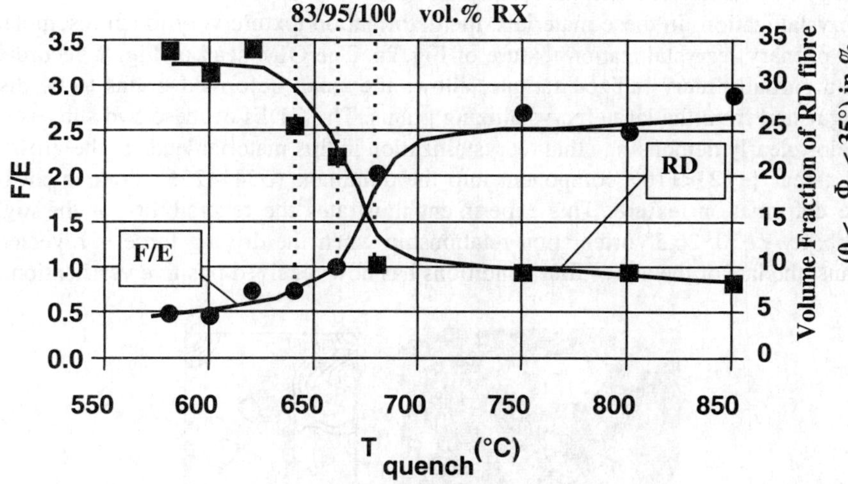

Fig. 6. Dependence on the quench temperature of the intensity ratio of the {111}<211> (F) to {111}<110> (E) component and of the volume fraction of the RD fibre.

The early nucleation texture can also be measured by OIM. In order to obtain statistically reliable data, the results of four OIM scans were compiled into a single ODF, displayed in Fig. 7. This texture represents the orientation distribution of almost 400 nuclei scattered over a sample area of approximately 1.5 mm x 0.15 mm when the recrystallized volume fraction was less than 5%. It is evident that the components that are significantly intensified during the late stages of recrystallization are far less dominant in the early nucleation texture; the maximum in the latter texture is located at the {111}<110> instead of at the {554}<225> component. This indicates that growth selection during the late stages of recrystallization is unlikely to be attributable to a size advantage due to early nucleation. Hence, these findings verify the third condition of the above list.

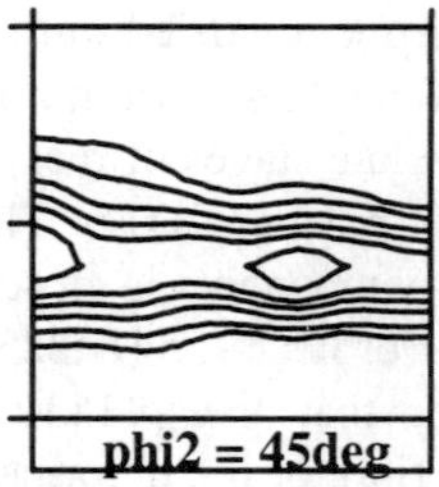

Fig. 7. ODF of nearly 400 recrystallization nuclei compiled from 4 OIM scans on a sample containing a recrystallized volume fraction of <5%. Levels: 1-2-4-6-8-11.

Some grades of electrical steel were rolled slightly (4 to 8% reduction) after primary recrystallization and then annealed again to produce an additional cycle of recrystallization. In these materials, the deformation texture very much resembled the primary recrystallization texture of Fig. 2b. The OIM scan of Fig. 8, recorded on a partially recrystallized sample, allows the small deformed grains to be distinguished from the large recrystallizing grains. The ODFs of these two subsets of grains clearly demonstrate that recrystallization in this material leads to the growth of a near {112}<110> component into the dominant {554}<225> component of the deformation texture. This experiment illustrates the reversibility of the high mobility <110>26.5° orientation relationship when the driving force is reversed. Thus, the last of the above four conditions has now received positive verification.

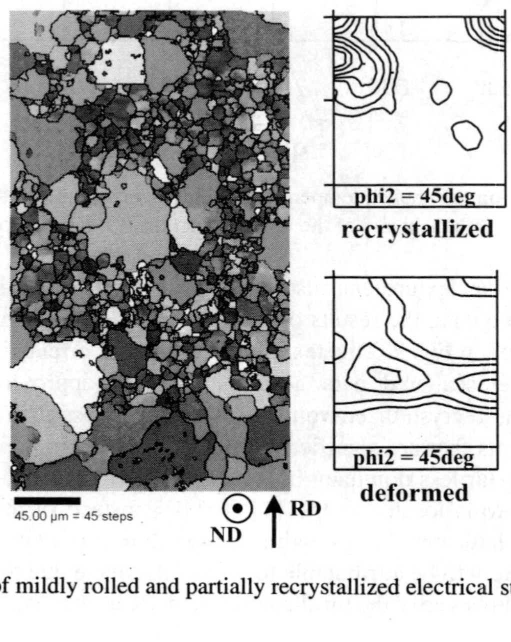

Fig. 8. OIM scan of mildly rolled and partially recrystallized electrical steel sample. Levels: 1-2-4-6-8-11-14.

4 Summary and Conclusions

The latest publications regarding orientation selection during the recrystallization of cold deformed low carbon steels were reviewed. The model of Kestens and Jonas [10] for high stored energy nucleation was reconsidered and evaluated, for the case of a Ti-Nb interstitial free steel, by comparing the predicted nucleation texture with the experimental one. The small discrepancies between the two textures can be attributed to the occurrence of heterogeneous nucleation at particular sites in the microstructure.

Using the *Orientation Imaging Microscopy* technique it was demonstrated that the dominant annealing components of a severely cold rolled ultra low carbon steel develop most strongly during the late stages of recrystallization, when the recrystallized volume fraction sluggishly reaches 100%. In the fully annealed condition, the {554}<225> oriented grains are among the largest in size and their growth correlates well with the disappearance of the last remaining {211}<110> orientations of the deformed matrix. These results, together with the observation that {554}<225> orientations were virtually absent in the early recrystallization texture, strongly support the view that selective growth during the late stages of recrystallization modifies the early texture produced by oriented nucleation. It is important to note that, although these data did not reveal the operation of a specific mechanism of selective growth, they were consistent with a number of common features of various selective growth models (as opposed to oriented nucleation mechanisms).

References

[1] W.B. Hutchinson, *Int. Metall. Rev.*, 29, 1984, p. 25.
[2] R.K. Ray, J.J. Jonas and R.E. Hook, *Int. Mater. Rev.*, 39, 1994, p. 129.
[3] F.J. Humphreys and M. Hatherly, "Recrystallization and Related Annealing Phenomena", Elsevier Science Ltd., Oxford, 1995, pp. 188-191.
[4] W.B. Hutchinson, *Scripta Met.*, 27, 1992, p. 1471.
[5] H. Hu, *Trans. Metall. Soc. A.I.M.E.*, 224, 1962, p. 75.
[6] C.J.E. Smith and I.L. Dillamore, *Met. Science J.*, 4, 1970, p. 161.
[7] P.A. Beck and P.R. Sperry, *J. Appl. Phys.*, 21, 1950, p. 150.
[8] R.H. Goodenow, *Trans. ASM*, 59, 1966, p. 804.
[9] H. Inagaki, *ISIJ Int.*, 34, 1994, p. 313.
[10] L. Kestens and J.J. Jonas, *Metall. and Mater. Trans.*, 27A, 1996, p. 155.
[11] A. Miroux, L. Badea, D. Arizmendi, B. Bacroix and L.J. Raphanel, *Archives of Metallurgy*, 44, 1999, pp. 241-252.
[12] D. Vanderschueren, N. Yoshinaga and K. Koyama, *ISIJ*, 36, 1996, p. 1046.
[13] L. Kestens, K. Verbeken and J.J. Jonas, "Orientation Selection by Nucleation and Growth in Highly Strained Low Carbon Steels", this proceedings.
[14] D.N. Lee, *Scripta Met.*, 32, 1995, p. 1689.

[15] Y.B. Park, D.N. Lee and G. Gottstein, *Acta Met.*, 46, 1998, p. 3371.
[16] J.H. Choi, S.Y. Kang and D.N. Lee, *J. Mat. Sci.*, 35, 2000, pp. 4055-4066.
[17] Y.B. Park, L. Kestens and J.J. Jonas, in "Proc. ReX'99", eds. T. Sakai and H.G. Suzuki, The Japan Institute of Metals, 1999, pp. 751-756.
[18] Y. Hayakawa and J.A. Szpunar, *Acta Met.*, 45, 1997, p. 3721.
[19] Y. Hayakawa and J.A. Szpunar, *Acta Met.*, 45, 1997, p. 1285.
[20] D. Juul Jensen, "Orientation Aspects of Growth During Recrystallization", Risø National Laboratory, Risø-R-987 (EN), 1997, p. 49.
[21] D. Juul Jensen, *Acta Met.*, 43, 1995, p. 4117.
[22] F.J. Humphreys and M. Hatherly, "Recrystallization and Related Annealing Phenomena", Elsevier Science Ltd., Oxford, 1995, pp. 85-126.
[23] G. Gottstein, D.A. Molodov and L.S. Shvindlerman, in "Proc. Third Int. Conf. on Grain Growth, ICGG-3", eds. H. Weiland, B.L. Adams and A.D. Rollett, 1998, p. 373.
[24] D. Wolf, *J. Appl. Phys.*, 69, 1991, p. 185.
[25] G. Ibe and K. Lücke, *Archiv für das Eisenhüttenwesen*, 39, 1968, p. 693.
[26] L. Kestens, N. Yoshinaga, J.J. Jonas and Y. Houbaert, in "Proc. Third Int. Conf. on Grain Growth, ICGG-3", eds. H. Weiland, B.L. Adams and A.D. Rollett, 1998, p. 621.
[27] L. Kestens, N. Yoshinaga, D. Vanderschueren and B.C. De Cooman, in "Proc. ReX'99", eds. T. Sakai and H.G. Suzuki, The Japan Institute of Metals, 1999, pp. 745-750.
[28] L.S. Tóth and J.J. Jonas, *Scripta metall.*, 27, 1992, p. 359.
[29] T. Urabe and J.J. Jonas, *ISIJ Int.*, 34, 1994, p. 435.
[30] P. Van Houtte, "The MTM-FHM Software System Version 2, Users Manual", 1995.

Recrystallization and Grain Growth
Proceedings of the First Joint International Conference
Eds. G. Gottstein and D.A. Molodov, Springer-Verlag, 2001

High-Temperature Structure and Properties of Grain Boundaries by Molecular-Dynamics Simulation

D. Wolf[1], V. Yamakov[1,2], P. Keblinski[3] , S. R. Phillpot[1] and H. Gleiter[2]

[1]Materials Science Division, Argonne National Laboratory, Argonne, IL, 60439
[2]Institute for Nanotechnology, Forschungszentrum Karlsruhe, 76021 Karlsruhe
[3]Dept. of Materials Science and Engineering, RPI, Troy, NY 12180

Keywords: grain boundary, structure, diffusion, mobility, diffusion creep.

Abstract. Recent molecular-dynamics simulations of the high-temperature structure of high-energy grain boundaries (GBs) in Si and fcc metals reveal a universal, highly confined, liquid-like structure. High-temperature properties, such as the GB diffusivity and mobility, therefore involve liquid-like mechanisms, with activation energies related to the melt. These GBs are therefore rate controlling in many high-temperature properties of polycrystalline materials. Consistent with experiments, at lower temperatures they undergo a reversible structural and dynamical transition from a confined liquid to a more or less disordered solid structure.

1 Introduction

Grain growth and deformation of polycrystalline materials are strongly influenced by the types and spatial distribution of the grain boundaries (GBs) in the microstructure. It is well known that the evolution of polycrystalline microstructures under the driving forces of temperature and stress involves three distinct types of GB processes, namely GB diffusion, migration and sliding which, in turn, are governed by the underlying GB structure and energetics. Achieving a better understanding of the evolution of polycrystalline microstructures therefore requires fundamental insights on the interrelation among high-temperature GB structure, the underlying GB energy and these three types of high-temperature GB processes.

A polycrystal contains GBs with a wide spectrum of energies, with very different structures and properties. Much work of recent years has suggested the usefulness of distinguishing among the following three different types of GBs: special high-angle, dislocation and general high-angle GBs. (For a recent overview, see [1].)

Special high-angle GBs contain no dislocations, similar to "special", i.e., step-free or flat surfaces. Giving rise to energy cusps in the GB-misorientation phase space, the absence of dislocations in their atomic structure is the reason that their properties are very much perfect-crystal like; i.e., their diffusivity, sliding resistance and mobility are very low while their cohesion is very high. In practice, a polycrystalline microstructure therefore evolves as though these GBs are not present at all, with a correspondingly larger grain size.

Dislocation boundaries are "low-angle" or "vicinal" interfaces whose atomic structure consists of periodic arrays of dislocations accommodating the deviation from the perfect crystal or a "special", i.e., cusped, GB misorientation [2,3]. For small vicinal deviations from a special GB-plane orientation or a particularly small GB planar unit cell [1,3], dislocations separated by elastically strained perfect-crystal regions are formed to accommodate the misfit between the two grains. Analogous to vicinal surfaces, the energy and physical properties of dislocation boundaries can be expressed in terms of the misorientation angle and axis with respect to the special misorientation, i.e., in terms of the structure and properties of isolated lattice dislocations and their elastic interactions [1-4].

General high-angle boundaries, the least understood - yet most important - GBs, control the evolution of polycrystalline microstructures. There has been much debate on their distinction from dislocation GBs, particularly on whether a dislocation picture is appropriate at high angles where the dislocation cores overlap completely. For example, in 1952 C. S. Smith [5] suggested that "...there is a continuous transition from the pure dislocation boundary to whatever exists at higher angles...", and he suggested that the experimental evidence "...does not in any way preclude a nearly horizontal curve of energy vs. misorientation once the maximum value has been achieved", implying that the energy of these high-angle GBs is independent of the misorientation between the grains. That Smith's intuition was correct was confirmed by extensive simulations of GB energies [3], revealing that there are, indeed, many high-angle GBs whose energy is entirely misorientation-independent. Because their atomic structure involves completely overlapping dislocation cores, by contrast with dislocation boundaries the structural disorder is distributed more or less homogeneously along these interfaces. We think of these boundaries as "true" high-angle GBs. A structural model that accounts for their unusual properties, including a high GB mobility and GB diffusivity coupled with a low sliding resistance, and also connects with other types of structural disorder, such as the amorphous state, is clearly needed.

In this paper we review recent molecular-dynamics (MD) simulations investigating the high-temperature structure and diffusion behavior of these high-energy, high-angle GBs. Many of the earlier simulations of high-temperature GB structure remain controversial (for a critical review, see [6]) in that they predict "premelting" at the GBs, i.e., GB disordering below the bulk melting point, T_m, with a width of the liquid-like GB layer that diverges as $T \rightarrow T_m$, where the GB is completely wetted by the liquid. It was not until the simulations had provided a clearer understanding of the melting process itself [7,8], and of the related superheating limit [8,9], that a clearer picture of high-temperature GB structure started to emerge.

The MD simulations reviewed here expose the structural reasons for the unique high-temperature properties of these GBs. The picture that emerges suggests that upon heating from 0K to melting, the highest-energy GBs in the polycrystal undergo a reversible structural and dynamical transition from a low-temperature, solid to a highly confined, liquid-like structure. Above the GB-energy dependent transition temperature, T_c, small bubbles of liquid are formed within the solid GB until, at T_m, the entire – yet still highly confined - GB region is consumed by the liquid. This continuous transition between T_c and T_m profoundly affects the high-temperature properties of these GBs, including GB diffusion, migration and sliding. These unique high-temperature properties are responsible for the rate-controlling role of these GBs in the evolution of polycrystalline microstructures under the driving forces of stress and temperature.

2 High-Angle, High-Energy Grain Boundaries in Silicon

Recent simulations of Si GBs equilibrated at high temperatures and then cooled to zero temperature revealed a highly disordered equilibrium structure of uniform thickness and energy of *all* the high-energy GBs, while low-energy GBs remained crystalline. [10] These simulations also revealed that above the glass-transition temperature, T_g, the high-energy GBs undergo a reversible structural and dynamical transition from a confined amorphous solid to a confined liquid [11]. By contrast with the bulk glass transition, however, this equilibrium transition is continuous and thermally activated, starting at T_g and completed only at the melting point, T_m, at which the entire amorphous film has become liquid. The coexistence for $T_g<T<T_m$ of the confined amorphous and liquid phases in this two-phase region less than 1 nm thick was shown to have a profound effect on GB diffusion [11].

Detailed structural analysis revealed short-range order in the center of this "universal" high-energy GB structure that is strikingly similar to that seen in bulk amorphous Si, with a virtually complete absence of crystalline order. Moreover, the simulations demonstrated that GB disordering during high-temperature annealing is driven by a lowering of the GB energy, suggesting that the highly disordered GB structure represents the thermodynamic ground state [10]. Qualitatively identical results were obtained via a tight-binding description of interatomic bonding in Si [12], suggesting the robustness of these conclusions [13].

3 Evidence for Structural Transition in fcc-Metal Grain Boundaries

The Si results raise the intriguing possibility that, in the spirit of Rosenhain's "amorphous-cement" model [14], even in fcc metals the high-energy GBs might exhibit a universal, highly disordered structure that is liquid-like above a certain critical temperature and solid below. Unfortunately, compared to Si in which the coordination and density of the melt differ significantly from those in the amorphous and crystalline phases, in fcc metals the structural fingerprint of such a transition is much more subtle because the coordination and density of the melt and the glass (producible by simulations only, by rapidly quenching the melt) are essentially the same.

The existence of a structural transition from a confined low-temperature solid to a confined high-temperature liquid implies a sudden decrease in the activation energies for GB migration, sliding and diffusion accompanied by a change in the under-lying atomic-level mechanisms [11]. Recent self- and impurity-diffusion experi-ments by Budke et al. [15] on Cu <001> tilt GBs near the Σ5 misorientation revealed exactly such a crossover, from a strongly misorientation dependent low-temperature diffusion mechanism with a high activation energy to a high-temperature mechanism with a 60% lower, misorientation-dependent activation energy. The transition was interpreted as a GB structural transition [15]. Recent MD simulations of GB migration and GB diffusion [16] in a (001) ϕ=43.60° (Σ29) twist bicrystal described by a Lennard-Jones potential fitted to Cu also revealed the existence of such a crossover. As seen in Fig. 1, above about T~750K (or ~0.62 T_m) the activation energies for both GB processes decrease significantly from their low-temperature values. Detailed analysis of the MD simulations revealed a related change in the underlying mechanisms, from a low-temperature mechanisms involving solid-like atom hopping to one involving liquid-like reshuffling of the GB atoms [16].

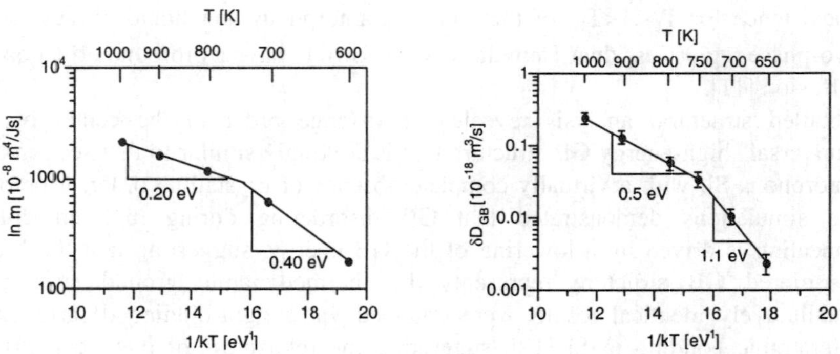

a) b)

Fig. 1. Arrhenius plots for (a) the GB mobility, m (in units of 10^{-8} m^4/Js) and (b) the GB diffusivity, δD^{GB} for the (001) ϕ=43.60° (Σ29) twist GB in Cu as described by a Lennard-Jones potential [16].

Consistent with the observation of a liquid-like high-temperature GB structure, Mott [17] had suggested that local disordering, or "melting", of small groups of atoms at the boundary is necessary to enable atoms belonging to one grain to reshuffle collectively while aligning themselves with the other grain. According to Mott's idea the activation energy for migration, $Q=nH_f$, should be governed by the latent heat of fusion, H_f, and the average number of atoms, n, involved in the melting and subsequent reshuffling. To test this idea, H_f was determined by MD simulation, yielding a value of 0.13eV/atom [16], remarkably close to the experimental value of 0.135eV for Cu. This value is, indeed, lower than the activation energy for GB migration, E_{mig}=0.20eV, in the high-temperature regime in Fig. 1a. This yields a value of $n=Q/H_f \sim 2$ that is consistent with Mott's idea of GB migration by collective reshuffling of atoms during local melting.

To gain a perspective on the rather low activation energies in the high-temperature regimes in Figs. 1a and 1b, lattice-statics relaxations were used to determine the activation energy for self-diffusion via mono-vacancies in the perfect crystal, yielding a value of E_{SD}=1.91eV [16], close to the experimental value of 2.01eV for Cu. By comparison, the high-temperature activation energy for GB migration (Fig. 1a) thus amounts to only about $0.1E_{SD}$ while that for GB diffusion (Fig. 1b) is only about $0.26E_{SD}$. Further, MD simulations of self-diffusion in the supercooled melt in the range 1300 - 1000K yielded an activation energy of 0.39eV [16], close to the experimental value of 0.42eV, but still much higher than the value of 0.20eV for GB migration in this temperature range. The authors therefore concluded [16] that the only energy scale consistent with the extremely low activation energy for GB migration is the heat of fusion, confirming that the underlying mechanism, indeed, involves the collective, liquid-like reshuffling of GB atoms.

4 Self-Diffusion in High-Energy, High-Angle fcc-Metal GBs

To further investigate the possible existence of a crossover from a solid low-temperature to a liquid-like high-temperature GB structure, Keblinski et al. have recently performed extensive simulations of self-diffusion in several carefully chosen high-energy tilt and twist GBs in Pd [18]. The Arrhenius plots thus obtained are shown in Fig. 2. Consistent with the experiments [15], these results strongly suggest that a GB structural transition, indeed, takes place in these GBs.

As clearly seen from Fig. 2, as a consequence of the transition the two tilt boundaries exhibit a crossover in the diffusion behavior, at some transition temperature, T_c (see Table I), from a high-temperature regime with a relatively low, universal activation energy to a low-temperature regime with a significantly higher activation energy that varies with the GB geometry. The fact that the twist boundaries in Fig. 1 do not show the transition indicates that T_c is lower than 700 K, where the mean-square displacement of the GB atoms is too small to be detectable by MD simulation.

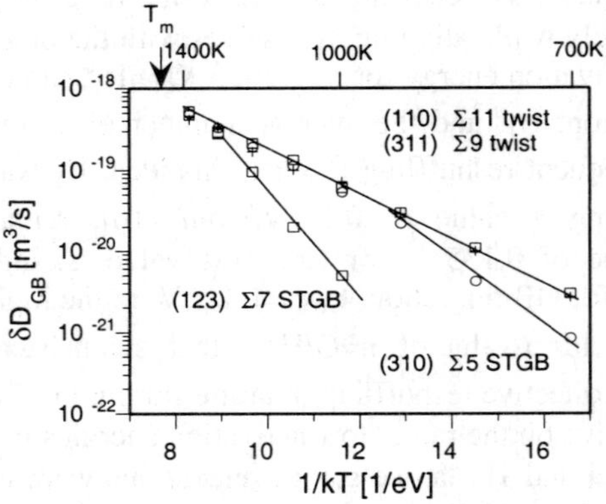

Fig. 2. Arrhenius plots for the diffusion flux, $D_{GB}\delta_D$, between 700K and 1400K for all four Pd GBs considered in the in Ref. [18]. D_{GB} is the GB diffusion constant; δ_D is the strongly temperature dependent diffusio-nal width of the GB, which is to be distinguished from its much less strongly temperature-dependent structural width, δ_S (for details see below).

Table I. GBs chosen in the study in Ref. [18], listed in order of decreasing GB energy. (Σ is the familiar density of coincident sites.) Also listed are the activation energies in the high- and low-T regimes and the crossover temperature, T_C.

GB (twist/tilt)	Energy [mJ/m^2]	High-T E_{GB} [eV]	Low-T E_{GB} [eV]	T_C[K]
(110) φ=50.48° (Σ11) tw.	1027	0.61±0.05	?	<700
(113) φ=67.11° (Σ9) tw.	1021	0.56±0.05	?	<700
(310) (Σ5) STGB	952	0.65±0.05	0.88±0.05	900
(123) (Σ7) STGB	881	0.70±0.10	1.50±0.10	1300

That, indeed, a structural transition takes place, for example, in the (110) Σ11 twist GB somewhere between 1400K and 0K is seen from the radial distribution functions in Fig. 3. Whereas the 0K structure in Fig. 3a is virtually indistinguishable from that of the glass characterized by the familiar split second peak, the high-temperature structure in 3b is identical to that of the bulk, supercooled melt [18].

For the case of the STGB on the (123) plane (see also Fig. 2), a direct analysis of the jump vectors of the diffusing atoms revealed a crossover in the underlying diffusion mechanism. Specifically, the simulations revealed that the low-temperature mechanism involves jumps in discrete crystallographic directions, whereas the high-temperature mechanism is liquid-like, characterized by a virtually complete absence of angular correlations between successive jump vectors.

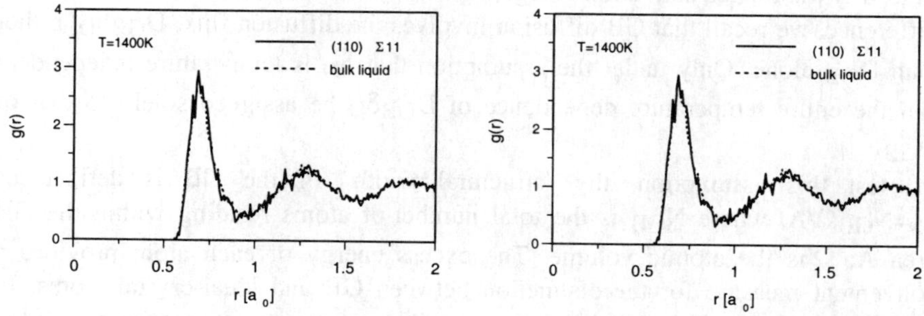

Fig. 3. Radial distribution functions, g(r), for the atoms in the two central planes of the (110) ϕ=50.48° (Σ11) twist GB in Pd at (a) T=1400K compared to that of the bulk, supercooled melt and (b) T=0K compared to that of the bulk glass produced by MD simulation, by rapidly quenching the melt. [18] The arrows in (a) indicate the familiar split second peak associated with the glass.

Similar to the earlier study of Si GBs [11], the comparison in Fig. 4 of Pd GB diffusion in the high-temperature regime with self-diffusion in the supercooled Pd melt demonstrates that the universal activation energy for GB diffusion (of E_{GB}=0.60±0.05 eV) is distinctly higher than that in the melt (of E_{melt}=0.40±0.03 eV). It is also worth noting that this value of E_{GB} amounts to only about 25% of that for self-diffusion via monovacancies in bulk perfect-crystal Pd (with a simulated activation energy of 2.41 eV [19]).

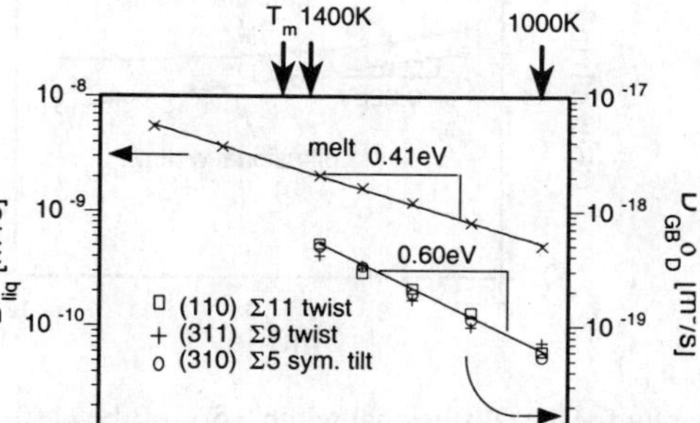

Fig. 4. Comparison of GB diffusion with self-diffusion in the supercooled melt [18]. Right-hand scale: high tempe-rature regime (1000 - 1400K) from Fig. 2 for the diffusion flux, $D_{GB}\delta_D$ (see also Table I). Left-hand scale: self-diffusion constant in the supercooled melt. The melting point for the Pd potential is estimated to be T_m~1500K [18].

As discussed above, the high-temperature GB structure and diffusion mechanism are virtually indistinguishable from those in the bulk melt. This raises the question as to why the related activation energies are not the same as well. To elucidate this difference, we recall that GB diffusion involves the diffusion flux, $D_{GB}\delta_D$, rather than D_{GB} alone. Only under the assumption that δ_D is temperature independent can the entire temperature dependence of $D_{GB}\delta_D$ be assigned solely to that of D_{GB}.

To test this assumption, the "structural width" of the GB is defined as $\delta_S=N_{GB}\Omega/A$, where N_{GB} is the total number of atoms residing within the GB area A; Ω is the atomic volume. The excess energy of each atom provides a convenient measure for the distinction between GB and ideal-crystal atoms. In practice, the half width at half maximum of the excess-energy profile across the GB provides a useful quantitative measure for δ_S [18]. This width has to be distinguished from the diffusional width, δ_D, which is governed by the number of *mobile* GB atoms, N_D, according to $\delta_D=N_D\Omega/A$, i.e., by only those atoms actually contributing to the mean-square displacement. N_D has been determined by following the diffusion of individual GB atoms over a certain time interval, revealing that the fraction of *mobile* GB atoms increases exponentially with increasing temperature (see Fig. 5), with an activation energy of E_δ=0.22±0.02 eV. As observed earlier for Si [11], the structural width increases much less strongly, indicating that the GB remains very narrow all the way up to T_m, where all GB atoms contribute to the diffusion flux.

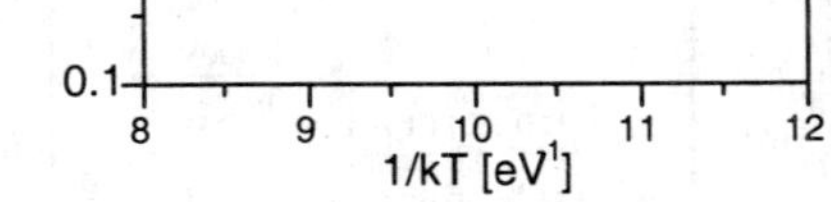

Fig. 5. Arrhenius plot for the "diffusional width", δ_D, of the (110) Σ11 twist GB, indicating that the number of diffusing GB atoms is thermally activated. For comparison, the "structural width", δ_S, with a much weaker temperature dependence, is also shown.

Adding the activation energies for δ_D and for D_{liq} (E_{melt}=0.41±0.03 eV; Fig. 4) yields 0.63±0.05 eV, a value in remarkable agreement with the universal activation energy of 0.60±0.05 for $D_{GB}\delta_D$ (see Fig. 4 and Table I). Given that the diffusion mechanism has already been identified as liquid-like, it seems reasonable to assume that GB diffusion involves two distinct processes: first the thermally activated formation of clusters of liquid from the solid GB (similar to Mott's suggestion for the mechanism of GB migration [17]) followed by liquid-like diffusion in these clusters, with an activation energy similar to that for diffusion in the melt; i.e.,

$$D_{GB}\delta_D = D^0_{GB}\,\delta^0_D\,\exp[-(E_\delta+E_{melt})/kT]\ . \tag{1}$$

In their MD simulations of both Si and Pd Keblinski et al. verified that these clusters, indeed, exist within the solid GB matrix [18].

5 Grain-Boundary Diffusion Creep

An important high-temperature plastic-deformation process in polycrystalline materials involves homogeneous grain elongation via GB diffusion, a process also known as Coble creep [20]. The related creep rate is given by [20],

$$\dot{\varepsilon} = A\,\frac{\sigma\Omega_D}{k_B T}\,\frac{\delta_D D_{GB}}{d^3} \tag{2}$$

where A is a proportionality constant depending on the grain geometry and k_B the Boltzmann constant. $\delta_D D_{GB}$ represents the GB diffusion flux discussed above, and $\sigma\Omega_D$ is the work performed by the stress σ during an elementary diffusion jump in the GB; the activation volume, Ω_D, depends on the GB-diffusion mechanism.

Due to the d^{-3} increase of the strain rate with decreasing grain size, d, Coble creep should be particularly important in nanocrystalline (nc) materials, i.e., polycrystalline materials with a grain size of nm dimensions. Moreover, recent MD simulations of nc Si [21] and Pd [22] demonstrated that under the assumption of random grain orientations, most of the GBs are high-energy, high-angle GBs of the type discussed above. All the GBs in such a material should therefore have liquid-like structures at elevated temperatures, giving rise to an even higher creep rate than one would expect for a course-grained polycrystal (in which all three types of GBs discussed in Sec. 1 are present); the Coble-creep rate of such a nc material should therefore have an activation energy that is identical to the universal value for high-energy *bicrystalline* GBs discussed in Sec. 4.

To test this hypothesis, MD simulations of 3d periodic nanocrystalline Pd microstructures were performed recently. [23] To prevent grain growth and thus to enable steady-state diffusion creep to be observed, model microstructures were tailored to have a uniform grain size and shape, with random orientations. Similar to our earlier Si simulations [24], these microstructures, indeed, exhibit steady-state diffusion creep that is homogenous, with a strain rate that agrees quantitatively with Eq. 2. In spite of the high stresses required to make the effects observable on an MD time scale (of typically $<10^{-8}$ s), the creep rate was found to vary linearly as a function of σ (see Eq. 2) Moreover, these stresses were verified to lie below the threshold value for dislocation nucleation from the GBs [23]; the entire plastic strain and related creep rate can therefore be attributed to the effects of GB diffusion.

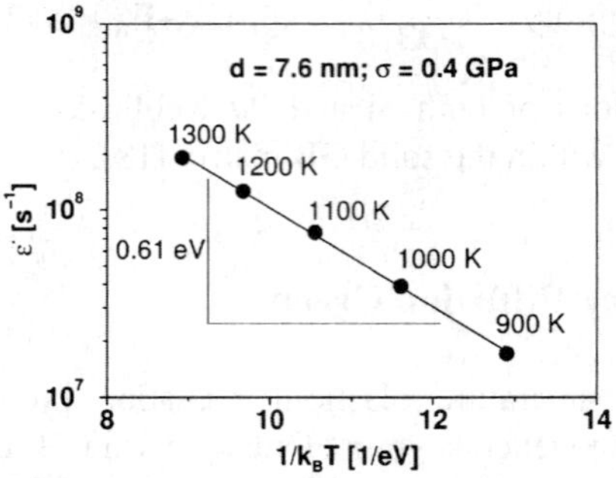

Fig. 6. Arrhenius plot for the creep rate of a tailored, 3d periodic nanocrystalline Pd microstructure consisting of 16 grains of identical shape and grain size (of 7.6 nm) arranged on a bcc lattice [23].

Figure 6 shows The temperature dependence of the strain rates thus obtained between 900K and 1300K. The resulting activation energy of 0.61±0.04 eV is in remarkable agreement with the universal high-temperature activation energy of 0.60±0.05 eV for GB diffusion in bicrystalline interfaces (see Fig. 4).
Another important characteristic of Coble creep is the grain-size dependence of the simulated creep rate shown in Fig. 7. For the three largest grain sizes, the slope of 1:3 agrees well with Eq. 2. However, for the two smallest values of d, deviations from the dashed line are clearly visible. An extension of the Coble derivation in which the GB width is treated as non-negligible compared to the grain size leads to a decrease of the grain-size exponent from 3 to 2 with decreasing grain size [23].

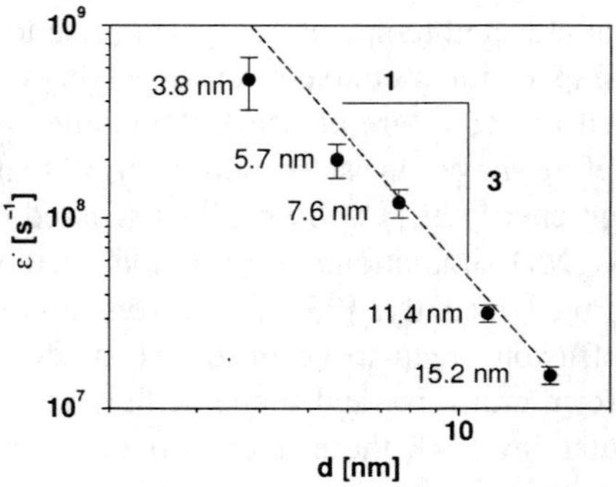

Fig. 7. Log-log plot of the strain rate as a function of the grain size. The dashed line represents a fit of the Coble-creep equation (2) to the large-grain-size data [23].

6 Conclusions

The larger picture that emerges from these simulations suggests that upon heating from zero temperature through the melting point, above a critical temperature T_c the highest-energy tilt and twist GBs in a polycrystalline material undergo a continuous, reversible transition from a low-temperature, solid GB structure to a highly confined liquid-like structure consisting of liquid bubbles embedded in the solid GB region. As the melting point is reached, the entire GB region is consumed by the liquid until, above T_m, the highly confined liquid GB layer spreads to induce thermodynamic melting [8,9]. The temperature at which this structural and dynamical transition occurs increases with decreasing GB energy, with the effect that the highest-energy GBs in the polycrystal play a rate-controlling role in high-temperature properties, such as grain growth and GB-diffusion creep.

Indeed, as first suggested by Mott [17], above T_c the atomic-level mechanisms for GB diffusion, sliding and migration are therefore liquid-like, with activation energies related to the diffusion in the melt and the heat of fusion. By contrast, lower-energy GBs never reach the critical level of (static and dynamic) structural disorder required to trigger the transition below T_m; their physical properties are therefore solid-like all the way up to T_m, with activation energies that increase with decreasing GB energy until the respective perfect-crystal limit is reached.

One should keep in mind, however, that this characterization of GB properties based on the GB energy alone applies only to *high-energy, high-angle* GBs, in which the structural disorder is uniformly distributed all along the GB plane. By contrast, in dislocation boundaries the structural disorder is localized in the dislocation cores separated by elastically strained perfect-crystal like regions. Two GBs with the same energy, one low-angle and the other high-angle, therefore exhibit rather different GB properties. For example, the low-angle GB shows fast diffusion only along the highly disordered dislocation pipes but no diffusion at all in the perfect-crystal regions; by contrast, the diffusion in the high-angle GB is homogeneously distributed all over the GB.

Acknowledgment

This work was supported by the US Department of Energy, BES-Materials Science under Contract No. W-31-109-Eng-38.

References

[1] D. Wolf, in "The Encyclopedia of Materials, Science and Technology", principal editor Robert Cahn, Pergamon Press, 2001.

[2] W. T. Read and W. Shockley, *Phys. Rev.*, 78, 1950, pp. 275-89.

[3] D. Wolf and K. L. Merkle, in "Materials Interfaces, Atomic-level Structure and Properties", eds. D. Wolf and S. Yip, Chapman and Hall, 1992, pp. 87-150.
[4] D. Wolf and J. A. Jaszczak, in Ref. [3], pp. 662-90.
[5] C. S. Smith, in "Metal Interfaces", ASM, Cleveland, OH, 1952, p. 64.
[6] V. Pontikis, *J. de Physique* (Paris), 49, 1988, C5-327 - C5-336.
[7] S. R. Phillpot, J. F. Lutsko, D. Wolf and S. Yip, *Phys. Rev.* B, 40, 1989, pp. 2831-2840.
[8] S. R. Phillpot, S. Yip and D. Wolf, *Computers in Physics*, 3, 1989, pp. 20-31.
[9] D. Wolf, P. Okamoto, S. Yip, J. F. Lutsko and M. Kluge, *J. Mater. Res.*, 5, 1990, pp. 286-301.
[10] P. Keblinski, S. R. Phillpot, D. Wolf and H. Gleiter, *Phys. Rev. Letters,* 77, 1997, 2965-2968; *J. Amer. Ceram. Soc.,* 80, 1997, pp. 717-32.
[11] P. Keblinski, S. R. Phillpot, D. Wolf and H. Gleiter, *Phil. Mag. Letters*, 76, 1997, pp. 143-151.
[12] F. Cleri, P. Keblinski, L. Colombo, S. R. Phillpot and D. Wolf, *Phys. Rev.*, B57, 1998, pp. 6247-6250.
[13] See the commentary on this work by R. W. Cahn, *Nature,* 390, 1997, p. 344.
[14] W. Rosenhain and D. Ewen, *J. of the Institute of Metals*, 10, 1913, pp. 119-148; ibid. 149-185; for a more recent review of this work, see K. T. Aust and B. Chalmers, in "Metal Interfaces", ASM, Cleveland, OH, 1952, pp. 153-178.
[15] E. Budke, T. Surholt, S. I. Prokofjev, L. Shvindlerman and C. Herzig, *Acta Mater.*, 47, 1999, pp. 385-395.
[16] B. Schoenfelder, P. Keblinski, D. Wolf and S. R. Phillpot, in *Intergranular and Mater.*, 47, 1999, pp. 385-395.
[17] N. F. Mott, *Proc. Phys. Soc.* 60, 1948, pp. 391-393.
[18] P. Keblinski, D. Wolf, S. R. Phillpot and H. Gleiter, *Phil. Mag., A* 79, 1999, pp. 2735-2761.
[19] J. B. Adams, S. M. Foiles and W. G. Wolfer, J. *Mater. Res.*, 4, 1989, pp. 102-112.
[20] R. L. Coble, *J. Appl. Phys.,* 34, 1963, pp. 1679-1682.
[21] P. Keblinski, D. Wolf, S. R. Phillpot and H. Gleiter, *Acta Mat.,* 45, 1997, pp. 987-998.
[22] P. Keblinski, D. Wolf, S. R. Phillpot and H. Gleiter, *Scripta Mat.,* 1999, 41, pp. 631-636.
[23] V. Yamakov, D. Wolf, S. R. Phillpot and H. Gleiter, *Acta Mater.*, (in press).
[24] P. Keblinski, D. Wolf and H. Gleiter, *Interface Science.* 6, 1998, pp. 205-212.

Recrystallization and Grain Growth
Proceedings of the First Joint International Conference
Eds. G. Gottstein and D.A. Molodov, Springer-Verlag, 2001

Experimental Measurements of Nucleation and Growth Rates

D. Juul Jensen

Materials Research Department, Risø National Laboratory, DK-4000 Roskilde, Denmark

Keywords: experimental methods, 3DXRD microscopy, synchrotron radiation, nucleation, growth rates.

Abstract. Methods frequently used to quantify critical nucleation and growth parameters are reviewed with focus on potentials and limitations of the methods. The parameters to be addressed here are:

- nucleation rates
- spatial distribution of nucleation sites
- orientation correlation between nuclei and matrix
- growth rates

Although not yet so frequently used, a new synchrotron based method: 3D X-ray diffraction (3DXRD) microscopy, is also included in this review. This is because 3DXRD microscopy can give some information on the critical parameters, which has not been accessible till now.

1 Introduction

Although computer modelling and simulation are increasingly applied as a key approach to understand many topics within materials science, most publications on recrystallization relies on experimental characterization. This is considered a sound approach to obtain more fundamental understanding of the basic recrystallization mechanisms, which in turn can form the basis for advancing recrystallization modelling.

Many experimental methods are available and are used in recrystallization studies. Very significant among these methods are the automatic electron back scattering pattern (EBSP) method operating in scanning electron microscopes. In the proceedings of the previous International Conference on Recrystallization and related Phenomena, Rex' 99 [1], 33% of the papers contains EBSP data. However, irrespective of the strength of a method, EBSP or other, it is generally true that a single method is not enough to give a complete precise description of a given recrystallization mechanism. It is in most cases necessary to use several complementary methods and combine the results.

The aim of the present paper is discuss the use of various experimental methods for quantification of a series of critical recrystallization parameters covering aspects of both nucleation and growth. As a speciality, the potentials of the new method, 3 dimensional x-ray diffraction (3DXRD) microscopy, to contribute new

information on the selected parameters will be included in this review. As this method is new, a few experimental details shall be given in the following.
The 3DXRD microscope is operating on the Materials Science beam line ID11 at the European Synchrotron Radiation Facility (ESRF) in France [2-5]. It is designed and developed to allow non-destructive measurements of the crystallographic orientation and the elastic strain within small volume elements in the bulk of typically mm^3-cm^3 sized samples. It is thus possible to determine the position (x,y,z) and size of e.g. a nucleus or a grain inside a sample. The 3D grain shape can also be mapped. Various furnaces and a uniaxial stress rig are available for in-situ experiments. The microscope typically operates at 50-100 keV and the flux in the focal point (typically at the sample) is 10^{11}-10^{12} photons/s. At present the mapping precision of e.g. grain boundaries or nuclei is 25 μm. The minimum size of observable nuclei or grains are 0.3 μm^3 if they have a crystallographic orientation difference to the majority of the sample, but the position of these small volume elements can not be determined better than the 25 μm mentioned above.
A sketch of the 3DXRD microscope is shown in Fig. 1, but for information on the microscope and on applications related to this paper see [2-11].

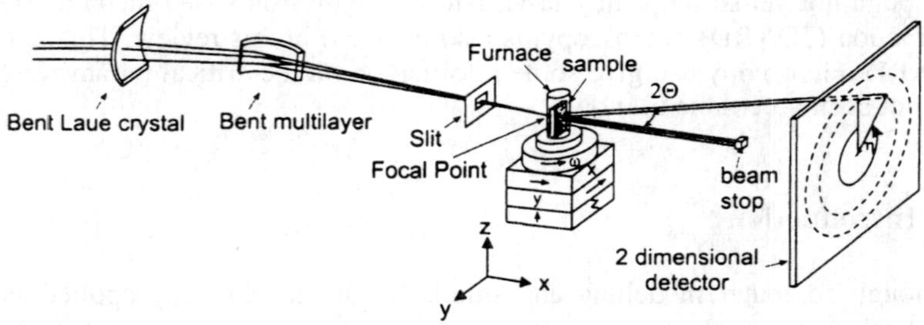

Fig. 1. Sketch of the 3DXRD microscope.

2 Nucleation Rates

The nucleation rate $\dot{N}$ is defined as the number of new nuclei appearing in unit volume and time; ie $\dot{N} = dN_v/dt$ where N_v is the number of nuclei per unit volume. Experimental measurements of nucleation rates are tedious and thus not very frequent in litterature. Instead various assumptions are generally used.
Typical assumption are: i) site saturation - all nuclei are pre-existing in the deformed structure and start growing at t = 0, ii) constant nucleation rate [12,13]. Because of the complexity of experimental nucleation rate measurements, it might be an efficient approach first to test if one of these two nucleation assumptions is valid. This may be done analytically by the microstructural path method [14,15]. For information on this method see [15]. In the following two experimental methods for direct measurements of nucleation rates shall be discussed.

2.1 Inspection of 2D sections

Counting the number of grains in two dimensional (2D) sample sections for a series of samples annealed to various degrees of recrystallization is the most frequently used method to determine the nucleation rate. Typically this is done by optical microscopy or the EBSP technique. If it is observed that the number of nuclei increase with increasing time, it does not necessarily mean that new nucleation occurs. This will also be observed for site saturation simply because nuclei already nucleated at $t = 0$ grow large enough to intersect the inspected sample section [16].

A correction for this growth effect may be done stereologically for specific cases [17]. If it for example can be assumed that the nuclei are spherical, continue to be spherical during growth and all are of the same size then

$$N_v = N_s/(2r) \tag{1}$$

where N_s is the average number of nuclei per unit area and r is the radius of the nuclei spheres.

If the nuclei are of different sizes, the variation in size has to be taken into account [17]. Still assuming spherically shaped nuclei

$$N_v = \frac{2\overline{m}N_s}{\pi} \tag{2}$$

where $\overline{m}$ is the average value of m_i, and m_i are measured directly in the 2D sample section (see Fig. 2).

Also for other nuclei shapes, formulas for N_v have been derived [17]. Depending on the actual nucleation conditions one of these formulas may be applied or new ones have to be derived matching the conditions. When N_v is determined, the nucleation rate is given as $\dot{N} = dN_v/dt$.

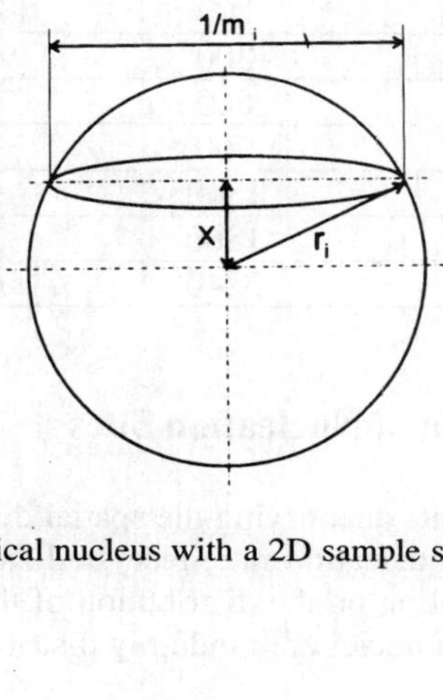

Fig. 2. Intersection of a spherical nucleus with a 2D sample section. defines the parameters x, r_i and $1/m_i$.

2.2 In-situ measurements of nucleation times for single nuclei

With the 3DXRD microscope it is possible to detect nuclei in the bulk of large samples and follow their development and growth in-situ while annealing the deformed sample. This is described in [10,18]. The noise limit for detecting a nucleus is approximately 2 μm (diameter). Its development and early growth can be followed with a time resolution of the order of seconds.

Applying the 3DXRD method, no correction for the appearance of "old" nuclei in the inspected area is needed (as discussed above in 2.1). This is because a full gauge volume (for example 70·70·1000 μm^3) is characterized and because a simple experimental checking procedure is applied to ensure that only nuclei fully inside the selected gauge volume is included in the analysis.

A series of experiments of this type has been performed for aluminium (AA1050) cold rolled 90% and annealed at 270°C [10,18]. For each observed nucleus, a full curve of volume versus annealing time is obtained. From such curves it is straight forward to determine the nucleation time. And by doing so for a sufficient number of nuclei, nucleation rates are directly obtained. In table 1, the nucleation time for 10 randomly selected nuclei are given.

Table 1 Nucleation times for 10 randomly selected nuclei in aluminium (AA1050) cold rolled 90 % and annealed at 270°C. The nucleation time t_{nucl} is defined as the time when the radius of a given nucleus has reached 1.3 μm which is slightly above the detection limit. Part of the data is from [10].

Nucleus number	t_{nucl} [sec]	t_{nucl}/t_{rex} [%]
1	0	0
2	0	0
3	130	1.1
4	200	1.7
5	300	2.5
6	600	5.0
7	900	7.5
8	1810	15.1
9	1860	15.5
10	3240	27.0

3 Spatial Distribution of Nucleation Sites

Experimental measurements quantifying the spatial distribution of nucleation sites are very few. As for the nucleation rate, recrystallization modelling is almost always based on an assumption on the distribution of the nucleation sites. And it is typically assumed that the nuclei are randomly distributed. However clustering of nucleation sites can have very large effects on the recrystallization kinetics, the recrystallized grain size and the texture. Numerical simulations on this are presented in [19].

As shall be described below it is fairly tedious work to measure the spatial distribution of nucleation sites. A first step could therefore be to test if the assumption of a random distribution applies or not [20]. A sensitive test is to plot the free unimpinged surface area of the nuclei/grains S_v as a function of the fraction of recrystallized material, V_v. An example is shown in Fig. 3. In the random nucleation case, the typical curve shape is very nearly symmetrical with a maximum at $V_v = 0.487$. In a linear clustered nucleation case, the average impingement starts earlier in the recrystallization process and the maximum is reached therefore at a lower V_v value. The actual position of the maximum depend on the assumed distance between the nuclei on the nucleation lines. A further discussion of this is given in [20]. It has to be underlined that this simple test method only can verify but *not prove* given nucleation assumptions.

Another fairly simple approach to differentiate clustered nucleation from random nucleation has been developed based on the microstructural path method [21].

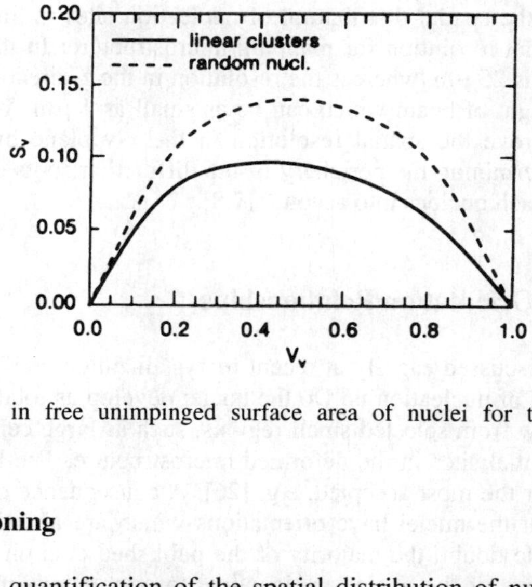

Fig. 3. Evolution in free unimpinged surface area of nuclei for the two nucleation assumptions [20].

3.1 Serial Sectioning

An experimental quantification of the spatial distribution of nucleation sites requires a full 3D analysis of the microstructure. The analysis of planar 2D sections does not allow such a quantification. However by careful serial sectioning and characterization of each plane, the necessary information is obtained. The sectioning step has to be chosen such that all nuclei above the critical size is intersected. For typical metals this means the sectioning steps in the μm-range.

Various techniques have been developed for best performing the serial sectioning [22-23], but only in few cases the method has been used to determine nucleation sites: In 1959 Vandermeer and Gordon [16] used the method to document prefer-

ential nucleation at triple lines and in 1995 Weiland [24] investigated correlations between second phase particles and nuclei.
The method seems powerful but is very labour intensive and extreme care is required when performing the sectioning and in aligning the sections relative to each other.

3.2 In-situ measurements of the spatial distribution of nucleation sites

With the 3DXRD microscope it is possible to determine the positions (x,y,z) of nuclei in-side bulk samples as soon as the nuclei sizes are above the detection limit (see section 2.2). The fastest procedure for doing this is the so-called tracking technique [5,8], by which a whole sample layer is characterized by one or a few series of recordings which in total typically takes less than 15 minutes. Information about a whole sample is then obtained by moving the sample position vertically (z-direction) relative to the X-ray beam and repeating the procedure. This method is fully non-destructive and the sample is thus maintained for further or supplementary investigations.
At present the potential of the tracking technique in the 3DXRD microscope for determination of the spatial distribution of nucleation sites is limited, due to the relative poor spatial resolution for mapping microstructure: In the x, y plane the spatial resolution is 25 μm, whereas the resolution in the z-direction is determined by slits to the height of beam which can be as small as 1 μm. Work is presently underway to improve the spatial resolution in the x-y plane by optimising the procedure for determining the periphery of the diffraction spots and by taking all reflections from each nucleus into account [7,8].

4 Nucleation Orientation Relationships

A key question discussed eagerly at recent recrystallization conferences [1,25] in the understanding of nucleation is: Do the nuclei develop as totally new grains or do the nuclei grow from selected small regions, such as large cells or cell blocks, at certain preferential sites in the deformed microstructure. The latter mechanism is presently by far the most accepted, e.g. [26]. A consequence of this nucleation mechanism is that the nuclei have orientations which are already present in the deformed state. No doubt, the majority of the published data on nucleation show that the orientations of the nuclei are identical to orientations present in the deformed state. A recent, general example is cube nuclei developing in cube bands present in the deformed microstructure e.g. [27]. Also the classical nucleation mechanisms: nucleation by strain induced boundary migration (SIBM) [28], nucleation by coalescence [29] nucleation within transition bands [30] and particle stimulated nucleation [31], all give rise to nuclei with orientations identical or almost identical to those present in the deformed state. Remaining are, however, several observations of nuclei which have orientations away from the orientations present in the deformed state. In the following subsections, methods used to study such "new nuclei" are discussed.

4.1 Electron microscopy of single crystals

The development of nuclei with new orientation relative to the deformed state is most convincingly documented for single crystals, e.g. [32-39]. For single crystals it is possible based on a detailed characterization of the deformation microstructure to establish an orientation relationship between a nucleus and the site in the deformed microstructure where it formed.

Typically the microstructure and the orientations of the deformed single crystals are characterized by TEM/Kikuchi lines and/or SEM/EBSP. An important question is always if this characterization has been done well enough to really prove that the nuclei have new orientations, i.e. for TEM if a sufficiently large sample area has been covered and for SEM if the spatial resolution was sufficient to find small areas with uncharacteristic orientations in the deformed microstructure.

An estimate of how large a sample area which has to be characterized can be found by a probability calculation. Such a calculation has been performed for the development of nuclei with new orientations in an S-oriented single crystal deformed by plane strain compression to $\varepsilon = 1.5$. From an experimental determination of the density of nuclei with new orientation for different annealing times, the volume fraction of cells/subgrains/cell-blocks leading to these nuclei is calculated to be in the range $3{\cdot}10^{-8}$ - $4{\cdot}10^{-7}$.

Number as low as this indicates that random (blind) TEM/SEM characterizations are most likely to fail finding the uncharacteristic orientations in the deformed microstructure even if they are there. A more potential experimental method would be 3DXRD microscopy (see section 4.3), but in general it is considered a better approach to try to relate the development of the nuclei with the new orientations to the deformation microstructure and deformation mechanism occurring there. This may lead to observations which can guide the experimental characterization, thus avoiding the mega-scans necessary if the search is random (blind).

4.2 Electron microscopy of polycrystals

For polycrystals, in particular for heavily deformed polycrystals it is experimentally not straight forward to study possible orientation relations between nuclei and deformed matrix. Since a given region in the bulk cannot be characterized before and after annealing and since there are significant variations in the deformed microstructure from region to region depending on the initial grain orientations, a direct comparison of the deformed and "nucleated" state is not possible.

One approach has been to relate the orientations of nuclei to the surrounding deformed structure for a large number of nuclei and thus get statistical information on orientation relationships [40]. Here it was found that some nuclei develop with orientation different from those in the deformation microstructure.

Another more direct method is to study surface behavior instead of bulk behavior (which may or may not differ) and characterize selected areas before and after deformation by SEM/EBSP. This has for example been done looking at nucleation at triple junctions in weakly deformed aluminium [41,42]. It was found that nucleation occurred almost exclusively at the triple junctions and that most junctions lead to at least one nuclei; it was thus only necessary to characterize in details the areas around the triple junctions and not the whole deformed surface area. When the orientation of each nucleus was compared to the measured orientations within that triple junction area, before deformation it was found that approximately half the nuclei had orientations as the "parent" grains, whereas the other half had orientations which were not within the scatter of deformation orientations but were further rotated about a common <111> axis [41,42].

This method is nice because it is direct. But besides only characterizing the surface behavior, a limitation is that it is uncertain in which depth the nuclei originated: If a nucleus develop somewhere below the surface and grow to coinside with the position of the triple junction seen in the surface, wrong orientation relationships may be reported. An investigation of this would require detailed serial sectioning.

4.3 In-situ measurements of nucleation orientation relationships

As discussed in [5] in-situ bulk observations of nucleation may be performed with the 3DXRD microscope.

The ideal experiment would be to characterize in detail in 3D the deformation microstructure. Then anneal the sample *in-situ* to the beginning of nucleation and see where and with what orientations the nuclei develop. However, for typical deformation microstructures, this is considered experimentally not possible. The spatial resolution of the 3DXRD microscope is not fine enough for the detailed microstructural characterization of reasonably sized samples. If the investigation is limited to measurements of possible relationships between the orientation of a nucleus and that/those present there (in a volume element of a size given by the spatial resolution) before annealing, it might be done. However, the necessary acquisition time for characterization of the orientations within a reasonably sized sample volume of the deformed microstructure would be substantial, and depending on this size, one may risk that no nucleation occurs within the characterized volume, and thus the work may be in vain. So far no work of this type has been performed, but is planned.

Another issue when trying to quantify nucleation orientation relationships, is experimental determination of uncharacteristic orientations (i.e. orientation away from the main texture components) in the deformed microstructure (see section 4.1). For this the 3DXRD microscope is ideal. It has been shown that isolated intensity peaks corresponding to volume fractions of the order of $3 \cdot 10^{-6}$ can be seen on the detector [43]. With a gauge volume of for example $10 \cdot 10 \cdot 1000\ \mu m^3$ this corresponds to a volume of $0.3\ \mu m^3$ which is typically less than the cell size in the deformed microstructure.

5 Growth Rates

Of current interest is the question whether grains of certain orientations grow faster than others or not. In this chapter methods that can determine growth rates of grains with different orientations are reviewed.
The most classic method to determine growth rates during recrystallization is based on measurements of the size of the largest unimpinged grain in 2D sample section of samples annealed for different times [12]. By for example EBSP, this method may easily be extended to include orientations, but still the well known limitation apply namely that the method is applicable only during the earliest stages of recrystallization before substantial impingement between the grains occurs, and that the resulting growth rate is only statistically sound if very large sample areas are inspected.

5.1 Average size and orientation measurements

Experimentally it is straightforward using the EBSP technique to measure grain sizes in a 2D plane and to categorize the grains according to orientation into a limited number of classes, for example cube, rolling and random. If such data are measured for samples annealed for various times to various fractions of recrystallization, the evolution in average size of grains within the selected orientation classes as a function of annealing time can be seen. Such plots often show that grains of a specific orientation, for example cube, grow to become larger than the other grains e.g. [44].
In the interpretation of such curves, the result that grains of certain orientations grow to become larger than the others, are sometimes used to prove faster growth rates of these grains. This is however not necessarily so. Only if the nucleation conditions, i.e. nucleation rate and spatial distribution of nucleation sites, are identical for different types of grains, this interpretation is correct.

5.2 The Cahn-Hagel Method

The basic idea of this method was originally suggested by Cahn and Hagel [45] for the formation of pearlite during austenite decomposition of steel. It was later shown that the method is equally successful for characterizing recrystallization [46]. The latest development of the Cahn-Hagel method has been to extend it to include crystallographic orientations [47]. Thereby, average growth rates of grains of various specified orientations $\langle G_i \rangle$ can be determined. This only requires that the crystallographic orientations of the individual nuclei/grains are determined. Then

$$\frac{dV_{vi}}{dt} = \langle G_i \rangle \cdot S_{vi} \quad (3)$$

Where "i" refers to a specific orientation (or orientations within x degrees of a specific orientation). Experimental values of V_{vi} and S_{vi} are determined by stereologial methods directly from 2D inspection of the microstructure of a series of samples annealed for different times, e.g. [48]. This method can be applied throughout the entire recrystallization process and is not restricted to certain nucleation conditions.

There are two issues requiring attention when applying the Cahn-Hagel method to determine average growth rates:

The first issue relates to the grain shape. For certain grain shapes, namely plates and rods, the Cahn-Hagel method will result in average growth rates including the whole unimpinged surface area including the non-moving areas and not the growth rate of the migrating areas only. This was pointed out by Gokhale et al. [49]. One may discuss which of the two growth rates are of interest and how frequent such growth conditions are for recrystallization. A plot of $S_{vi}/(1\text{-}V_{vi})$ versus $-\ln(1\text{-}V_{vi})$ can be used to test if such shapes apply for a specific investigation. If the plot shows a non-linear dependency, the grain shape is not plates or rods.

The second issue relates to determination of the time derivative of V_{vi}. As discussed in [50] this is done most precise if a suitable function describing $V_{vi}(t)$ can be deduced from which the time derivative can be calculated. Frequently, the Avrami equation [13] is used. There are, however, many recrystallization studies where it has been observed that this equation does not describe $V_{vi}(t)$ sufficiently well over the entire annealing time period. An example is shown in Fig. 4a. Other functions have to be chosen. Often the following equation fits better

$$\log \ln\left(\frac{1}{1-V_{vi}}\right) = A + B\exp\left[-C(\log t - D)^{E}\right] \tag{4}$$

where A, B, C, D and E are fitting constants. The same data as in Fig. 4a is fitted to this equation in Fig. 4b and a good fit is obtained. One concern could be the many fitting constants compared to the often fairly low number of experimental points (typically 5 to 20). But as the fitting constants are used only to gain a precise description of the experimental data not for any interpretation of physical phenomena, this is generally not important for the resulting growth rates.

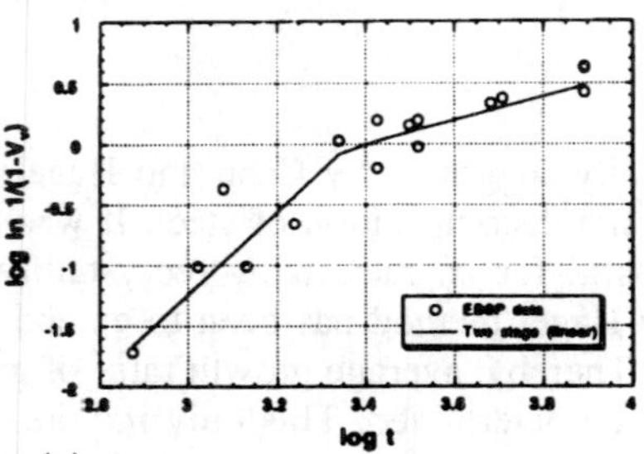

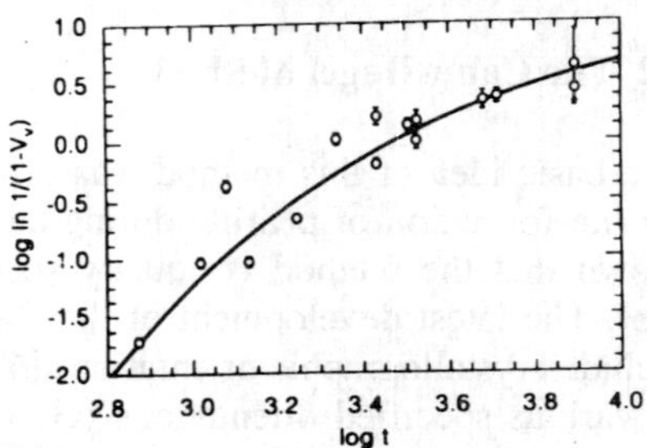

Fig. 4. Cu cr 92 % annealed at 121°C. (a) Fit to two Avrami functions. (b) Fit to double exponential equation.

5.3 In-situ measurements of the growth of single grains

As discussed in section 2.2, the 3DXRD microscope allows in-situ measurements of the growth of individual grains. An example of such results is given elsewhere in these proceedings [18]. Such data are unique as they for the first time allow us to see the behavior of single grains of known orientations in the bulk during annealing. The data have already shown that even grains within one group of orientations have very different growth behavior [18], and suggests that future work on recrystallization should address growth rate distributions.

The direct output of the measurements is intensity versus annealing time. The intensity is relative easily converted into volume of the grain [10]. However what is not known is the possible impingement with other grains, i.e. S_v is not known. This means that the migration rate of the "recrystallization boundaries" surrounding a grain are not directly available from this type of data.

Alternative approaches may be i) to focus the X-ray beam on a selected boundary (in the bulk) using focusing analyser optics [51] and simply follow its migration, somewhat similar to the approach used in Aachen [52] and ii) to use the tracking technique to map the complete recrystallized structure as a function of annealing time. The potentials of latter method is limited by the present spatial resolution in the x, y plane to about 25 μm.

6 Concluding Remarks

Many detailed measurements of nucleation and growth parameters have advanced significantly the field of recrystallization. However many questions still excist and answers are required to advance the field further. As discussed in this paper, 3DXRD microscopy is expected in the coming years to provide a number of such answers.

References

[1] "Proc. ReX'99, The Fourth Int. Conf. on Recrystallization and Related Phenomena", eds. T. Sakai and H.G. Suzuki, Japan Inst. Metals, Sendai, Japan, 1999, pp. 1-928.

[2] H.F. Poulsen, S. Garbe, T. Lorentzen, D. Juul Jensen, F.W. Poulsen, N.H. Andersen, T. Frello, R. Feidenhans'l, H. Graafsma, *J. Synch. Rad*, 4, 1997, pp. 147-154.

[3] U. Lienert, H.F. Poulsen, Å. Kvick, in "Proc. 40th Conf. AIAA Structures, Structural Dynamics and Materials", 1999, pp. 2067-2075.

[4] D. Juul Jensen, Å. Kvick, E.M. Lauridsen, U. Lienert, L. Margulies, S.F. Nielsen, H.F. Poulsen, in "Mat. Res. Soc. Symp. Proc.", 5990, 2000, pp. 227-240.

[5] D. Juul Jensen and H.F. Poulsen, in "Recrystallization - Fundamental Aspects and Relations to Deformation Microstructure", eds. N. Hansen et al., Risø Nat. Lab., Roskilde, Denmark, 2000, pp. 103-124.
[6] U. Lienert, H.F. Poulsen, V. Honimäki, C. Schulze, O. Hignete, *J. Synch. Rad.*, 6, 1999, pp. 979-984.
[7] E.M. Lauridsen, S. Schmidt, R.M. Suter, H.F. Poulsen, submitted for publication.
[8] S.F. Nielsen, W. Ludwig, D. Bellet, E.M. Lauridsen, H.F. Poulsen, D. Juul Jensen, in "Recrystallization - Fundamental Aspects and Relations to Deformation Microstructure", eds. N. Hansen et al., Risø Nat. Lab., Roskilde, Denmark, 2000, pp. 473-478.
[9] O.V. Mishin, E.M. Lauridsen, N.C. Krieger Lassen, G. Brückner, T. Tschentscher, B. Bay, D. Juul Jensen, H.F. Poulsen, *J. Appl. Cryst.*, 3, 2000, pp. 364-371.
[10] E.M. Lauridsen, D. Juul Jensen, H.F. Poulsen, U. Lienert, *Scripta mater.*, 43, 2000, pp. 561-566.
[11] L. Margulies, G. Winther, H.F. Poulsen, *Science*, 291, 2001, pp. 2392-2394.
[12] W.A. Anderson and R.F. Mehl, *Trans AIME*, 161, 1945, pp. 140-172.
[13] M. Avrami, *J. Chem. Phys.*, 1, 1939, pp. 1103-1112; ibid, 8, 1940, pp. 212-224; ibid, 9, 1941, pp. 177-184.
[14] R.A. Vandermeer, R.A. Masumura, B.B. Rath, Acta Metall. Mater., 39, 1991, pp. 383-389.
[15] R.A. Vandermeer, "Recrystallization - Fundamental Aspects and Relations to Deformation Microstructure", eds. N. Hansen et al., Risø Nat. Lab., Roskilde, Denmark, 2000, pp. 179-200.
[16] R.A. Vandermeer and P. Gordon, *Trans TMS-AIME*, 215, 1959, pp. 577-588.
[17] R.L. Fullman, *Trans AIME*, 197, 1953, pp. 447-452.
[18] E.M. Lauridsen, S. Schmidt, L. Margulies, H.F. Poulsen, D. Juul Jensen, in these Proceedings.
[19] D. Juul Jensen, *Met. Mat. Trans.*, 28A, 1997, pp. 15-25.
[20] D. Juul Jensen and R.A. Vandermeer, in “Proc. ICOTOM 11”, eds. Z. Liang et al., 1996, pp. 490-496.
[21] R.A. Vandermeer, in "Microstructural and Crystallographic Aspects of Recrystallization", eds. N. Hansen et al., Risø Nat. Lab., Roskilde, Denmark, 1995, pp. 193-213.
[22] M.V. Kral and G. Spanos, *Acta Mater.*, 47, 1999, pp. 711-724.
[23] A.D. Rollett et al., in “Proc. Microscopy and Microanalysis”, 5, 1999, pp. 230-231.
[24] H. Weiland, in "Microstructural and Crystallographic Aspects of Recrystallization", eds. N. Hansen et al., Risø Nat. Lab., Roskilde, Denmark, 1995, pp. 215-228.
[25] "Recrystallization - Fundamental Aspects and Relations to Deformation Microstructure", eds. N. Hansen et al., Risø Nat. Lab., Roskilde, Denmark, 2000, pp.1-677.
[26] F.J. Humphreys and M. Hatherly, “Recrystallization and Related Annealing Phenomena”, Pergamon Press, Oxford, 1995.

[27] R.D. Doherty, I. Samajdar, C.T. Necker, H.E. Vatne, E. Nes, in "Microstructural and Crystallographic Aspects of Recrystallization", eds. N. Hansen et al., Risø Nat. Lab., Roskilde, Denmark, 1995, pp. 1-23.
[28] P. Beck and P.R. Sperry, *J. Appl. Phys.*, 21, 1950, pp. 150-152.
[29] H. Hu, in "Recovery and Recrystallization of Metals", ed. H. Himmel, AIME Interscience, NY, 1993, pp. 311-362.
[30] I.L. Dillamore, P.L. Morris, C.J.E. Smith, W.B. Hutchinson, in "Proc. R. Soc. Lond.", 329A, 1972, pp. 405-420.
[31] F.J. Humphreys, M. Ferry, C. Johnson, P. Paillard, in "Microstructural and Crystallographic Aspects of Recrystallization", eds. N. Hansen et al., Risø Nat. Lab., Roskilde, Denmark, 1995, pp. 87-104.
[32] S. Kikuchi, E. Kimura, M. Koiwa, *J. Mat. Sci.*, 27, 1992, pp. 4927-4934.
[33] T. Kamijo, S. Kataoka, H. Inagaki, *Acta Metall. Mater.*, 41, 1993, pp. 1713-1720.
[34] A. Godfrey, D. Juul Jensen, N. Hansen, in "Microstructural and Crystallographic Aspects of Recrystallization", eds. N. Hansen et al., Risø Nat. Lab., Roskilde, Denmark, 1995, pp. 365-370.
[35] X. Huang, T. Okada, J.A. Wert, F. Inoko, in these Proceedings.
[36] K. Kashihara, M. Tagami, T. Okada, F. Inoko, *Mater. Sci. Engng. A*, 291, 2000, pp. 207-217.
[37] J.H. Driver, H. Paul, J-C. Glez, C. Maurice, in "Recrystallization - Fundamental Aspects and Relations to Deformation Microstructure", eds. N. Hansen et al., Risø Nat. Lab., Roskilde, Denmark, 2000, pp. 35-48.
[38] A. Godfrey, D. Juul Jensen, N. Hansen, submitted to *Acta mater.*
[39] M.G. Ardakani, F.J. Humphreys, *Acta Metall. Mater.*, 42, 1994, pp. 763-780.
[40] D. Juul Jensen, in "Proc. ReX'99", eds. T. Sakai and H.G. Suzuki, Japan Inst. Metals, Sendai, Japan, pp. 3-14.
[41] T.J. Sabin, D. Juul Jensen, in "Recrystallization - Fundamental Aspects and Relations to Deformation Microstructure", eds. N. Hansen et al., Risø Nat. Lab., Roskilde, Denmark, 2000, pp. 539-544.
[42] T.J. Sabin, G. Winther, D. Juul Jensen, in these Proceedings.
[43] H.F. Poulsen, D. Juul Jensen, T. Tschentscher, L. Wcislak, E.M. Lauridsen, L. Margulies, S. Schmidt, to be published.
[44] D. Juul Jensen, in "Microstructural and Crystallographic Aspects of Recrystallization", eds. N. Hansen et al., Risø Nat. Lab., Roskilde, Denmark, 1995, pp. 119-137.
[45] J.W. Cahn and W.C. Hagel, in "Decomposition of Austenite by Diffusional Processes", eds. V.F. Zackay and H.I. Aaronson, Interscience, NY, 1962, pp. 131-196.
[46] J.E. Hilliard, in "Recrystallization, Grain Growth and Textures", ed. H. Margolin, ASM, 1966, pp. 267-286.
[47] D. Juul Jensen, *Scripta Metall. Mater.*, 27, 1992, pp. 1551-1556.
[48] R.A. Vandermeer and D. Juul Jensen, *Acta Metall. Mater.*, 42, 1994, pp. 2427-2436.
[49] A.M. Gokhale, C.V. Iswaran, R.T. Dehoff, *Metal. Trans,* 10A, 1979, pp. 1239-1245.

[50] R.A. Vandermeer, D. Juul Jensen, *Scripta Metall. Mater.*, 30, 1994, pp. 1575-1580.
[51] U. Lienert, H.F. Poulsen, V. Honkimäki, C. Schulze, O. Hignette, *J. Synch. Rad.*, 6, 1999, pp. 979-984.
[52] U. Czubayko, D.A. Molodov, B-C. Petersen, G. Gottstein, L.S. Shvindlerman, *Meas. Sci. Techn.*, 6, 1995, pp. 947-952.

Recrystallization and Grain Growth in Minerals: Recent Developments

Janos L. Urai[1], and Mark Jessell[2]

[1] Geologie-Endogene Dynamik, Fachgruppe Geowissenschaften, Lochnerstrasse 4-20, D-52056 Aachen, Germany.

[2] Department of Earth Sciences, Monash University, Box 28E, Melbourne, Victoria, 3800, Australia.

Keywords: dynamic recrystallization, grain growth, minerals, geoscience.

Abstract. Recent developments in the study of recrystallization and grain growth in rocks and minerals are briefly reviewed and key issues discussed.

1 Introduction

The study of deformation mechanisms and microstructure evolution in geosciences is different from that in engineering [1]. Firstly, we aim to reconstruct the processes and conditions in the geologic past from observations on rocks which are brought to the Earth's surface by geologic processes or by deep drilling [2, 3] or to infer such processes operating at present in the deep subsurface using measurements of acoustic or electric response of the lithosphere [4]. The second aim is to provide constitutive models which can be used as input for the modelling of geodynamic evolution [5-7].
The very low strain rates in geologic deformation (10^{-13} s^{-1} is a common value) mean that these natural processes are not directly accessible by experiment. Therefore Earth materials scientists combine extrapolations of laboratory data [8-14] with observations of both naturally and experimentally deformed rocks and theoretical modelling [15-19].
In the sixties these studies were strongly influenced by the metallurgical and materials science literature [20-22]. In the past four decades understanding of the subjects has increased dramatically, and the limits of applicability of concepts from engineering materials were clarified. With the increase of data available and the improvement of analytical tools, communication between experimentalists and structural geologists has become increasingly challenging. Laboratory studies focus more on quantifying the behaviour of simple, pure or synthetic rocks under well defined conditions [23-27]; studies addressing the behaviour of polyphase rocks [28-30] are less frequent. On the other hand, studies of naturally deformed and recrystallized materials [31-37] emphasize the complexities of the processes

occurring in nature which often involve changes in chemical composition and complex deformation paths under changing temperature and pressure.
The aim of this paper is to give a brief overview of the developments and issues in the study of recrystallization and grain growth in minerals over the past fifteen years, using a selection of papers from the large body of literature in this rapidly evolving field.

2 Dynamic Recrystallization

Displacements of the Earth's crust and upper mantle are accommodated at least in part by shear zones in which very large strains are accumulated in zones of localized noncoaxial deformation [38, 39]. During localization of strain in polyphase rocks softening mechanisms involve the rearrangement of phases by plastic flow and dynamic recrystallization, producing a change of the load-supporting phase [28, 40] or in some cases a change in deformation mechanism to grain size sensitive flow by a grain-scale mixing of the two phases [41].
Dynamic recrystallization is a common process in ductile shear zones [42-45], in fact the vast majority of recrystallization processes in the Earth are dynamic in nature.
The microscale mechanisms of dynamic recrystallization are rotation of subgrains until new high angle grain boundaries are formed [46, 47], and grain boundary migration [48, 49]. Fig. 1 is an illustration of dynamic recrystallization in biotite, with kinking due to the presence of only one easy slip system and grain boundary migration in the presence of a small amount of partial melt in the grain boundaries.
As a result of dynamic recrystallization, a transition to grain size sensitive flow may occur, during phases of transiently low grainsize after a previous period of dynamic recrystallization at high shear stress [50] or caused by stabilization of the grainsize by second phases [51]. The conditions under which the changes in grain size during dynamic recrystallization can lead to a switch in deformation mechanism to grain size sensitive creep and consequent localization of strain has been discussed extensively in the Earth Science literature. Recently, an interesting new model was presented by de Bresser and coworkers [52, 53]. According to this model, dynamic grain growth and grain size reduction during dynamic recrystallization lead to a balance at steady state such that the strain rate contributions of dislocation and diffusion creep become equal. This means that during steady state dynamic recrystallization there will be a significant contribution of diffusionally accommodated grain boundary sliding. It also means that dynamic recrystallization in a pure material cannot produce a switch in deformation mechanism to dominantly grainsize sensitive creep. During geological deformation under changing temperature and strain rate however, it is possible that steady state is not reached and large part of the rock's history are dominated by transient effects. Research into these transient effects is still in its infancy.

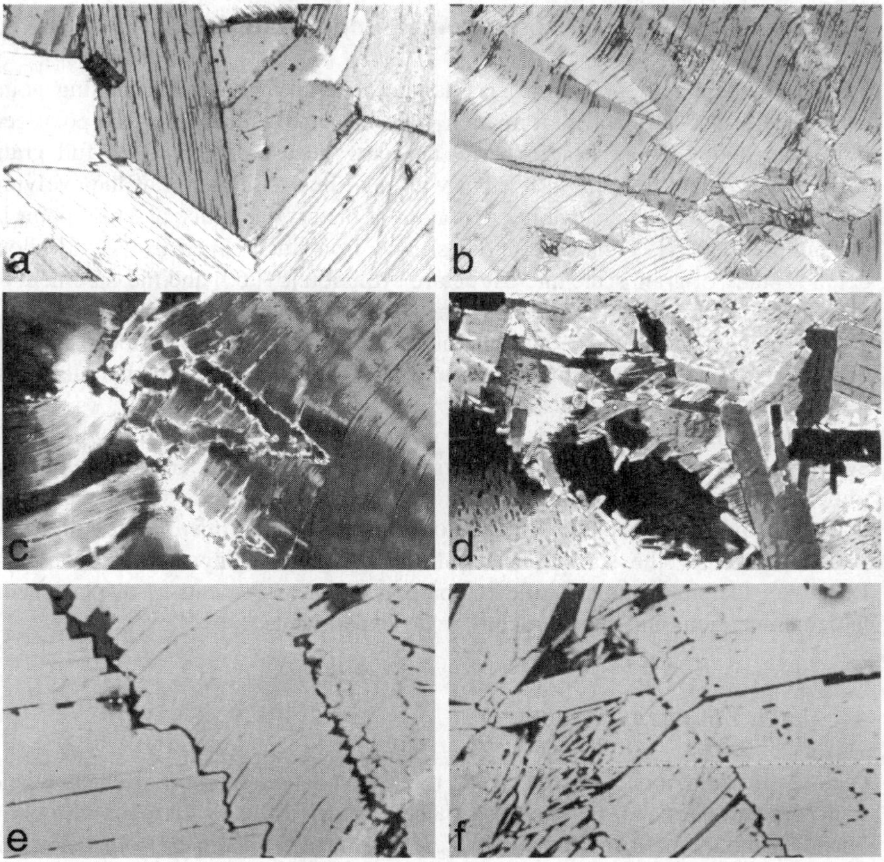

Fig. 1. Experimentally deformed biotite, a highly anisotropic sheet silicate [54]. Samples were deformed at 1 GPa pressure with about 3% H2O-rich fluid added. (a) Starting material, a coarse grained natural biotite rock. Transmitted light micrograph, horizontal edge of photo is 5 mm. (b) Deformed grain showing kinks developed to accommodate deformation by basal glide which is the only easy slip system in Biotite. Black lines are microfractures along the basal plane. Transmitted light micrograph, horizontal edge of photo is 1 mm. (c) Kinks in a grain deformed at 10^{-6} s^{-1} at 800 C, for 90 hours. Incipient recrystallization by kink boundary migration. Transmitted light micrograph, horizontal edge of photo is 1 mm. (d) Dynamically recrystallized grains in a sample deformed 20,000 hours at 10^{-7} s^{-1} at 820 C. Anisotropic growth of new grains occurs along former grain boundaries. Transmitted light micrograph, horizontal edge of photo is 1 mm. (e, f) SEM BSE micrographs of the sample shown in (d), showing kink boundary migration and growth of new grains which have a significantly different composition from the parent grains. Dark grey areas are the small amounts of melt which formed along preexisting grain boundaries and facilitated grain boundary migration. Horizontal edge of photos is 0.1 mm.

3 Transmitted Light Deformation of Model Materials

Transparent, optically anisotropic crystalline materials with a low melting point can be deformed in small deformation cells in plane strain under the polarized light microscope [11]. These experiments allow measurement of the full grain scale displacement field [55], together with real time observation of the evolving microstructure and the change in orientation of the crystal lattice at every point in the sample [56, 57]. Time lapse movies made from such experiments [58] allow detailed study of grain scale microstructure evolution [59, 60] and measurement of grain boundary velocity [61, 62]. In some cases the experiments are scaled models of much higher temperature deformation of minerals because the model material belongs to the same isomechanical group [62]. An example of a transmitted light deformation experiment is shown in Fig. 2.

With the rapid development in the numerical modelling of microstructural evolution [63] transmitted light deformation experiments are becoming an important tool for the validation of such models. Using the properties of the model material and the boundary conditions of the experiment, the digitized microstructure of the starting material can be used as input for simulations. Results of the simulations can be compared with measurements of displacement, lattice orientations and microstructure in the experiments.

4 Grain Boundary Structure

Changes in grain boundary structure in rocks have large effect on mechanical and transport processes [64, 65] and fast grain boundary diffusion coupled with grain boundary migration is an effective process to change the chemical composition of grains, with the difference in composition between old and new grains providing a possible additional driving force for grain boundary migration [66].

Our understanding of the structure and mobility of grain boundaries in minerals is poorly constrained. This partly due to the experimental difficulties and also to the large variability in grain boundary structure in recrystallizing rocks, from clean, solid-state boundaries [67] to micron-wide zones of fluid or melt between two grains. Such fluid-filled grain boundaries migrate by dissolution of the grain with the higher free energy, diffussion across the fluid and crystallization on the other side [21]. The morphology of the fluid in these grain boundary films has been proposed to be different from the morphology in static equilibrium [68-73], but more research is needed for a satisfactory resolution of this issue.

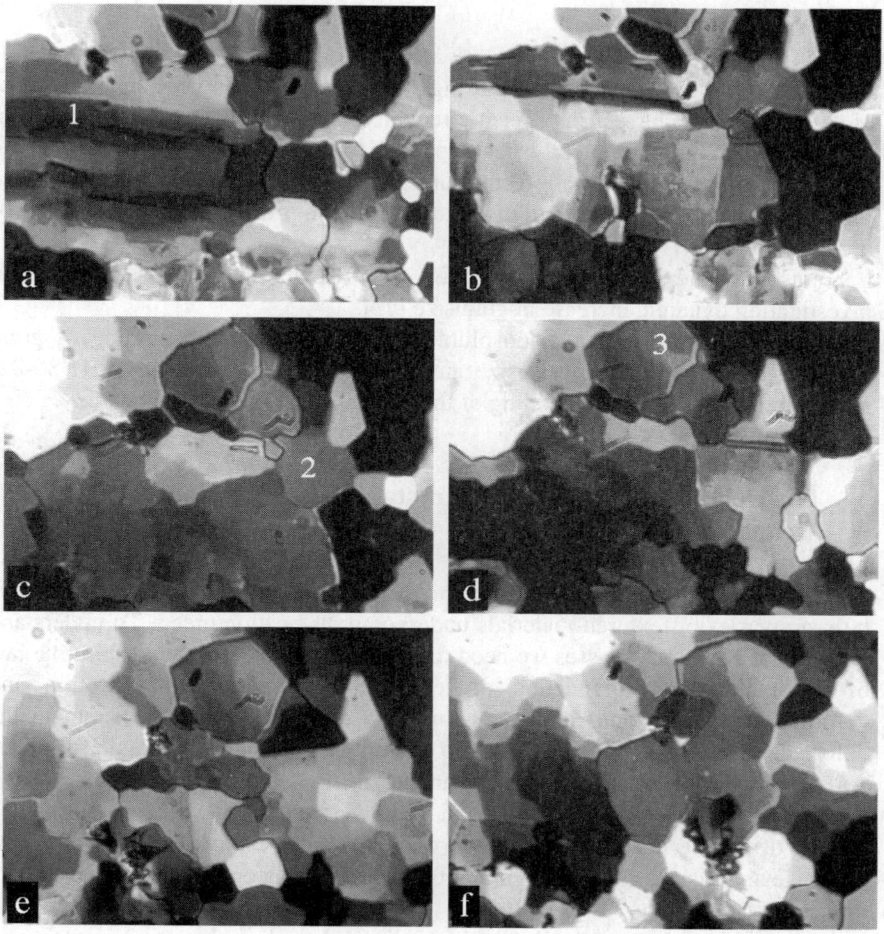

Fig. 2. Sequence of images from a transmitted light deformation experiment of polycrystalline $NaNO_3$ at 0.98 of its melting temperature [22]. The images were taken during simple shear deformation at a strain rate of 10^{-5} s^{-1}, over a period of about one hour. The microstructural evolution consists of the rapid growth [62] of large new grains (1, 2 and 3) nucleated from pre-existing small grains, followed by rotation recrystallization by progressive misorientation of subgrains, and slow migration of these new high angle boundaries. Crossed polarizers, width of each image is 0.5 mm.

5 Grain Growth

Because of the low driving forces during static grain growth, very small amounts of deformation can have a significant effect on the evolving microstructures. Grain growth in nature is frequently accompanied by deformation [74, 75] and the perfect foam textures which form during static grain growth of metals are infrequent in rocks. Examples of close-to static settings are the contact aureoles of granite intrusions [76].

Laboratory studies of grain growth [27, 77, 78] are in part also directed to investigating dynamic increase in grainsize [79].

The well known effects of an insoluble second phase on the pinning of grain boundaries and stabilizing grainsize are also common in rocks [24, 51, 80-82] although quantitative understanding of their effect needs much further study.

6 The Relationship between Recrystallization, Mechanical Properties and CPO

The highly anisotropic crystal structure of most minerals results in anisotropic bulk properties of textured materials undergoing dislocation creep. To understand the flow of such rock masses we need to quantify the anisotropic constitutive law of the material as a function of temperature, impurity concentration and deformation history, and implement this in a numerical model to simulate the deformation in nature. Sufficient data to attempt this are rarely available for rocks. An example is the flow of ice in polar ice sheets. Here an unusually complete dataset is available consisting of well characterized cores with a well known history of simple coaxial deformation over 400,000 years. Measurement of the grainsize and dislocation density along the cores shows good correspondence with modelling [83-85]

During burial to several hundred metres depth the ice undergoes grain growth during slow deformation, driven by the reduction in grain boundary energy. The rate of grain growth correlates with small variations in impurity levels caused by climate cycles in the past [86]. At further burial and deformation the ice undergoes rotation recrystallization which prevents further increase of the grainsize [87]. A strong CPO is developed, with an associated increase in mechanical anisotropy which in turn influences the displacement field in the ice sheet.

In the bottom 100 m of the ice sheet temperatures are sufficiently high (above -10 C) to cause a dramatic increase of grain boundary mobility and a three orders of magnitude increase in grain boundary migration rate [88]. In this regime grain boundary migration recrystallization is the dominant process, producing large lobate grains and a reduction of the strength of the CPO.

7 Interpretation of Structures in Naturally Recrystallized Rocks

Building on the rapidly growing body of knowledge of which this paper gives a brief overview, modern studies of naturally deformed rocks are capable of reconstructing considerable detail of the thermal and deformational history using detailed microstructural and mineralogical study [89-94].

On the other hand, some of these interpretations are not accepted by all, as quite different microscale processes can produce strikingly similar microstructures [95], and the complex overprint on a microstructure during a rock's journey to the surface is not well understood.

One of the methods to reconstruct the stress state in the past from microstructural information is based on the well known relationship between dynamically recrystallized grainsize and shear stress. Based on laboratory calibrations of this relationship for minerals, paleopiezometers are now in use for several minerals [53, 96, 97]. Recent work has shown that, in contrast to previous models, the temperature dependence of the grainsize-stress relationships cannot be neglected, and for improved paleostress estimates more extensive characterization of the deformation conditions is required.

An important development in analytical techniques is the possibility to create orientation images, using EBSD [98, 99] or optical techniques [100, 101]. This allows rapid access to the full richness of microstructural information in a rock, and has the promise to bring new diagnostic microstructures to improve our interpretation capabilities. The first studies in this field use boundary hirarchy characteristics to distinguish the dominant deformation mechanisms [102-104].

References

[1] S. M. Schmid, R. Heilbronner, H. Stünitz, *Tectonophysics,* 303, 1999, pp. vii-ix.

[2] R. J. Knipe, *Journal of Structural Geology,* 11, 1989, pp. 127-146.

[3] P. J. O'Brien, J. Duyster, B. Grauert, B. Stöckhert, K. Weber, *Journal of Geophysical Research,* 102 B8, 1998, pp. 18203-18220.

[4] S. Zhang, S.-i. Karato, J. Fitz Gerald, U. H. Faul, Y. Zhou, *Tectonophysics,* 316, 2000, pp. 133-152.

[5] R. N. Pysklywec, C. Beaumont, P. Fullsack, *Geology,* 28, 2000, pp. 655-658.

[6] S. Ellis, S. Wissing, A. Pfiffner, *Int. J. Earth Sci.,* 2001, in press.

[7] M. R. Handy *et al., Int. J. Earth Sci.,* 2001, in press.

[8] M. S. Paterson, *Tectonophysics,* 133, 1987, pp. 33-43.

[9] G. Hirth, J. Tullis, *Journal of Structural Geology,* 14, 1992, pp. 145-159.

[10] L. N. Dell'Angelo, J. Tullis, *Tectonophysics,* 169, 1989, pp. 1-21.

[11] W. D. Means, *Journal of Structural Geology,* 11, 1989, pp. 163-174.

[12] M. S. Paterson, *International Journal of Earth Sciences,* 2001, in press.

[13] G. Hirth, C. Teyssier, W. J. Dunlap, *Int. J. Earth Sci,* 2001, in press.
[14] W. Skrotzki, R. Tamm, C.-G. Oertel, J. Röseberg, H.-G. Brokmeier, *Journal of Structural Geology,* 22, 2000, pp. 1621-1632.
[15] J. P. Busch, B. A. van der Pluijm, *J. Struct. Geol.,* 17, 1995, pp. 677-688.
[16] C. Pauli, S. M. Schmid, R. P. Heilbronner, *Journal of Structural Geology,* 18, 1996, pp. 1183-1203.
[17] R. L. M. Vissers, M. R. Drury, J. Newman, T. F. Fliervoet, *Tectonophysics,* 279, 1997, pp. 303-325.
[18] H. R. Wenk, G. Canova, Y. Bréchet, L. Flandin, *Acta Materialia,* 45, 1997, pp. 3283-3296.
[19] D. L. Olgaard, *Geol. Soc. Spec. Pub.,* 54, 1990, pp. 175-181.
[20] Poirier, "Creep of Crystals" , Cambridge University Press, 1985.
[21] J. L. Urai, W. D. Means, G. S. Lister, in "AGU Geophysical Monograph" 36, 1986, pp. 161-199.
[22] M. R. Drury, J. L. Urai, *Tectonophysics,* 172, 1990, pp. 235-253.
[23] S.-I. Karato, *Phys. Earth Planet. Interiors,* 51, 1988, pp. 107-122.
[24] D. L. Olgaard, B. Evans, *Contrib. Min. Petrol.,* 100, 1988, pp. 246-260.
[25] G. C. Gleason, J. Tullis *J. Struct. Geol.,* 15, 1993, pp. 1145-1168.
[26] S. J. Covey-Crump, *Journal of Structural Geology,* 19, 1997, pp. 225-241.
[27] G. Dresen, Z. Wang, Q. Bai, *Tectonophysics,* 258, 1996, pp. 251-262.
[28] L. Dell'Angelo, J. Tullis, *Tectonophysics,* 256, 1996, pp. 57-82.
[29] J. Tullis, H. Stünitz, C. Teyssier, R. Heilbronner., http://www.virtualexplorer.com.au/VEjournal/Volume2, 2000
[30] P. D. Bons, J. L. Urai, *Tectonophysics,* 256, 1996, pp. 145-164.
[31] R. J. Cumbest, M. R. Drury, H. L. M. v. Roermund, C. Simpson, *Contributions to Mineralogy and Petrology,* 101, 1989, pp. 339-349.
[32] R. Kruse, H. Stünitz, *Tectonophysics,* 303, 1999, pp. 223-249.
[33] T. F. Fliervoet, S. H. White, *J. Struct. Geol.,* 17, 1095-1109 (1995).
[34] D. J. Prior, J. Wheeler, *Tectonophysics* 303, 1999, pp.29-49.
[35] J. C. White, C. K. Mawer, *J. Struct. Geol.,* 8, 1986, pp. 133-143.
[36] J. Newman, W. M. Lamb, M. R. Drury, R. L. M. Vissers, *Tectonophysics,* 303, 1999, pp. 193-222.
[37] U. Altenberger, S. Wilhelm, *Tectonophysics,* 320, 2000, pp. 107-121.
[38] R. L. M. Vissers, M. R. Drury, *Tectonophysics,* 249, 1995, pp. 155-171.
[39] E. H. Rutter, *Tectonophysics,* 303, 1999, pp. 147-158.
[40] K. Schulmann, B. Molcoch, *J. Struct. Geol.,* 18, 1996, pp.719-733.
[41] J. H. Behrmann, D. Mainprice, *Tectonophysics,* 140, 1987, 297-305.
[42] D. Jin, S.-I. Karato, M. Obata, *J. Struct. Geol.,* 20, 1998, pp. 195-209.
[43] G. Molli, P. Conti, G. Giorgetti, M. Meccheri, N. Oesterling, *Journal of Structural Geology,* 22, 2000, pp.1809-1825.
[44] M. Bestmann, K. Kunze *J. Struct. Geol.,* 22, 2000, pp.1789-1807.
[45] N. S. Mancktelow, *Tectonophysics,* 135, 1987, pp. 133-153.
[46] O. Nishikawa, T. Takeshita, *J. Struct. Geol.,* 22, 2000, pp. 259-276.

[47] F. Heidelbach, K. Kunze, H.-R. Wenk, *J. Struct. Geol.,* 22, 1999, pp. 91-104.
[48] B. Lafrance, B. E. John, B. R. Frost, *J. Struct. Geol.,* 20, 1998, pp. 945-955.
[49] T. Takeshita, I. Hara, *Journal of Structural Geology,* 20, 1998, pp. 417-431.
[50] L. Dell'Angelo, D. Olgaard *J. Geophys. Res.,* 100, 1995, pp. 15425-15440.
[51] M. Krabbendam, J. L. Urai, *J. of Structural Geology,* 2001, in press.
[52] J. H. P. de Bresser, C. J. Peach, J. P. J. Reijs, C. J. Spiers, *Geophysical Research Letters,* 25, 1998, pp. 3457-3460.
[53] J. H. P. de Bresser, J. H. ter Heege, C. J. Spiers, *International Journal of Earth Sciences,* 2001, in press.
[54] J. L. Urai, R. L. M. Vissers, *Abstract, Leeds Conf 1988.,* 1989.
[55] P. D. Bons, M. W. Jessell, *J. of Structural Geology,* 21, 1999, pp. 323-334.
[56] M. Herwegh, M. R. Handy, R. Heilbronner, *Tectonophysics,* 280, 1997, pp. 83-106.
[57] R. Heilbronner, M. Herwegh, *J. of Struct. Geol.,* 19, 1997, pp. 861-874.
[58] J. L. Urai, F. J. Humphreys, www.virtualexplorer.com.au/ VEjournal/Volume2, 2000
[59] M. W. Jessell, *Journal of Structural Geology,* 8, 1986, pp. 527-542.
[60] J. L. Urai, *Tectonophysics,* 135, 1987, pp. 251-263.
[62] P. D. Tungatt, F. J. Humphreys, *Acta Met.,* 32, 1984, pp.1625-1635.
[63] M. Jessell, in This Proceedings, 2001.
[64] D. L. Kirschner, C. Teyssier, R. T. Gregory, Z. D. Sharp, *Journal of Structural Geology,* 17, 1995, pp. 1407-1423.
[65] S. de La Chapelle, H. Milsch, O. Castelnau and P. Duval, *Geophysical Research Letters,* 26, 1999, pp. 251-254.
[66] H. Stuenitz, *Contrib. to Mineralogy and Petrology,* 131, 1998, pp. 219-236.
[67] D. L. Kohlstedt, *AGU Geophysical Monograph,* 56, 1990, pp. 211-218.
[68] J. L. Urai, C. J. Spiers, H. J. Zwart, G. S. Lister, *Nature,* 324, 1986, pp. 554-557.
[69] Z. M. Jin, H. W. Green, Y. Zhou, *Nature*, 372, 1994, pp.164-129.
[70] M. R. Drury, J. D. FitzGerald, *Geoph. Res. Lett.,* 23, 1996, pp. 701-704.
[71] H. J. M. Visser, *Geologica Ultraiectina,* 174, 1999, p. 242.
[72] S. H. Hickman, B. Evans, *J. Geophy. Res.,* 100, 1995, 13, pp.113-132.
[73] W. K. Heidug, *J. Geophy. Res.,* 100 B4, 1995, pp. 5931-5940.
[74] S. J. Covey-Crump, E. H. Rutter, *Contributions to Mineralogy and Petrology,* 101, 1989, pp. 69-86.
[75] K. A. Hickey, T. H. Bell, *Europ. J. of Mineralogy,* 8, 1996, pp. 1351-1373.
[76] R. Joesten, *American Journal of Science,* 283-A, 1983, pp. 233-254.
[77] S. J. Covey-Crump, *Contrib. Min. Petrol.,* 129, 1997, pp. 239-254.
[78] J.-H. Ree, Y. Park, *Journal of Structural Geology,* 19, 1997, pp. 1521-1526.
[79] S. I. Karato, T. Masuda, *Geology,* 17, 1989, pp. 695-698.

[80] T. Masuda, T. Koike, T. Yuko *J. Metam. Geology,* 9, 1991, pp.389-402.
[81] S. J. Nichols, S. J. Mackwell, *Phys. Chem. Min.,* 18, 1991, pp. 269-278.
[82] D. L. Olgaard, B. Evans, *J. Am. Ceramic Soc.,* 69, 1986, pp. C272-C277.
[83] M. Montagnat, P. Duval, *Earth Planet. Sci. Let.,* 183, 2000, pp. 179-186.
[84] N. Azuma, K. Goto-Azuma, *Ocean. Lit. Review,* 44, 1997, p. 548.
[85] P. Duval, O. Castelnau, *Journal of Physics C,* 3, 1995, pp. 197-205.
[86] R. B. Alley, G. A. Woods, *Journal of Glaciology,* 42, 1996, pp. 255-260.
[87] S. de la Chapelle, O. Castelnau, V. Lipenkov, P. Duval, *Journal of Geophysical Research,* 103, 1998, pp. 5091-5105.
[88] C. J. L. Wilson, B. Marmo, www.virtualexplorer.com.au/ VEjournal/Volume2, 2000
[89] D. van der Wal, R. L. M. Vissers, M. R. Drury, E. H. Hoogerduijn-Strating, *Journal of Structural Geology,* 14, 1992, pp. 839-846.
[90] B. Stöckhert, J. Duyster, *J. of Struct. Geol.,* 21, 1999, pp. 1477-1490.
[91] B. Stöckhert, M. R. Brix, R. Kleinschrodt, A. J. Hurford, R. Wirth, *Journal of Structural Geology,* 21, 1999, pp. 351-369.
[92] S. van der Klauw, T. Reinecke, B. Stöckhert, *Lithos,* 41, 1997, pp. 79-102.
[93] C. W. Passchier, R. A. J. Trouw, "Microtectonics", Springer, ed. 1, 1996.
[94] J. L. Urai, R. D. Schuiling, J. B. H. Jansen, *Geol. Soc. Special Publication,* 54, 1990, pp. 509-522.
[95] J. Hippertt, M. Egydio-Silva, *J. Struct. Geol.,* 18, 1996, pp. 1345-1352.
[96] A. Post, J. Tullis, *Tectonophysics,* 303, 1999, pp.159-173.
[97] D. van der Wal, P. Chopra, M. Drury, J. FitzGerald, *Geophysical Research Letters,* 20, 1993, pp.1479-1482.
[98] P. W. Trimby, M. R. Drury, C. J. Spiers, *Journal of Structural Geology,* 22, 2000, pp. 81-89.
[99] P. W. Trimby, M. R. Drury, C. J. Spiers, *Journal of Structural Geology,* 22, 2000, pp. 1609-1620.
[100] R. Heilbronner, *Journal of Structural Geology,* 22, 2000, pp. 969-981.
[101] M. van Daalen, R. Heilbronner, K. Kunze, *Tectonophysics,* 303, 1999, pp. 83-107.
[102] P. W. Trimby, D. J. Prior, J. Wheeler, *J. Struct. Geol.,* 20, 1998, pp. 917-935.
[103] T. F. Fliervoet, M. R. Drury, *Tectonophysics,* 303, 1999, pp. 1-27.
[104] B. Neumann, *Journal of Structural Geology,* 22, 2000, pp. 1695-1711.

Grain Growth

Invited Papers

Recrystallization and Grain Growth
Proceedings of the First Joint International Conference
Eds. G. Gottstein and D.A. Molodov, Springer-Verlag, 2001

Topological Relationships in 2D Trivalent Mosaics (Topological Metallography)

R. Brandt[1], K. Lücke[1], G. Abbruzzese[2], and J. Svoboda[3]

[1] Institut für Metallkunde und Metallphysik, RWTH Aachen, D-52056 Aachen, Germany
[2] Centro Sviluppo Materiali, Terni, Italy
[3] Academy of Sciences of the Czech Republic, Brno, Czech Republic

Keywords: topology, mosaics, simulation of grain growth, grain growth kinetics

Abstract. A statistical theory for a general description of 2-dimensional (2D), trivalent mosaics is derived from first principles. The main parameters of this theory are introduced and values are discussed for the grains of different mosaics. The arrangement of grains with respect to their next neighbours is described by topological relationships (Topological Metallography). As main results the topological equations $n_i = n(r)$ and $r_n = r(n)$ Abbruzzese-Lücke equations, ALEs) are derived which are necessary for deriving the statistical theory of grain rowth (GG) kinetics. The validity of the topological law n = 6 for all trivalent mosaics is shown analytically and a rivation of the "Weaire - Aboav equation" is found in a correct way.

1 Introduction

Although homogeneous microstructures eem to look in cases rather random, they are not formed in a random fashion but show rather strong correlations. However, only few investigations being applicable to metallographical purposes exist on this subject. By the present authors such topological aspects have recently been derived in connection with the attempt to formulate a statistical theory of GG in analytical terms. Up to now the topological aspects were mostly taken care of by some heuristic assumptions or rather rough models, but, here for the 2D case these neighbourhood relations are considered correctly. The present lecture gives a brief overview of this new field here being called "Topological Metallography" for only 2D and only trivalent mosaics. For the description of each grain 4 parameters are considered (grain size R, grain shape f, number of sides n, crystallographic orientation g) and five examples of 2D microstructures are used (s. Fig. 1):

M-I: A Voronoi mosaic of N = 10000 grains.
M-II: A GG mosaic of N = 2488 grains starting from N_0 = 9000 grains.

M-III: A hard-core mosaic of N = 4900 grains. (No figures are given.)
M-IV: A hard-core GG mosaic of N = 3039 grains with N0 = 4900 grains.
M-V: A 2D section through an 3D Al3%Mg sample after GG (N = 1988).

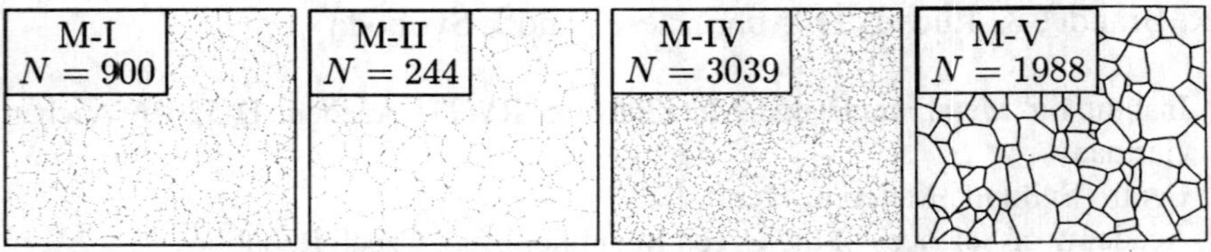

Fig. 1. GB networks: Voronoi mosaic M-I with straight GBs, M-II after simulation of GG, M-IV after simulation of GG, and M-V, a 2D section of a 3D Al3%Mg specimen after GG.

2 Individual Grains and their Distributions (One-Grain Model)

2.1 Definition of Grain Parameters

The main parameter of a grain μ is its perimeter L_μ equivalent grain size R_μ The shape of the grain shall be described by a second parameter f_μ giving the ratio of the grain size R_μ and the area A_μ equivalent radius R_μ^A Related to the mean radius $\overline{R}$ a reduced radius *r* is defined. Thus one obtains

$$R_\mu = R_\mu^L = \frac{L_\mu}{2\pi}; \quad f_\mu = \frac{R_\mu^L}{R_\mu^A}; \quad R_\mu^A = \sqrt{\frac{A_\mu}{\pi}} = \frac{R_\mu}{f_\mu}; \quad r_\mu = \frac{R_\mu}{\overline{R}} = f_\mu r_\mu^A . \tag{1}$$

The value $f_\mu = 1$ characterizes a circular grain shape. A third parameter is the coordination number n_μ and a fourth parameter is the grain orientation g. R and g will be represented in a discrete way by the size class indices i, j, k,... and by the orientation class indices H, K, L,...

Disregarding first the parameter f, the whole microstructure is described in a 3D phase space (r, n, g) by the three-parametric distribution function φ_{in}^H wich denotes the fraction of the number of grains in size class i having the number of sides n and being in orientation class H. One thus has

$$\sum_{i,n,H} \varphi_{in}^H = \sum_i \varphi_i = \sum_n \varphi_n = \sum_H \varphi^H = 1. \tag{2}$$

φ_i represents the grain size distribution function (SDF), φ_n the coordination distribution function (CDF), and φ^H the grain orientation distribution function (ODF) with the average values $r = 1$ and $n = 6$ [1].

2.2 Distributions of Grain Parameters

In Fig. 2 values f of the grains are plotted as dots. The open circles are representing the average values f_i. For the simulated mosaics M-II and M-IV f_i is approximately constant given by f_i = 1:08. The lower limits of f are plotted as solid lines showing almost free zones between the lower limit and f = 1 for circular grain shapes.

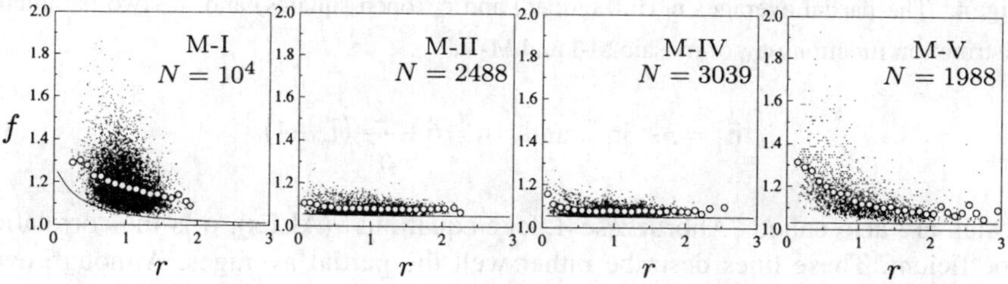

Fig. 2. Shape factor f vs. r for the individual grains. The mean shape factors f_i (open cirles) and a lower limit to f (solid line) are printed, too.

Practically most often applied for describing microstructures is the SDF. In Fig. 3 the measured values are plotted as full circular symbols and they are compared with best fits of 4 different types of distributions often applied in literature.
The frequency of grains possessing a size r and the coordination number n is given by a two-parametric distribution φ_m plotted into a 2D phase space (r, n). In Fig. 4 for mosaics M-I and M-V the individual grains are plotted as dots (for mosaics M-II and M-IV the dots are omitted). For the sake of visibility, they are not positioned at n = 3; P4;S5..., but they are randomly distributed between these integer values. The partial average number of sides n_i (closed cirecles) and r_n (open squares) were plotted, too, and one recognizes that these average values are situated near a solid and a dashed line

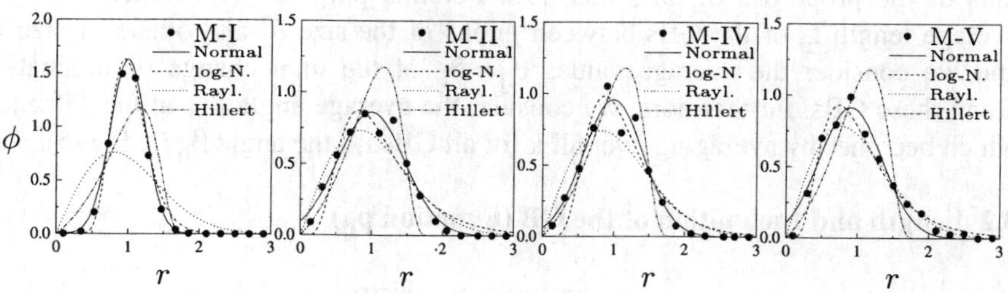

Fig. 3. Size distribution functions (SDF) for the considered mosaics.

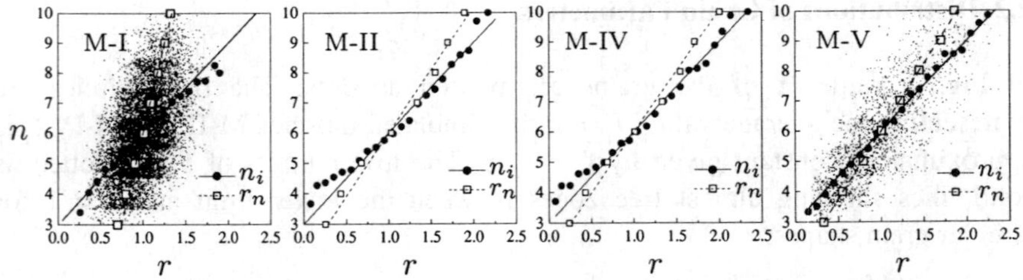

Fig. 4. The partial averages n_i (full circles) and r_n (open squares) and the two parametric distribution functions φ_m of mosaic M-I and M-V.

$$\tilde{n}_i = 3 + 3r_i \quad \text{and} \quad n = 6 + \frac{3}{\rho^2}(\tilde{r}_n - 1) \tag{3}$$

which are also called "Abbruzzese–Lücke equations" (ALEs). ρ is the correlation coefficient. These lines describe rather well the partial averages. Although even the partial averages of the simulated GG mosaics show particular deviations of it the ALEs are representing a kind of idealised behaviour being well suited for giving a general description. As has been shown in [2] the ALEs forces a particular symmetry on the phase space n vs. r. For the case of symmetry one can calculate the CDF from the SDF an vice versa. This has been shown by the results of simulated mosaics and for experimentally gained mosaics, too.

3 Topological Relationships of Trivalent Mosaics

3.1 Grain Boundaries (GBs) and Triple Points (TPs)

While in Sec. 2, as a first approximation for describing a polycrystalline microstructure, a one-grain model, i.e. only individual grains, were considered, now, in order to include the topology, a two-grain model is applied. This means that we take care of the arrangement of grains with respect to their neighbours and thus of the properties of GBs and TPs. For this purpose we consider now the average length l_{ij} of the GBs between grains of the size R_i and others of size R_j and we consider the average values $\theta_{ij} = \vartheta_{ij}$ of the total change of inclination along these GBs. Furthermore, we consider the average angle β_{ijk} at the TP i; j; k which becomes by averaging over all k for all GBs i; j the angle β_{ij} (s. Fig. 5).

3.2 Length and Inclination of the GB (Function p_{ij})

Here use is made of the fact that the individual GBs of a grain μ always form a closed loop leading to a total change of inclination of 2π by going once around clockwise (by convention) along its perimeter L. The perimeter L is composed of

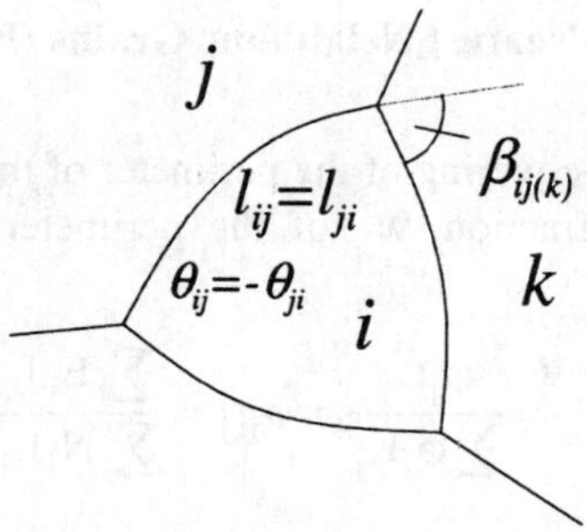

Fig. 5. Grain i in contact to other grains j and k. The average length l_{ij} of all GBs ij its average total change of inclina tion θ_{ij} and the angle $p_{ij(k)}$ at the common TP are indicated.

the lengths l_λ of the $\lambda = 1,..., n_\mu$ GBs and the total change of inclinaton of 2π by the changes of inclination ϑ_λ (by convention negative sign) of the n_μ GBs and β_λ of the n_m TPs at the end of the GBs, i.e.

$$\sum_{\lambda=1}^{n_\mu} l_\lambda = L; \quad \sum_{\lambda=1}^{n_\mu} \beta_\lambda - \vartheta_\lambda = 2\pi. \tag{4}$$

These equations lead to the fraction w_{ij} of the perimeters $N_i L_i$ of all grains i which are covered by grains j and to the fraction w^*_{ij} of the total change of inclination $2\pi N_i$ of all grains i which is made up by grains j

$$w_{ij} = \frac{E_{ij}}{N_i n_{ij}} \Rightarrow \sum_j w_{ij} = 1; \quad w^*_{ij} = \frac{E_{ij}}{N_i n^*_{ij}} \Rightarrow \sum_j w^*_{ij} = 1. \tag{5}$$

E_{ij} is the total number of contacts between the number N_i grains of size R_j and others of size R_j and $n_{ij} = L_i / l_{ij}$ and $n^*_{ij} = 2\pi \;/\left(\beta_{ij} + \vartheta_{ij}\right)$ are coordination numbers with respect to the length l_{ij} and the total change of inclination $\beta_{ij} + \varphi_{ij}$, respectively. The ratio between these quantities gives the topological function p_{ij}

$$p_{ij} = \frac{n_{ij}}{n^*_{ij}} = \frac{w^*_{ij}}{w_{ij}}. \tag{6}$$

The function p_{ij}, therefore, indicates wether a number of n_{ij} averaged GBs ij together with a number of n_{ij} averaged TPs ij can form a closed loop of total inclination 2π.

3.3 Size Correlations of Nearest Neighbour Grains (Functions k_{ij} and k^*_{ij})

For a completely random covering of the perimeter of the grains i, the quantity w_{ij} would be equal to the fraction w_j of the perimeter of the grains j and the perimeter of all grains:

$$w_j = \frac{\varphi_j L_j}{\sum\limits_j \varphi_j L_j} = \varphi_j r_j = \frac{\sum_i E_{ij} l_{ij}}{\sum_i N_i L_i}. \tag{7}$$

Correspondingly one obtains for a completely random covering of the total inclination 2π of the grains *i*, the fraction w^*_j with

$$w^*_j = \frac{\sum_i E_{ij}\left(\beta_{ij} - \vartheta_{ij}\right)}{\sum_i 2\pi N_i} = \frac{\varphi_i}{3}\left(\sum_i w_{ji} n_{ji} q_{ji} - 3\right) \tag{8}$$

since

$$\sum_i E_{ij}\left(\beta_{ij} - \vartheta_{ij}\right) = \sum_i E_{ji}\left[\beta_{ij} + \beta_{ji} - \left(\beta_{ji} - \vartheta_{ji}\right)\right] 2\pi / 3N_j \sum_i w_{ji} n_{ji} q_{ji} - 2\pi N_j.$$

For the non random cases correlation functions (coefficients) must be introduced which characterise the deviations from the random bahaviour:

$$k_{ij} = \frac{w_{ij}}{w_j}; \quad k^*_{ij} = \frac{w^*_{ij}}{w^*_j}. \tag{9}$$

Since the part w_{ij} of the perimeters $N_i L_i$ of the N_i grains of class i which is covered by grains j is equal to that part w_{ji} of the perimeters $N_j L_j$ of the N_j grains j which is covered by grains i one obtains the symmetric relationship $k_{ij} = k_{ji}$. A generally valid relationship $k^*_{ij} = k^*_{ji}$, however, does not exist.

3.4 Space Filling (Function q_{ij})

Space filling along the GBs requires that the pair of boundaries confining two adjacent grains μ and ν are identical, i.e. one obtains the principal GB symmetry conditions:

$$l_{ij} = l_{ji}; \vartheta_{ij} = -\vartheta_{ji} \tag{10}$$

This leads to

$$\frac{n_{ij}}{n_{ji}} = \frac{L_i / l_{ij}}{L_j / l_{ji}} = \frac{r_i}{r_j} \text{ and } \frac{\vartheta_{ij}}{\vartheta_{ji}} = \frac{\beta_{ij} - 2\pi p_{ij} / n_{ij}}{\beta_{ji} - 2\pi p_{ji} / n_{ji}} = -1 \tag{11}$$

and with a further topological function q_{ij}

$$q_{ij} = \frac{\beta_{ij} + \beta_{ij}}{2\pi / 3} \quad \text{where } \bar{q} = \frac{\sum_{i,j} \varphi_i w_{ij} n_{ij} q_{ij}}{\sum_{i,j} \varphi_i w_{ij} n_{ij}} = 1 \tag{12}$$

to

$$n_{ij} = \frac{3}{q_{ij}} \left(p_{ij} + \frac{r_i}{r_j} p_{ji} \right) \tag{13}$$

By averaging the relative number of contacts $E_{ij} / N_i = w_{ij} n_{ij}$ over all j one gets

$$n_i = \sum_j w_{ij} n_{ij} = 3 \left(\sum_j \varphi_j r_j \frac{k_{ij} p_{ij}}{q_{ij}} + r_i \sum_j \varphi_i \frac{k_{ji} p_{ji}}{q_{ji}} \right). \tag{14}$$

This relationship which only depends on the function $k_{ij} p_{ij} / q_{ij}$ of the above topological functions is very important since n_i which gives the average number of sides of grains of the size r_i, for example, is absolutely needed for the derivation of the statistical theory of GG kinetics.

4 Cases of Randomness

4.1 "Abbruzzese-Lücke Equations" (ALE, Full Randomness)

One recognizes that the 2D networks are characterized by the above topological functions k_{ij}, k^*_{ij}, p_{ij} and q_{ij} (Secs. 3.2, 3.3 and 3.4) which, in general, have to be experimentally determined. Of special interest, however, is the case that all four topological functions are constant and that_they_are_replaced by their "expectation values", i.e. by their average values $\bar{k} = \overline{k^*} = \bar{p} = \bar{q} = 1$. These as well as the functions which are derived from it are always indicated by a tilde, e.g. $\tilde{k}_{ij}$ or $\tilde{n}_{ij}$. By substitution of $\tilde{k}_{ij} = \tilde{p}_{ij} = \tilde{q}_{ij} = 1$ into Eq. (14) and by averaging over all j one finally gets

$$\tilde{n}_i = 3 \left(\sum_j \varphi_j r_j + r_i \sum_j \varphi_j \right) = 3(1 + r_i) \tag{15}$$

and from Eq. (8) one gets

$$\tilde{w}_j^* = \varphi_j \frac{n_j - 3}{3} = w_j \frac{n_j - 3}{3r_j} \Rightarrow \frac{\tilde{w}_j^*}{w_j} = \frac{n_j - 3}{3r_j} \tag{16}$$

Together with Eqs. (6) and (15) one finally obtains

$$w_j = \tilde{k}_{ij}^* \tilde{w}_{ij}^* = \tilde{k}_{ij}^* w_j \frac{\tilde{n}_j - 3}{3r_j} = \tilde{k}_{ij}^* w_j \tag{17}$$

so that $\tilde{k}_{ij}^* = 1$ follows automatically leading to full randomness. Eq. (15) is denoted as "Special Linear Relationship" (SLR). For a linearly correlated set of pairs of data (r; n) a second, corresponding equation for $\tilde{r}_n$ can be found by linear regression analysis. This leads together with the SLR (Eq. (15)) again to the "Abbruzzese - Lücke Equations" (ALE)

$$\tilde{n}_i = 3(1 + r_i),\ \tilde{r}_n = 1 + \frac{\rho^2}{3}(n - 6) \qquad 18)$$

which were so far only experimentally found [2]. The ALEs can thus be considered as the idealised case which proves that most mosaics are nearly random.

4.2 Simulated GG Mosaics (Partial Randomness)

Beside the fully random case also cases of partial randomness are existing where only some of the functions are replaced by their "expectation values", e.g. as observed for simulated GG mosaics [3]. This special case of partial randomness is characterized by a hat over the corresponding topological function, i.e. $\hat{k}_{ij}^* = 1$, and the functions which are derived from it, e.g. $\hat{w}_{ij}^* = w_j^*$.All angles are $\beta_{\mu\nu\lambda} = \pi/3$ by applying the mechanical equilibrium condition so that one obtains $\hat{w}_j^* = \hat{w}_j^*$ (s. Eq. (16)).

This case of partial randomness has been observed for mosaics generated by simulation of GG. The product $k_{ij}p_{ij} = \hat{w}_{ij}^* / w_j$ is found to be independent of r_i (Fig. 6) and it is fitting well to $\hat{w}_j^* / w_j$ which is given in analogy to Eq. (16) by substitution of n_j by a parabolic relation ship (s. Eq. (23)) as a best fit to n_j of mosaic M-II. It is indicated by a full line in Fig. 6. This yields for the example of mosaic M-II by use of Eq. (6) a correlation coefficient $k_{ij}^* = \hat{k}_{ij}^* = 1$ which means that one observes here completely random covering of the circumference of a grain of total inclination 2π by other grains j. A similar case of partial randomness is found for mosaic M-IV, too. The Voronoi mosaics M-I and M-III show a non random behavior with respect to this definition.

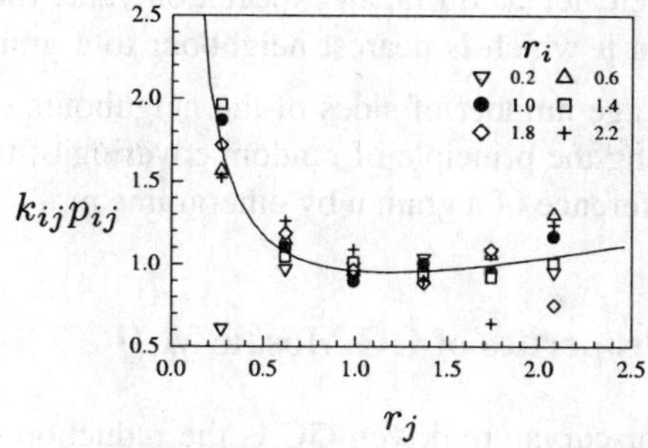

Fig 6. The product $k_{ij}p_{ij}$ as function of r_j. The measured values of mosaic M-II (symbols) are fitted by $\hat{w}_j^* / w_j$.

Since the fraction w_{ij}^* of the total inclination $2\pi N\varphi_i$ of the grains i is made up by the number $E_{ij} = E_{ji} = N\varphi_j w_{ji}^* n_{ji}^*$ of GBs ij and their TPs, i.e.

$$2\pi\varphi_i w_{ij}^* = \varphi_j w_{ji}^* n_{ji}^* (\beta_{ij} + \beta_{ji}) \tag{19}$$

one gets for the coordination number $\hat{n}_{ij}^*$ of this case of and $\hat{q}_{ij} = 1$ the definition of n_{ij}^* (Eq. (6)). Substituting w_{ij}^* by $\hat{w}_j^*$ finally yields

$$\hat{n}_{ij}^* = 3\left(1 + \frac{n_i - 3}{n_j - 3}\right) \tag{20}$$

4.3 "Weaire - Aboav Equation" (WAE, Partial Randomness)

For the derivation of the WAE [4] instead of the grain size r_i, the coordination number n is of main interest. By application of $\hat{k}_{mn}^* = q_{mn} = 1$ one gets

$$\hat{w}_{0mn}^* = w_n^* = \varphi_n \frac{n-3}{3} \quad \text{and} \quad \hat{n}_{mn}^* 3\left(1 + \frac{m-3}{n-3}\right). \tag{21}$$

The result of this special case of randomness can be applied for the correct derivation of the "Weaire - Aboav equation" which gives the average number $\hat{m}_{n\mu}$ of sides for the grains which are next neighbour to the n-sided grains by

$$\hat{m}_n = 5 + \frac{6 + \sigma_n^2}{n} \quad \text{with} \quad \sigma_n = \bar{n} k_n . \tag{22}$$

For the derivation we calculated first an expectation value for the mean number of sides $\hat{m}_{n\mu}$ of a grain μ which is nearest neighbour to a grain n and, finally, in a second step the average number of sides of the neighbours of grain n. This is the result of only applying the principle of random covering of the total inclination of 2π along the circumference of a grain n by othergrains m.

5 Topological Properties of GG Mosaic M-II

The driving force for curvature driven GG is the reduction of GB energy so that GG shall here be understood as the reduction of GB length. The microstructure can be changed in two ways to gain this reduction:
(i) By reduction of the shape factor f (microscopic GG) in the average, i.e. that at constant area A the grain enclosing perimeter L is reduced (s. Fig. 2). (ii) By reduction of the total number N of grains (macroscopic GG) so that the average grain size $\overline{R}$ increases, but, simulataneously the total GB length and the total amount of GB energy is reduced.
Before a grain vanishes leading to macroscopic GG (ii) it has to shrink and it is exposed to a cascade of reduction of sides $... \rightarrow 5 \rightarrow 4 \rightarrow 3 \rightarrow 2$. However, grains of 2 sides are rarely observed in microstructures probably since they are very small and their life is very short. But also grains of 3 sides are observed seldomly. Instead the grains of 4 sides are predominant among the small grains. This topological stability of 4-sided grains shows up in Fig. 4 where for $r_i = 0$ instead of $n_i = 3$ approximately the value $n_i = 4$ is observed. With this value $n_i(r_i = 0) = 4$ the topological relationship n_i can well be approximated by the parabola

$$n_i = 4.0 + 1.1 r_i + 0.7 r_i^2 \text{ with } \overline{n} = \sum_i \varphi_i n_i = 6. \tag{23}$$

Considering here microstructures of only uniform GBs, this topological relationship $n_i(r_i)$ is important for the description of GG by a statistical theory. Here one gets for the growth rate G_i by use of the averaged von Neumann-Mullins law $dA_i / dt = m\gamma\pi(n_i - 6)/3$ and the assumption of an uniform shape factor $f_i = f \approx 1.08$

$$G_i = G(r_i) = \frac{dR_i}{dt} = \frac{f^2}{2\pi R_i}\frac{dA_i}{dt} = \frac{m\gamma f^2}{6R_i}(n_i - 6). \tag{24}$$

A linear topological relationship

$$n_i = 6 + a(r_i - 1) \tag{25}$$

leads by applying the term of Eq. (24) to a continuity equation (s. [5]) to

$$\frac{\delta N\phi}{\delta t} = -\frac{\delta}{\delta R}(N\phi G) = -c\frac{\delta}{\delta R}\left[\phi\left(\frac{1}{\overline{R}} - \frac{1}{R}\right)\right] \text{ with } c = \frac{m\gamma f^2 Na}{6}. \tag{26}$$

For a long run of simulation one obtains a self similar, reduced SDF $\phi(R;t)dR=\phi^*(r,t)dt$ which means that, by applying r instead of R to the SDF, the reduced SDF becomes independent of time, i.e.

$$\frac{\delta\phi^*(r,t)}{\delta t}=0. \tag{27}$$

As was shown first by Hillert [6] the corresponding SDF can be calculated yielding the Hillert distribution $\phi^*(r)=\phi_H(R)$ In contrast a non-linear relationship n_i supplies another self similar, reduced SDF [5]. This significantly differs from the Hillert distribution. For example for the parabolic relationship n_i of Eq. (23) a self similar, reduced SDF is obtained which, moreover, fits very well to the measured data of mosaic M-II. A Hillert distribution, however, is only found for linear relationships given by Eq. (25).

6 Summarizing Remarks

The main parameters (R, f, n, and g) have been introduced for a topological description of homogeneous 2D trivalent microstructures (s. Sec. 2.1). Although these microstructures seem to look rather random it could be shown that there exist well-defined relationships between these parameters of one grain and its nearest neighbour grains (one-grain model). This model has been extended by a two-grain and a three-grain model where two or three next neighbour grains and thus GBs and TPs are included into the considerations. The points are thoroughly discussed at the example of five different types of mosaics (s. Sec. 1).
By this way a statistical theory of microstructures has been developed which is well suited for deriving various types of topological relationships.
The topological functions k_{ij}, k^*_{ij}, p_{ij} and q_{ij} have been introduced where k_{ij} represents the fraction of the perimeters of grains of the size class i which is covered by other grains of size class j. k^*_{ij} represents the analoguous relationship for the total changes of inclination. (For p_{ij} and q_{ij} see Sec. 3.)
This theory is leading to a new and elegant derivation of the fundamental topological law $\bar{n}=6$ which had been derived so far mainly by geometrical means [1]. By substitutions of Eqs. (13), (9), (12), and $k_{ij}=k_{ji}$ and finally by averaging the terms of Eq. (14) over all i one gets

$$\bar{n}=\sum_{i,j}\varphi_i w_{ij} n_{ij}=\sum_{i,j}\varphi_i w_{ij} n_{ij} q_{ij}=6. \tag{28}$$

By means of these functions k_{ij}, k^*_{ij}, p_{ij} and q_{ij} a well founded definition of randomess has been established. Since the total average values characterizes the randomness for the special case of fully randomness one gets $k_{ij} = k^*_{ij} = p_{ij} = q_{ij} = 1$. These equations yield the "Abbruzzese - Lücke Equations" (ALEs)

$$\tilde{n}_i = \tilde{n}(r_i) = 3(1 + r_i),\ \tilde{r}_n = \tilde{r}(n) = 1 + \frac{\rho^2}{3}(n - 6). \tag{29}$$

which are shown to be a reasonable approximation for most 2D mosaics (e.g. Fig. 4). This means that these equations are a kind of ideal topological relationships. However, the exact fully random case is not found for any of the mosaics. The mosaic M-V comes close to it since the 2D section intersects the 3D body randomly so that k_{ij}, k^*_{ij}, p_{ij} and q_{ij} are given by their average values. For simulated GG mosaics a kind of partial randomness $k^*_{ij} = q_{ij} = 1$ has been observed.

However, there do not only exist correlations between grains of different size classes but also between grains of different coordination classes. The topological function $k^*_{mn} = 1$ represents thus also a quantity giving randomness. This provides a new interpretation and derivation of the "Weaire - Aboav equation":

$$\hat{m}_n = 5 + \frac{6 + \sigma_n^2}{n}. \tag{30}$$

The partial mean values m_n of simulated GG mosaics, e.g. M-II, and of 2D sections of microstructures after GG, e.g. mosaic M-V, are described in good approximation by this equation. The corresponding values m_n of a Voronoi mosaic M-I, however, are badly fitted by this equation since strong correlations occur leading to $k^*_{mn} \neq \hat{k}^*_{mn}$. A derivation of the "Weaire – Aboav equation" has also been given in [7], but there it is based on the heuristic assumption that there exist no correlation between the coordination number of a grain and any one of its second nearest neighbours which, however, does not agree to the measurements [2].

The statistical theory of microstructures allows an understanding of the mechanisms of actual formation of the microstructure which is by C.S. Smith [8] "a result of the interplay between the requirements of the physical forces operating between the individual parts and the mathematical requirements of space filling". The present theory has been applied also to numerical simulations of curvature driven GG. The introduction of the effect of textures, the effect of particles, and the effect of TP mobilities on GG into this theory will be given in [5, 3, 9, 10]. In contrast to other statistical models on GG, here no heuristic assumptions nor assumptions on the shape of grains vave been made. This theory has been derived by means of first principles including all geometrical and topological requirements. It will be of great advantage to apply such a theory to the extensive and precise data which are delivered by modern, automatically working experimental methods, like OM (Orientation Microscopy [11]). This gives a rather realistic description of microstructures by the topological functions k_{ij}, k^*_{ij}, p_{ij} and q_{ij}

which often can be well approximated by use of only few parameters for its fit (Fig. 6). This helps to understand the mechanisms of GG by analytical means.

Acknowledgments

The authors gratefully acknowledge the financial support from the MaTech program of the "Bundesministerium für Bildung, Wissenschaft, Forschung und Technologie" (BMBF) under number 03K8008 2.

References

[1] W.C. Graustein. *Annals of Math.*, 32, 1931, pp. 149-153.
[2] I. Heckelmann, PhD thesis, RWTH Aachen, 1992.
[3] K. Lücke, R. Brandt, G. Abbruzzese. to be published, 1999.
[4] D.A. Aboav, *Metallography*, 3, 1970, pp. 383-390.
[5] K. Lücke, R. Brandt, G. Abbruzzese, to be published, 1999.
[6] M. Hillert, *Acta metall.*, 13,:1965, pp 227 - 238.
[7] M. Blanc, A. Mocellin. *Acta metall.*, 27, 1979, pp. 1231-1237.
[8] C.S. Smith, "Metal Interfaces". ASM, Cleveland 1952, p. 65.
[9] K. Lücke, R. Brandt, G. Abbruzzese, J. Svoboda. to be published, 1999.
[10] K. Lücke, R. Brandt. to be published, 2000.
[11] B.L. Adams, S.I. Wright, K. Kunze. *Met. Trans.A*, 24A, 1993, pp. 819 - 831.

Recrystallization and Grain Growth
Proceedings of the First Joint International Conference
Eds. G. Gottstein and D.A. Molodov, Springer-Verlag, 2001

Abnormal Grain Growth – Some Theoretical and Practical Considerations

P.R. Rios

Universidade Federal Fluminense, Escola de Engenharia Industrial Metalúrgica de Volta Redonda, Av. dos Trabalhadores, 420, Volta Redonda, RJ, 27255-125, Brasil

Keywords : grain growth, abnormal grain growth.

Abstract. Abnormal grain growth development is examined. Several different circumstances in which abnormal grain growth might happen are considered. Effects of pinning force, mobility advantage and preexistence or not of large grains are discussed. In particular, situations in which abnormal grain growth takes place after a precursor normal grain growth are highlighted.

1 Introduction

Abnormal grain growth is often distinguished from normal grain growth. In the latter one can say that all grain boundaries can move freely so that as grain growth takes place the grain size distribution remains uniform. In contrast during abnormal grain growth a few large grains, abnormal grains, dominate the grain growth process. The boundaries of the abnormal grains can migrate much faster than those of the matrix grains. In this case the grain growth proceeds heterogeneously. The grain distribution does not remain uniform and a bimodal distribution can develop until the abnormal grains consume all the matrix grains. Recently, abnormal grain growth has received significant theoretical attention [1-20]. Based on previous work this paper examines the development of abnormal grain growth in a variety of situations.

2 Criterion for Abnormal Grain Growth

With regard to the grain growth of an abnormal or of a "candidate" grain, R_A, relative to R_{Cr}, that is the critical grain radius in the sense of Hillert's theory [1], its generally accepted that there are two possibilities:

$$\frac{d\frac{R_A}{R_{Cr}}}{dt} \leq 0 \qquad (1)$$

or

$$\frac{d\dfrac{R_A}{R_{Cr}}}{dt} > 0 \tag{2}$$

In the first case, Eq. 1, the increase in the growth rate of the candidate grain is compensated by the normal grain growth of the matrix. No abnormal grain growth will take place as the candidate grain cannot increase its relative size and the grain size distribution remains uniform. However in the second case, Eq. 2, the candidate grain increases its relative size and moves away from the grain size distribution. This latter condition is considered to correspond to abnormal grain growth. The problem with the above criterion are those situations in which a large grain is present but:

$$\frac{d\dfrac{R_A}{R_{Cr}}}{dt} \approx 0 \tag{3}$$

In this case it does not matter too much if the derivative is slightly greater or slightly smaller than zero. The point is that a given heterogeneity will persist. Neither the large grain will break away from the distribution nor it will join the grain size distribution.

Expanding the derivatives one obtains:

$$\frac{d\dfrac{R_A}{R_{Cr}}}{dt} = \frac{1}{R_{Cr}}\frac{dR_A}{dt} - \frac{R_A}{R_{Cr}^2}\frac{dR_{Cr}}{dt} \tag{4}$$

Since $dR_A/dt>0$ it can be seen from Eq. 4 that generally speaking abnormal grain growth will be favored when dR_{Cr}/dt is small. So for abnormal grain growth to take place it is necessary that the grain growth rate of the matrix is small and this normally happens by some sort of restriction acting on the matrix. This restriction is usually by Zener pinning but orientation pinning is also possible if the matrix texture is very strong.

3 Abnormal Grain Growth Without a Pinning Force

When the matrix grain growth can proceed free from any restraining forces and all grains have the same grain boundary mobility abnormal grain growth does not occur from an initially uniform distribution [8,15].

One then considers the growth of a candidate grain of size R_A possessing a mobility, M_A, greater than the mobility of the matrix, M [1]:

$$\frac{dR_A}{dt} = K_A\left(\frac{1}{R_{Cr}} - \frac{1}{R_A}\right) \tag{5}$$

where $K_A=\alpha M_A\gamma$, γ is the grain boundary energy per unit of area and α is a geometric factor approximately equal to one for three dimensional grains; R_{Cr} is the critical grain radius. It is assumed here that the matrix has Hillert's distribution so that [1]:

$$\frac{dR_{Cr}^2}{dt}=\frac{K}{2} \tag{6}$$

where $K=\alpha M\gamma$. Moreover for a Hillert's distribution the maximum grain size is $R_{MAX}=2R_{Cr}$ so that the maximum relative size is $\rho_{MAX}=R_{MAX}/R_{Cr}=2$.
Abnormal growth is said to occur if the relative size, $\rho_A=R_A/R_{Cr}$, increases or $d(R_A/R_{Cr})/dt>0$ this condition together with Eqs. 5, 6 results in [4,7,12]:

$$\frac{d\rho_A}{d\ln R_{Cr}}=4\kappa\left(1-\frac{1}{\rho_A}\right)-\rho_A>0 \tag{7}$$

where $\kappa=K_A/K$.
For $\kappa=K_A/K>1$ Eq. 7 has two real roots: $\rho_{AMIN}<2$ and $\rho_{AMAX}>2$. It can be easily shown that a grain with relative size between ρ_{AMIN} and ρ_{AMAX} will grow abnormally. However, this growth is limited since when its relative size approaches ρ_{AMAX} its relative growth rate tends to zero [4,7,12]. One important point to notice here is that since $\rho_{AMIN}<2$ grains with relative size smaller than the maximum size can grow "abnormally". These grains will outgrow other grains in the same size class and eventually grow beyond 2 towards the upper limit ρ_{AMAX}. This possibility is a direct consequence of the mobility advantage.
Another point is that if the local environment changes and K_A decreases becoming equal to K then the abnormal growth cannot continue and the abnormal grain will start to decrease its relative size [8,15]. It is important to point out that grains that acquired a size advantage due to this mechanism will remain the largest grains even though their relative size might decrease. So if on further processing grain growth either normal or abnormal can happen these grains are going to dominate the final microstructure.

4 Abnormal Grain Growth Without Normal Grain Growth

One considers in the first place the more simple case of a matrix which is stagnant. No normal grain growth can occur because of a pinning force exerted by stable particles. One can write for the grain of maximum radius, R_{MAX}:

$$\frac{dR_{MAX}}{dt}=K\left(\frac{1}{R_{Cr}}-\frac{1}{R_{MAX}}-\frac{Z}{\alpha}\right)\leq 0 \tag{8}$$

where Z is a pinning force or an opposing pressure related to the size and volume fraction of the particles. In this case it is common to distinguish two levels of pinning:

a) ormal pinning: $Z/\alpha=1/R_{MAX}$; all grains $R<R_{MAX}$ will be pinned so no normal grain growth can occur but abnormal grain growth can for grains $R_A>R_{MAX}$.
b) bnormal pinning: $Z/\alpha=1/R_{Cr}$; no grain growth either normal or abnormal can occur.

In this case, abnormal grain growth depends on the preexistence of favorable conditions such as large preexistent grains of size $R_A>R_{MAX}$. If Eq. 7 is always obeyed abnormal grain growth does not take place at all.

If possible abnormal grain growth would take place immediately after the specimen was heated to a temperature high enough to make grain boundaries sufficiently mobile. If no abnormal grain growth occurs immediately then no grain growth, normal or abnormal, is likely to occur up to the melting point. It is worthy of note that if abnormal grain growth does occur it is going to be extremely severe since the matrix is pinned by stable particles and no normal grain growth takes place [17]. This suggests that the overall kinetics should be "site-saturated" in this case. Evidence for this was found in a study of the overall kinetics of abnormal grain growth in $BaTiO_3$ [21,22].

5 Abnormal Grain Growth With Precursor Normal Grain Growth - Stable Particles

Abnormal grain growth development might also take place from a matrix partially pinned by stable particles. This situation can arise in practice by different routes. One possibility is illustrated by the work of Tweed et al. [23] in which they had an aluminum matrix pinned by stable alumina particles. Their initial grain size was below the limiting grain radius so that normal grain growth would initially take place. In some samples in the later stages of grain growth they could observe abnormal grain growth. Another interesting possibility is illustrated by the work of Cabrera et al. [24] who studied the grain growth of austenite in the presence of two particle dispersions: aluminum nitride and vanadium/titanium carbonitrides. Cabrera et al. found that aluminum nitride would dissolve at a lower temperature than the more stable vanadium/titanium carbonitrides. As a consequence, at the dissolution temperature of aluminum nitride the pinning force was suddenly lowered leaving only the more stable vanadium/titanium carbonitrides to pin the matrix and abnormal grain growth took place. This sudden decrease in the pinning force due to the dissolution of one particle dispersion might have lowered the pinning force to the extent of allowing some normal grain growth to take place, leading to a situation essentially similar to that of Tweed et al.. A similar situation could of course also occur in a material with a single particle dispersion heated at a temperature is high enough to dissolve part of the precipitates.

The first possibility is that the grain boundaries are uniform and there is no difference in grain boundary mobility among them. This case has been studied in detail. Both the situation in which large grains were initially present and the situation in which the abnormal grain had to develop from an initially uniform distribution were considered. The main result was that abnormal grain growth can take place

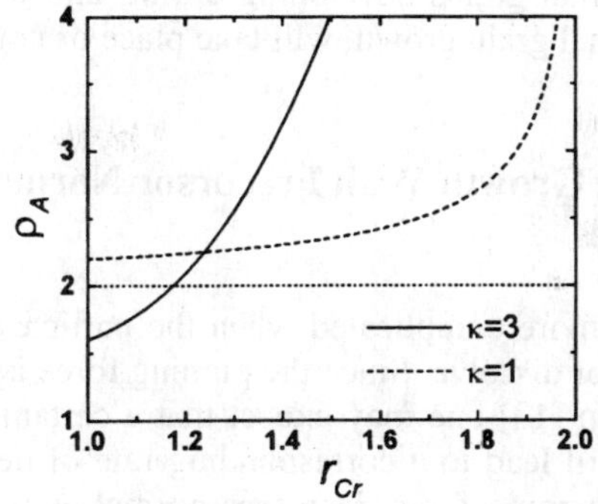

Fig. 1. Relative size of abnormal grain, ρ_A, plotted against normalized critical radius, r_{Cr}, for different mobility advantages, κ, for z=0.25.

when matrix grain growth slows down significantly as it approaches the fully pinned state [16,21]. A more interesting situation is that in which a mobility advantage is present [19]. Both cases are illustrated in Fig. 1.

Fig. 1 exhibits a plot of $\rho_A=R_A/R_{Cr}$ as a function of $r_{Cr}= R_C/R_{Cr0}$ in a matrix initially possessing Hillert's distribution so that $\rho_{MAX}=2$. The solid line corresponds to the abnormal grain growth of a grain possessing a mobility advantage, $\kappa=M_A/M=3$ with an initial size $\rho_{A0}=1.6<2$ in a matrix with a pinning force given by $z=ZR_{Cr0}=0.25$. One can see that abnormal grain growth would continue unbounded if κ remained unchanged. The dashed line corresponds to the growth of a grain not possessing a mobility advantage, κ=1 with an initial size $\rho_{A0}=2.2$. The growth rate is much lower than that of the grain with κ=3 but still abnormal grain growth takes place albeit slowly and its growth rate increases rapidly only as $r_{Cr}\rightarrow 2$. Suppose that at the point in which the curves cross the abnormal grain with κ=3 finds a local environment which no longer favors a mobility advantage so that κ decreases from 3 to 1. In this case the abnormal grain would decrease its growth rate but would still continue to grow abnormally following the dashed line.

It is apparent in Fig. 1 a fundamental difference between abnormal grain growth development in a matrix with uniform boundaries and one in which some grain may have a mobility advantage. In the former abnormal grain growth can only happen if either there is a preexistent large grain [ρ_A >2 in Fig. 1) or if such a grain develops from the largest grains from the distribution [(ρ_A=2 in Fig. 1) as consequence of a favorable local environment. On the other hand in a matrix in which certain grain have a mobility advantage the abnormal grain can develop from a grain with an initial size below the maximum grain size of the distribution [ρ_A=1.6<2 in Fig. 1). Such a grain due to its mobility advantage can outgrow grains of the same or even larger size class (crossing line ρ_A=2 in Fig. 1) and develop into an abnormal grain. It is interesting to point out that for the development of abnormal grains in this way the precursor stage in which normal grain growth takes place [from r_{Cr}=1 up to r_{Cr}=1.2 in Fig. 1) is essential. In this particular case the possibility of abnormal grains developing during this stage will probably determine whether abnormal grain growth will take place or not.

6 Abnormal Grain Growth With Precursor Normal Grain Growth - Unstable Particles

The situation becomes more complicated when the particles are unstable, that is, when they can coarsen or dissolve. Since the pinning force is a function of particle size and volume fraction [13] one may expect that a certain rate of particle coarsening or dissolution will lead to a corresponding rate of decrease in the pinning force dZ/dt<0. As the pinning force decreases normal as well as abnormal grain growth may become possible [14]. One possibility is that normal grain growth follows particle coarsening. Both the situation in which large grains were initially present and the situation in which the abnormal grain had to develop from an initially uniform distribution were considered in detail in previous works [14,18]. When there is a large preexistent grain $R_A>R_{MAX}$ the magnitude of the parameter η was found to be critical:

$$\eta = \frac{3Kr_0 f^{2m}}{kb^2} = \frac{3M\alpha\gamma_g r_0 f^{2m}}{\left(\frac{8D\gamma_p\Omega^2 C_\infty}{9kT}\right) b^2} \tag{9}$$

The parameter η is very complex and depends on several factors such as: grain boundary mobility [M), diffusion coefficient [D) and solubility [C_∞) of solute in the matrix, the size, r_0, and volume fraction of the particles, f, grain boundary and particle/matrix interfacial energy per unit of area, γ and γ_g, among others. The parameter η is essentially a measure of the ability of the matrix to "resist" abnormal growth. One may interpret η as the ratio of the rate of grain boundary movement to the coarsening rate, if grain boundary can move fast relative to particle coarsening η is high and normal grain growth of the matrix will be controlled by coarsening. Whereas if coarsening is fast relative to grain boundary movement η will be low and normal grain growth of the matrix will be controlled by grain boundary movement. In terms of normal/abnormal growth, a low value of η indicates that normal grain growth of the matrix is fast relative to abnormal grain growth which means that the matrix behaves as if it were free of particles. on the other hand high values of η indicate that normal grain growth of the matrix takes place slowly so that matrix behavior approaches that of matrix containing stable particles.

In summary a matrix containing coarsening particles in which the critical grain radius increases together with the limiting grain radius can have a behavior ranging from a particle free matrix to that of a matrix containing stable particles. The abnormal grain growth depends on this and one would expect high values of η to favor abnormal grain growth of a large preexistent grain [14]. Such a grain would not grow abnormally for low values of η [14].

In the above it is considered that a large grain $R_A > R_{MAX}$ is already present. However one can also have the situation in which large preexistent grains are not present [18]. In these circumstances they must develop from an initially uniform matrix. It is possible to show that the development of such a grain requires that a large grain is surrounded by a locally lower pinning force or by a locally higher driving force. The final expression is:

$$\frac{\Delta Z}{Z} > R_{Cr}^3 \frac{n^3}{(n-1)^2}\left(-\frac{d(Z/\alpha)}{dt}\right)\frac{1}{M\gamma\alpha} \tag{10}$$

If a certain "candidate grain", $R_A = R_{MAX}$, grows into a region in which the pinning force is lower, Z-ΔZ, than the average pinning force, Z, then this grain will become unstable and grow abnormally if Eq. 10 is satisfied. One can modify the above equation to account for the effect of a mobility advantage:

$$\frac{\Delta Z}{Z} > \left(\frac{1}{\kappa}\right) R_{Cr}^3 \frac{n^3}{(n-1)^2}\left(-\frac{d(Z/\alpha)}{dt}\right)\frac{1}{M\gamma\alpha} \tag{11}$$

where n=2 for a Hillert distribution.

The difference between Eq. 10 and Eq. 11 is the factor 1/κ. So if a candidate grain has a mobility advantage, κ>1, over other candidate grains then a locally lower fluctuation in the pinning force ΔZ is required. As a consequence it will become much more probable that the higher mobility grains will grow abnormally in preference to other candidate grains without a mobility advantage. As above, once the abnormal grain develops it is not necessary that it retains a mobility advantage since abnormal growth can be maintained without it in the first place. So one reaches the same conclusion as above: a mobility advantage can favor abnormal grain growth to develop but it is not necessary that the mobility remains high for abnormal grain growth to continue. It can also be seen in this case, similarly to the situation describe in the section 5, that the precursor stage of slow normal grain growth can be important and perhaps essential for the development of abnormal grain growth.

6 Summary and Conclusions

Abnormal grain growth has been shown to develop either from an uniform matrix or when there are already large preexistent grains in a variety of circumstances. The ideas and the theoretical modeling can be taken as a basis for the qualitative understanding of abnormal grain growth in practical cases. However quantitative modeling and prediction of abnormal grain growth in practice still seems to be a considerable problem. One factor is probably the difficulty to obtain data in certain model variables, for example, the grain boundary mobility. In any case future works will hopefully attempt to fill the gap of significant theoretical advancement and scarce practical quantitative verification.

Acknowledgement

The author is grateful to Conselho Nacional de Desenvolvimento Científico e Tecnológico [CNPq) of the Brazilian Government for the financial support.

References

[1] M. Hillert, *Acta Metall.*, 13, 1965, pp. 227-238.
[2] T. Gladman, *Proc. R. Soc.*, 294, 1966, pp. 298-309.
[3] D.J. Srolovitz, G.S. Grest and M.P. Anderson, *Acta Metall.*, 33, 1985, pp. 2233-2246.
[4] D. Rollet, D.J. Srolovitz and M.P. Anderson, *Acta Metall.*, 37, 1989, pp. 1227-1240.
[5] G.S. Grest, M. P. Anderson, D.J. Srolovitz and A.D. Rollet, *Scripta Metall. Mater.*, 24, 1990, pp. 661-665.
[6] S. Ling and M.P. Anderson, *JOM*, 44, 1992, pp. 30-36.
[7] A.D. Rollet and W.W. Mullins, *Scripta Mater.*, 36, 1997, pp. 975-980.
[8] C.V. Thompson, H.J. Frost and F. Spaepen, *Acta Metall.*, 35, 1987, pp. 887-890.
[9] H.J. Frost, C.V. Thompson, and D.T. Walton, *Acta Metall. Mater.*, 40, 1992, pp. 779-793.
[10] I. Andersen, Ø. Grong and N. Ryum, *Acta Metall. Mater.*, 43, 1995, pp. 2689-2700.
[11] T. Nakayama and Y.Ushigami, *Nippon Steel Technical Report*, 47, 1990, pp. 8-17.
[12] C.H. Worner, S. Romero, and P. M.J. Hazzledine, *Mater. Research*, 6, 1991, pp. 1773-1777.
[13] P.R. Rios, *Acta Metall.*, 35, 1987, pp. 2805-2814.
[14] P.R. Rios, *Acta Metall. Mater.*, 40, 1992, pp. 649-651.
[15] P.R. Rios, *Acta Metall. Mater.*, 40, 1992, pp. 2765-2768.
[16] P.R. Rios, *Acta Metall. Mater.*, 42 , 1994, pp. 839-843.
[17] P.R. Rios, *Mater. Sci. Forum*, 204-206, 1996, pp. 247-256.
[18] P.R. Rios, *Acta Mater.*, 45 ,1997, pp. 1785-1789.
[19] P.R. Rios, *Scripta Mater.*, 38, 1998, pp. 1359-1364.
[20] P.R. Rios, *Scripta Mater.*, 39, 1998, pp. 1725-1730.
[21] P.R. Rios, T. Yamamoto, T. Kondo, T. Sakuma, *Acta Mater.*, 46, 1998, pp. 1617-1623.
[22] T. Yamamoto, T. Sakuma and P.R. Rios, *Scripta Mater.,* 39, 1998, pp. 1713-1717.
[23] C.J. Tweed, N. Hansen and B. Ralph, *Metall. Trans.* 14A, 1983, pp. 2235-45.
[24] J.M. Cabrera, A. Al Omar and J. M. Prado, *Z. Metallkd.*, 89, 1998, pp. 47-55.

Numerical Modelling of Grain Growth : Current Status

Claire Maurice

Ecole Nationale Supérieure des Mines de Saint-Etienne - Centre SMS -
CNRS URA 1884 - 158 cours Fauriel - 42023 Saint-Etienne Cedex 2 - France

Keywords : computer simulation, normal grain growth, topological models.

Abstract. Eight network and Monte Carlo numerical simulation techniques of 2D ideal normal grain growth are compared in terms of kinetics, size and side number distributions. This initial study demonstrates a surprisingly similar behaviour for the distributions but major differences in kinetics.

1 Introduction

Normal grain growth is a microstructure evolution process driven by the reduction of interfacial grain boundary energy leading to an increase of grain size. When all boundaries are considered to have the same properties (i.e. energy and mobility), this process is referred to as *ideal* normal grain growth. In this case, all analytical theories (e.g. [1]) predict parabolic kinetic laws for the mean grain radius as given by equation (1)

$$\langle R(t) \rangle^2 - \langle R(0) \rangle^2 = k\,t \tag{1}$$

It is also accepted that the structure reaches a "scaling regime", characterised by the time invariance of the normalised grain size distribution. Real life, however, seriously departs from this highly idealised situation : complications such as solute drag, second phase particle pinning or anisotropic boundary properties are thought to be responsible for the observed experimental behaviour. Such complications are not easily amenable to analytic models, and numerical simulations are necessary to investigate complex situations. A wide variety of computer simulations of grain growth have therefore been proposed over the years (see [1, 2 and 3] for exhaustive reviews).

The purpose of this paper is to give a preliminary account of a comparative study of different computer simulation techniques of grain growth. We are here primarily concerned with those that explicitly take full account of the topological arrangement of grains (hereafter designated as topological models), since as recognised by C. S. Smith [4] "Normal grain growth results from the interaction between the topological requirements of space-filling and the geometrical needs of surface tension equilibrium". Moreover, this type of model is more suitable to deal with "real world complications" than those based only on distribution functions. However, the extension of numerical simulations to more complex situations requires sound physical bases, which can be assessed via the results obtained for the simple case of ideal normal grain growth.

2 Topological models - A Brief Summary

Among the topological models, one can distinguish two distinct classes : the discrete models and the continuous models. This distinction refers not only to the nature of the space used to describe the structure, but also to the time scale. It is not the purpose of this paper to describe the details of these numerical simulations, and the reader is referred to the original papers for technical descriptions.

2.1 Discrete models

Discrete models are characterised by mapping the grain structure onto a regular array of cells, each cell being assigned some state variables. The structural evolution is obtained by modifying these state variables according to specific rules, designed so as to mimic grain boundary displacement.
The most frequently used discrete model of grain growth is the Monte-Carlo model (see for example [5, 6]). Cellular automata [7] and phase field models [8] have also been applied to the simulation of grain growth. Recent developments of discrete models are described in [9]. The major advantage of discrete models is the relatively easy computational effort required, even in three dimensions, but the questions of space and time scales are still open.

2.2 Continuous models

On the other hand, continuous models describe the granular structure as interconnected grain boundaries, whose spatial position can be specified by points lying on the boundary, together with triple junctions. They will be termed "network" models in the following. Starting with Fullman [10] in 1952, a relatively large number of "network" models have been proposed. They principally differ by the way the boundary dynamics are taken into account. A common feature of all these models is the need to explicitly provide a solution to the topological transformations. The computer implementation is therefore significantly more challenging than for the discrete models, where such topological transformations are implicitly taken care of. The major part of this paper being devoted to the comparison of a number of "network" models, it is worth spending some time describing the different techniques.

3 Various "Network" Models

It is easy to classify the "network" models according to the type of points (or vertices) used to describe the structure. Two broad categories then emerge : the so-called "vertex" models where only triple junctions are used and the "front tracking" techniques where points lying on the boundaries are added.

3.1 Vertex models

In this type of model, the boundary migration is concentrated in the vertex motion equations. Their reliability is therefore very much dependent on the way vertex motion equations are derived from the migration of the adjacent interfaces. Several different solutions have been proposed over the years, starting with the graphical resolution of bubble soap froth evolution due to Fullman [10]. He considered the interfaces to be flat and proposed that the direction of vertex migration was the resultant of the surface tensions, the rate of migration being proportional to this driving force divided by a "drag coefficient" calculated as the sum of the projected boundary lengths on the normal to the direction of motion. Soares et al. [11] claimed this drag coefficient to have no thermodynamical basis and simply assumed the vertex velocity to be proportional to the sum of the line tensions at this vertex. More sophisticated formulations were proposed by Kawasaki and co-workers [e.g. 12] : the vertex motion equations are derived from a variational approach that involves the total surface energy as a potential energy term, and a dissipative term due to the viscous drag opposing the boundary motion. It should be noted that all the aforementioned solutions neglect the boundary curvature by considering straight lines, thus violating the mechanical equilibrium at the triple junctions. Explicit curvature-driven vertex models have also been developed [13-16]. In these curvature-driven vertex models, local equilibrium is assumed at the triple junction ; the mean boundary curvature is estimated from the tangent direction and two end-points, and is used to calculate the normal velocity of the boundary in the vicinity of the triple point.

3.2 "Front tracking" techniques

In this type of model, boundaries are described as sets of points and intersect at triple junctions. The points lying on boundaries are moved in the normal direction according to the boundary local curvature, and then the position of the triple junctions is separately adjusted to satisfy mechanical equilibrium requirements. The major contribution to the development of front tracking techniques is due to Frost and co-workers [17] ; the local boundary curvature is obtained by local fitting of circular or parabolic arcs. Another contribution is due to Saetre and Ryum [18], who approximate the grain boundaries by cubic spline functions fitted to a series of discrete points along the boundary. Local curvature is readily obtained by standard differential geometry applied to the spline approximation.

The accuracy of the "front tracking" algorithms depends on the number of points used to discretise the boundary, and therefore seem more flexible than the "vertex" models which totally ignore the boundary shape between two triple junctions. However, "front tracking" techniques suffer from the more or less arbitrary adjustment of triple junctions at their equilibrium position. Recently, two distinct techniques have been proposed to overcome this difficulty. The first one, due to Weygand et al. [19] uses Kawasaki's formulation applied to the triple junctions *plus* a number of so-called "virtual" vertices located along the boundaries. The two types of vertices are therefore treated in the same way. The second technique proposed by Brandt et al. [20], follows [13, 15] for the motion of triple points but also uses discrete points along the boundaries. The motion of these points is governed by the local curvature estimated by fitting circular arcs between three successive points.

4 Comparing "Network" Models - Results

With such a plethora of "network" models that share the same framework and differ only by the way vertices, real or virtual, are moved, it appears that a comparative study would be useful to assess their respective qualities. We have therefore undertaken such a study. At the time being ***seven*** models have been tested on the same initial structure, using throughout the same methodology to perform the topological transformations and to control the time integration scheme. Results are therefore directly comparable. Table 1 summarises the models that have been tested up to now.

Table 1. Summary of "network" models used for the comparative study

Key	GB description	Motion	Reference
A	Triple junctions	Curvature	[15]
B	Triple junctions	Curvature	[14]
C	Triple junctions + 7 virtual vertices/boundary	Curvature	[20]
D	Triple junctions	Dissipation	[12]
E	Triple junctions + 7 virtual vertices/boundary	Dissipation	[19]
F	Triple junctions	Line tension	[10]
G	Triple junctions	Line tension	[11]

In order to get good statistics the initial structure consisted of 25000 grains, and was obtained by grain growing a 50000 Voronoi tessellation using model A. Periodic boundary conditions were used. The initial structure should therefore be representative of the grain growth scaling regime. Grain growth was then simulated until 1000 grains were left in the structure. All simulations were run on a standard desktop personal computer (Pentium® III 600 MHz - 256 Mo RAM).

4.1 Kinetics

Figure 1 shows the evolution of the normalised mean grain area as a function of time for all seven models. It is seen that all models yield the expected linear relationship (with the possible exception of model G) :

$$\frac{\langle A(t)\rangle}{\langle A(0)\rangle} = \alpha \frac{M\gamma}{\langle A(0)\rangle} t \tag{2}$$

where M is the boundary mobility, γ is the interfacial energy.

However, the slope α varies significantly from one model to the other. In fact, four groups of kinetics can be distinguished. Models A, B and C yield almost exactly the same kinetics, with a slope of ≈ 2.0. Models D and E have kinetics which are twice as slow (slope ≈ 1.0 for D and 1.08 for E). Note that the slope obtained for model E is very close to the one reported in [19] (1.12 ± 0.04). It is to be noticed that both model E and D are here resolved in their complete form, without using the simplifications introduced by their respective authors (see [12, 19] for details). It can be shown that omitting the coupling term in the Lagrange equation of models E and D yields kinetics with a slope of about 2.0, similar to the kinetics found for models A, B and C where coupling is not introduced. The fact that models A, B and C have the same kinetics, despite the addition of virtual points for model C, means that the kinetics are entirely governed by the triple junction motion. Two extreme behaviours are obtained with the line-tension models : model F has the slowest kinetics of all with a slope of ≈0.5, while model G has an extremely fast kinetics, not quite linear, with a slope of ≈20.

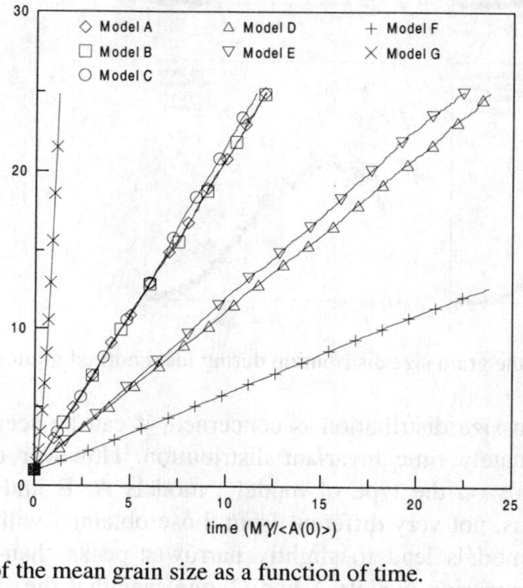

Fig. 1. Evolution of the mean grain size as a function of time.

4.2 Distributions

If a scaling regime does exist for ideal normal grain growth, it can be revealed by the time invariance of distribution functions such as the grain size distribution and the side number distribution. These distributions are shown in Figs. 2 and 3 respectively. Each one of the figures represents the distributions obtained at four successive steps of the grain growth simulation, i.e. when 20000 (respectively 15000, 10000 and 5000) grains were left in the structure.

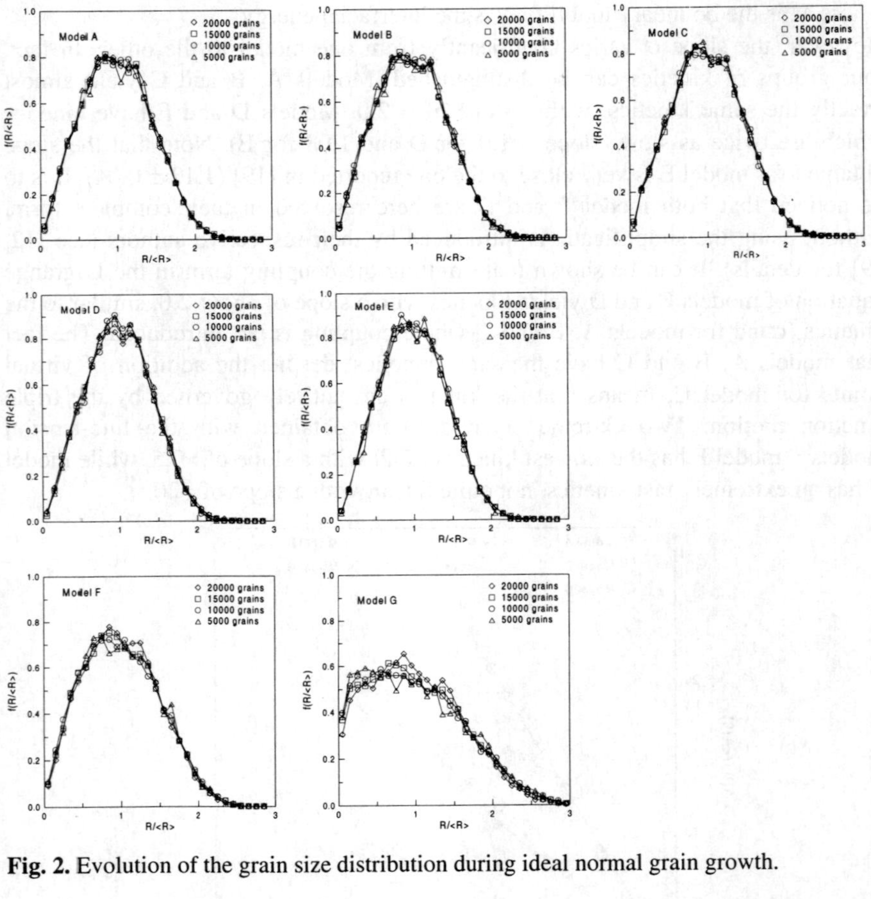

Fig. 2. Evolution of the grain size distribution during ideal normal grain growth.

As far as the grain size distribution is concerned, it can be seen that all models yield an approximately time invariant distribution. However the shape of the distribution depends on the type of model : models A, B and C produce very similar distributions, not very different from those obtained with models D or E. Still, dissipation models lead to slightly narrower peaks than curvature based models. Again, comparing models A and C, one sees that the addition of virtual vertices along the boundaries does not modify the results in a significant way (this conclusion can also be drawn from a comparison of models D and E). On the other hand, line-tension models tend to produce somewhat broader distributions ; this is particularly clear for model G, where one sees a much higher proportion of either very small or very big grains.

Similar conclusions are drawn from a close examination of side number distributions (Fig. 3). Time invariance (i.e. scaling regime) is obtained for all models, and all models, except line-tension models, produce approximately the same side number distribution. Between the two line-tension models studied so far, the original one proposed by Fullman seems to give results closer to those obtained by more sophisticated ones compared with the results of the Soares model.

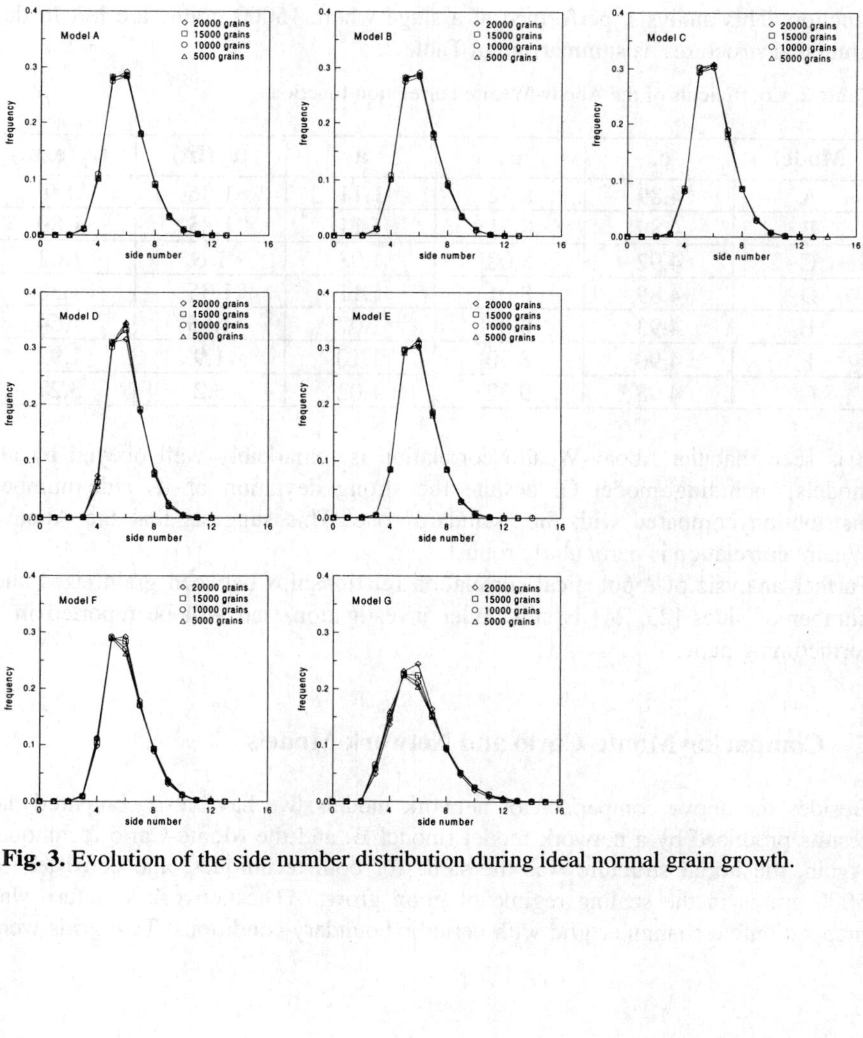

Fig. 3. Evolution of the side number distribution during ideal normal grain growth.

4.3 Topology

A specific advantage of topological models over analytical ones is the possibility to investigate topological correlations, such as the so-called Aboav-Weaire law [21, 22]. This law establishes a correlation between the number of vertices n of a grain and the mean number of vertices of the neighbouring grains, according to equation (3)

$$m(n) = 6 - a + \frac{6a + \mu_2}{n} = c_1 + \frac{c_2}{n} \tag{3}$$

where a is close to 1 and μ_2 is the second moment of the side distribution function. This analysis, performed at a stage where 15000 grains are left in the simulated structures, is summarised in Table 2.

Table 2. Coefficients of the Aboav-Weaire correlation function.

Model	c_1	c_2	a	μ_2 (fit)	μ_2 (exp.)
A	4.89	8.52	1.11	1.86	1.9
B	4.89	8.51	1.11	1.85	1.89
C	4.92	8.08	1.08	1.6	1.62
D	4.89	8.01	1.11	1.35	1.38
E	4.93	8.07	1.07	1.65	1.64
F	4.90	8.50	1.10	1.9	1.92
G	4.98	9.32	1.02	3.2	3.27

It is seen that the Aboav-Weaire correlation is remarkably well obeyed by all models, including model G, despite the strong deviation of its side number distribution compared with the "standard" one. This suggests that the Aboav-Weaire correlation is particularly robust.

Further analysis of topological correlation relationships between grain sizes and number of sides [23, 24] is still under investigation, and will be reported in a forthcoming paper.

5 Comparing Monte-Carlo and Network Models

Besides the above comparison of network models, we have also compared the results produced by a network model (model E) and the Monte-Carlo technique. Again, the initial structure was the same for both techniques, and consisted of 5000 grains in the scaling regime of grain growth. The network structure was mapped onto a triangular grid with periodic boundary conditions. Two grids were used to discretise the structure : a 1000x1000 site grid (MC1) and a 2000x2000 site grid (MC2). Simulations were run until 500 grains were left in the structure. Figure 4 compares the time evolution of the normalised mean area. It is to be noticed that the definition of time is not the same for the network model and for the Monte-Carlo runs : in the latter, the usual MCS unit (Monte-Carlo step) is normalised by NbSites/NbG(0), where NbSites is the number of sites contained in the grid and NbG(0) is the initial number of grains. Using this normalisation procedure, it is seen that both Monte-Carlo runs produce the same linear kinetics. Moreover, it appears that the Monte-Carlo time unit is twice the network time unit. Further tests are needed to check and interpret this relationship, but it is obvious that such a comparison between the network model, where the space and time scales are explicitly defined, and the Monte-Carlo technique would help to identify the Monte-Carlo space and time scales.

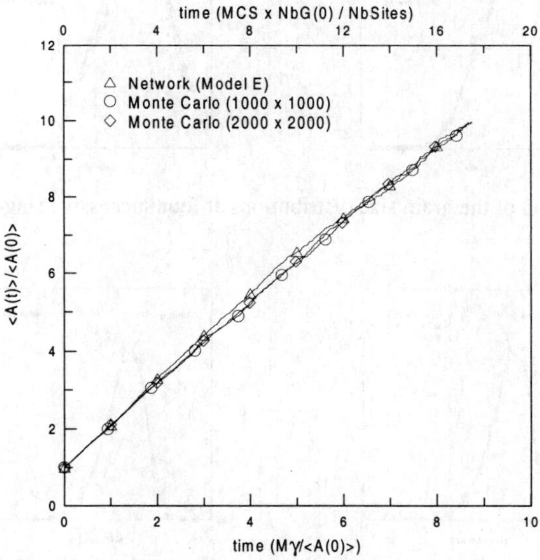

Fig. 4. Evolution of the mean grain size as a function of time. Lower time scale for the Network model and upper time scale for the Monte-Carlo model.

Normalised grain size and number of side distributions are compared in Figs. 5 and 6 respectively, at four successive stages of the coarsening process (i.e. 4000, 3500, 3000 and 2500 grains left). Although the number of grains is now too low to give as reliable results as for Figs. 2 and 3, the similarity of both distributions for the two types of model is remarkable at all four coarsening stages. The time invariance of the distributions (i.e. the scaling regime) is difficult to assert, but this is thought to be due to the relatively low number of grains.

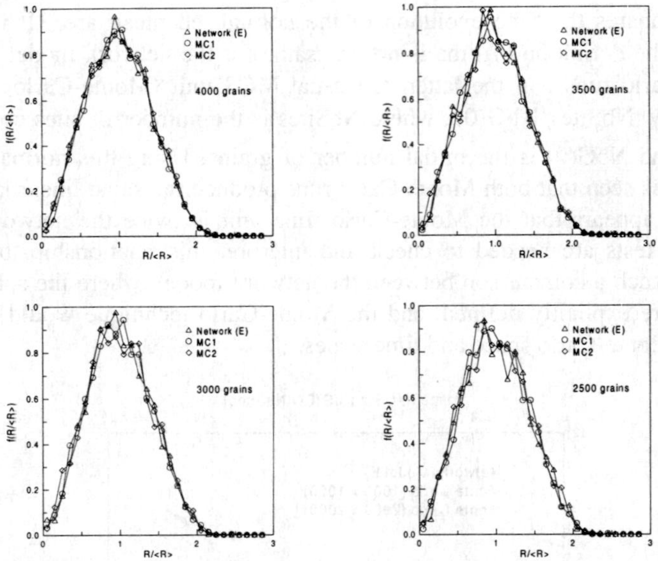

Fig. 5. Comparison of the grain size distributions at four successive stages of coarsening.

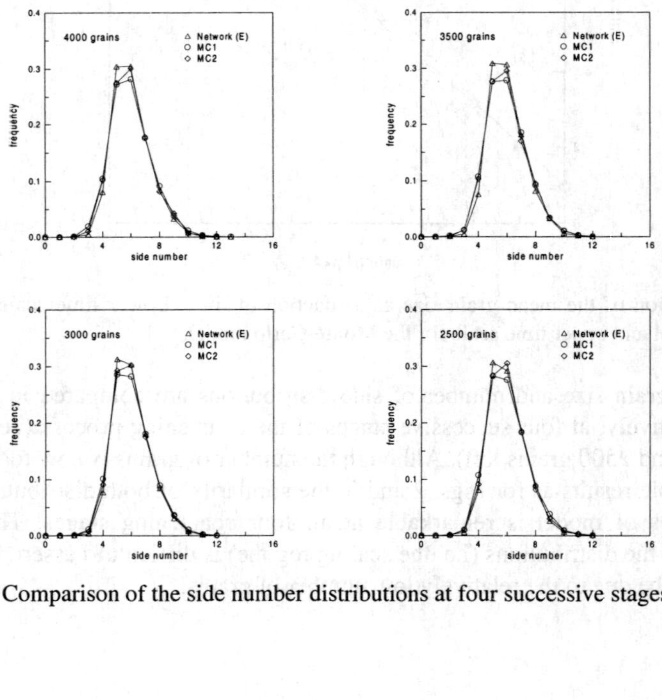

Fig. 6. Comparison of the side number distributions at four successive stages of coarsening.

6 Concluding Remarks

This study aims at comparing the results of published topological models of ideal grain growth. At the time of writing, only two-dimensional models have been tested, and the comparison is not exhaustive since some major contributions have not yet been incorporated yet in the study (e.g. [7, 8, 17, 18]).

Also, the results presented here pertain to the global behaviour of aggregates of grains. The analysis at the individual grain level is still underway and will be reported in due course. However, to the author's knowledge, it is the first time that such a wide-ranging comparison is undertaken.

Preliminary results presented in this paper show in an unambiguous way that all models are very good at producing a scaling regime of grain growth, with parabolic kinetics.

From a statistical point of view, all the network models tested here produce quasi-identical distributions, with the exception of the line-tension network model [11] (it is to be remarked that the oldest network model [10] produces quite reasonable results compared with more sophisticated ones). The direct comparison between one network model [19] and the Monte-Carlo technique [5, 6] also shows a striking similarity of distributions. Moreover, a kind of relationship between the Monte-Carlo time scale and the physical one, implicit in Network models, has been revealed.

From a computational point of view, it has already been said that network models are more challenging to code than discrete models, and this is even worse in 3D. However, for a given amount of RAM, network models can handle much greater numbers of grains and are therefore worth the pain !

References

[1] H.V. Atkinson, *Acta Metall.*, 36, 1988, pp. 469-491.

[2] F.J. Humphreys, *Mat. Sci. Tech.*, 8, 1992, pp. 135-143.

[3] H.J. Frost and C.V. Thomson, *Current Opinion in Solid State & Materials Science*, 1, 1996, pp. 361-368.

[4] C.S. Smith, in "Metal Interfaces", ASM, Cleveland, 1952, pp. 65-108.

[5] M.P. Anderson, D.J. Srolovitz, G.S. Grest, *Acta Metall. Mater.*, 32, 1984, pp. 783-791.

[6] D.J. Srolovitz, M.P. Anderson, P.S. Sahni and G.S. Grest, *Acta Metall. Mater.*, 32, 1984, pp. 793-802.

[7] Y. Liu, T. Baudin and R. Penelle, *Scripta Mater.*, 34, 1996, pp. 1679-1683.

[8] D. Fan and L.-Q. Chen, *Acta Mat.*, 45, 1997, pp. 611-622.

[9] E.A. Holm, M.A. Miodownik and J.W. Cahn, *in these proceedings*.

[10] R.L. Fullman, in "Metal Interfaces", ASM, Cleveland, 1952, pp. 179-207.

[11] A. Soares, A. C. Ferro and M. A. Fortes, *Scripta Metall.*, 19, 1985, pp. 1491-1496.

[12] K. Kawasaki, T. Nagai and K. Nakashima, *Phil. Mag. B*, 60, 1989, pp. 399-421.
[13] J. Svoboda, *Scripta Metall. Mater.*, 28, 1993, pp. 1589-1590
[14] S.P. Marsh, R.A. Masumura and C.S. Pande, in "Modeling of Coarsening and Grain Growth", eds. C. S. Pande and S. P. Marsh, TMS, Chicago, 1993, pp. 339-346.
[15] C. Maurice and F.J. Humphreys, in "Thermomechanical Processing in Theory, Modelling & Practice", eds. B. Hutchinson et al., The Swedish Society for Materials Technology, Stockholm, 1997, pp. 201-209.
[16] C. Maurice and F.J. Humphreys, in "Grain Growth in polycrystalline materials III, eds. H. Weiland et al. TMS, 1998, pp. 81-90.
[17] H.J. Frost, C.V. Thompson, C.L. Howe and J. Whang, *Scripta Metall. Mater.*, 22, 1987, pp. 65-70.
[18] T.O. Saetre and N. Ryum, in "Modeling of Coarsening and Grain Growth", eds. C.S. Pande and S.P. Marsh, TMS, Chicago, 1993, pp. 281-294.
[19] D. Weygand, Y. Bréchet and J. Lépinoux, *Phil. Mag. B*, 78, 1998, pp. 329-352.
[20] R. Brandt, J. Svoboda and K. Lücke, in "Grain Growth in Polycrystalline Materials III, eds. H. Weiland et al. TMS, 1998, pp. 131-136.
[21] D.A. Aboav, *Metallography*, 3, 1970, pp. 383-390.
[22] D. Weaire, *Metallography*, 7, 1974, pp. 157-160.
[23] F.T. Lewis, *Anat. Rec.*, 38, 1928, pp. 341-362.
[24] N. Rivier, *Phil. Mag. B*, 52, 1985, pp. 795-819.

Vacancy Effects in Grain Growth

Y. Estrin

IWW, TU Clausthal, Agricolastr. 6, D-38678 Clausthal-Zellerfeld, Germany
E-mail: juri.estrin@tu-clausthal.de

Keywords: grain growth, vacancies, thin films, ultra fine grained materials.

Abstract. The excess free volume of grain boundaries that is liberated as a result of grain growth is assumed to be accommodated in the form of vacancies. This kinetically necessary vacancy generation is shown to give rise to inhibition of grain growth in fine grained bulk materials and in thin films. Critical conditions for the occurrence of this effect, particularly with respect to the initial grain size and temperature, are obtained. Some predictions of the model, e.g. the occurrence of *linear* grain growth controlled by the coefficient of self-diffusion have been recently confirmed by experiment. Other salient features, such as a pronounced vacancy supersaturation during linear grain growth, await experimental validation.

1 Introduction

In this paper we address the role of vacancies produced as a result of grain growth in the kinetics of the very process in which they are generated. In the approach taken, details of the vacancy generation mechanism are not considered, the focus being on the energetics of the grain growth process. It will be shown that even with such a generalised treatment, important conclusions with regard to the grain growth kinetics can be drawn. The overview of the vacancy effects presented below is a result of recent work [1-4] done in collaboration with G. Gottstein, E. Rabkin and L.S. Shvindlerman.

The philosophy developed in those papers is as follows. During grain growth, the total amount of the grain boundary 'phase' contained in the material decreases, which is accompanied with release of excess free volume. Indeed, the density of matter is lower in the grain boundaries than in the bulk by a couple of percent, cf. e.g. [5,6]. (Some estimates even suggest a two-digit difference in the densities.) In any case it can be assumed that this excess volume that has to be generated as a 'by-product' of grain growth is 'digested' by the bulk of the material in the form of individual vacancies or, perhaps, larger agglomerates of vacancies. The idea that grain boundary motion may result in vacancy production has, in fact, been around for quite some time, cf. [7-10]. Recent computer simulations of grain boundary motion provide substantiation to it [11]. While in light of these considerations vacancy generation is a kinematically necessary process accompanying grain growth, it may not be favourable thermodynamically. It can thus turn out that the kinematic necessity of vacancy generation has an inhibiting effect on grain growth.

2 Vacancy Inhibited Kinetics of Grain Growth – the Case of a Bulk Material

We start by looking at the kinetics of grain growth accompanied by vacancy generation in a bulk material. To look at this effect quantitatively, let us consider the various contributions to the Gibbs free energy of the system. For small deviations of the vacancy concentration c from its thermal equilibrium value, c^{eq}, the vacancy contribution can be written [1,2] as

$$G_{vac} = \frac{1}{2}\frac{NkT}{c^{eq}}\left(c - c^{eq}\right)^2. \tag{1}$$

Here N is the number of atomic sites per unit volume, T the absolute temperature and k the Boltzmann constant. The portion of the Gibbs free energy of the system that is associated with the grain boundary energy is given by

$$G_{GB} = \frac{3}{2}\frac{\gamma}{R} \tag{2}$$

where γ is the grain boundary energy per unit area and R is the average grain radius. It is this part of the free energy that provides the driving force for grain growth in the first place. As suggested in Ref. [3], one can write an equation expressing the balance of energy associated with an increment of grain size, dR within a time increment dt:

$$-\frac{d}{dR}\left(\frac{3\gamma}{2R}\right)dR = \frac{3}{2}\cdot\frac{1}{R}\frac{\left(dR/dt\right)^2}{m}dt + \frac{NkT}{c^{eq}}\left(c - c^{eq}\right)\frac{6\delta\beta}{R^2}dR. \tag{3}$$

The left hand side of this equation represents the energy release associated with the reduction of the total grain boundary area. Part of this energy is dissipated due to the drag forces determining the grain boundary mobility m (the first term on the right hand side of the equation). The rest – represented by the second term – is the energy spent to generate the kinematically necessary vacancy concentration. The form of that second term is suggested by the fact that the vacancy concentration increment dc associated with the grain radius increment dR is given by $6(\delta\beta/R^2)dR$. (Here δ and β are, respectively, the width of the grain boundary and the relative excess volume, or the relative density 'defect' associated with grain boundaries.)

Eq. (3) can be seen as an 'equation of motion' for the average grain radius R. The time evolution of the system would be complete if the variation of the vacancy concentration was known. The following equation can be proposed to describe the variation of the vacancy concentration with time [3]:

$$\dot{c} = \frac{6\delta\beta}{R^2}\cdot\frac{dR}{dt} - \frac{D_v}{d^2}\left(c - c^{eq}\right) \tag{4}$$

Here the average spacing between vacancy sinks, d, and the vacancy diffusivity, D_v, determine the characteristic vacancy annihilation time d^2/D_v. Eqs. (3) and (4) that furnish a full description of the evolution of the average grain radius and the vacancy concentration in the bulk can be rewritten in a non-dimensional form:

$$\xi\frac{d\xi}{dt} + A\Psi C - A = 0 \tag{5}$$

$$\frac{dC}{dt} = \frac{p}{\xi^2}\frac{d\xi}{dt} - C \tag{6}$$

Here the time t is now non-dimensional and is measured in the units of d^2/D_v. Numerical solution of the above set of equations suggests simple analytical expressions that reveal some interesting features of the system under consideration. As seen from the numerical results presented in Fig. 1a, for sufficiently small initial grain size $\xi_o = R_o/d$ the grain growth uninhibited by vacancies is preceded by an incubation time during which the growth rate is substantially reduced, the time dependence of the grain size exhibiting a plateau-like behaviour. For large values of ξ_0 no incubation time is observed. The incubation time can be defined [3] as the time at which the grain growth rate is a maximum. It corresponds to a transition from the plateau-like behaviour to uninhibited, parabolic grain growth, cf. Fig. 1a. Over the incubation time, the vacancy concentration stays at an approximately constant, increased level, cf. Fig. 1b. (Note that in the example shown, the value of c^{eq} was 10^{-8}.)

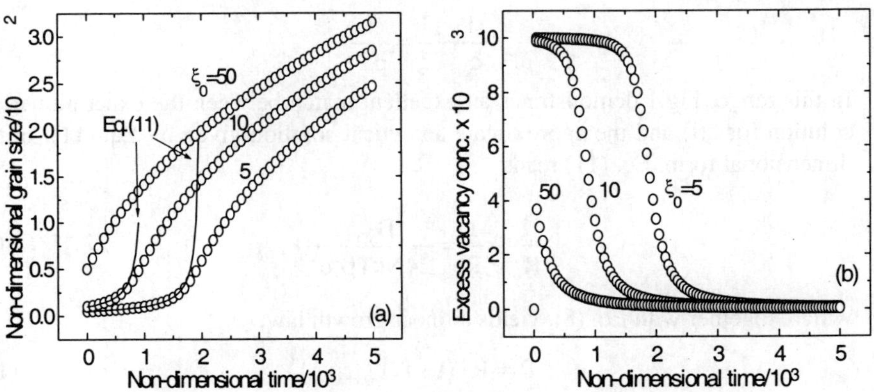

Fig. 1. Dependence of non-dimensional grain size (a) and of excess vacancy concentration (b) on the non-dimensional time calculated from Eqs. (5),(6) for the following values of parameters: A=10, Ψ=100 and p=100.

A fortunate circumstance is that an approximate analytical expression for the incubation time can be found from the numerical calculation results. It can be shown [3] that the non-dimensional incubation time is inversely proportional to ξ_o and can be represented by the following formula:

$$\tau_{incubation} = \frac{p\Psi}{\xi_o} \tag{7}$$

or, in dimensional form:

$$t_{incubation} = 24\frac{NkT(\delta\beta d)^2}{\gamma R_o D_{SD}} \tag{8}$$

where $t_{incubation} = 24\frac{NkT(\delta\beta d)^2}{\gamma R_o D_{SD}}$ is the coefficient of self-diffusion.
For times larger than the incubation time, a common parabolic grain growth law follows:

$$\xi^2 - \xi_o^2 = 2At \tag{9}$$

while within the incubation time, an approximate analytical solution of the set of Eqs. (5) and (6) reads [3]:

$$C = 1/\Psi, \tag{10}$$

$$\frac{1}{\xi_o} - \frac{1}{\xi} = \frac{t}{p\Psi} \tag{11}$$

In this range, Fig.1 demonstrates an excellent match between the exact numerical solution for $\xi(t)$ and the approximate analytical solution given by Eq. (11). In the dimensional form, Eq. (11) reads

$$\frac{1}{R_o} - \frac{1}{R} = \frac{D_{SD}\gamma}{24NkT\beta^2 d^2}t \tag{12}$$

which, together with Eq. (8), yields a linear growth law,

$$R = R_o\left(1 + t/t_{incubation}\right) \tag{13}$$

for times that are small compared to $t_{incubation}$. A striking observation is that in the grain growth inhibition regime considered, the growth rate is *controlled by the coefficient of self-diffusion*, rather than by the grain boundary mobility. It is noted that Eq. (10) suggests constancy of the vacancy concentration during the incubation time, in which the vacancy inhibited linear grain growth prevails.
A comparison between vacancy inhibited grain growth and the conventional parabolic grain growth behaviour was used in Ref. [1] to suggest a criterion for the

efficiency of the vacancy-induced stabilisation against grain growth. Inhibition of grain growth means that its rate described by Eq. (11) is much smaller than that for uninhibited parabolic growth given by Eq. (9). According to Ref. [3], this condition reads

$$R_o << R_c^* = \left[24NkT(\delta\beta)^2 d^2 \frac{m}{D_{SD}} \right]^{1/3} \tag{14}$$

It should be noted that this condition is not very material sensitive, as the ratio of the grain boundary mobility and the coefficient of bulk self-diffusion is commonly not strongly material dependent. A possible material dependence is further weakened due to the smallness of the exponent of 1/3. One source of material dependence may be the sink spacing d, for instance if it is related to the inverse of the square root of the dislocation density or the dislocation cell size.
In the above considerations, it was tacitly assumed that the concentration of vacancies released during grain growth and retained in the material is controlled by sinks in the bulk of the material implying that d is not related to the grain size. A particular case needs to be considered separately, though. It is the case when the initial grain size is so small that no other vacancy sinks but the grain boundaries themselves are available, so that d is to be identified with the grain size R or is at least proportional to it. This may particularly be the case with ultra fine grained (notably nanostructured) materials. A picture behind this assumption is that a vacancy produced at a particular grain boundary does not have a special affinity for it, but would rather have to migrate a typical distance scaling with R until it can be efficiently removed from the grain. Eq. (6) has to be replaced for this case with

$$\frac{dC}{dt} = \frac{1}{\xi^2}\left(6\frac{d\xi}{dt} - C\right) \tag{15}$$

Eq. (5) remains unchanged. However, the meaning of the non-dimensional grain size and the non-dimensional time are different now: $\xi = R/(\delta\beta)$, and time is measured in units of $(\delta\beta)^2/D_v$. A different type of grain size variation is found for this case by numerically solving the set of Eqs. (5) and (15), cf. Fig. 2. After a short initial period of time t_{trans} the rate of vacancy-inhibited growth slows down considerably, cf. Fig. 2a. After that, the grain growth effectively stays inhibited for very long times. The growth rate is thus always lower than the respective rate for parabolic growth (cf. Fig. 2b). An analytical expression by which the numerical solution of the set of Eqs. (5) and (15) can be approximated reads [3]:

$$R - R_o = \frac{1}{24} \cdot \frac{\gamma D_{SD} t}{NkT(\delta\beta)^2} \tag{16}$$

From Fig. 2b it can be seen that Eq. (16) indeed provides a good analytical approximation for sufficiently short times, which nevertheless should be longer than t_{trans}. However, with increasing initial grain size t_{trans} increases, too, and Eq. (16) becomes less accurate.

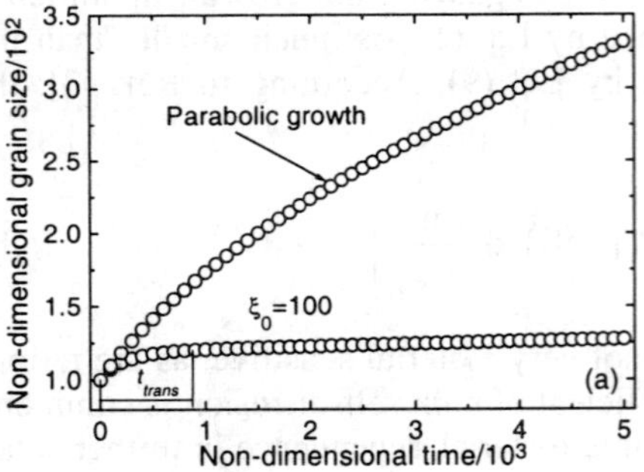

a)

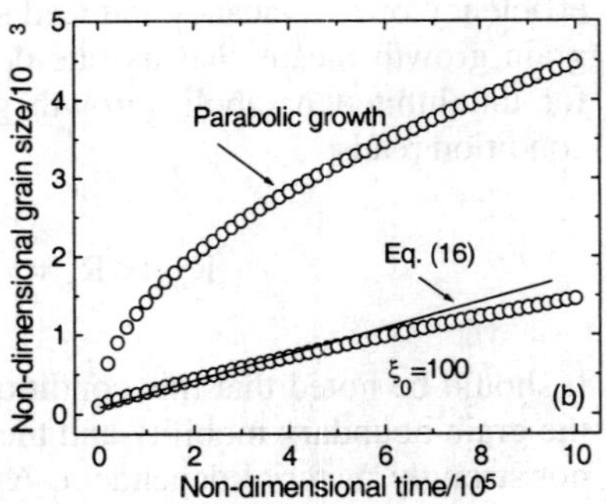

b)

Fig. 2. Time dependence of the grain size for the case of d = R.. The following values of parameters were used in the calculation: A=10, Ψ=100. (Note the different time scales in (a) and (b).)

Stability of a nanocrystalline material against grain growth can be checked by requiring, as above, that the growth rate given by Eq. (16) be much smaller than that corresponding to the parabolic law. The stability condition is then expressed by the inequality

$$R_o << R_c = 24NkTZ(\delta\beta)^2 \frac{m}{D_{SD}} \tag{17}$$

which is different from Eq. (14). These results have an interesting interpretation: a nanocrystalline material cannot be stable against grain growth if condition (14) or (17) (whichever is relevant) is violated. In that sense, the critical radius R_c or R_c^* can be regarded as a limiting stable grain size above which grain growth is no longer inhibited by vacancy generation.

An estimate of the critical radius and an assessment of its temperature dependence is fairly difficult. Indeed, the grain boundary mobility may depend very strongly on the type of the boundary, the texture in the material and its purity [12]. Also, the self-diffusion coefficients entering the critical radius may be different from that for bulk diffusion in a coarse-grained material, as the self-diffusion characteristics of the grain interiors of an ultra fine grained material may be affected by the proximity of grain boundaries. Still, Eqs. (14) and (17) make an estimate possible in principle. For the case of Al - the only material for which m is known with some confidence – this was done for the temperature of 300°C. Taking m=2x10^{-14} $m^4J^{-1}s^{-1}$ and D_{SD}=1.3x10^{-17} m^2/s both R_c and $R_c^* = (R_c d^2)^{1/3}$ were estimated to be of the order of 100 nm [3]. (In the latter case it was assumed that vacancy sinks are provided by dislocations, so that d^2 could be identified with the inverse of the dislocation density ρ taken to be in the range of 10^{13} – 10^{15} m^{-2}.) These estimates explain the relative stability of nanocrystalline materials against grain growth, as they typically fall in the subcritical range of R where vacancy induced grain growth inhibition is prevalent.

The qualitative results of the above analysis can be summarised as follows:

- Within an incubation time, or below a critical grain size, the grain structure is stabilised in that the grain growth rate is lower than what one would expect on the basis of parabolic grain growth models.
- In that range, the grain growth rate is controlled by the self-diffusion coefficient, rather than by the grain boundary mobility. The grain growth rate in this range is constant, i.e. linear grain growth occurs.
- A change of mechanism to that leading to uninhibited parabolic grain growth is expected when the critical grain size is reached. Here the grain boundary mobility becomes the rate controlling quantity.

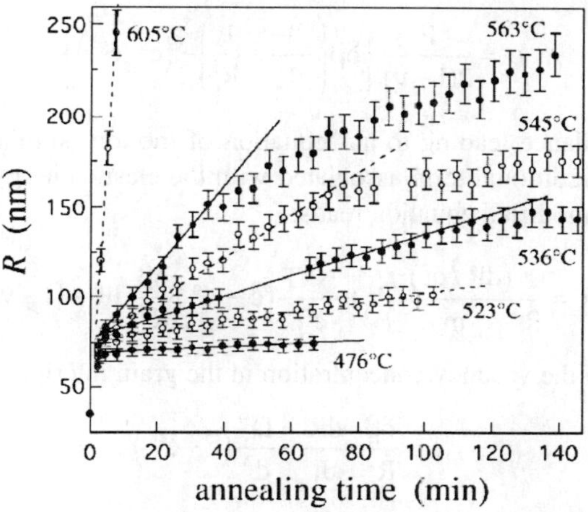

Fig. 3. Time dependence of the average grain size $\bar{R}$ in high purity iron with nanoscale grain structure. The straight lines indicate the slopes of the linear portions of the curves [14]. (Courtesy Carl Krill, Saarbrücken.)

3 Grain Growth in Thin Films

Grain growth in thin films, though similar to that in the bulk in principle, has certain specifics though. They are mainly connected with the stresses developing in thin films as well as with thermal grooving [15,16] whose effect on grain boundary mobility increases with decreasing film thickness. In a recent paper [4], tensile stresses associated with grain growth were included in the analysis of the grain growth kinetics. Here we shall present the results of this analysis disregarding the effect of grooving on grain growth.

Following ref. [4], we consider a polycrystalline film deposited on a substrate. It is assumed that the film was initially in a stress-free state. However, removal of the excess volume associated with the grain boundary 'phase' that occurs during grain growth should give rise to a tensile stress in the film. Indeed, shrinkage of the film due to this loss of excess volume is prevented by its cohesion with the substrate. According to Chaudhari [17], this stress is given by

$$\sigma = \frac{E}{2(1-\nu)} \delta\beta \left(\frac{1}{R_o} - \frac{1}{R} \right) \quad 18)$$

where E is Young's modulus and ν Poisson's ratio of the film material. A similar equation was recently used by Brückner et al. [18]. However, Chaudhari's equation implies that the excess volume release is instantaneous, while we claim that part of it is redistributed and is contained in the vacancies present in the film. Eq. (18), which thus overestimates the tensile stress magnitude, is to be replaced with [4]

$$\sigma = \frac{E}{2(1-\nu)} \left\{ \delta\beta \left(\frac{1}{R_o} - \frac{1}{R} \right) - \left(c - c^{eq} \right) \right\} \quad (19)$$

In the energy balance leading to an 'equation of motion' similar to Eq. (3) there now will be an additional term associated with the elastic energy due to the stress in the film. The modified equation reads:

$$\frac{\gamma}{R^2} dR = \frac{1}{R} \frac{(dR/dt)^2}{m} dt + \frac{NkT}{c^{eq}} \left(c - c^{eq} \right) \frac{\delta\beta}{R^2} dR + (1-\nu) \frac{\sigma}{E} d\sigma \quad (20)$$

The evolution of the vacancy concentration in the grain interior is given by

$$\dot{c} = \frac{\delta\beta}{R^2} \cdot \frac{dR}{dt} - \frac{D_v}{d^2} \left(c - c^{eq} \right) \quad (21)$$

an equation similar to Eq. (4). The differences in the vacancy generation term, also seen in the form of the first term of equation of motion, stem from geometry: in the case consider here columnar cylindrical grains normal to the film surface are considered.

The set of Eqs. (20) and (21), with σ given by Eq. (19), provide a full description of the time evolution of the grain size, vacancy concentration and stress. Solving the equations numerically yields results very similar to those for the bulk case. The same trends as in Fig. 1 were found [4]. As with grain coarsening in a bulk material, for thin films there exists an incubation time $t^*_{incubation}$ during which grain growth is markedly inhibited by vacancies generated in the process. The two incubation times, i.e. $t^*_{incubation}$ and the corresponding quantity for the bulk case, $t^*_{incubation}$ are proportional to each other, but are different in magnitude due to differences in the grain topology:

$$t^*_{incubation} = \frac{1}{24} t_{incubation} \quad (22)$$

Over the incubation time, a significant vacancy supersaturation prevails, while the stress in the film falls well behind the level estimated by Chaudhari [17] and has little effect on the rate of grain growth. It is only after a period of nearly linear increase of σ (close to the incubation time in duration) that the Chaudhari stress level,

$$\sigma = \frac{E}{2(1-\nu)}\beta\frac{\delta}{R_o} \tag{23}$$

is reached. Depending on the initial grain size R_o, this stress can be very appreciable in magnitude and, for sufficiently small R_o, can even lead to rupture of the film.

4 Concluding Remarks

The above consideration shows that redistribution of the excess volume in the form of vacancies generated during grain growth is a process stabilising a grain structure against coarsening. Some salient features of the model considered were validated experimentally, but more experimental work is needed to fully verify its predictions. This particularly refers to the thin film case where a measurable vacancy supersaturation should occur simultaneously with the inhibited grain coarsening during the incubation time. Linear variation with time of the tensile stress in the film that should occur at the same time can also be measured.

The incubation time – a crucial quantity in the grain coarsening kinetics discussed – is inversely proportional to the coefficient of self-diffusion, the grain boundary energy and the initial grain size and thus has a huge range of variation. The excess free volume efficiency of grain boundaries, given by the product $\beta\delta$, that enters quadratically, does not provide for much additional variability. Estimates show that for the grain size below about 100 nm, the incubation time in the temperature range of interest in grain coarsening problems is such that the grain structure is stabilised against grain growth, cf. the case of Fe [14]. Obviously, the smaller the initial average grain size, the more pronounced is the inhibiting effect of vacancies on grain growth. Regions of a fine grained polycrystal (be it a bulk material or a thin film) with the average grain size smaller than the overall average will thus be more stagnant, which may promote selective local coarsening, or perhaps even abnormal grain growth.

The notion of vacancy inhibition of grain growth may appear to hinge on the assumption that the excess free volume release occurs via diffusion of vacancies through the bulk. However, if the obvious alternative channel through which the system can dispose of the excess free volume was considered instead, the situation would change quantitatively, but not qualitatively. Indeed, dynamic equilibrium between grain boundaries and grain interiors will have to be established, and this will require an increase of the vacancy concentration in the grain interiors, thus producing a thermodynamic retarding force acting against grain growth – in much the same way as in the model considered above. A more detailed study of the role

of the alternative channel of excess free volume removal is needed, of course, to assess its effect on the grain growth inhibition effect.
Similar effects of vacancy-inhibited process kinetics can be expected whenever a solid state transformation is accompanied with release of excess free energy.

Acknowledgments

The work reported here was done in collaboration with Profs. G. Gottstein, E. Rabkin and L. S. Shvindlerman. The author is indebted to them for intense interaction and many an illuminating insight into the role of vacancies in grain growth they provided. Generous support of the AvH Foundation through an Alexander von Humboldt Award is gratefully acknowledged.

References

[1] Y. Estrin, G. Gottstein, and L.S. Shvindlerman, *Scripta mater.*, 41, 1999, pp. 385-390
[2] Y. Estrin, G. Gottstein and L.S. Shvindlerman, *Acta mater.*, 47, 1999, pp. 3541-3549.
[3] Y. Estrin, G. Gottstein, E. Rabkin and L.S. Shvindlerman, *Scripta mater.*, 43, 2000, pp. 141-146.
[4] Y. Estrin, G. Gottstein, E. Rabkin and L.S. Shvindlerman, *Acta mater.*, 49, 2001, pp. 673-681.
[5] Ya.E. Geguzin, „Physik des Sinterns", VEB Deutscher Verlag für Grundstoffindustrie, Leipzig, 1973.
[6] D. Wolf and K.L. Merkle, in "Material Interfaces", eds. D. Wolf and S. Yip, Chapman & Hall, London, 1992, p. 87.
[7] W. in der Schmitten, P. Haasen and F. Häßner, *Z. Metallk.*, 51, 1960, pp. 101-108.
[8] H. Gleiter, *Acta metall.*, 27, 1979, pp. 1749-1754.
[9] K. Lücke and G. Gottstein, *Acta metall.*, 29, 1981, pp. 780-790.
[10] Y. Estrin and K. Lücke, *Acta metall.*, 29, 1981, pp. 791-802.
[11] M. Upmanyu, D.J. Srolovitz, G. Gottstein and L.S. Shvindlerman, *Interface Sci.*, 6, 1998, p. 289.
[12] G. Gottstein and L.S. Shvindlerman, "Grain Boundary Migration in Metals: Thermodynamics", Kinetics, Applications, CRC Press, Boca Raton, USA, 1999.
[13] T.R. Malow and C.C. Koch, *Acta mater.*, 45, 1997, p. 2177.
[14] C.E. Krill, L. Helfen, D. Michels, H. Natter, A. Fitch, O. Masson and R. Birringer, *Phys. Rev. Lett.*, 86, 2001, pp. 842-845.
[15] W.W. Mullins, *J. Appl. Phys.*, 28, 1957, p. 333.
[16] F.Y. Genin, *J. Appl. Phys.*, 77, 1995, p. 5130
[17] P. Chaudhari, *J. Vac. Sci. Techn.*, 9, 1972, p. 520.
[18] W. Brücknerm, V. Weihnacht, W. Pitschke, J. Thomas and S. Baunack, *J. Mater. Res.*, 15, 2000, pp. 1062- 1068.

Grain Growth Phenomena during Steel Processing

I. Gutierrez, B. Lopez, and J.M. Rodriguez-Ibabe

CEIT and Univ. of Navarra (ESI), Pº Manuel de Lardizabal, 15, 20018 Donostia-San Sebastián Basque Country, Spain

Keywords: low carbon steel, grain growth, particle pinning, solute drag, texture.

Abstract. The ferrite grain size is one of the main parameters controlling the mechanical properties in low carbon steels. Consequently, the control of the microstructure at different stages during the steel processing is of paramount importance. Grain refinement can only be obtained on the base of the allotropic transformation or the recrystallisation, both of them producing the nucleation of new grains. During the periods the steel remains at high temperature, the natural tendency of the grains is to coarsen by the effect of the grain growth. Particle pinning or solute drag effects can be used as a tool in structural steels to limit or inhibit the grain growth. In soft formable steels, the opposite happens and these effects have to be avoided in order to promote the grain growth.

1 Introduction

The most important part of the steel production concentrates on low carbon steels with a final ferrite-pearlite microstructure. From all the production, two broad groups, in terms of the final application, can be established: structural steels and formable strip steels. Each group comprises a large set of qualities ranging from standard to high performance properties, giving tailored solutions to particular requirements. The main microstructural parameter directly related to the mechanical properties in ferrite-pearlite steels is the ferrite grain size. Consequently, all the efforts for producing high performance steels, passes, at a stage or another of the production sequence, through the control of the grain size.
There are different metallurgical factors able to increase the strength of structural steels: grain refinement, precipitation, solution strengthening and cold working [1]. Among all, only the first of them allows to improve, at the same time, both the tensile properties and the toughness [2]. Consequently, the production of fine grained steels becomes of a paramount importance when a high strength combined to a good toughness are required directly after hot rolling.
When formability is an important property, grain size also plays a significant role, but now the challenge is to get large grain sizes [3]. In general, steel strips for cold forming are hot rolled, cold rolled and annealed. Traditional batch annealing is

able to produce appropriate ferrite grain sizes. On the contrary, the short times involved in continuous annealing lines tend to produce a lower extent of grain growth [4,5]. The main scope here is to overcome the possible restrictions to the grain growth in order to get soft and formable steel sheets.

2 Metallurgical Factors Allowing the Control of the Grain Growth

Recrystallisation and transformation involve the complete rebuilding of a new grain microstructure and are the only mechanisms able to produce a drastic grain refinement. Outside them, as far as the diffusivity of the atoms involved in the process is high enough, the natural tendency of the material is to grain coarsen, as a way to reduce the internal energy associated to the grain boundary interfaces. The grain growth retardation or inhibition can be produced by the effect of the particle and/or solute pinning [6,7]. Grain growth is driven by the curvature of the grain boundaries and can be expressed as being proportional to the inverse of the grain radius, R. The pinning pressure, P_z exerted by particles of radius, r, and volume fraction, f_v, can be expressed as

$$P_z \propto \frac{f_v}{r} \tag{1}$$

When the driving pressure, P_G, equals the pinning one exerted by second phase particles, the grain growth ceases and a limiting grain size is reached, D_{lim}. However, when using second phase particles as grain growth inhibitors, their stability is of prime importance. The thermal cycle able to produce grain growth can also alter their distribution through the dissolution/precipitation or the particle coarsening due to the Ostwald ripening effect [8]. Both processes alter P_z and consequently the effect the particles exert on the grain boundaries.

3 Grain Growth Control during Hot Rolling

In industrial hot rolling practice, static recrystallisation between passes is an effective way to refine austenite grain size. However, there are different factors acting against an effective grain refinement:

- Grain growth during reheating
- Grain growth after static recrystallisation
- Coarse initial grain sizes in thin slab casting and direct charging technologies.

The normal sequence in the production of the steels is the continuous casting, cooling under the A_1 temperature, and reheating followed by hot rolling. The austenite to be hot rolled is produced by the transformation $\alpha \rightarrow \gamma$ during reheating and relatively fine grain sizes are obtained. However, during the long times involved in reheating to temperatures high enough to allow hot rolling, grain growth can be very important. Large grains at the beginning of the hot working will require further refinement during the passes, which is not always possible depending on the deformation sequence. Second phase particles can be used to control, at least to a certain extent the grain size. As seen before, second phase particles, if finely dispersed, can exert a pinning effect on the grain boundaries and inhibit or restrict the grain growth. This effect depends on the volume fraction and size of the particles, Eq. 1, and their thermodynamic and size stabilities are very important.

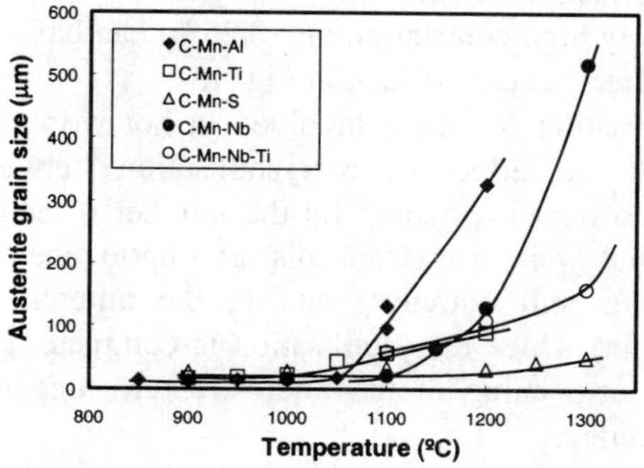

Fig. 1. Austenite grain size evolution during reheating for different steels C-Mn-Al [9, 10], C-Mn-Ti, C-Mn-Nb and C-Mn-Nb-Ti [10], C-Mn-S [11].

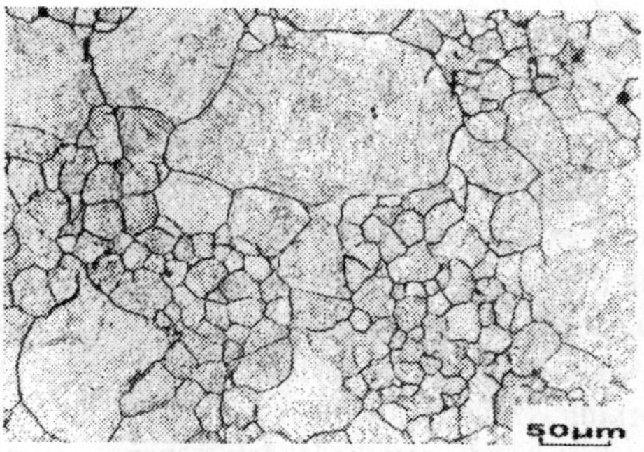

Fig. 2. Region showing abnormal grain growth in a C-Mn-Nb steel. 20 min at 1100ºC [12].

Fig. 1 shows the evolution of the austenite grain size as a function of the reheating temperature for different element additions in steel. It can be seen that until about 1050ºC, the grain size remains stable in all the steels. In the case of the C-Mn-Al steel, the AlN particles exert an effect in controlling the grain size on reheating until about 1050ºC, a similar effect is observed in the C-Mn-Nb steel until about 1100ºC. At higher temperatures in both steels, an intense increase of the mean grain size is observed due to abnormal grain growth, taking place, as illustrated in

Fig. 2. The result is, in both cases, that some grain boundaries become able to migrate very fast, while the rest of them remain pinned by the undisolved particles. This difference in the boundary ability to migrate produces the very fast growth of some grains that reach a size advantage with respect to the remaining fine grained matrix. Abnormal grain growth is very deleterious because it produces much coarser grains than those obtained in the non stabilised steels [13]. Additionally, this effect can be very heterogeneously distributed in the steel, depending on the local alloy concentration. Other alloying elements produce more stable precipitate particles. Among then, Ti which produces TiN precipitates is one of the most effective in controlling the grain size. However, an element like sulphur in relatively high concentrations (0.25%) can have a very important grain size controlling effect, as can be seen in Fig. 1.

During the deformation sequence involved in hot working, grain refinement is possible, through the effect of recrystallisation between passes [14]. The recrystallised grain depends mainly on the number of active nuclei. This last is related to the initial grain size (from this the importance of controlling the grain size at the rolling mill entrance) and to the amount of the stored energy (accumulated strain). Once recrystallisation is complete, grain growth is able to produce the fast coarsening of the microstructure, depending on the interpass temperature and time.

There are different equations reported in the literature to describe the grain growth in austenite [15]. All of them are of the general form which is generally accepted to describe the evolution of the grain size, D, after recrystallisation, as a function of time, t, and temperature, T, [6]:

$$D^n - D_{rex}^n = Bt\exp(-\frac{Q_G}{RT}) \tag{2}$$

where Q_G is the activation energy for grain growth and n is a constant adopting different values depending on the steel composition. The statically recrystallised grain size, D_{rex}, can be expressed as a function of the initial grain size before the pass and the strain in the pass [16].

In Al-killed C-Mn steels, the AlN enter into solution during reheating. The relatively short interpass times do not allow the precipitation of these particles during the hot working sequence. However, these times are long enough for recrystallisation and fast grain growth. In consequence, significant grain growth takes place between passes. In these conditions, the control of the microstructure is only possible if the temperatures of the last passes are low enough to restrict the extent of austenite grain growth, as can be seen in Fig. 3. In the case of thin slab casting and direct charging technologies, the first deformation pass, is applied to an austenite with the coarse (~1mm) microstructure coming directly from the cast. The thin slab enters directly into the finishing mill, without passing the previous roughing. Due to the coarse initial austenite grain size, the static recrystallisation [10] between the first passes is noticeably delayed. As a result, the time available for grain growth is shortened and an important grain refinement due to

recrystallisation and limited growth is observed to take place after the first deformation pass, as can be seen in Fig. 3. This compensates for the large initial starting grain size and, at the end, approximately the same final austenite grain size is obtained than in the case of the conventional process route.
Stable particles after reheating can continue to be effective during hot working. This can be the case of TiN which completely inhibit grain growth during the hot working sequence, leading to an effective recrystallisation control rolling [17]. Nb becomes into solution during reheating and is used to perform non-recrystallisation control rolling [18] by promoting the strain induced precipitation of Nb(C,N) particles which are able to stop the static recrystallisation in the last deformation passes of the sequence. Before this happens, Nb produces a solute drag effect which slows down the grain growth and helps in the control of the microstructure [19]. Nevertheless, in the case of thin slab rolling, elements able to retard the static recrystallisation can induce the presence in the final microstructure of "refractory grains" coming directly from the cast that are very deleterious in terms of the final mechanical properties.

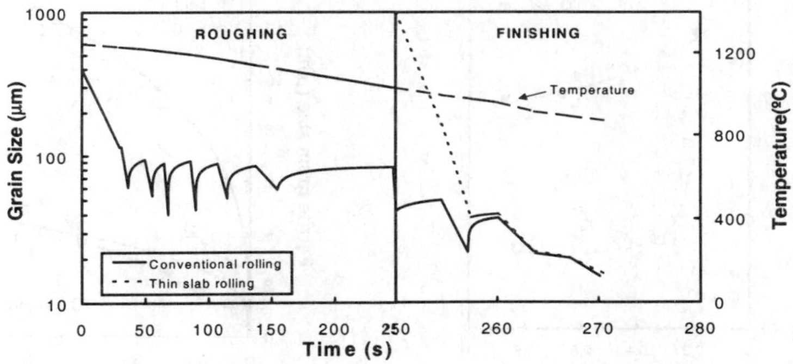

Fig. 3. Evolution of the temperature and the grain size in a rolling sequence applied to a C-Mn conventional slab or a thin slab and direct charging.

4 Ferrite Grain Size Produced by Transformation.

The austenite to ferrite transformation implies nucleation and growth of new grains and, like recrystallisation, is a powerful tool in terms of microstructure refinement. It is clear that to produce fine grained ferrite, small prior austenite grain sizes and high cooling rates within the range ensuring the formation of equiaxed ferrite are required. The range of grain sizes industrially achievable are, by applying recrystallisation control rolling in the range 8-10 μm and by non recrystallisation control rolling about 4-5μm. Efforts done to get finer equiaxed ferrite microstructures, require non conventional production routes [20,21].

The number of grains per unit volume nucleated at the beginning of the transformation is strongly dependant on the austenite initial grain size and accumulated strain. But the final ferrite microstructure will also depend on the evolution of the transformation. A reduction in the number of the ferrite grains, as the transformation progresses, has been observed by different authors [22-25]. This effect is shown in Fig. 4a for a recrystallised and a strained initial austenite. In Fig. 4b the mean ferrite grain size has been plotted as a function of the volume fraction of transformation. The dark line corresponds to the evolution predicted using the following equation:

$$d_{\alpha_i} = d_\alpha \left(\frac{f_{\alpha i}}{f_\alpha} \right)^{1/3} \tag{3}$$

where d_α and f_α are the mean ferrite grain size and the volume fraction of polygonal ferrite at the end of the transformation and d_{α_i} and f_{α_i} represent the corresponding instantaneous values during the progress of it.

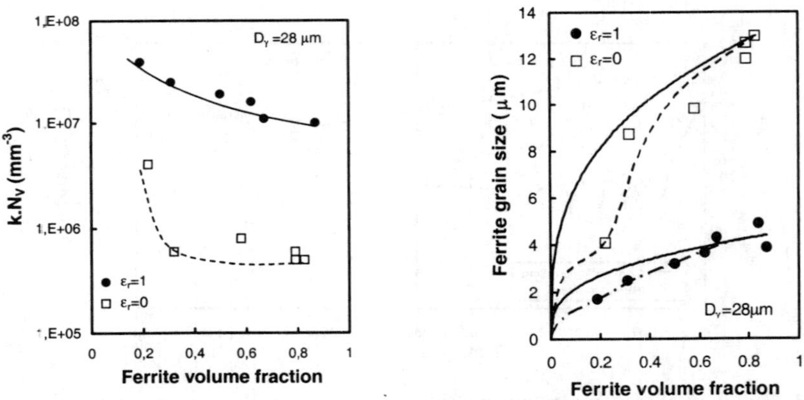

(a) (b)

Fig. 4. Austenite to ferrite partial transformation: (a) Evolution of the number of ferrite grains per unit volume and (b) Change in the ferrite grain size [26].

It can be seen that, at the initial stages of the transformation, the experimental grain sizes are lower than those predicted by Eq. 3. This indicates that the conditions on which this last is applicable, i.e, all the grains nucleated at the beginning of the transformation remain in the final microstructure and the grain size is uniform, are not fulfilled. Site saturation at austenite grain boundaries is observed to occur during the initial stages of transformation, as can be seen in Fig. 5a and, after impingement, ferrite grain coarsening takes place, resulting in the mentioned decrease in the number of the grains per unit volume. The result is that the final ferrite grain size is significantly higher than that expected considering the number of grains nucleated from austenite.

Coarsening of ferrite grains can be due to different mechanisms. Priestner et al. [22,23] suggested that the elimination of ferrite grains is produced by normal growth after full impingement on the austenite grain boundary plane. In the work by Bengochea et al. [24] some TEM work was carried out to determine the orientation relationship between adjacent grains at an early stage of transformation. It was observed that the transformation produced a mixture of low and high angle boundaries. The Fig. 5b shows this type of misorientation distribution between neighbour ferrite grains, as obtained by EBSD in a partially transformed sample. The presence of low angle boundaries opens the possibility to some coalescence being the responsible of the observed ferrite grain coarsening during transformation.

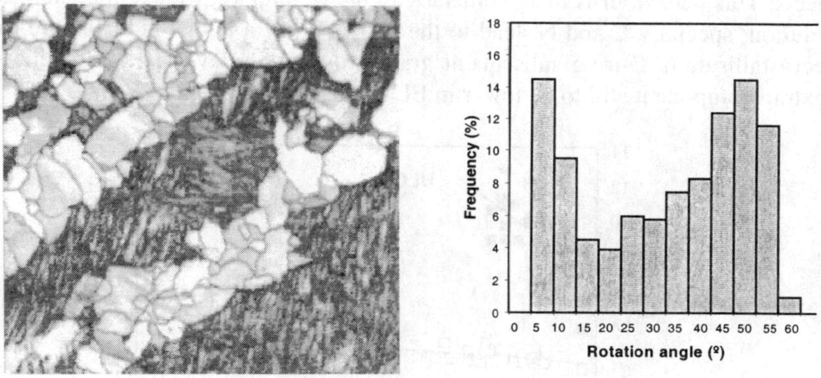

Fig. 5. Partial transformation from a strained austenite a) EBSD image of the ferrite grains on prior grain boundaries b) grain boundary misorientation distribution in ferrite behind the transformation front [27].

5 Grain Growth Control during Annealing after Cold Rolling

In the case of low carbon cold rolled and annealed sheet steels, for many applications, the ability they present to deep drawing is an important property which is related to the texture of the material [3]. A large amount of work has been carried out on low carbon steels and the quotient between the intensity of the {111} and {001} texture component is used very often to have a first picture on the ability of the steel to be deep drawed [28]. The main efforts in developing deep drawable steels concentrate on the control of the steel chemistry and of the entire production process from hot rolling trough cold rolling to annealing [5-31]. The {111} texture components develop during recrystallisation and continue to intensify during grain growth [32]. It is generally accepted that grains with a {111} orientation nucleate early during recrystallisation and have a size advantage that helps them to grow preferentially [30]. However, different factors can have an important effect on grain growth and consequently on the final texture. The carbon content, [%C], of the steel is one of them. It has been deduced [33] that the recrystallised grain size depends on the carbon content of the steel in the way:

$$D_{rec'} = F\,[\%C]^{-0.2} \tag{4}$$

with %C in weight per cent and F=2.6 µm in the work by Petite et al. [34] and applicable to the data of Ushioda et al. in ref. [5] with a value of F=2 µm. The carbon content in the steel also plays a role with respect to the evolution of the texture produced by the grain growth after recrystallisation, as can be seen in Fig. 6. Two steel families have been considered ELC steels with carbon contents between 0.03 and 0.09% and ULC steels containing around 0.003%C. It can be seen that the $P_{\{111\}}/P_{\{100\}}$ is more intense after recrystallisation in ULC than in ELC steels. This behaviour is in agreement with the general knowledge that elements in solution, specially C and N, lead to the formation of a poor {111} texture during recrystallisation. During subsequent grain growth, the development of this same texture component still to be lower in ELC than in ULC steels.

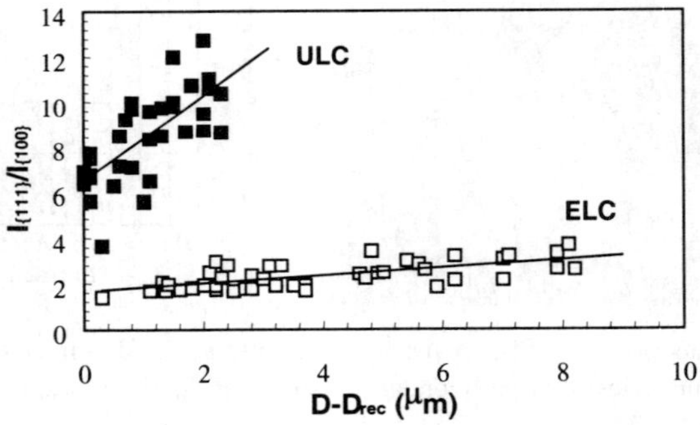

Fig. 6. Evolution of the texture component intensity relation as a function of the grain growth after recrystallisation [33].

The small solubility in ferrite of the main part of the alloying elements leads to the precipitation taking part during annealing of the elements solutioned during hot working processing. Specially important in this field is the nitrogen. It is important in continuous annealing of ULC and ELC steels to minimise the presence of free nitrogen in the steel after coiling in order to avoid the precipitation of fine AlN particles in the annealing line. These particles have a very strong grain growth inhibition effect, as can be seen in Fig. 7. As discussed elsewhere [35], the heterogeneous precipitation occurring preferentially at grain boundaries, in a fully recrystallised microstructure, can be the cause for this strong effect.

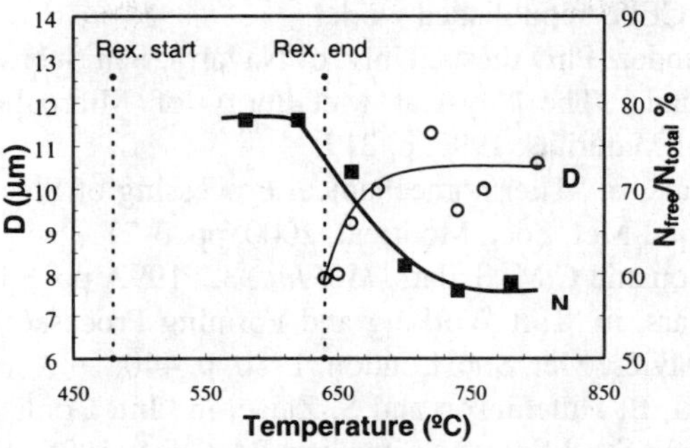

Fig. 7. Grain size and free nitrogen evolution, as a function of the annealing temperature under continuous annealing conditions in a ULC-steel [35].

6 Acknowledgements

The authors want to acknowledge the finantial support of part of this work by the "Consejería de Educación del Gobierno Vasco" under the project: EX-1999-89

7 References

[1] T. Gladman, D. Dulieu and I.D. McIvor, "Microalloying´75", ed. M. Korchynsky, Union Carbide Corporation, New York, 1977, pp. 32-55.

[2] F.B. Pickering, "Microalloying´75", ed. M. Korchynsky, Union Carbide Corporation, New York, 1977, pp. 9-31.

[3] M. Abe, in "Material Science and Techonology a Comprehensive Treatment", eds. Cahn et al., Vol. 7 (ed. F.B. Pickering), pp. 285-333.

[4] P.R. Mould, in "Metallurgy of Continuous-annealed Sheet Steel", eds. B.L. Bramfitt and P.L. Mangonon, Met. Soc. AIME, Warrendale, 1982, pp. 3-34.

[5] R.C Hudd, as in Ref. [3], pp. 219-284.

[6] F.J. Humphreys and M. Hatherly, "Recrystallisation and Related Annealing Phenomena", Elsevier, 1995, pp. 73-120.

[7] A. J. DeArdo, in "Thermomechanical Processing of Steels", IOM Communications, 2000, pp. 309-312.

[8] J.W. Martin and R.D. Doherty, "Stability of Microstrcuture in Metallic Systems", Cambridge Univ. Press, 1980, pp. 154-243.

[9] L.E. Cepeda, J.M. Rodríguez-Ibabe, J.J. Urcola and M. Fuentes, *Materials Science and Technology*, 5, 1989, pp. 1191-1199.

[10] A.I. Fernandez, P. Uranga, B. Lopez and J.M. Rodriguez-Ibabe, *ISIJ International*, 40, 2000, pp. 893-901.

[11] J. Ochoa, CEIT unpublished work.
[12] A.I. Fernandez, PhD thesis, Univ. of Navarra, San Sebastian, 2001.
[13] T. Galdman, "The Physical Metallurgy of Microalloyed Steels", The Institute of Materials, 1997, p. 217.
[14] C.M. Sellars, in "Thermomechanical Processing of Steel", eds. S. Yue and E. Essadiqui, Met. Soc., Montreal, 2000, pp. 3-21.
[15] J.H. Beynon and C.M. Sellars, *ISIJ Int.*, 32, 1992, pp. 359-367.
[16] C.M. Sellars, in "Hot Working and Forming Processes", eds. C.M.Sellars and G.J.Davies, Met. Soc. London, 1980, p. 440.
[17] T. Siwecki, B. Hutchinson and S. Zajac, in "Int. Conf. Microalloying 95", The Iron and Steel Society, Pittsburg, PA, USA, 1995, pp. 197-211.
[18] L.J. Cuddy, in "Proc. of Int. Conf. Thermomechanical Processing of Microalloyed Austenite", eds. A.J. DeArdo and al., AIME, Warrendale, PA, 1982, p. 129.
[19] P.S. Hodgson and R.K. Gibbs, *ISIJ Int.*, 32, 1992, pp. 1329-1338.
[20] A.K. Ibraheem, R. Priestner, J.R. Bowen, P.B. Prangnell and F.J. Humphreys, as in Ref. [7], pp. 446-455.
[21] P.D. Hodgson, M.R. Hickson and R.K. Gibbs, in "Microalloying in Steels", *Materials Science Forum*, 284-286, eds. J.M. Rodriguez-Ibabe et al., 1998, pp. 63-72.
[22] R. Priestner and P.D. Hodgson, *Mat. Sci. and Technol.*, 8, 1992, pp. 849-854.
[23] R. Priestner and L. Ali, *Materials Science and Technology*, 9, 1993, p. 135.
[24] R. Bengochea, B. López and I. Gutierrez, *Metall. Trans.*, 29A, 1998, pp. 417-426.
[25] R. Bengochea, B. López and I. Gutierrez, as in Ref. [22], pp. 201-208.
[26] E. Cotrina, R. Bengochea, A. Argüelles, B. Lopez and I. Gutierrez, as in Ref. [7], pp. 255-264.
[27] E. Cotrina and A. Iza, to be published.
[28] J.F. Held, "Mechanical Working and Steel Processing IV", AIME, Warrendale, 1965, pp. 3-38.
[29] U. Schlippenbach, F. Emren and K. Lücke, *Acta Metall.*, 1986, 34, p. 1289.
[30] W.B. Hutchinson, *Int. Met. Rev.*, 29, 1984, pp. 25-42.
[31] R.K. Ray, J.J. Jonas and R.E. Hook, *Int. Mater. Rev.*, 39, 1994, p. 129.
[32] W.B. Hutchinson and E. Nes, in "Grain Growth in Polycrystalline Materials", eds. Abbruzzese and Brozzo, Trans. Tech. Publications, Rome, 1992, p. 385.
[33] I. Gutierrez et al., Final Report ECSC project 7210.EC/940, Nov. 1998.
[34] M.M. Petite, PhD. thesis, Univ. of Navarra, San Sebastian, 1999.
[35] M.M. Petite, I. Gutierrez and M. Fuentes, in "Recrystallization and Related Phenomena", ed. T.R. McNelley, Monterey Institute of Advanced Studies, California, 1997, pp. 389-396.

Aspects on Grain Growth in Cemented Carbides and Cermets

T. Yamamoto[1], Y. Ikuhara[2], and T. Sakuma[1]

[1] Department of Advanced Materials Science, The University of Tokyo, 7-3-1, Hongo, Bunkyo-ku, Tokyo 113-8656, Japan
[2] Engineering Research Institute, The University of Tokyo, 2-11-16, Yayoi, Bunkyo-ku, Tokyo 113-8656, Japan

Keywords: cemented carbide, cermet, grain boundary structure, HRTEM, EDS.

Abstract. High resolution electron microscopy study was carried out for VC-doped WC-Co and TiC-Mo_2C-Ni alloys with a special interest in WC/Co, WC/WC boundaries and core-rim interfaces. It was found that the morphology of carbide grains largely changed by an addition of VC. The WC/Co interfaces tend to facet with fine multi-steps of (0001) and $\{10\bar{1}0\}$ habits by VC-doping. In addition, EDS analysis with sub-nano scale probe revealed that doped V was strongly concentrated to the two habits. Thus it can be supposed that the retardation of the grain growth of WC grains caused by VC-doping is due to the concentration of V at the migration front of the two habits. Meanwhile, it was revealed that the interface of core-rim structure was highly coherent, and the chemical composition discretely changed across the core-rim interface at a nano-scale level in TiC-Mo_2C-Ni cermet. The discrete change in molybdenum content at the interface suggests the existence of a miscibility gap between TiC and Mo_2C system at the sintering temperature.

1 Introduction

WC-Co based cemented carbides or TiC-Mo_2C-Ni based cermets are one of the most useful materials for mechanical machining tools because they have high hardness, strength and wear-resistance [1,2]. In general, the microstructures of both alloys have similar feature that hard carbide grains of WC or (Ti, Mo)C are surrounded by metallic binder phase mainly consisting of Co or Ni, respectively. The morphology of carbide grains, however, is very different between the two alloys, i.e., facetted shape in cemented carbides while round shape in cermets [3,4]. In particular, (Ti, Mo)C grains exhibit "core-rim" structure, i.e., a "core" region of TiC is surrounded by a "rim" of (Ti, Mo)C solid solution. Although the morphology of carbide grains in both alloys differs each other, their mechanical properties are known to be mainly depending on the microstructure, especially, the size of carbides. For example, transition metal carbides are often doped as inhibitors for carbide grain growth in WC-Co based alloys in order to improve their mechanical properties [5]. Such grain size control is generally closely related to the mechanism of the microstructure formation. Therefore, it will be important to examine the microstructure in atomic scale. So far, our group has revealed the microstructure in WC-Co based alloys and cermets by high resolution transmission electron microscopy study with nano-scale electron probe [6-9]. In this study, we summarize the results obtained from HRTEM study for VC-doped WC-Co alloys prepared by liquid-phase and solid-phase sintering, and also TiC-Mo_2C-Ni alloys.

2 Experimental Procedure

WC-20wt%Co, WC-20wt%Co-0.5wt%VC and TiC-20wt%Mo_2C-20wt%Ni alloys were prepared by ordinary vacuum-sintering process with raw powders of WC (Japan New Metal, nominal grain size of 0.8 μm), Co (United Metal, 1.2 μm) and VC (Starck, 1.6 μm) in cemented carbides and TiC, Mo_2C, and Ni (Japan New Metal Co. Ltd.) in cermets. Respective green compacts were sintered at 1380°C for 1h or quenched at 1200°C in cemented carbides, and 1390°C for 1h in cermets. Here, quenching temperature of 1200°C used for WC-Co alloys means that sintering was stopped in solid state [10]. Meanwhile, specimens for microstructural study were fabricated by ion-thinning method (PIPS, Gatan691). Microstructure observations were carried out with a high resolution transmission electron microscope (HRTEM, EM-002BF, TOPCON, H-9000NAR, HITACHI) with X-ray energy dispersive spectroscopy (EDS, Voyager System, Noran). Chemical analysis was made with a probe size of ϕ0.5 nm.

3 Results

3.1 WC-Co alloys

Liquid-phase sintered alloys. Fig. 1 shows typical microstructures in (a) undoped and (b) VC-doped alloys in as sintered state. In Fig. 1a, the grains with black contrast represent WC grains, and the phase with white contrast a Co-phase, respectively. As seen in the figure, the undoped alloy exhibits the well-known microstructure, i.e., facetted WC grains are present in a metallic binder phase consisting of mainly Co. On the contrary, the morphology of the WC/Co interface largely changes by an addition of VC as seen in Fig. 1b. Fine multi-steps generate along WC/Co interfaces in the VC-doped alloy as indicated by the arrows in Fig. 1b. This distinct feature of multi-stepped boundaries is observed only in the VC-doped alloys. On the other hand, the average grain size of carbide grains is estimated to be 0.67 μm in the undoped alloy while 0.13 μm in the VC-doped one. It is thus clear that VC-doping is effective to suppress the grain growth of the carbide grains as previously reported.

Fig. 1. Bright field images in (a) non-doped WC-Co and (b) VC-doped WC-Co. The arrows in (b) indicate multi-steps caused by VC-doping.

Fig. 2 is a HRTEM image of the VC-doped alloy in the vicinity of WC/Co interface. The $[1\bar{2}10]$ of the WC grain is parallel to the observation direction and the WC/Co interface is set at the edge-on condition. In the figure, the indications with circles represent the areas for EDS analyses carried out in Fig. 3. As seen in the HRTEM image, the fine multi-steps consist of two types of habits, i.e., (0001) and $(10\bar{1}0)$ habits, respectively. The width of (0001) habit tends to be longer than that of $(10\bar{1}0)$ one. In addition, a contrast can be seen along the habit A-A and at the corner of crossing the two habits as indicated by the arrows.

Fig. 2. A HRTEM image in VC-doped WC-Co. EDS areas described in Fig. 3 are numerically shown and the arrows indicate other compounds formed at the comer of steps.

Fig. 3 shows the EDS spectra taken from the areas as numerically indicated in Fig. 2. The area 1 and 3 are 2 nm apart from the habit of B-B outside or inside the WC grain. Comparing these three profiles, it is clear that the large difference exists in V content. The peak of V Kα-line clearly appears at the interface of the area 2 and exhibits slight intensity at the area 1 while it disappears in the WC grain interior as in the area 3. Doped V segregates at the WC/Co interface.
Fig. 4 shows a plot of V content against a distance from WC/Co boundaries. EDS analysis was conducted systematically at every 1nm across the two habits of (0001) and $(10\bar{1}0)$. The both habits exhibit strong V-segregation, but the V content differs between the two habits. V segregates more at the (0001) habit than at the $(10\bar{1}0)$ one. The behavior of V-segregation was similar in other WC/Co interfaces examined in this study.

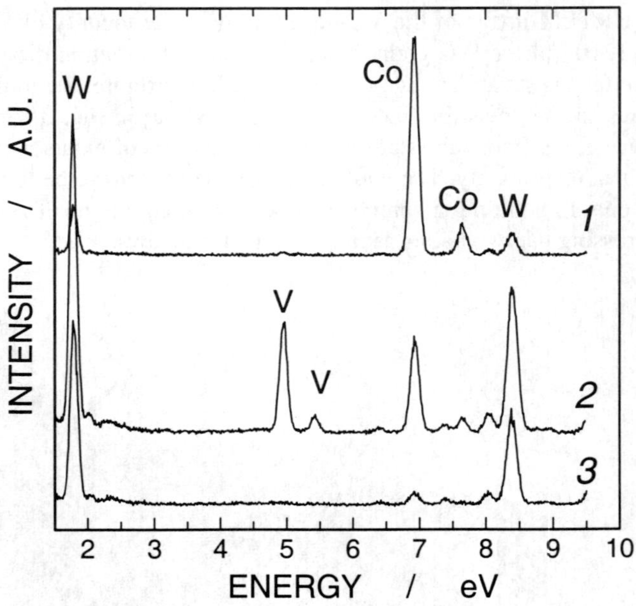

Fig. 3. EDS profiles taken from the area as indicated in Fig. 2. V-Kα peak clearly appears at WC/Co interface.

Solid-phase sintered alloys. Table 1 presents the grain sizes estimated from quenched alloys. In the table, the grain size of the fully sintered alloys are also shown for comparison. Comparing the grain sizes, it is noted that there is not so large difference in the grain size of VC-doped alloys, i.e., 0.13 μm in fully sintered alloy and 0.12 μm in the quenched one. This means that the grain growth of carbides in the VC-doped alloy is already retarded even in solid state.

Table 1. Grain sizes in fully sintered and quenched WC-Co alloys.

	Undoped alloy	Doped alloy
Fully-sintered	0.67μm	0.13μm
Quenched	0.36μm	0.12μm

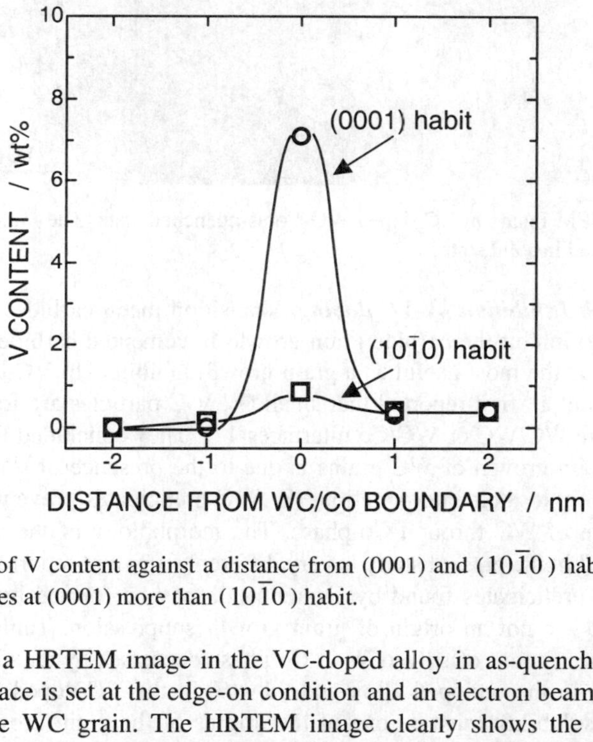

Fig. 4. A plot of V content against a distance from (0001) and $(10\bar{1}0)$ habits. It is noted that V segregates at (0001) more than $(10\bar{1}0)$ habit.

Fig. 5 shows a HRTEM image in the VC-doped alloy in as-quenched state. The WC/Co interface is set at the edge-on condition and an electron beam is parallel to $[1\bar{2}10]$ of the WC grain. The HRTEM image clearly shows the formation of stepped boundaries in an atomic scale, which is similar to that in the fully sintered alloy as shown in Fig. 2. Namely, the multi-stepped boundary with the two habits is already formed in the solid-state reaction, i.e., (0001) and $\{10\bar{1}0\}$, respectively. Comparing the width of the two habits in the multi-stepped boundaries, both (0001) and $(10\bar{1}0)$ habits in the quenched alloys have similar width whereas (0001) habit tends to be wider than $(10\bar{1}0)$ one in the fully sintered one. In addition, V-segregation at the two habits was also confirmed in the quenched alloys by EDS analysis.

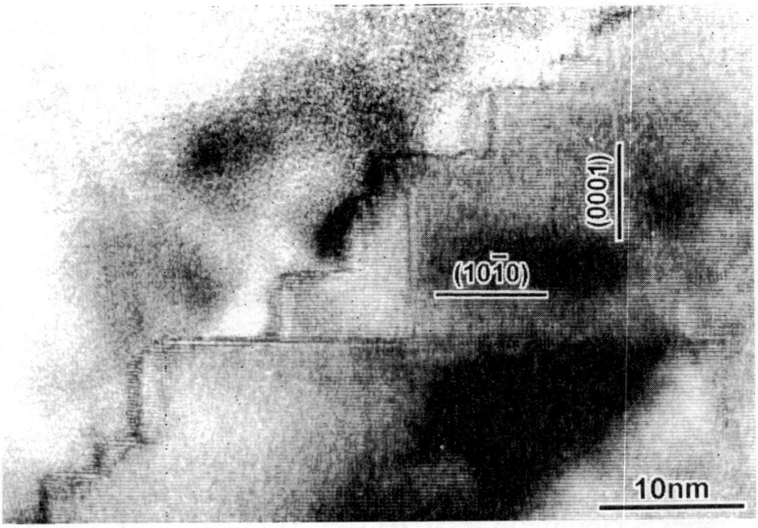

Fig. 5. A HRTEM image in VC-doped WC-Co as-quenched state. The multi-step interface is already formed in solid state.

Grain Growth Inhibition by VC doping. Transition metal carbides are known to be effective to inhibit the carbide grain growth in cemented carbides. Among the carbides, VC is the most useful as a grain growth inhibitor. In VC-doped WC-Co alloys, Egami et al. first reported that small (W,V)C particles are formed periodically along the WC/WC or WC/Co interfaces[11]. They concluded that the inhibition of the grain growth of WC grains is due to the presence of (W,V)C precipitates. According to their interpretation, (W,V)C preticipates prevent the solution-reprecipitation of WC through Co-phase. The morphology of the vanadium-rich phase detected in the present study is very different from that reported by Egami et al. Very fine preticipates found by Egami et al. may be formed by cooling after sintering, and are not an origin of grain growth suppression. Taniuchi et al. reported that the amount of solved W in Co phase decreases by an addition of VC, i.e., 25.2wt% in the undoped alloy and 9.8wt% in the VC-doped one [12]. The reduction in solubility must be one of the origins of the grain growth inhibition. The present result demonstrates that doped V tends to segregate in WC/Co interface, and probably form carbides. Although the composition of carbides cannot be determined precisely in the present analysis, segregated V is likely to form $(W,V)_2C$ precipitates at elevated temperature.

3.2 Discrete compositional change of Mo in TiC-Mo_2C-Ni cermet

Fig. 6 is a bright field image in TiC-20wt%Mo-20wt%Ni cermet in as sintered state. In the figure, typical core-rim structure can be clearly seen, i.e., the area of white contrast is a "core" and the surrounding region of gray contrast outside the cores is a "rim". A metallic phase mainly consisting of Ni surrounds core-rim carbides. In addition, small black grains are occasionally observed inside carbide grains as indicated by the arrows in the figure. They were determined to be Ni-phases from EDS analysis. They may be included during the formation of carbide grains.

Fig. 6. A bright field image in TiC-Mo_2C-Ni cermet in as-sintered state.

Fig. 7 shows a HRTEM image in the vicinity of the_core-rim interface. In the image, the carbide grain is oriented parallel to the [011] electron beam direction. The interface is highly coherent as recognized in the figure, and the interface can be confirmed only by the slight contrast difference as indicated by the arrows. The darker region is confirmed to be a rim from EDS analysis as described later. The high coherency is consistent with the previous report of X-ray diffraction result showing that core-rim phase is determined as a single phase.

Fig. 8 shows a plot of the result of chemical analysis carried out across the core-rim interface as indicated in Fig. 7 with a probe size less than 1 nm at every 2 nm interval. Note that the content of Ti and Mo discretely changes at the interface. The content of Ti does not change largely in the core region while it reduces to about 80wt% at the interface. On the other hand, the content of Mo increases from a few wt% to about 20wt% at the interface. This is probably the first report to detect the discrete change in the content at the core-rim interface in TiC-Mo-Ni cermets in nanometer scales.

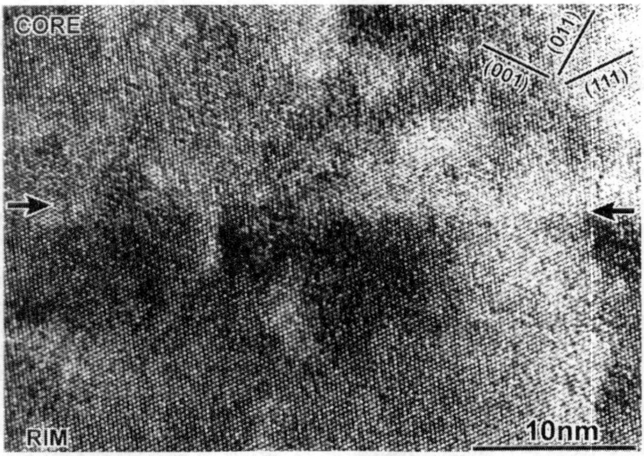

Fig. 7. A HRTEM image in the vicinity of a core-rim interface. The arrows indicate the interface. Note that the interface is highly coherent.

As for the formation mechanism of core-rim structure, several interpretations have been proposed. For example, Suzuki et al. described that the core-rim structure is due to the precipitation of (Ti, Mo)C solid solutions on TiC grains, which is equilibrium to Ni-rich eutectic liquid [13]. Their conclusion was obtained from the data of SEM-EDS analysis with less accuracy rather than the present study, and the interface was assumed to be diffuse after the formation of core-rim structure. However, the compositional change at the interface is not diffuse but discrete as clarified in this study. The present authors wish to point out that the presence of such discrete interface can be explained by assuming that a miscibility gap exists in (Ti, Mo)C solid solutions.

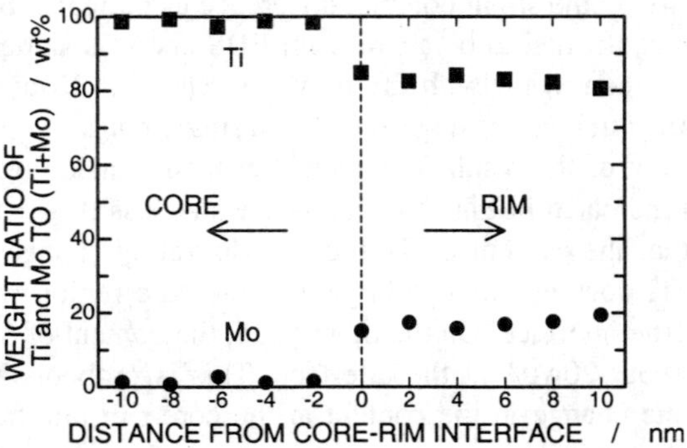

Fig. 8. A plot of Ti and Mo content as a function of a distance from the interface. It is noted that the composition discretely changes at the interface in nano-scale level.

4 Conclusion

The nano-probe analysis were carried out in VC-doped WC-Co and TiC-Mo_2C-Ni alloys, and the following conclusions were obtained.
The WC grains are facetted and sometimes form multi-steps with the two habits of (0001) and $\{10\bar{1}0\}$. EDS analysis with subnano-scale probe revealed that doped-V segregates in the two habits. The suppression of the grain growth by VC-addition into WC-Co alloys is due to the reduction of the solubility of W in Co phase and also by the V-segregation in WC/Co boundaries.
HRTEM study in TiC-Mo_2C-Ni cermets revealed that the core-rim interface in TiC carbides is highly coherent, and the Mo content changes discretely at the interface in a nano-scale level. The result can be explained by the presence of a miscibility gap in TiC and (Ti, Mo)C system at high temperatures.

Acknowledgement

A part of this study was financially supported by Grant-in-Aid for Scientific Research on Priority Areas (B) (No. 12130202). The authors wish to thank Dr. T. Tanase, Dr. T. Taniuchi and Dr. K. Okada (Mitsubishi Materials Co. Ltd.) for useful discussions and Dr. A. Jaroenworaluck for experimental assistance.

References

[1] F.V. Lenel, "Powder Metallurgy and Its Applications", Metal Powder Industries Federation, Princeton, NJ, 1980.
[2] H. Doi, in "Proceedings of 2nd Int. Conf. Science Hard Mater.", Adam Hilger Ltd., Rhodes, 1986, pp. 489-523.
[3] J. Vicens, M. Benjdir, G. Nouet, A. Dubon and J.Y. Laval, *J. Mater. Sci.*, 29, 1994, pp. 987-994.
[4] A.H. Heuer, J.S. Sears and N.J. Zaluzec, in "Sience of Hard Metals", eds. by C.A. Brookes and R. Warren, Adam Hilger Ltd., Bristol, 1986, pp. 321-326.
[5] S. Luyckx, C. Osborne, L.A. Cornish and D. Whitefield, *Powder Metallurgy*, 39, 1996, pp. 210-216.
[6] T. Yamamoto, Y. Ikuhara and T. Sakuma, *Science and Technology of Advanced Mater.*, 1, 2000, pp. 97-104.
[7] T. Yamamoto, Y. Iuhara, T. Watanabe, T. Sakuma, Y. Taniuchi, K. Okada and T. Tanase, *J. Mater. Sci.*, in printing.
[8] A. Jaroenworaluck, T. Yamamoto, Y. Ikuhara, T. Sakuma, T. Taniuchi, K. Okada and T. Tanase, *Ceramic Transactions*, 99, 1998, pp. 501-509.
[9] T. Yamamoto, A. Jaroenworaluck, Y. Ikuhara and T. Sakuma, *J. Materials Research*, 14, 1999, pp. 4129-4131.

[10] E.J. Sandford and E.M. Trent, *Spec. Rep.*, Iron and Steel Inst, 38, 1947, pp. 84.

[11] A. Egami, T. Kusaka, M. Machida and K. Kobayashi, in "Proc. 12th Int. Plansee Seminar", eds. H. Bildstein and H.M. Ortner, Metallwerk Plansee Reutte, 2, 1989 pp. 822-826.

[12] A. Egami, M. Ehira and M. Machida, in "Proc. 13th Int. Plansee Seminar", eds. H. Bildstein and R. Eck, Metallwerk Plansee Reutte, 3, 1993 pp. 639-648.

[13] T. Taniuchi, K. Okada and T. Tanase, in "Proc. of 14th Int. Plansee Seminar" eds. H. Bildstein and R. Eck, Metallwerk Plansee Reutte, 3, 1993 pp. 639-645.

[14] H. Suzuki, K. Hayashi and O. Terada, *Jpn. Inst. Metal.*, 35, 1971, pp. 146-151.

[15] L. Lindau and K.G. Stjernberg, *Powder Metall.*, 19, 1976, pp. 210-218.

[16] D. Moskowitz and M. Humenik, Jr., in "Science of Hard Materials", eds. R.K. Viswanadham, D.J. Roweliffe and J. Gurland, Plenum Press, New York, 1983, pp. 299-315.

Recrystallization and Grain Growth
Proceedings of the First Joint International Conference
Eds. G. Gottstein and D.A. Molodov, Springer-Verlag, 2001

Grain Boundary Property Determination through Measurement of Triple Junction Geometry and Crystallography

A.D. Rollett[1], C.-C. Yang[1], W.W. Mullins[1], B.L. Adams[5], C.T. Wu[1], D. Kinderlehrer[2], S. Ta'asan[2], F. Manolache[2], C. Liu[2], I. Livshits[2], D. Mason[4], A. Talukder[3], S. Ozdemir[3], D. Casasent[3], A. Morawiec[1], D. Saylor[1], G.S. Rohrer[1], M. Demirel[1], B. El-Dasher[1], and W. Yang[1]

[1] Department of Materials Science and Engineering, Carnegie Mellon University, Pittsburgh, PA 15213-3890
[2] Department of Mathematical Sciences, Carnegie Mellon University
[3] Department of Electrical and Computer Engineering, Carnegie Mellon University
[4] Department of Materials Science and Mechanics, Michigan State University
[5] Department of Mechanical Engineering, Brigham Young University

Keywords: grain boundary properties, polycrystals, energy, mobility, triple junctions, aluminum.

Abstract. Microstructure controls the properties of most useful materials. Thus an ability to control microstructure through the processing of materials is a key to optimization of materials performance. Most materials are polycrystalline and their grain structure is a very important aspect of their microstructure. Thanks to their complexity there is a great variety of grain boundary types even in relatively isotropic materials such as the cubic metals. Simply describing the crystallography requires five (macroscopic) parameters (e.g. disorientation and inclination). Evidently, acquiring a knowledge of the variation of properties such as energy and mobility as a function of grain boundary type would be of great value in predicting properties and optimizing processing. This paper outlines the methods being employed to extract such properties from the geometry and crystallography of triple junctions between grain boundaries.

1 Introduction

1.1 Approach

The aim of this paper is to lay out a procedure, with an example, for measuring the properties of interfaces as a function of their crystallographic character that is capable of characterizing the full range of boundary type. The scope is limited to determination of excess free energy and mobility of homophase interfaces, i.e. grain boundaries in single phase materials. The approach relies on measurement of the geometry and crystallography of grain boundary junctions i.e. triple points [1], Fig. 1.

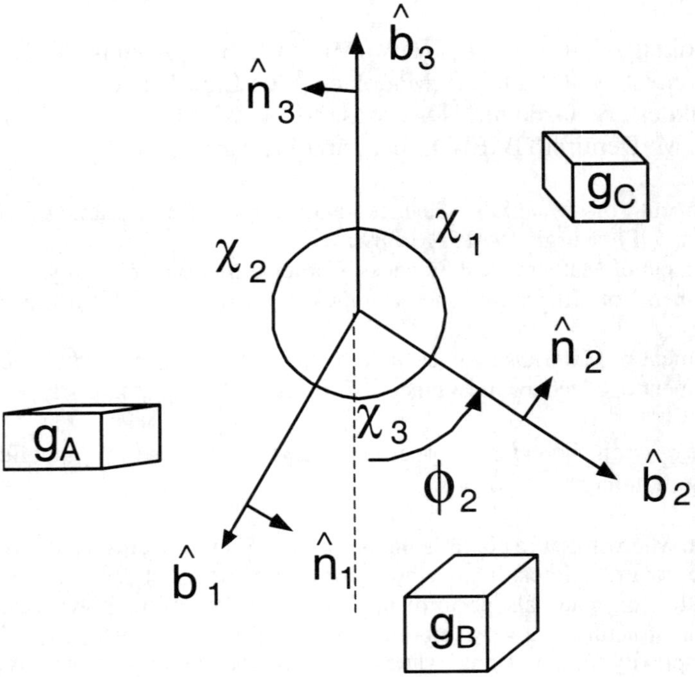

Fig. 1. Diagram of a labeling convention for grain boundary character, showing dihedral angles, χ, inclination angles, ϕ, boundary tangent vectors, b, boundary normals, n, and grain orientations, g. The triple junction line is perpendicular to the plane of the diagram.

In the general case, the crystallographic orientations of each grain, the grain boundary tangents and the triple line tangent are measured with the aid of serial sectioning. For samples in which columnar structures can be generated (typically thin foils or films), the geometry may be measured directly on a single section perpendicular to the column axis. The analytical procedure for extracting properties is statistical and multiscale which means that the number of measurements must be large enough to ensure that each boundary type has been adequately sampled. It is assumed that local equilibrium exists at each triple junction so that the Herring relations (for force balance) may be applied [2]. Each triple junction provides two independent relations between boundary energies. Analysis of the set of relations generates a distribution of boundary energies that (simultaneously) satisfies the Herring relations at each triple junction. The Herring Eq.s [3], describing equilibrium at a triple junction, result from the requirement that a virtual displacement of the triple junction in any direction causes no first order change in energy. The equations are

$$\sum_{j=1}^{3}\left\{\sigma_j\hat{b}_j+\left(\partial\sigma_j/\partial\phi_j\right)\hat{n}_j\right\}=\bar{0} \tag{1}$$

where the sum is conducted over the three interfaces intersecting at the triple junction, σ_j is the excess free energy of the jth boundary, $\hat{n}^{(j)}$ is the unit boundary normal of the jth boundary, $\hat{b}^{(j)}$ is a unit vector lying in the jth boundary and perpendicular to the triple line, $\hat{l}=\hat{n}^{(j)}\wedge\hat{b}^{(j)}$ which is common to all three adjacent boundaries and ϕ_i is defined to be the right handed angle of rotation about triple line of the jth boundary from a reference direction. The derivative terms are referred to as torque terms and reflect the dependence of interface energy on orientation about the triple junction.

If the interface energy is independent of interface orientation, Herring's equations reduce to the Young equations:

$$\sigma_1/\sin(\chi_1)=\sigma_2/\sin(\chi_2)=\sigma_3/\sin(\chi_3) \tag{2}$$

Further, a geometrical relationship between boundary migration rates has been described by Adams et al. [2] based on the assumption that boundaries migrate only due to their capillarity and that other driving forces are negligible. Then the relation

$$\sum_{i=1,3} M_i[\sin(\chi_i)\sigma_i\kappa_i]=0 \tag{3}$$

holds, where κ_i is the sum of the two principal curvatures at a triple junction associated with the ith grain boundary. The assumptions underlying Eq. (3) are that the boundary migration rate is linearly related to the driving force f ($=\sigma\kappa$) by a proportionality factor M, representing mobility, and that M and the energy are independent of the boundary inclination. Thus each triple junction geometry yields two energy-related equations and one mobility-related equation. After characterizing a sufficient number of triple junction geometries, relative boundary energies and mobilities as a function of grain boundary misorientation can be extracted through a statistical/multiscale analysis [4] of Eqs. (2) and (3).

The current size of the data sets is only large enough, however, to allow for the analysis to be executed on the basis of a three-parameter description of grain boundary character. A data set for predominantly low angle boundaries in an aluminum of moderate purity (99.98%) has been analyzed based on misorientation angle and misorientation axis type for both energy and mobility. The results of the one-parameter analysis (angle) are in good agreement with the literature. The results of the two-parameter analysis (axis) show mild variations in energy with type but strong variations in mobility. The results for energy measurements in magnesia indicate that the conventional view of uniform boundary energy except for low angle boundaries and for well matched lattices, as indicated by a high fraction of coincident lattice sites (low sigma values) is not likely to be true [5]. For a special case of grain boundaries in equilibrium with solid-vapor surfaces in MgO for which one of the torque terms could be neglected, a reconstruction of the surface energy has been performed that includes torque terms for the solid-vapor surfaces [6]. The results from this analysis indicate that the torque terms are very important for determination of the map of surface energy. Torque terms arising from the inclination dependence of the boundary energy for internal triple junctions (i.e. between three grain boundaries) will be included in the analysis of (five parameter) grain boundary energy once larger data sets are available.

1.2 Current knowledge of grain boundary properties

Current knowledge of the properties of grain boundaries is limited to special boundary types. Furthermore much of the available literature concerns computer simulation. Thorough reviews by Wolf and Yip and, more recently by Sutton and Baluffi [7] provide excellent overviews of boundary structure and properties. Most experimental determinations of grain boundary energy have relied on a force balance between surface energy and grain boundary energy, e.g. [8]. In contrast to the literature on metals, one of the few contributions on oxides is that of Dhalenne et al. who measured thermal grooves on NiO for boundaries in <110> tilt boundaries [9]. Their results indicated little variation in energy for <100> high angle tilt boundaries. <110> tilt boundaries, however, showed cusps at the $\Sigma 9$ (221 plane), $\Sigma 11$ (311 plane) and the (expected) $\Sigma 3$ positions. The grain boundary energies were higher for <110> misorientations than for <100> boundaries, although this result assumed constant surface energy. The methods outlined above have been applied by Saylor and Rohrer to measurement of (relative) grain boundary energy in MgO [5], also using thermal grooves. As discussed elsewhere, they also found surface energy to be orientation (inclination) dependent in MgO [6]. More recently Otsuki [10,11] has conducted a series of experiments on aluminum in which intersections of grain boundaries with a solid-liquid surface were characterized. By assuming a constant solid-liquid interfacial energy and examining a variety of interfaces based on <100> and <110> rotation axes, information has been obtained on symmetric tilt, asymmetric tilt, twist and mixed character boundaries. The results also showed cusps at the $\Sigma 11$ (311 plane) and the (expected) $\Sigma 3$ positions, though not at the $\Sigma 9$ (221 plane) position. Their results also confirmed previous results for metals, i.e. that twist boundaries tend to have lower energies than the corresponding tilt boundaries. A concern with this approach is that a concentrated alloy is used as the liquid in order to avoid melting the (unalloyed) solid aluminum: solute will diffuse into the solid, especially along boundaries and the presence of the solute may perturb the boundary energy.

The mobility of grain boundaries has been extensively studied since the late 1950s [12-16] and an excellent review is available in the book by Gottstein and Shvindlerman [17]. Although this introduction cannot do justice to the topic, there is general agreement that low angle boundaries have much lower mobilities than high angle boundaries as a consequence of their discrete dislocation structure. This difference in behavior is also apparent in the generally higher activation energies for low angle boundary mobilities, corresponding to bulk diffusion mechanisms, than for high boundary mobilities. There is also considerable variation in mobility with boundary type for high angle boundaries. As with the case of energy, mobility has been mostly studied for bicrystal specimens containing a single boundary whose type generally close to a low index misorientation axis such as <100> or <111>. For the series of boundaries based on rotations about a <111> axis, simulation using molecular dynamics suggests that the CSL-related boundaries have low energy and high mobility relative to general boundaries. These simulations have been carried out with potentials for aluminum [18,19]. This limited range of boundary type for which information is available has motivated the present study of grain boundary properties over the whole fundamental zone.

2 Experimental Procedure: Al Foil

An Al foil sample of purity 99.98% was annealed at 550°C for 9 hours in a N_2 environment. This resulted in a columnar grain structure as observed by optical microscopy. The texture was dominated by a strong cube component,

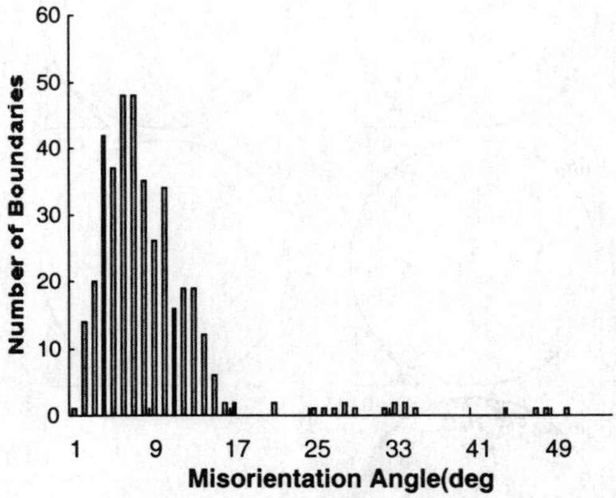

Fig. 2. Frequency of boundaries versus misorientation (°)

{100}<001>, with occasional randomly oriented grains. Since the columnar structure results in triple junctions that are nearly straight and perpendicular to the surfaces, the difficulty and error of measuring true dihedral angles and principal curvatures for each grain boundary by the serial sectioning technique is effectively eliminated. Crystallographic information for the grains adjacent to each triple junction was obtained by using orientation imaging microscopy in a scanning electron microscope. Analysis of the grain boundaries showed a predominance of low angle boundaries, Fig. 2. The full five-parameter nature of the boundaries was further analyzed by locating the boundary plane in a pole figure after determining the disorientation, Fig. 3. In this figure, the disorientation is plotted in the small triangular section in Rodrigues space in the lower right corner of each subplot. The first subplot, fig 3a, contains most of the boundaries because of the predominance of low angle boundaries. The boundary planes are plotted in the full circle pole figure (equal area projection), together with the disorientation axis in a unit triangle on the right hand side of the circle. Note that the full range of disorientation vectors is not plotted here because there were so few boundaries with higher values of R_3. The strong cube texture means that most of the misorientations are close to the origin in Rodrigues space and therefore plot in the first section, 3a.

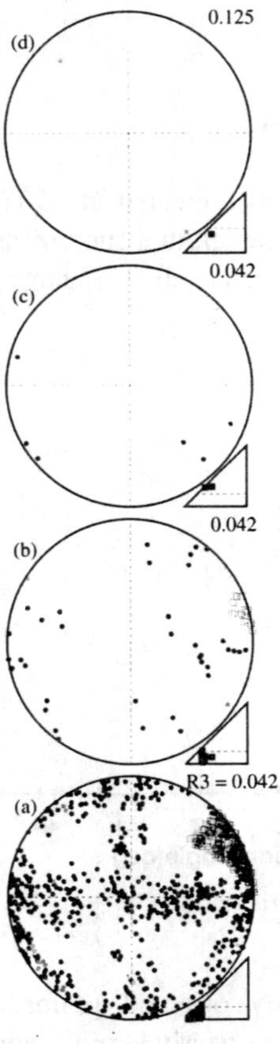

Fig. 3. Plot of the grain boundary plane (full circle pole figure in equal area projection) with disorientation (section of Rodrigues space). Most boundaries are low angle boundaries

in the first section, 3a, and the strong {001}<100> texture means that the boundary planes are confined to <010> zones.

Also the boundary planes are confined to the 001 zone(s) with very few {111} planes present in the sample. In order to extract the dihedral angles and curvatures, each triple junction was recorded on an individual scanning electron micrograph, Fig. 4a. Image processing methods [20] were used to skeletonize the grain boundaries, Fig. 4b, and curve fitting of a conic section to each boundary [21], Fig. 4c, was performed in order to extract the required quantities.

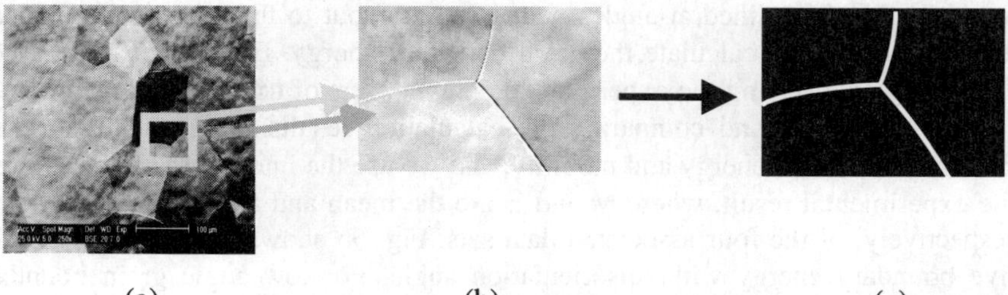

(a) (b) (c)

Fig. 4. (a) Back scatter contrast scanning electron micrograph of Al foil in plan view. (b) Magnified view of a triple junction. (c) Skeletonized image of boundaries comprising the triple junction in (b).

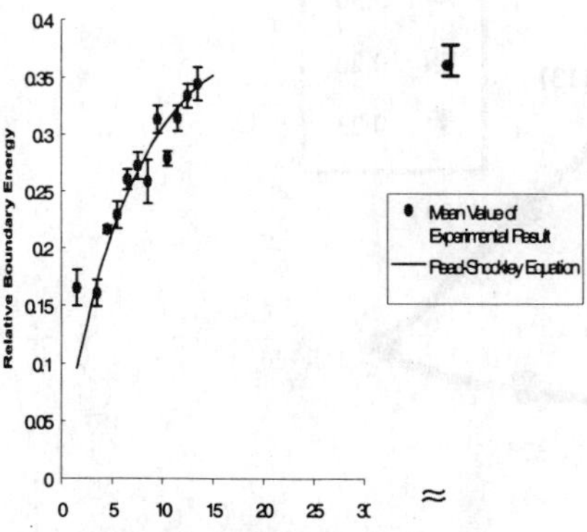

Fig. 5a. Plot of grain boundary energy versus misorientation angle (points) with Read-Shockley equation fit (line).

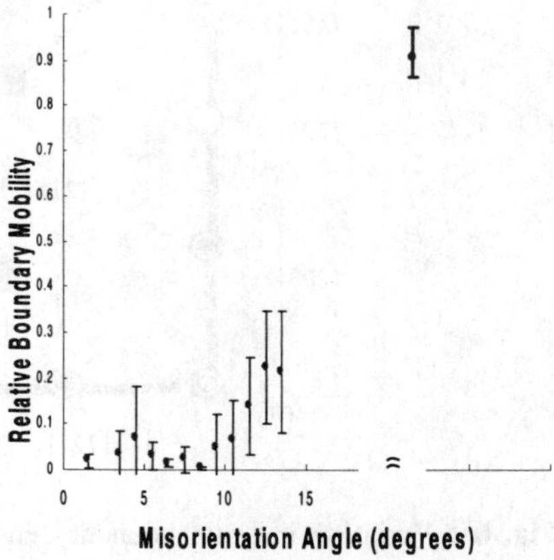

Fig. 5b. Plot of grain boundary mobility versus misorientation angle showing sharp transition for angles greater than 10°. The mobilities are scaled by a high angle mobility of one.

3 Results

3.1 Measurement of grain boundary energy and mobility in Al: as a function of misorientation angle

402 grain boundaries were characterized at 134 triple junctions in the Al foil. The boundaries were sorted into the 13 different types based on their misorientation angles. The misorientation angle distribution of the boundaries is shown in Fig. 5(a), which shows that most are low angle grain boundaries, i.e. misorientation angle $\theta < 15°$. The dihedral angles were used as input to the statistical/multiscale analysis [4] to first calculate the grain boundary energy. Error analysis was performed by subdividing the experimental data, i.e. sets of triple junctions, into two batches randomly, and comparing the calculation results. Four calculations are performed for both energy and mobility, and we use the interval $M \pm S$ to represent the experimental result, where M and S are the mean and the standard deviation, respectively, of the four associated data sets. Fig. 5b shows the variation of relative boundary energy with misorientation angle. For low angle grain boundary energies, the experimental data is fit well with the Read-Shockley equation; the

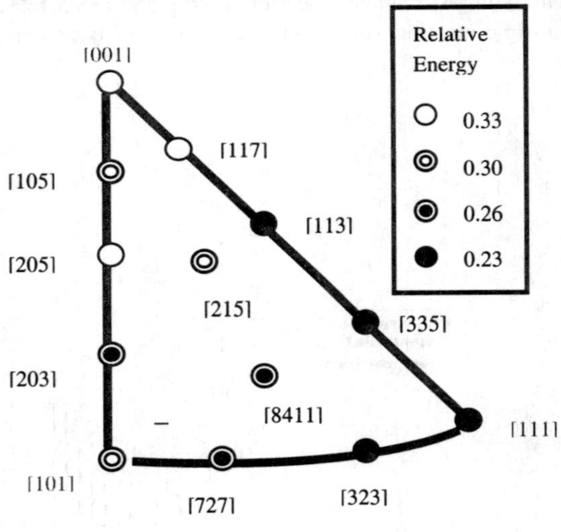

Fig. 6a. Variation in grain boundary energy for low angle boundaries as a function of misorientation axis

average relative energy for high angle grain boundaries is 0.37. A similar statistical calculation was performed based on Eq. (3) to obtain grain boundary mobility as a function of misorientation angle as shown in Fig. 5c. The results show that high angle grain boundaries are much more mobile than low angle boundaries and that there is a sharp transition in mobility in the range 10-15°.

3.2 Measurement of grain boundary energy and mobility in Al: as a function of misorientation axis

In a second analysis, only low angle grain boundaries, i.e. boundaries with misorientation angles less than 15°, were included, and the boundaries were relocated to their physically equivalent positions within the fundamental zone, i.e. the standard stereographic triangle, 001-101-111. All sorted into 13 different types, chosen to give uniform coverage of the standard stereographic triangle. Each boundary was assigned to a particular type by finding the nearest axis from the list of 13 types. As for the previous analysis, the resolution was limited by the size of the experimental data set. The same statistical analysis for the calculation of boundary energies and mobilities was performed as used in the previous section

Fig. 6a shows the misorientation axis dependence of the boundary energy. Comparing the three low Miller index axes in the corners of the triangle, <001>, <101>, and <111>, it was found that $\sigma_{[001]} \approx \sigma_{[101]} > \sigma_{[111]}$. Fig. 6b shows the variation of low angle boundary mobility with the boundary misorientation axis. <111> type boundaries are much more mobile than <101> and <001> type boundaries, i.e. $M_{[111]} > M_{[101]} \gg M_{[001]}$. Bauer and Lanxner also theoretically predicted and experimentally demonstrated [22] that boundaries with <111> and <101> much more mobile than <001> type boundaries by a factor of approximately 150, which is consistent with the results of the current work. If

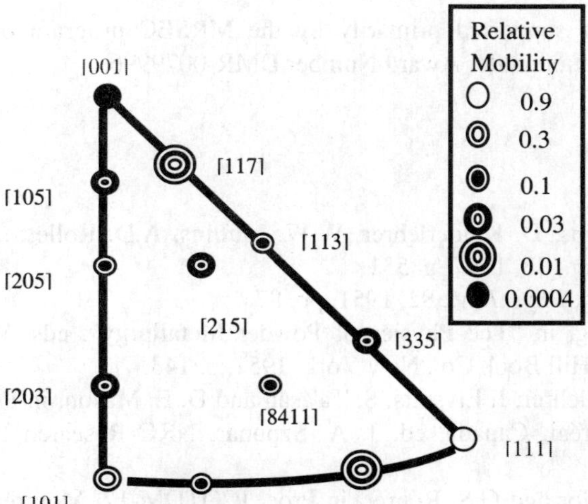

Fig. 6b. Variation in grain boundary mobility for low angle boundaries as a function of misorientation axis

symmetric tilt boundaries are considered whose mobility is limited by diffusion between adjacent (parallel) dislocations comprising the boundary, then <001> tilt boundaries are seen to be less mobile than either <110> or <111> tilt boundaries because the Burgers vectors are inclined at approximately 45° to the plane in the former case. In the latter types (<110> and <111> tilts) the Burgers vectors are nearly normal to the boundary plane which means that much less diffusion is required in order to advance the boundary and maintain the equilibrium dislocation spacing.

4 Summary

The method of extracting grain boundary energy and mobility from triple junction geometry and crystallography has been demonstrated for a sample of aluminum foil. The foil was annealed in order to produce a columnar grain structure which allowed all the required information to be obtained from a single section. The strongly (cube) textured material contained a large fraction of low angle boundaries. The variation in energy of these boundaries was found to follow the Read-Shockley equation, with a mild dependence on the misorientation axis. The mobility of the boundaries was found to depend strongly on both misorientation angle and on misorientation axis.

Acknowledgement

This work was supported primarily by the MRSEC program of the National Science Foundation under Award Number DMR-0079996.

References

[1] B.L. Adams, D. Kinderlehrer, W.W. Mullins, A.D. Rollett and S. Ta'asan, *Scripta mat.*, 38, 1997, p. 531.
[2] C. Herring, *Phys. Rev.*, 82, 1951, pp. 87.
[3] C. Herring, in "The Physics of Powder Metallurgy", eds. W.E. Kingston, McGraw-Hill Book Co., New York, 1951, p. 143.
[4] D. Kinderlehrer, I. Livshits, S. Ta'asan and D. E. Mason, in Proc. ICOTOM-12, Montréal, Canada, ed. J. A. Szpunar, NRC Research Press, 2, 1999, p. 1643.
[5] D.M. Saylor and G.S. Rohrer, in Proc. ICOTOM-12, Montréal, Canada, ed. J.A. Szpunar, NRC Research Press, 2, 1999, p. 1690.
[6] D.M. Saylor, D.E. Mason and G.S. Rohrer, *J. Am. Ceram. Soc.*, 83, 2000, p. 1226.
[7] P. Sutton and R.W. Balluffi, in "Interfaces in Crystalline Materials"., ed. R.J. Brook, et al., Clarendon Press, Oxford, UK, 1995.
[8] N.A. Gjostein and F.N. Rhines, *Acta metall.*, 7, 1959, p. 319.

[9] G. Dhalenne, M. Dechamps and A. Revcolevschi, in Proc. Advances in Ceramics, eds. M.F. Yan, A.H. Heuer, American Ceramic Society, Columbus, OH, 6, 1983, p. 138.

[10] A. Otsuki, *Mater. Sci. Forum*, 207-209, 1996, p. 413.

[11] A. Otsuki, H. Yato and I. Kinjyo, *Mater. Sci. Forum*, 126-128, 1993, p. 285.

[12] J.W. Rutter and K.T. Aust, *Trans. Met. Soc. AIME*, 218, 1960, p. 682.

[13] J.W. Rutter and K.T. Aust, *Acta metall.*, 13, 1960, p. 181.

[14] B.B. Rath and H. Hu, *Trans. Met. Soc. AIME*, 236, 1966, p. 1193.

[15] R. Viswanathan and C.L. Bauer, *Acta metall.*, 21, 1973, p. 1099.

[16] R.C. Sun and C.L. Bauer, *Acta metall.*, 18, 1970, p. 639.

[17] G. Gottstein and L.S. Shvindlerman, "Grain Boundary Migration in Metals"., CRC Series in Materials Science & Technology CRC Press, Boca Raton, FL, 1999.

[18] D. Srolovitz, M. Upmanyu, L.S. Shvindlerman and G. Gottstein, *Acta mater.*, 47, 1999, p. 3901.

[19] M. Upmanyu, D. Srolovitz and R. Smith, *Int. Sci.*, 6, 1998, p. 41.

[20] A. Talukder, D. Casasent and S. Ozdemir, in Proc. ICOTOM-12, Montréal, Canada, eds. J. A. Szpunar, NRC Research Press, 1, 1999 p. 243.

[21] C. Yang, Extraction of Grain Boundary Energies and Mobilities from Triple Junction Geometry, Ph.D., Carnegie Mellon 2000.

[22] C.L. Bauer and M. Lanxner, in Proc. JIMIS-4, Minakami, Japan, Y. Ishida, Eds., Trans. Japan Inst. Metals, 27, 1986, p. 411.

Triple Junction Motion in Metals

L.S. Shvindlerman[1,2], G. Gottstein[1], D.A. Molodov[1], and V.G. Sursaeva[2]

1 Institut für Metallkunde und Metallphysik, RWTH Aachen, Kopernikusstr. 14, D-52056 Aachen, Germany
2 Institute of Solid State Physics, Russian Academy of Sciences, Chernogolovka, Moscow Distr., 142432 Russia

Keywords: grain boundary, triple junction, migration, mobility.

Abstract. The motion of grain boundary systems with triple junctions and the interaction of these structural elements of a polycrystal are considered. The experimental results demonstrated that the motion of grain boundary systems with triple junctions can be controlled by slowly moving triple junctions. In a certain temperature interval these structural elements of a polycrystals can drag each other. The experiments reveal a drastic difference between activation enthalpy of grain boundary and triple junction motion. The motion of the same triple junction in opposite directions was found to be essentially different as well. Computer experiments carried out by molecular dynamics showed that the mobility of the triple junction is reduced when the straight grain boundary is a low Σ boundary. The process of grain growth in 2D systems is analyzed with respect to the controlling kinetics: from solely boundary kinetics, when grain growth in a polycrystal is determined by the Von Neumann-Mullins relation [1,2], to exclusively triple junction kinetics, when grain growth is governed by the mobility of triple junctions.

1 Introduction

Although the number of triple junctions in polycrystals is comparable in magnitude with the number of boundaries, all peculiarities in the behaviour of polycrystals during grain growth were solely attributed to the motion of grain boundaries so far. It was tacitly assumed in theoretical approaches, computer simulations and interpretation of experimental results that triple junctions do not disturb grain boundary motion and that their role in grain growth is reduced to preserve the thermodynamically prescribed equilibrium angles at the lines (or the points for 2D systems) where boundaries meet. The most prominent example of how this assumption determines the fundamental concepts of grain structure evolution gives the Von Neumann-Mullins relation, which defines the rate of change of the grain area in the course of grain growth. No doubt this relation forms the basis for practically all theoretical and experimental investigations as well as computer simulations of microstructure evolution in 2D polycrystals in the course of grain growth [3-6]. This relation is based on three essential assumptions, namely: (i) all grain boundaries possess equal mobility and surface tension, irrespective of their misorientation and crystallographic orientation of the boundaries; (ii) the mobility of a grain boundary is independent of its velocity; (iii) triple junctions do not affect grain boundary motion; therefore, the contact angles at triple junctions are in equilibrium and, due to the first assumption, are equal to 120°.

For 2D systems the rate of change of the grain area S can be expressed by

$$\frac{dS}{dt} = -\oint V d\ell \qquad (1)$$

where V is the grain boundary migration rate, $d\ell$ is an element of grain perimeter. For grain growth

$$V = \sigma m_b K \equiv A_b K \qquad (2)$$

where m_b is the grain boundary mobility, σ is the grain boundary surface tension, K is the local curvature of the grain boundary and equal to

$$K = \frac{d\varphi}{d\ell} = -\frac{y''}{\left[1+(y')^2\right]^{3/2}} \qquad (3)$$

where φ is the tangential angle at any given point of the grain boundary and y(x) is the shape of the boundary . From Eqs. (1)-(3) follows

$$\frac{dS}{dt} = -A_b \oint d\varphi \qquad (4)$$

If the grain were bordered by a smooth line, the integral in Eq. (4) would equal 2π. However, owing to the discontinuous angular change at every triple junction, the angular interval $\Delta\varphi=\pi/3$ is subtracted from the total value 2π for each triple junction. Consequently

$$\frac{dS}{dt} = -A_b\left(2\pi - \frac{n\pi}{3}\right) = \frac{A_b\pi}{3}(n-6) \qquad (5)$$

where n is the number of triple junctions for each respective grain, i.e. the topological class of the grain. Thus, the rate of area change is independent of the shape of the boundaries and determined by the topological class n only. Grains with n > 6 will grow and those with n < 6 will disappear [2].

The existence of triple junctions drastically affects the kinetics of grain growth. To discuss this problem quantitatively the mobility of a triple junction should be measured.

2 How to Measure the Triple Junction Mobility

There are two possible ways to measure the triple junction mobility. Similar to grain boundaries, reliable data of triple junction mobility can be obtained in the course of steady-state motion. However a steady-state motion of a grain boundary system with a triple junction is only possible in a small set of geometrical configurations. Two of them are shown in Figs. 1 and 2 [7].

These grain boundary systems can move steadily. In the framework of a 2D uniform grain boundary model the shape and velocity of moving grain boundaries with a triple junction can be calculated analytically [7,8].

These special boundary systems were investigated in [8-10] under three main assumptions. Two of them comply with the assumptions (1) and (2) of the Von Neumann-Mullins consideration, while the third one is determined by Eq. (2): the normal grain boundary displacement rate v is proportional to the grain boundary curvature K. As shown in [1, 8], the model grain boundary systems (Figs. 1 and 2) can move steadily, and the analysis of their motion permits us to understand the influence of the finite mobility of a triple junction on the migration of grain boundaries.

The strict solution for the shape of a boundary-triple junction configuration (Fig. 1) can be expressed as

$$y(x) = \xi \arccos\left(e^{-x/\xi + c_1}\right) + c_2$$

$$\xi = a/2\Theta;\ c_1 = \ln(\sin\Theta);\ c_2 = -\xi(\pi/2 - \Theta) \tag{6}$$

The velocity V of steady-state motion of the system is

$$V = \frac{2\Theta\, m_b \sigma}{a} \tag{7a}$$

On the other hand the velocity of steady-state motion of the triple junction can be represented as:

$$V = m_{tj}\sigma(2\cos\theta - 1) \tag{7b}$$

where m_{tj} is the triple junction mobility.

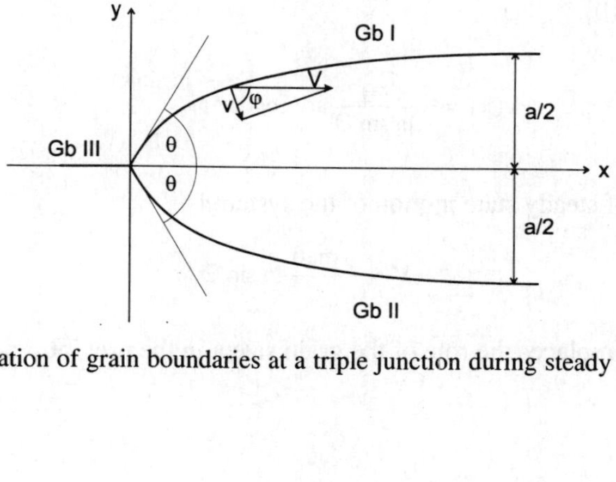

Fig. 1. Configuration of grain boundaries at a triple junction during steady state motion for n < 6 [7].

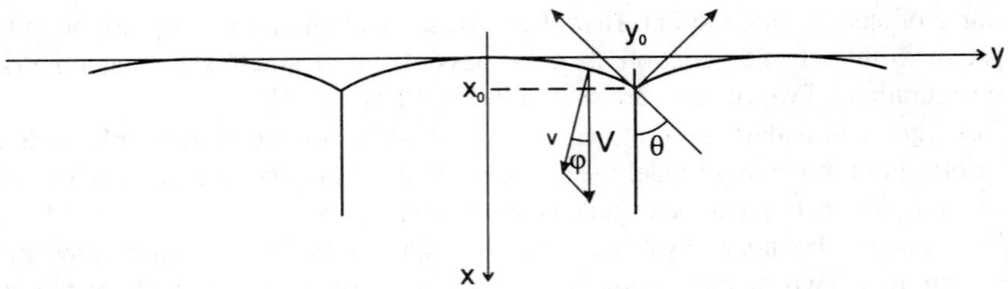

Fig. 2. Configuration of grain boundaries at triple junctions during steady state motion for n > 6 [7].

An important relation which links the steady-state value of the angle θ at the vertex of a triple junction with the dimensionless criterion $\Lambda = m_{tj}a/m_b$ results from the Eqs. (7a) and (7b)

$$\frac{2\Theta}{2\cos\Theta - 1} = \frac{m_{tj}a}{m_b} = \Lambda \tag{8}$$

If a triple junction is mobile and does not drag grain boundary motion, the criterion $\Lambda \to \infty$ and $\Theta \to \pi/3$, i.e. the equilibrium angular value at a triple junction in the uniform grain boundary model. In contrast, however, when the mobility of the triple junction is relatively low (strictly speaking, when $m_{tj}\ a \ll m_b$) then $\Theta \to 0$ (Fig. 3).

It should be stressed that the angle Θ is strictly defined by the dimensionless criterion Λ, which, in turn, is a function of not only the ratio of triple junction and grain boundary mobility, but of the grain size as well.

As can be seen from Fig. 1, the model configuration correlates to grains in a polycrystal with less than 6 neighbours (adjacent grains), in other words, the topological class of the grain is smaller than 6.

The grain boundary configuration outlined in Fig. 2 corresponds to grains with topological class greater than 6 [1,10]. The steady state motion of this system can be described by the same set of equations, only with different boundary and initial conditions [1,10]:

$$y(x) = -\frac{x_0}{\ln\sin\Theta}\arccos\left(e^{\frac{x}{x_0}\ln\sin\Theta}\right) \tag{9}$$

The velocity of steady state motion of the system is

$$V = -\frac{m_b\sigma}{x_0}\ln\sin\Theta \tag{10}$$

The length x_0 replaces the role of the grain size a in the previous case (Fig. 1) or

$$y_0 = y(x_0) = -\frac{x_0}{\ln \sin \Theta} \arccos\left(e^{\ln \sin \Theta}\right) = -\frac{x_0}{\ln \sin \Theta}(\pi/2 - \Theta) \tag{11}$$

The criterion Λ, which describes the influence of the triple junction mobility on grain boundary migration now reads

$$-\frac{\ln \sin \Theta}{1-2\cos \Theta} = \frac{m_{tj}\, x_0}{m_b} = \Lambda 2 \tag{12}$$

Obviously, for $\Lambda >> 1$, when the boundary mobility determines the kinetics of the system the angle Θ tends to its equilibrium value ($\pi/3$).
Again, the angle Θ changes when a low mobility of the triple junction starts to drag the motion of the boundary system. However, as can be seen from Eq. (12) and Fig. 3, in this case the steady state value of the angle Θ increases as compared to the equilibrium state. (Otherwise the triple junction would move in the negative direction of the x-axis, increasing the free energy of the system.)

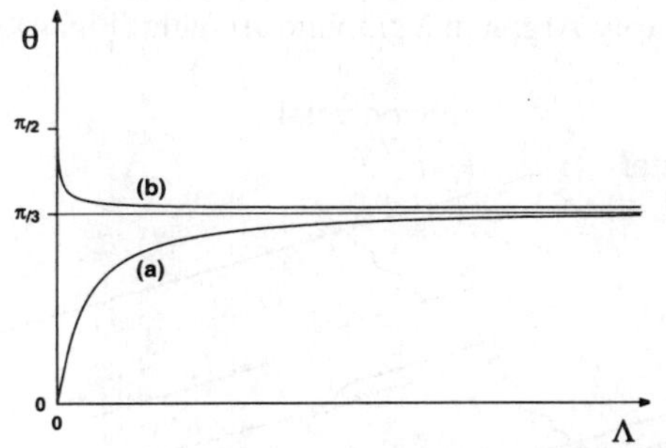

Fig. 3. Angle Θ as a function of Λ. (a) for $n < 6$, Eq. (8); (b) for $n > 6$, Eq. (12).

For $\Lambda << 1$ the angle Θ (Eq. (12)) tends to approach $\pi/2$. The dependency $\Theta = \Theta(\Lambda)$ for both $n < 6$ and $n > 6$ are shown in Fig. 3.
One can see that such a steady-state motion will take place for the so-called "symmetrical" configuration as well (Fig. 1), where both curved grain boundaries I and II are the same and different from the straight boundary III. In this case the relations (7b) and (8) yield

$$V = m_{tj}\left(2\sigma\cos\theta - \sigma_{III}\right) = m_{tj}\sigma\left(2\cos\theta - \frac{\sigma_{III}}{\sigma}\right) \tag{13}$$

$$\Lambda = \frac{m_{tj}a}{m_b} = \frac{2\theta}{2\cos\theta - \frac{\sigma_{III}}{\sigma}} \tag{14}$$

The grain boundary-triple junction configurations considered provide a basis for experimental investigations of triple junction mobility.

3 Experimental

In order to study the triple junction motion we must not only fix the displacement but also the angles at the tip of the half-loop. So a device was designed which makes it possible to observe and record the motion of the grain boundary system with triple junction. Experiments on Zn tricrystals were performed in an optical microscope with polarized light and a hot stage for observation of grain boundary motion at high temperatures. A video camera attached to the microscope recorded the motion of the boundary system. To protect the sample from oxidation, it was kept in a nitrogen atmosphere during the experiment. The experiments were carried out in the temperature range 300°C-410°C. For each temperature the velocity of the triple junction, the angle Θ and the width a of the half loop were continuously measured. For this single video frames of the recorded video tapes were grabbed by a computer and the shape of the triple junction system was determined.
Zn tricrystals with a triple junction were grown by a directed crystallization technique in high purity Argon in a graphite crucible (Fig. 4).

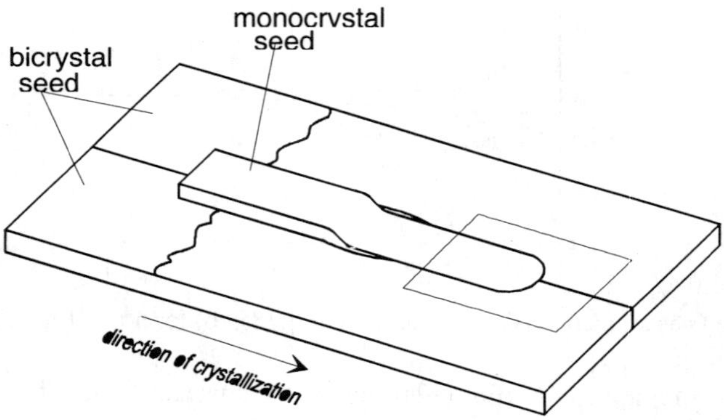

Fig. 4. Scheme of growing a Zn tricrystal; the inserted square indicates the used tricrystal.

4 Results and Discussion

Let us consider briefly the results of grain boundary and triple junction motion in specially grown tricrystals of Al and Zn [7, 9-12]. The motion of different triple junctions with a grain boundary configuration as in Fig. 1 formed by two high-angle and one low-angle boundary were investigated in the temperature range between 300°C and 410°C (Zn) and 400–600°C (Al). The shape of the moving grain boundary system was similar to the shape predicted by theory [10].
In [10] grain boundary motion in tricrystals (configuration Fig. 1) of pure (99.999%) Zn was studied. Fig. 5 reveals the a strong temperature dependence of

the angle Θ at the apex of the triple junction. This cannot be attributed to the temperature dependence of the surface tension. Actually, the surface tension temperature coefficient is much higher for a high-angle grain boundary than for a low angle boundary, like in the case shown in Fig. 5. Thus the change of Θ for the investigated sample would have to be opposite to that observed in the experiment, i.e. the angle Θ has to decrease with increasing temperature. Also, the order of change of Θ is too large to be explained by thermodynamical reasons in terms of different temperature coefficients of the surface tension. The function Θ(T) consequently is the result of a change of the kinetics of motion, indicated by the temperature dependence of Λ (Fig. 5) [10]. As can be seen at low temperatures Λ is of the order of unity (Fig. 6), and the motion of the boundaries is dragged by the hardly mobile triple junction. Accordingly, the value of Θ is lower than in the equilibrium state (Fig. 5), and the motion of the grain boundary system is controlled by the mobility of the triple junction. With increasing temperature the triple junction becomes more mobile with respect to the grain boundary as indicated by an increasing value of Λ (Fig. 5,6). In other words, two different branches of grain boundary motion were observed, indicated by the change of the angle Θ with temperature. Qualitatively similar results were obtained in experiments on high purity Al [11], where a <111> tilt grain boundary system with triple junction was formed by two curved high-angle boundaries and one low-angle straight boundary. In Fig. 6 the temperature dependence of the criterion Λ and the reduced grain boundary (A_b) and triple junction (A_{tj})mobility:

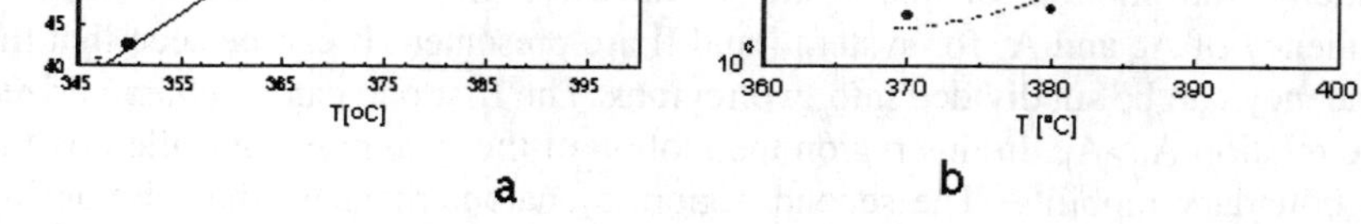

Fig. 5. Temperature dependence of the angle Θ in the vertex of a triple junction (configuration Fig. 1) (a) and of the criterion Λ (b) in Zn tricrystals investigated [10].

$$A_b \equiv m_b \sigma = \frac{V \cdot a}{2\theta} = A_{0b} e^{-H_b/kT} \qquad (15)$$

$$A_{tj} \equiv m_{tj}\sigma a = \frac{V \cdot a}{2\cos\theta - \sigma_{III}/\sigma} = A_{0tj} e^{-H_{tj}/kT}$$

for the boundary systems discussed are presented [11]. Fig. 6, in particular, reveals clearly the defined transition from triple junction kinetics at low temperatures to grain boundary kinetics at elevated temperatures, and an essential difference in the activation enthalpy for triple junction motion (H_{ij}) and grain boundary migration (H_b). The behavior of the grain boundary controlled branch in Fig. 7 compares well with measurements of the reduced grain boundary mobility obtained from literature data of independent bicrystal experiments [7]. Actually, the configuration (Fig. 1) without a triple junction constitutes a grain boundary half-loop, the theory and experimental feature of which are well elaborated [7,13]. Essentially, the problem is as follows: the motion of the grain boundary half-loop in Zn with 99.995% purity was studied. Independently, the motion of practically the same half-loop but now attached to the low-angle straight boundary in configuration Fig. 1 was investigated. "Practically the same" means that the incorporation of the low-angle grain boundary changes the misorientation of curved grain boundaries only very moderately. Hence, we can conduct a direct comparison between the motion of a grain boundary half-loop and the motion of the same half-loop with a triple junction. In other words, we can compare the mobility of a moving grain boundary system with and without the triple junction. Two tilt grain boundary systems were studied: $<11\bar{2}0>$ with misorientation of the curved boundaries (half-loop) 84°, and $<10\bar{1}0>$, the misorientation of the curved boundaries was 62°, the misorientation of the straight boundary in both cases was about 3°. Firstly, we would like to discuss the dependency of the reduced displacement $a \cdot \ell(t)$ on time for grain boundary half-loop and half-loop with triple junction (Fig. 7). Since half-loops (and half-loops with a triple junction) of different widths were studied experimentally, the comparison of the product $a \cdot \ell(t)$ compensates for the discrepancy between the major parameter of driving force – the width a. The experiment proves that the triple junction drags essentially the motion of the grain boundaries. In Fig. 8 the temperature dependency of A_b and A_{tj} for system I and II are presented. It can be seen that the dependency can be subdivided into two regions. The first one can be characterized by the relation $A_{tj}>A_b$. In this region the motion of the system is controlled by the grain boundary mobility. The second region is characterized by the relationship $A_{tj}<A_b$. In this region the motion of the grain boundary system is determined by the triple junction mobility. In Fig. 8a the positions of some experimental points that determine the motion of the grain boundary system with triple junction were recalculated according to the equation for boundary kinetics (Eq. (7)). They are the points located above the intersection point with the line $A_b(1/T)$, the mobility of the half-loop. Obviously the recalculated points open circles - are in good agreement with the half-loop mobility (solid circles).

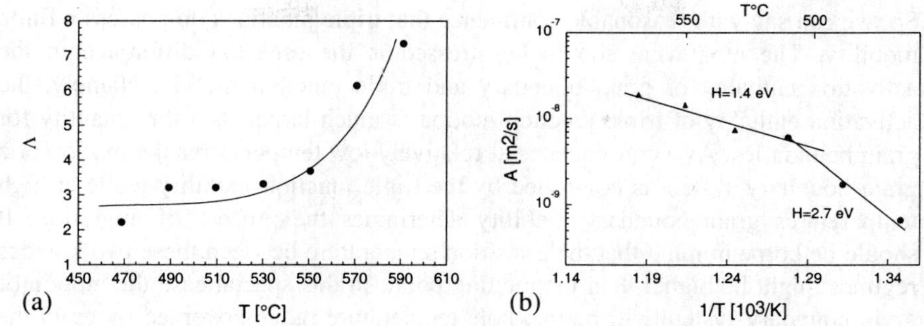

Fig. 6. Temperature dependence of the criterion Λ (a) and of triple junction (●) and grain boundary mobility (▲) (b) for a <110> tilt grain boundary system in pure Al [11].

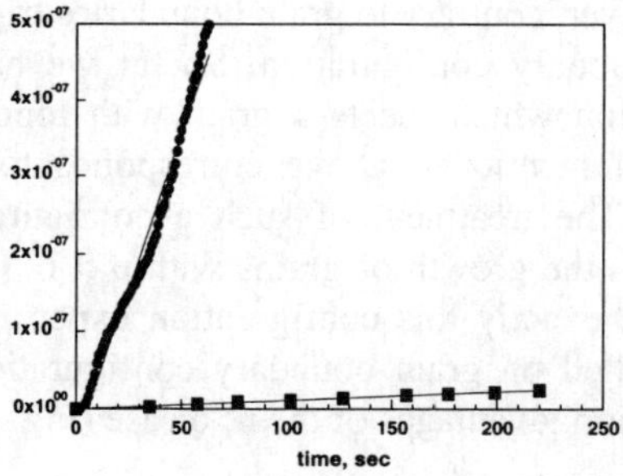

Fig. 7. Time dependence of the reduced displacement $a \cdot \ell(t)$ for a $<11\bar{2}0>$ tilt grain boundary half-loop (solid circles) and triple junction configuration (solid squares) at T=390°C [12].

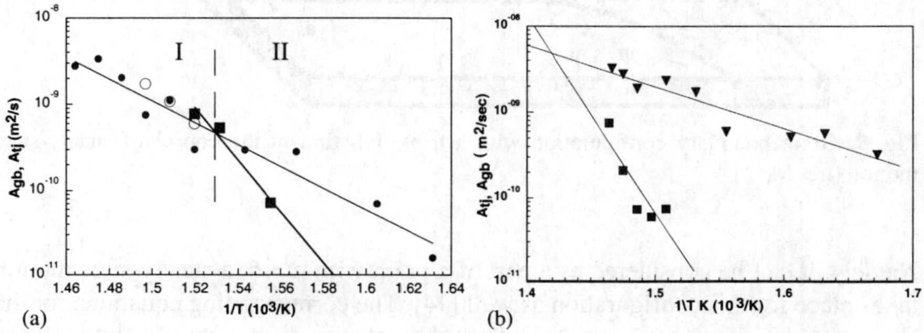

Fig. 8. Temperature dependence of the reduced grain boundary mobility A_h (solid circles) and reduced triple junction mobility A_{tj} (solid squares) for the system $<10\bar{1}0>$ (a) and $<11\bar{2}0>$ (b): reduced grain boundary mobility A_h (solid triangles) and reduced triple junction mobility A_{tj} (solid squares) [12].

So, we can say with reasonable confidence that triple junctions do possess a finite mobility. The next what should be stressed is the essential difference in the activation enthalpy of grain boundary and triple junction motion. Namely, the activation enthalpy of triple junction motion is much larger then this quantity for grain boundaries. As a consequence at relatively low temperatures the motion of a grain boundary system is controlled by the triple junction mobility while at high temperatures grain boundary mobility determines the process of migration. It should be borne in mind that the transition temperature between these two kinetics regimes might be higher than the melting point. In this special case the motion of grain boundary system will be in whole temperature range governed by grain the boundary mobility.
Similar to grain boundaries, one might expect a dependence of triple junction mobility on its structure, on the structure of the grain boundaries which form the triple junction. However, contrary to grain boundaries triple junction mobility may depend on grain boundary configuration. So far we have considered the grain boundary configuration which meets a grain with topological class n < 6. The second configuration mentioned above corresponds to grains with topological class n>6 (Fig. 2). The treatment of such a configuration is of great interest because it determines the growth of grains with n > 6, however there are serious difficulties to realize exactly this configuration experimentally. That is why the authors of [14] dwelled on grain boundary configuration represented in Fig. 9, which retains the major advantages of the n>6 case (Fig. 2).

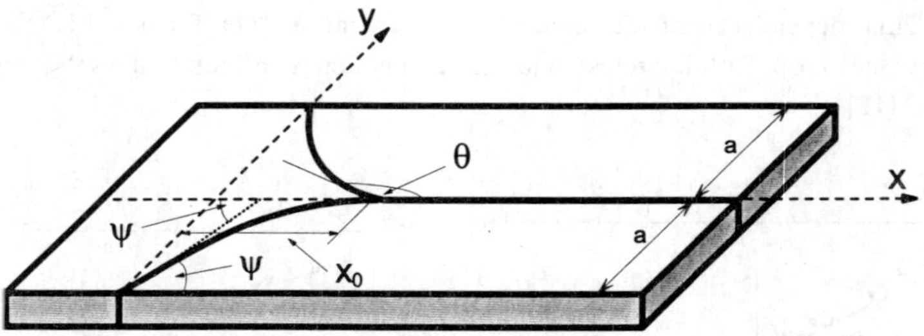

Fig. 9. Grain boundary configuration with a triple junction in the course of steady-state motion (n > 6).

Namely, it can be considered as a part of a grain with n > 6, a steady-state motion takes place for this configuration as well [14]. The corresponding equations for the shape y(x) of a moving grain boundary, the velocity V of a steady-state motion and the criterion Λ read

$$y(x)=\xi\arccos\left(e^{-\frac{x}{\xi}+C_1}\right)+C_2,$$

$$\xi=-\frac{x_0}{\ln\dfrac{\sin\theta}{\sin\psi}};\quad C_1=\ln\sin\psi;C_2=\frac{x_0}{\ln\dfrac{\sin\theta}{\sin\psi}}\left(\frac{\pi}{2}-\theta\right)\tag{16}$$

$$V=-\frac{m_b\sigma}{x_0}\ln\frac{\sin\theta}{\sin\psi}\tag{17}$$

$$\Lambda=\frac{m_{tj}x}{m_b}=-\frac{\ln\dfrac{\sin\theta}{\sin\psi}}{1-2\cos\theta}\tag{18}$$

A special feature of Eqs. (16)-(18) lies in the fact that they are determined not only by the angle at the apex of the triple junction but on the angle ψ as well. The angle ψ is defined by the equilibrium between the surface tension of the crystal and grain boundary at the face of the crystal (Fig. 9) [7]. One can see that such a grain boundary and triple junction configuration exist within the region of the angles θ qual to ($(\pi/3-\pi/2)$). At $\theta\to\pi/3$ on the source side of large angles $\Lambda\to\infty$, at $\theta\to\pi/2$, $\Lambda\to\ln\sin\psi$.

Thus, the configuration considered shows that the transition from boundary kinetics to triple junction kinetics can be dictated not only by grain boundary and triple junction mobility and the grain size but by the "situation" in the vicinity of the triple junction as well. It is significant that such a situation can be caused by adjacent grain boundary triple junctions. This is a radically new conclusion which have refined our understanding of triple junction behavior [14]. As it follows from (17), (18) at $\theta\to\psi$, $\Lambda\to 0$ the system transfers to triple junction kinetics. The situation corresponding to $\theta>\psi$ is specific for configuration Fig. 1. In [14] the motion of <100> tilt grain boundary system with a triple junction in two configurations - Fig. 1 and Fig. 9 respectively - was studied. It was found that the grain boundary system in configuration Fig. 9 is immobile while the same system in configuration Fig. 1 yields a high mobility [14]. In other words, to mobility of a triple junction is determined also by its configuration.

A comprehensive investigation of grain boundary and triple junction migration by atomistic simulations was undertaken in [15-18]. It was shown that triple junctions do possess a finite mobility, and this mobility is a function of misorientation of grain boundaries which form the triple junction and of the geometry of grain boundary configuration.

5 The Effect of Triple Junction Drag on Von Neumann-Mullins Relation.

It was shown that triple junctions do possess a finite mobility. As repeatedly mentioned the Von Neumann-Mullins relation is the reflection of the assumption that triple junctions are mobile. The rate of area change for a grain with n < 6 can be expressed as

$$\frac{dS}{dt} = -m_b \sigma \oint d\varphi = -A_b\left(2\pi - n(\pi - 2\Theta)\right) = A_b(\pi - 2\theta)\left(n - \frac{2\pi}{\pi - 2\Theta}\right) \qquad (19)$$

Since the limited mobility of the triple junction reduces the steady state value of the angle Θ as compared to the equilibrium angle, the shrinking rate of grains with n < 6 decreases, as obvious for the case when the mobility of the triple junction becomes very low. In other words, for grains with n < 6 the influence of the triple junction mobility slows down the process of grain structure evolution, decreasing the vanishing rate of grains with small topological class (n < 6).

For grains with topological class greater than 6 (Fig. 2) the dimensionless parameter Λ which describes the influence of the triple junction mobility on grain boundary migration for such a system, is given by Eq. (12). When a low mobility of the triple junction starts to drag the motion of the boundary system, the angle Θ changes. However, in this case, the steady state value of the angle Θ increases as compared to the equilibrium state. Such an increase of the angle Θ also decreases the magnitude of (n - 2 Θ) in Eq. (19), in other words, it decreases the "effective" magnitude of the topological class of the growing grain with n > 6. Consequently, microstructural evolution will slow down due to triple junction drag for any n-sided grain.

The only exception holds for n = 6, since a hexagonal grain structure becomes unstable when the contact angle $2\Theta \neq 2\pi/3$. Since the actual magnitude of θ is determined by the triple junction and grain boundary mobility as well as the grain size and is independent of the number of sides of a grain, there is no unique dividing line between vanishing and growing grains with respect to their topological class anymore, like n = 6 in the Von Neumann-Mullins approach.

This section discusses the rate of change of a grain area S when the motion of grain boundaries is controlled by the motion (mobility) of triple junctions. As is shown in [19], in case when grain growth is determined by the mobility of grain boundary triple junctions (triple junction kinetics) all grains are transformed into regular polygons in the course of grain growth. The only exceptions are triangles: they collapse without transforming into a regular polygon.

The respective relation for the rate of a change of the grain area under triple junction kinetics will be discussed with regard to microstructure evolution. Firstly it was established that under triple junction control in the course of grain growth in 2D systems the grains will eventually be bordered by straight lines, i.e. they will assume a polygonal shape [19]. Actually, the expression for the curvature K can be found from (3), (6), (9). For the system shown in Fig. 1 we obtain [19]:

$$K=\frac{1}{\xi}e^{-x/\xi+\ln\sin\Theta}=\frac{2\Theta}{a}e^{-\frac{2\Theta}{a}x}\sin\Theta=\frac{\Lambda}{a}\sin\Theta(2\cos\Theta-1)e^{-\frac{2\Theta}{a}x} \qquad (20a)$$

while for the system shown* in Fig. 2

$$K=\frac{\ln\sin\Theta}{x_0}e^{\frac{x}{x_0}\ln\sin\Theta}=\frac{\Lambda}{x_0}(1-2\cos\Theta)e^{\frac{x}{x_0}\ln\sin\Theta} \qquad (20b)$$

Since for triple junction kinetics $\Lambda\to 0$, also the grain boundary curvature K approaches zero, i.e. the grain structure of 2D polycrystals comprises straight grain boundaries which extend between the triple junctions. In other words, the grains in a 2D polycrystal represent a system of polygons. More specifically, as shown in [19], in the framework of triple junction kinetics a polygon of arbitrary shape will be transformed in an equilateral polygon, and any deviation from an equilateral polygon will generate a force to restore the equilibrium. The only exception is a triangle. In other words, a grain with topological class $n = 3$ is always instable and must disappear. Finally, all other shrinking polygons must by necessity go through this stage eventually. We like to stress that we are considering triple junction kinetics, when grain growth is controlled by the motion of triple junctions.

We reason that this phenomenon has important consequences for the development of grain growth. Let us take a look at the evolution of a shrinking grain in the course of grain growth. The topological class of such a grain should be smaller than 6, naturally taking into account all corrections to the Von Neumann-Mullins relation. As shown above the transition between boundary and triple junction kinetics does not depend on grain boundary and triple junction mobility, but on the size of a grain as well. When the size of a grain diminishes progressively there comes a time where boundary kinetics becomes replaced by junction kinetics. This will happen to grains of the topological class $n = 4$ or $n = 5$ which are bound to shrink even after such a transition to triple junction kinetics. Grains of topological class $n = 3$ will collapse without transforming into a regular polygon. Since the kinetics of triple junctions is significantly slower than the boundary kinetics, the four- and five-side polygons will shrink, and eventually converge to a point although at a markedly smaller rate. Experimentally this phenomenon will manifest itself in the mean value of the topological class of vanishing grains. In Fig. 10 experimental data of grain growth in Aluminum foil with 2D (columnar) structure is presented, in terms of the grain size dependence of the mean topological class <n> [20,21]. Extrapolation of this experimental dependence to zero area yields the mean value of the topological class of vanishing grains. As

* We would like to note that for the grain boundary system presented in Fig. 2 triple junction control causes $\Theta\to\pi/2$.

can be seen <n>(0) = 4.2 [20] and <n>(0) = 4.0 [21] respectively., i.e. n = 4 is the smallest topological class to shrink in a stable manner. (Figs. 10a,b).
The change of its area dS can be represented as [19]

$$\frac{dS}{dt} = \int V_n d\ell = -V_n \int d\ell = V_n P \qquad (21)$$

where V_n is the boundary velocity parallel to the boundary normal, P is the perimeter of the polygon.

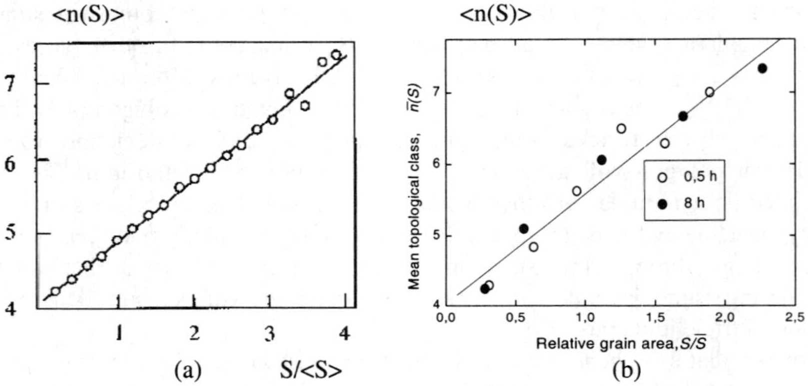

Fig. 10. Dependence of the mean topological class <n(s)> on "grain size" s/<s> (normalized by the average grain size) in pure Al [20] (a), [21] (b).

Finally, as shown in [19] we obtain the rate of change of the grain area S, when grain boundary motion is controlled by the displacement of the triple junctions

$$\frac{dS}{dt} = -m_{tj}\sigma[2\sin(\pi/n) - 1]\cos(\pi/n)\, P \qquad (22)$$

In essence, therefore, a limited triple junction mobility always slows down the evolution of grain microstructure of polycrystals, irrespective whether the topological class of the considered grain is smaller or larger than 6. Formally, for grains with n < 6, the sluggish motion of the triple junction "reduces" the effective topological class of growing grains, while for grains with n > 6 the triple junction behavior "increases" the topological class of vanishing grains.
The only exception holds for n = 6, since a hexagonal grain structure becomes unstable when the contact angle $2\Theta \neq 2\pi/3$. Since the actual magnitude of Θ is determined by the triple junction and grain boundary mobility as well as the grain size and is independent of the number of sides of a grain, there is no unique dividing line between vanishing and growing grains with respect to their topological class anymore, like n = 6 in the Von Neumann-Mullins approach.

As for the situation, when grain boundary migration is completely controlled by triple junction motion, it is interesting to compare the obtained expression (22) with the Von Neumann-Mullins relation (5), which gives the rate of area change of a grain with topological class n under the condition that the triple junctions do not affect grain boundary motion [2]. Evidently, the qualitative behaviour of Eqs. (22) and (5) is similar. Grains with n>6 will grow while those with n<6 disappear in accordance with both Eqs. (5) and (22). What distinguishes Eqs. (22) and (5) is, firstly, the dependence of the rate of change on the topological class n. According to Eq. (5), dS/dt increases infinitely fast with the number n, whereas for the triple junction kinetics

$$\lim_{n\to\infty}\frac{dS}{dt} = m_{tj}P\sigma$$

i. e., remains finite.

The second distinctive difference is connected with the dependence of the rate of area change on the grain size. The Von Neumann-Mullins relation (5) does not depend on grain size, whereas Eq. (22) relates the rate of area change to the grain size through the perimeter P of the grain.

References

[1] J. Von Neumann, in „Metal Interfaces", Americal Society for Testing Materials, Cleveland, 1952, p. 108.

[2] W.W. Mullins, *J. Appl. Phys.*, 27, 1956, p. 900.

[3] K. Lücke, I. Heckelmann, G. Abbruzzese, *Acta Metall. Mater.*, 40, 1992, p. 533

[4] M.P. Anderson, D.J. Srolovitz, G.S. Grest, P. Sahni, *Acta Metall.*, 32, 1984, p. 783.

[5] V.E. Fradkov, L.S. Shvindlerman, D.G. Udler, *Scripta Metall.*, 19, 1985, p. 1285.

[6] V.E.Fradkov, L.S.Shvindlerman, "Structure and Properties of Interfaces in Metals", Moscow, "Nauka", 1988, p. 213.

[7] G. Gottstein, L.S. Shvindlerman, "Grain Boundary Migration in Metals: Thermodinamic, Kinetics, Applications", CRC Press, 1999.

[8] A.V. Galina, V.E. Fradkov, L.S. Shvindlerman, *Phys. Met. Metall.*, 63, 1987, pp. 165-168.

[9] G. Gottstein, V. Sursaeva, and L.S. Shvindlerman, *Interface Sci.*, 7, 1999, pp. 273-283.

[10] U.Czubayko, V.G.Sursaeva, G.Gottstein, L.S.Shvindlerman, *Acta Mater.*, 46, 1998, pp.5863-5871.

[11] S.G. Protasova, G. Gottstein, D.A. Molodov, V.G. Sursaeva, and L.S. Shvindlerman, *Acta Mater.*, submitted.

[12] V.G. Sursaeva, G. Gottstein, L.S. Shvindlerman, *Scripta Mater.*, to be published.
[13] V.Yu. Aristov, V.E. Fradkov, L.S. Shvindlerman, *Sov. Phys. Solid State*, 22, 1980, pp. 1055-1059.
[14] D.A. Molodov, G. Gottstein, S.I. Prokofjev, L.S. Shvindlerman, to be published.
[15] M. Upmanyu, D.J. Srolovitz, G. Gottstein and L.S. Shvindlerman, *Interface Science*, 6, 1998, p. 289.
[16] M. Upmanyu, D.J. Srolovitz, L.S. Shvindlerman, and G. Gottstein, *Acta Mater.*, 47, 1999, pp. 3901-3914.
[17] M. Upmanyu, D.J. Srolovitz, L.S. Shvindlerman, and G. Gottstein, *Interface Science*, 7, 1999, pp. 307-319.
[18] M. Upmanyu, D. J. Srolovitz, L. S. Shvindlerman, and G. Gottstein, to be published.
[19] G. Gottstein, L. S. Shvindlerman, to be published.
[20] V.E. Fradkov, A.S. Kravchenko, L.S. Shvindlerman, *Scripta Metall.*, 19, 1985 pp. 1291-1296.
[21] V.G. Sursaeva, S.G. Protasova, to be published.

Grain Boundary Mechanics

Myrjam Winning

Institut für Metallkunde und Metallphysik, RWTH Aachen, 52056 Aachen, Germany

Keywords: grain boundary motion, X-ray diffraction, external mechanical stress field, grain boundary mechanics, aluminium.

Abstract. The reaction of grain boundaries to mechanical stresses is reviewed. It was found that the motion of low angle grain boundaries as well as of high angle grain boundaries can be induced by the imposed external stress. The observed activation enthalpies allow conclusions on the migration mechanism of the grain boundary motion. The motion of planar low and high angle grain boundaries under the influence of a mechanical stress field can be attributed to the movement of the grain boundary dislocations which comprise the structure of the boundary. A sharp transition between low and high angle grain boundaries was observed for different tilt axes. Also an influence of the mechanical stress on grain growth was investigated. The fact that boundaries can also be moved by mechanical forces sheds new light on microstructure evolution during elevated temperature deformation.

1 Introduction

Defects play an essential role in microstructure evolution, in particular 2D defects, i.e. internal surfaces between similar and dissimilar phases. Interfaces between equal phases but different crystallographic orientations are referred to as grain boundaries. Grain boundaries have the unique property that they react to exerted forces by a change of position. *Grain boundary mechanics*, i.e. the reaction and motion of grain boundaries under the influence of mechanical stresses is the main subject of this paper.

It has long been known that planar low angle grain boundaries can be driven by a mechanical stress field [1]. Most recently it was shown that even planar high angle grain boundaries in aluminium interact with a mechanical stress field and that the motion of the grain boundaries can be enforced by a respective stress [2,3]. The measured activation enthalpies for low angle grain boundaries were found to be similar to the self diffusion enthalpy and for high angle grain boundaries akin to the grain boundary diffusion enthalpy in aluminium. There is a distinct transition between low angle and high angle grain boundaries which can be identified by a conspicuous step in the measured activation enthalpies. The aim of this paper is to show the influence of mechanical stresses on the behaviour of grain boundaries, and the effect on recrystallization and grain growth.

2 Experimental Procedure

To observe the grain boundary motion of individual grain boundaries in-situ the method of continuous determination of the grain boundary position with X-rays [4] was used. Bicrystals of high purity aluminium (7.7 ppm and 1 ppm impurity content) with <112>-, <111>- and <100>-tilt boundaries were investigated in a misorientation angle range from 3° to 33°. We measured the motion of planar low and high angle grain boundaries with different tilt axes under the influence of a stress perpendicular to the boundary.

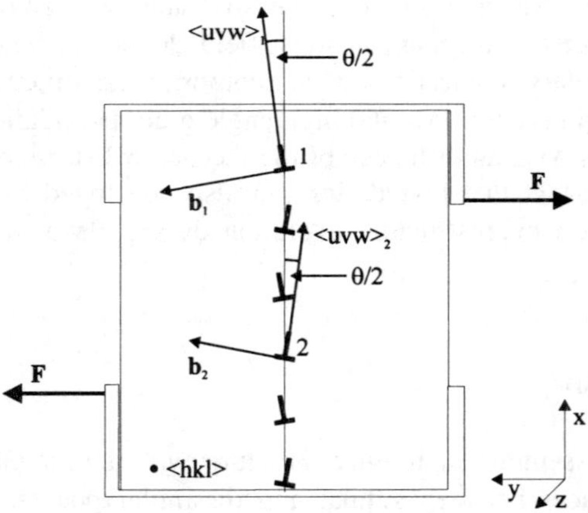

Fig. 1. Illustration of the loading on the sample for a symmetrical tilt boundary.

Every symmetrical tilt boundary can be described as an arrangement of a single set of edge dislocations which is schematically shown in Fig. 1. For a given stress tensor which is a result of the experimental stress state, Burgers vector **b** and line element **s** of the dislocation, the force per unit length on the dislocation can be calculated by the Peach-Koehler equation.
The result is a motion of the dislocations in x- and y-direction, i.e. parallel and perpendicular to the boundary. The external shear stress perpendicular to the boundary plane will cause a force on each dislocation:

$$F_{dis}^{y} = \tau \cdot b \cdot \cos\frac{\theta}{2}. \qquad (1)$$

and in summary a driving force on the boundary [4]

$$p_{gb} = F_{dis}^{y} \cdot \rho_{dis} = 2 \cdot \tau \cdot \sin\frac{\theta}{2} \tag{2}$$

results, where ρ_{dis} is the dislocation density in the boundary and can be calculated easily for each tilt boundary. To expose the samples to a shear stress in the elastic range the applied stress ranging from 10^{-1} MPa to 10^{-3} MPa corresponding to driving forces on the boundary ranging from 10^{-2} MPa to 10^{-4} MPa according to Eq. 2.

The same calculation can be made for asymmetrical tilt boundaries. Every asymmetrical grain boundary consists of at least two sets of perpendicular edge dislocations. The driving force on the grain boundary is then:

$$p_{gb}^{asym} = 2 \cdot \tau \cdot \sin\frac{\theta}{2} - 4 \cdot \tau \cdot \sin^2\frac{\theta}{2} \cdot \sin\phi \cdot \cos\phi \tag{3}$$

where θ is the angle of misorientation for the symmetrical boundary and ϕ is the deviation from the symmetrical position. The driving force on asymmetrical boundaries is nearly the same as for symmetrical boundaries, because the second term on the right hand side of Eq. 3 is small compared to the first term, which is the driving force on a symmetrical boundary.

The temperature dependence of the grain boundary mobility is given by:

$$\frac{v}{p} = m = \frac{m_0'}{kT} \cdot \exp\left\{-\frac{\Delta H}{kT}\right\} = m_0 \cdot \exp\left\{-\frac{\Delta H}{kT}\right\} \tag{4}$$

for the investigated temperature range (300°C to 600°C) the term 1/T is small compared to the exponential function and ist temperature dependence can be neglected.

3 Results

3.1 Motion of planar symmetrical low angle tilt boundaries

Fig. 2 shows the Arrhenius plot of the mobility of planar, symmetrical low angle tilt boundaries with different tilt axes. Apparently the stress driven grain boundary motion is a thermally activated process.

According to Eq. 4:

$$\ln\frac{v}{p} = \ln m_0 - \frac{\Delta H}{kT} \tag{5}$$

we can determine the activation enthalpy for grain boundary motion ΔH from the slope of the Arrheniusline and the pre-exponential factor m_0 from the extrapolated intersection with the mobility axis.

Obviously, it is possible to activate the motion of planar low angle grain boundaries. The activation enthalpies for the motion of the low angle boundaries are close to the self diffusion enthalpy in aluminium which is ΔH_{SD}=1.25eV [5], and in the range ΔH=1.19eV for the motion of the symmetrical <100>-tilt boundary and ΔH=1.29eV for the motion of the symmetrical <111>-tilt boundary. The pre-exponential factors for the motion of the different low angle grain boundaries are also similar, namely between $\ln m_0$=22.25 in the case of the <100>-boundary and $\ln m_0$=23.99 for the <111>-tilt boundary.

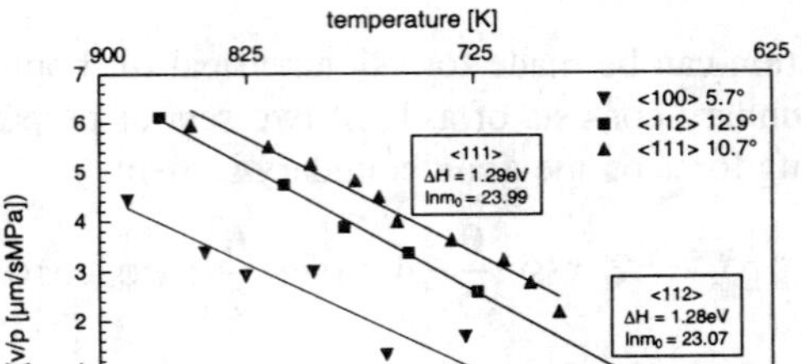

Fig. 2. Arrhenius plot of the mobility of planar, symmetrical low angle tilt boundaries with different tilt axes and 7.7 ppm impurity content.

3.2 Motion of planar symmetrical high angle-tilt boundaries

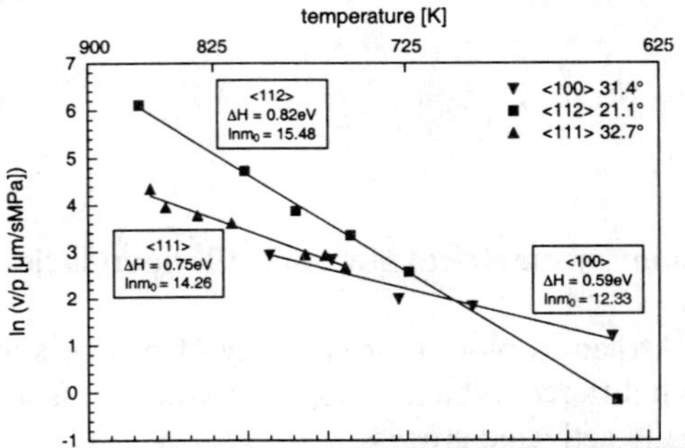

Fig. 3. Arrhenius plot of the mobility of planar, symmetrical high angle tilt grain boundaries with different tilt axes and 7.7 ppm impurity content.

The Arrhenius lines of the mobility of different planar high angle grain boundaries are shown in Fig. 3. Even for the planar high angle grain boundaries the migration of the planar boundary can be induced by the external shear stress and nearly the same activation enthalpies are measured which were close to the grain boundary diffusion enthalpy in aluminium (0.6-0.9eV [6]). Also the pre-exponential factors for the different tilt axes were similar between ln m_0=12.33 for the <100>-tilt boundary and ln m_0=15.48 for the <112>-boundary.

3.3 Motion of planar asymmetrical grain boundaries

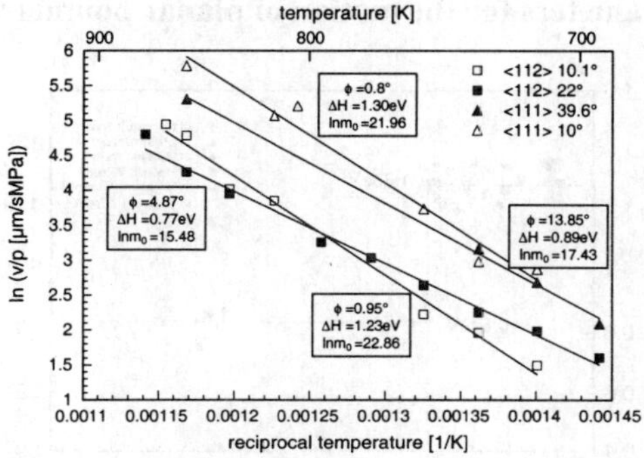

Fig. 4. Arrhenius plot of the mobility of planar, asymmetrical <112>- and <111>- tilt boundaries with 7.7 ppm impurity content.

The experiments on the asymmetrical tilt boundaries with <112>- and <111>-tilt axes were carried out in the same way as for the symmetrical boundaries. The Arrhenius lines of the grain boundary mobility for <112>- and <111>- tilt boundaries are shown in Fig. 4. The activation parameters for the low and high angle boundaries are in good agreement to the results of the symmetrical boundaries this holds for the <112>- as well as for the <111>-boundaries.

3.4 Motion of planar random boundaries

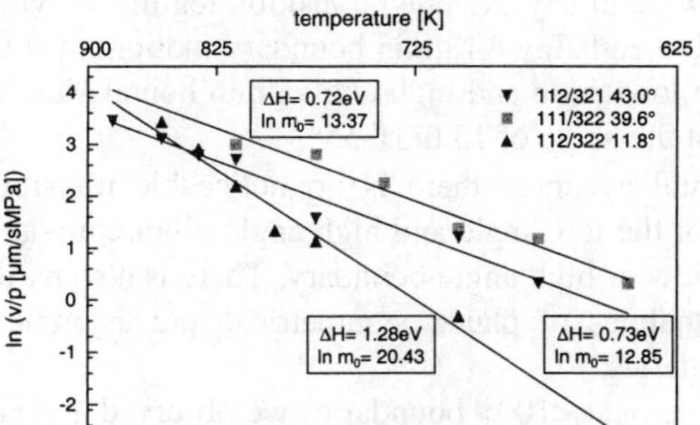

Fig. 5. Arrhenius plot of the mobility of planar (random) grain boundaries.

The results of the experiments on planar, random boundaries are given in Fig. 5. The activation enthalpies of the high angle boundaries were again in the range of the grain boundary diffusion enthalpy. The same good agreement was found for the low angle boundary for which the activation enthalpy was ΔH=1.28eV and very akin to the self diffusion enthalpy and the results for all other low angle grain boundaries.

3.5 Activation parameters for the motion of planar boundaries

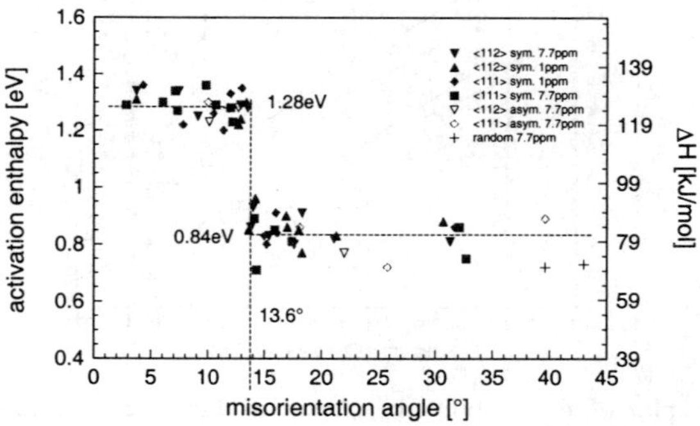

Fig. 6. Dependence of the activation enthalpy on misorientation for the <112>- and <111>-boundaries (symmetrical as well as asymmetrical boundaries) with different impurity contents.

Fig. 6 shows the dependence of the activation enthalpies on the misorientation angle for all measured symmetrical and asymmetrical <112>- and <111>-tilt boundaries. There is a distinct subdivision into two ranges. For angles smaller than 13.6° an average activation enthalpy ΔH_m=1.28eV±0.01eV was found, while for angles larger than 13.6° the average activation enthalpy was ΔH_m=0.84eV±0.01eV. The transition between these two ranges is marked by a conspicuous step without any extended transition regime. Obviously, the observed step in the activation enthalpy for grain boundary motion must be attributed to the transition between low angle and high angle grain boundaries. The corresponding transition occurs at the angle of 13.6°±0.55°.

Within experimental accuracy there is no noticeable transition range between these two levels for the low angle and high angle grain boundaries. The boundary behaves either as low or high angle boundary. There is also no difference between the activation enthalpies of planar symmetrical and asymmetrical <112>- and <111>-grain boundaries.

Even for the symmetrical <100> boundaries we observed a clear division into the two levels with the same activation enthalpies for low as well as for high angle boundaries. But the transition angle for the <100>-tilt boundaries is between 5.7° and 10.9° this means much smaller than for the <112>- and <111>-tilt boundaries. Since there is no influence of the misorientation or the impurity content on the the activation enthalpy, the pre-exponential factor depends on these parameters. In Fig. 8a the influence of the impurity content on the mobility for the <112>-

boundaries with 7.7 ppm and 1 ppm impurity content is shown. One recognizes a difference in the pre-exponential factor for the two impurity contents. The grain boundaries with the lower impurity content have a higher pre-exponential factor for grain boundary motion, i.e. at constant activation enthalpy a higher mobility. The same dependence of the pre-exponential factor on the impurity content was observed for the <111>-tilt boundaries.

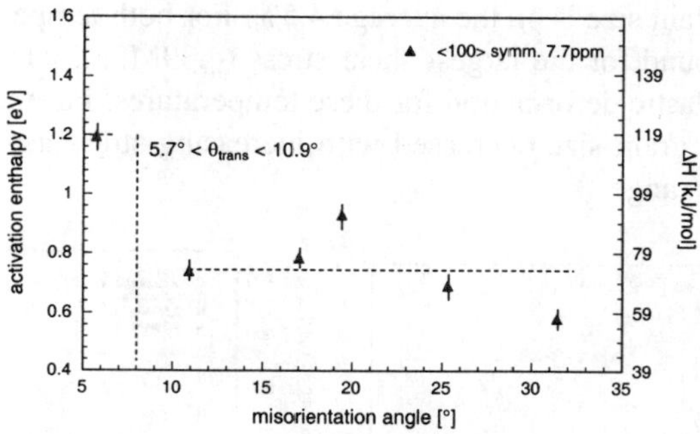

Fig. 7. Dependence of the activation enthalpy on misorientation for the symmetrical <100>-tilt boundaries.

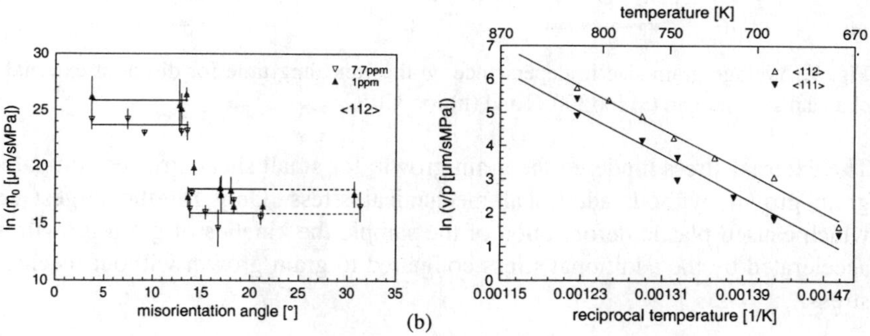

Fig. 8. (a) Influence of the impurity content on the grain boundary pre-exponential factor for the <112>-tilt boundaries. (b) Comparison of the mobilities for a <111>-tilt boundary with θ=7.3° and a <112>-tilt boundary with θ=7.4° and the same impurity content.

In Fig. 8b the mobility is shown for a <111>-tilt boundary with θ=7.3° and a <112>-boundary with θ=7.4°. The impurity content of both boundaries was 7.7 ppm. The mobility of the <112>-tilt boundary was higher than the mobility of the <111>-tilt boundary.

3.6 Influence of shear stresses on grain growth in polycrystals

To investigate the influence of an external shear stress on grain growth in polycrystals, high purity aluminium polycrystals were annealed under different shear stresses and the kinetics of grain growth was determined by measuring the grain size at different annealing times and several temperatures. The experiments were carried out on three samples for each combination of temperature, stress state and annealing time. The average grain size was determined for each sample by measuring the grain size of at least 200 grains (for large grain sizes) and 400 grains (for small grain sizes). In Fig. 9 the average grain size is shown in dependence on annealing time for 300°C and 350°C and four different stress states. The deviation from the given grain size is on the average 4.5%. For both temperatures the largest grain size was found for the largest shear stress (4.39MPa), which was definitely in the range of plastic deformation for these temperatures. For the same annealing time the average grain size decreased with increasing stress as long as the stress was in the elastic range.

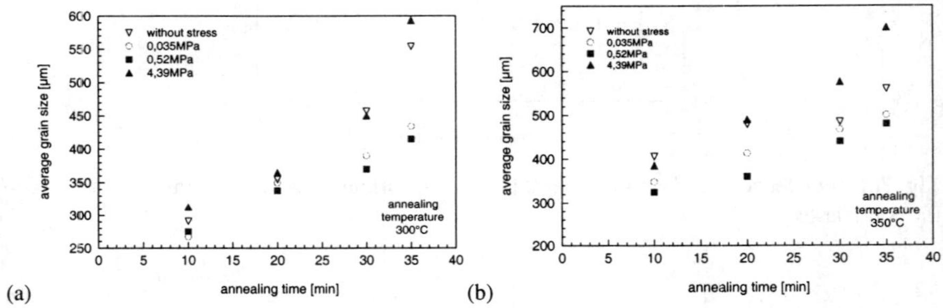

Fig. 9. Average grain size in dependence on the annealing time for different external mechanical stresses and (a) for 300°C and (b) for 350°C.

The external stress hindered the grain growth for small shear stresses compared to grain growth without additional mechanical stress. Only for the largest stress which caused plastic deformation of the sample the kinetics of grain growth were accelerated by the additional stress compared to grain growth without mechanical stress.

4 Discussion

The results prove that planar grain boundaries can be driven by a mechanical stress field irrespective of the angle of misorientation, i.e. irrespective whether high or low angle grain boundary.

Motion of planar boundaries under a mechanical stress can be conveniently associated with the motion of structural grain boundary dislocations [2]. The measured activation enthalpies for low angle grain boundaries (ΔH_m=1.28eV±0.01eV) are comparable to the volume self diffusion enthalpy ΔH_{SD}=1.25eV [5] in aluminium, and for the high angle grain boundaries the activation enthalpies

(ΔH_m=0.84eV±0.01eV) are comparable to the grain boundary self diffusion enthalpy ΔH_{GB}=0.6-0.9eV [6].
In fcc crystals the reaction of edge dislocations to an applied shear stress ought to be purely mechanical and not thermally activated. Obviously, the observed grain boundary motion is a thermally activated process controlled by diffusion. Moreover, real boundaries are never perfect symmetrical tilt boundaries but always contain structural dislocations of other Burgers vectors. These dislocations have to be displaced by nonconservative motion to make the entire boundary migrate. The climb process requires diffusion, which can only be volume diffusion for low angle grain boundaries but grain boundary diffusion for high angle grain boundaries according to the observed activation enthalpies.
Self diffusion in aluminium proceeds via vacancy site exchange. If this also holds for grain boundary migration the fundamental process of stress driven grain boundary motion could be the interaction of vacant lattice sites with edge dislocations. The fact that the motion of a dislocation arrangement in low angle grain boundaries depends in the same way on the temperature as self diffusion indicates that the dislocations during their movement encounter barriers which impede their glide motion. The ability of an edge dislocation to overcome a barrier depends on diffusion, because edge dislocations can climb perpendicular to their glide direction and thus bypass a barrier by absorption of vacancies.
Any grain boundary can be formally (geometrically) described by a dislocation structure. Correspondingly, the motion of planar high angle grain boundaries can also be formally described by the movement of structural grain boundary dislocations. Again grain boundary diffusion assists dislocation climb in the grain boundary plane to overcome obstacles for grain boundary motion. Since the climb process is much slower than the glide process it determines the velocity of boundary motion. Therefore, the measured activation enthalpies of grain boundary motion must be associated with the activation enthalpies of the climb processes. The results show that irrespective of the misorientation angle low angle grain boundaries as well as high angle grain boundaries can be driven by an externally imposed mechanical stress. Obviously, the motion of any grain boundary is due to the coupling of a respective external stress with the strain field of the boundary as expressed in terms of the movement of the dislocations which compose the grain boundary.
One major result is that the activation enthalpy of grain boundary motion does not change with the angle of rotation except for a conspicuous step at about 13.6° for the <111>- and <112>- tilt boundaries and between 5.7° and 10.9° for the <100>-tilt boundaries. This means that the transition angle depends on the tilt axis and therefore also the Brandon criterion is dependent on the tilt axis. This step in the activation enthalpy reflects the transition from the low angle grain boundary to the high angle grain boundary regime (Figs. 6, 7).
In this view, the transition from the migration behaviour of low and high angle tilt grain boundaries merely changes the diffusion path of the vacancies necessary for the climb of the structural dislocations, namely from volume self diffusion to grain

boundary diffusion. The elastic coupling of the dislocations with the applied shear stress remains virtually unchanged.

It is widely accepted that the grain boundary mobility increases with increasing angle of misorientation [7,8]. Especially the mobility of high angle grain boundaries should be larger than the mobility of low angle grain boundaries by several orders of magnitude [7]. The current results show a constant activation enthalpy and pre-exponential factor although different for low angle grain boundaries and high angle grain boundaries. From the measured activation parameters we can determine the grain boundary mobility for each temperature. Fig. 10a shows the dependence of grain boundary mobility of planar <112>- and <111>-tilt boundaries on the misorientation angle determined for 200°C.

The constant activation parameters lead to a constant mobility for low angle grain boundaries. The mobility of the high angle grain boundaries is also constant, but the difference between the mobilities of the planar high angle and low angle grain boundaries is nearly a factor 50-60 at 200°C. For the temperature 600°C the mobility of the low angle grain boundaries is even larger than the mobility of the high angle grain boundaries by a factor of 2-3 (Fig. 10b).

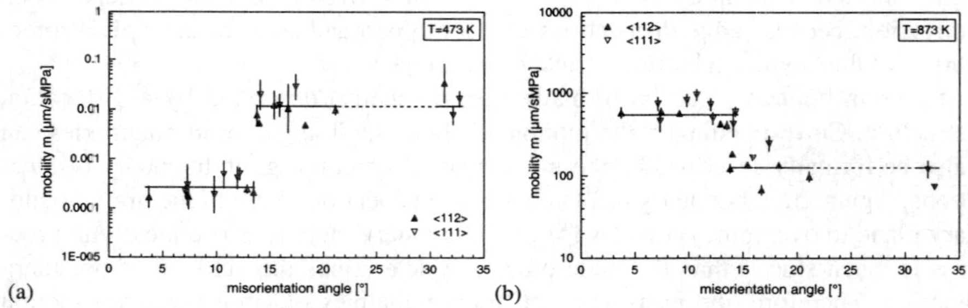

Fig. 10. Dependence of the grain boundary mobility on the misorientation angle for the motion of symmetrical planar <112> and <111>-tilt boundaries (a) at the temperature T=473K and (b) at T=873K.

Even more in contrast to common conception is the result that the difference of the mobilities of low angle grain boundaries and high angle grain boundaries depends on temperature such that at temperatures above 500°C the mobility of low angle grain boundaries even exceeds the mobility of high angle grain boundaries in the case of stress induced motion. This is due to the compensation effect [9] which predicts the process with the higher activation enthalpy to attain a higher mobility at temperatures above the compensation temperature. This may be of particular importance for subgrain growth and nucleation of recrystallization, and maybe a reason for the observation that recrystallisation at very high temperatures may be entirely dominated by low angle grain boundaries [10]. Obviously the presented results are important for recrystallization and grain growth. With regard to the simulations of these processes the results should be more taken into account. The assumption in such simulations is normally that low angle grain boundaries are nearly immobile and do not play a role for recrystallization and grain growth. But the results show that the motion of low angle grain boundaries can be activated and that the mobility can exceed the mobility of high angle grain boundaries at high temperatures. Also the dependence of the transition angle from the low angle to the high angle regime on the grain boundary orientation may have consequences on the Brandon criterion which defines the angular deviation of special boundaries from the exact coincidence orientation relationship. This may have an additional influence on microstructure and texture evolution during recrystallization and grain growth.

First experiments on grain growth showed an influence of the external shear stress on the kinetics of grain growth. The acceleration of the grain growth kinetics by a large stress is in good agreement with literature (e.g. [11]) where the same behaviour was observed in polycrystalline thin films under the influence of pressure. Applied shear stresses in the elastic range lead to a hindering of grain growth. This is a new experimental observation and maybe due to the interaction of grain boundaries with the external shear stress. One possible explanation is that the additional driving force on the grain boundaries can act in the same direction as the capillary force due to the grain boundary curvature or in the opposite direction. In the second case small grains would shrink more slowly compared to the case without additional force. To clarify the observed phenomenon it will be necessary to determine the kinetics of grain growth in dependence of external mechanical stresses in a more detailed way coupled with investigations of grain morphology and the texture evolution during grain growth.

Obviously there is a possibility to change the grain growth kinetics and thus also the microstructure evolution by an external mechanical stress field. This means the microstructure evolution and the properties of materials are not only influenced by dynamic properties of grain boundaries but also by *grain boundary mechanics* i.e. the interactions between grain boundaries and mechanical stresses. Another effect of this interaction can be expected for high temperature plasticity e.g. during cyclic deformation at high temperatures.

5 Summary

A method was presented which allows the activation of the motion of planar, symmetrical and asymmetrical tilt boundaries in aluminium under the influence of a mechanical stress field. The grain boundary motion is due to the movement of the structural dislocations in the boundary. Two different activation enthalpies were found: for low angle grain boundaries an activation enthalpy close to the self diffusion enthalpy and for high angle grain boundaries an activation enthalpy close to the grain boundary diffusion enthalpy. Besides these two activation enthalpies no additional dependence of the activation enthalpies for the grain boundary motion on the misorientation angle was found for the planar grain boundaries.

There is a sharp transition between the low and the high angle grain boundaries which is marked by a conspicuous step in the activation enthalpy. The transition angle is 13.6°±0.55° for the <112>- and the <111>-tilt boundaries, and between 5.7° and 10.9° for tilt boundaries with a <100>-tilt axis.

For asymmetrical as well as for random planar boundaries the same activation enthalpies were found as in the case of the symmetrical boundaries. The impurity content and the tilt axes seem to have only an influence on the pre-exponential factor but not on the activation enthalpy.

Also an influence of the external shear stress on grain growth could be observed. For small shear stresses the kinetics of the grain growth is hindered compared with the grain growth without additional mechanical force. For large shear stresses the grain growth is accelerated due to plastic deformation.

Acknowledgment

The author would like to express her gratitude to the Deutsche Forschungsgemeinschaft for financial support through contract Wi 1917/1.

References

[1] C.H. Li, E.H. Edwards, J. Washburn and E.R. Parker, *Acta Metall.*, 1, 1953, pp. 122-133.

[2] M. Winning, G. Gottstein and L.S. Shvindlerman, *Acta Mater.*, 49, 2001, pp. 211-219.

[3] M. Winning, G. Gottstein and L.S. Shvindlerman, *Mat. Science & Engineering A*, in press 2001.

[4] U. Czubayko, D.A. Molodov, B.-C. Petersen, G. Gottstein and L.S. Shvindlerman, *Meas. Sci. Technol.*, 6, 1995, pp. 947-952.

[5] J. Philibert, "Diffusion et Transport de Matière dans les Solides", Les Éditions de Physique, Les Ulis Cedex, France, 1985.

[6] T.E. Volin, K.H. Lie and R.W. Baluffi, *Acta Metall.*, 19, 1971, pp. 263-273.

[7] Y. Huang, F.J. Humphreys and M. Ferry, *Acta Mater.*, 48, 2000, pp. 2543-2556.

[8] Y. Huang and F.J. Humphreys, *Acta Mater.*, 48, 2000, pp. 2017-2030.

[9] G. Gottstein and L.S. Shvindlerman, "Grain Boundary Migration in Metals: Thermodynamics, Kinetics, Applications", CRC Press LLC, Boca Raton Florida, USA, 1999.

[10] M. Heinrich and F. Haider, *Philosophical Mag. A*, 74, 1996, pp. 1047-1052.

[11] C.A. Volkert and C. Lingk, *Appl. Physics Letters*, 73, 1998, pp. 3677-3679.

Grain-Growth Kinetics and Thermal Stability of Nanocrystalline Materials

C. E. Krill III and R. Birringer

FR 7.3 Technical Physics, Universität des Saarlandes, Postfach 151150, Geb. 43B, D-66041 Saarbrücken, Germany

Keywords: nanocrystalline materials, grain-growth kinetics, thermal stability, x-ray diffraction.

Abstract. An essential step in the development of nanocrystalline materials for technological application is understanding the growth kinetics of nanometer-sized grains. Although several studies of such materials find growth exponents and activation-energy values similar to those of polycrystalline materials of conventional grain size, theoretical considerations predict a change in the qualitative nature of the growth kinetics in the submicron size range. Experimental evidence for grain-size-dependent kinetics in nanocrystalline Fe is presented and discussed. The thermal stability of nanocrystalline materials can be improved using the same strategies for grain-boundary pinning that are successful at conventional grain sizes. Alternatively, one can exploit the thermodynamics of chemical segregation to the boundaries to reduce the driving force for grain growth. The effectiveness of this approach is demonstrated for nanocrystalline Pd containing 20 at.% Zr.

1 Introduction

Experimental studies of grain-growth kinetics have traditionally been performed on polycrystalline samples with an average grain size of ~5 μm or larger. The extent to which the concepts underlying our understanding of grain growth at micrometer length scales can be extended to much smaller grain sizes is of vital importance to the technological application of so-called nanocrystalline materials, in which the average grain size $\overline{D}$ lies between 5 and 100 nm. A variety of macroscopic sample properties—such as hardness, sinterability, electrical and ionic conductivities, corrosion resistance, magnetic coercivity and catalytic activity—have been found to depend on $\overline{D}$, varying sometimes dramatically in nanocrystalline specimens from their corresponding values in coarse-grained counterparts of the same material [1,2]. Of course, any property enhancements brought about via nanostructuring are lost upon the onset of significant grain growth. It is therefore crucial to develop effective methods for stabilizing nanocrystalline samples against coarsening, and this requires a fundamental understanding of the kinetics of grain growth in such materials.

There are, however, more than just technological reasons for investigating the kinetics of grain growth in nanocrystalline materials: owing to its inverse dependence on $\overline{D}$, the specific grain-boundary area (i.e., the grain-boundary area per unit volume) in such samples is several orders of magnitude larger than in conventional polycrystalline materials, resulting in a corresponding enhancement in the driving force for grain growth. In coarsening studies performed with conventional materials, it is the rather small value of this driving force that allows processes with similar or even larger energetic effects, such as recrystallization or Zener drag, to cause deviations from purely curvature-driven growth kinetics. Hence, nanocrystalline samples may constitute model systems for studying the "ideal" case envisioned by theories for curvature-driven boundary motion. Another consequence of an increased specific grain-boundary area is the potential for much cleaner grain boundaries: at a given total impurity concentration, the number of impurity atoms per unit grain-boundary area is orders of magnitude smaller in a nanocrystalline sample than in a conventional specimen. This could be advantageous for studying the migration kinetics of boundaries free of the pinning influence of impurities.

Moreover, nanocrystalline materials offer the intriguing potential for investigating grain-size-dependent aspects of grain-growth kinetics. The migration of a grain boundary is a far more complex process than the phenomenological picture of an idealized surface moving toward its center of curvature. For instance, grain boundaries are defined by the spatial arrangement of individual atoms, which can conceivably jump stochastically from one side of the boundary to the other under the influence of thermal energy [3,4]. Furthermore, significant grain-boundary motion can take place only with the simultaneous movement of triple junctions, the mobility of which may deviate considerably from that of the grain boundaries [5]. Finally, grain-boundary migration can lead to an increase in the average grain size only to the extent that the excess volume localized in the grain-boundary cores is transported out of the sample [6]. Recent theoretical considerations of these effects [7–11] suggest that, in the limit of nanometer-sized $\overline{D}$, each of these processes could potentially replace curvature-driven boundary migration as the rate-controlling step for grain growth. Hence, an investigation of growth kinetics in nanocrystalline materials may uncover aspects of the mechanism for grain growth—in the form of unexpected values for the growth exponent or the activation energy, for instance—that manifest themselves only at small $\overline{D}$.

Lastly, we note that the average size of the grains in a nanocrystalline material can be determined nondestructively using x-ray diffraction, which opens up the exciting possibility for performing *in situ* studies of grain growth in nanocrystalline materials with near-real-time resolution [12]. Owing to the rapid change in the specific grain-boundary area during growth at such small grain sizes, one can also use techniques like calorimetry [13] or measurements of electrical resistivity [14] to characterize the isothermal evolution of the grain size in nanocrystalline samples (albeit more indirectly than by x-ray diffraction).

2 General Features of Grain Growth in Nanocrystalline Materials

Because thermal stability is such an important parameter governing the processing and technological application of nanocrystalline materials, numerous experimental studies of grain growth have been performed on such samples, usually as isochronal annealing steps over a range of temperatures or, less frequently, in the form of isothermal measurements, as have traditionally been employed to characterize coarsening in conventional polycrystalline materials. The grain-growth data recorded in nanocrystalline samples have generally been interpreted in the classical manner, assuming that the isothermal evolution of the grain size $\overline{D}$ can be described by a power law of the form

$$\overline{D}^{n}(t) - \overline{D}^{n}(0) = kt, \tag{1}$$

where n is the so-called growth exponent, k is a temperature-dependent rate parameter, and t is the annealing time. In the case of ideal curvature-driven growth, n equals 2 [15], but, for reasons to be discussed below, this value is rarely found in fits of Eq. 1 to experimental data. Furthermore, one usually assumes that the rate constant k varies with temperature T according to an Arrhenius law [13],

$$k(T) = \frac{k_0}{T} \exp(-Q/RT), \tag{2}$$

where Q is the activation energy for grain-boundary migration, R is the universal gas constant, and k_0 is a temperature-independent prefactor. From isothermal growth curves measured at several different temperatures, one can deduce Q from a fit of Eq. 2 to the k values arising from Eq. 1 (with a fixed value for n at all temperatures).

The literature on grain growth in nanocrystalline materials has been reviewed by several authors (see [16,17] and references therein). Malow and Koch [16] tabulated the exponents n reported for grain growth in a variety of nanocrystalline samples, finding values ranging from 2 to well over 10, with a tendency toward lower values as the annealing temperature approached the melting point. Similar behavior has been reported for samples of conventional grain size [18]; however, high-purity polycrystalline specimens of conventional grain size always manifest exponent values close to 2, and it can be shown that the existence of pinning forces slowing boundary migration cause an increase in the apparent value of n [19]. Thus, it can be argued that exponent values larger than 2 are primarily caused by sample impurities. Impurity drag can be taken into account, for instance, through the use of Burke's growth equation [19], which has an underlying growth exponent of 2, in place of Eq. 1. Indeed, just such a procedure has repeatedly succeeded in explaining the origin of exponent values $n > 2$ for grain growth in nanocrystalline materials [20–25].

The activation energy Q for grain growth has been evaluated in a number of nanocrystalline materials, although primarily by calorimetric means rather than through isothermal studies. In most investigations of elemental nanocrystalline materials, Q is found to take on values close to those for grain-boundary self-diffusion [16,26–28], mirroring the situation in coarse-grained materials. A notable exception is nanocrystalline Fe prepared by ball milling (Sec. 3.1), in which grain growth manifests an activation energy close to that of lattice self diffusion [20,29]. (Similar coarsening behavior was observed in microcrystalline zone-refined Fe, although it was suggested for this sample that the unexpectedly high value for Q could be attributed to boundary drag arising from an unknown impurity [30].)

The similarity between the values found for the growth exponent n and the activation energy Q in nanocrystalline and coarser-grained polycrystalline materials suggests that the mechanism for grain growth is independent of the average grain size $\overline{D}$, despite the factors mentioned above that argue to the contrary. If this is true, then we can use Hillert's model for curvature-driven grain growth [31] to express the rate of boundary migration as

$$\frac{d\overline{D}}{dt} = \frac{4m\gamma}{\overline{D}}, \tag{3}$$

where m denotes the (temperature-dependent) grain-boundary mobility and γ the grain-boundary energy. At a given temperature, the rate of grain growth should scale as $\overline{D}^{-1}$ and therefore be several orders of magnitude higher in a nanocrystalline sample than in the same material of conventional grain size. In Fig. 1 the growth rates measured in nanocrystalline Fe at 438°C, 488°C and 535°C [21] are compared to the values expected for coarse-grained Fe [30] extrapolated to the nanometer size range using Eq. 3. Surprisingly, the coarsening rate in nanocrystalline Fe is several orders of magnitude slower than that of its coarse-grained counterpart, despite the fact that the boundary migration in the zone-refined Fe samples is thought to have been impeded by impurity drag [30].

Unexpectedly sluggish grain growth like that illustrated in Fig. 1 has frequently been observed in nanocrystalline samples—in particular, in those prepared by inert-gas condensation (IGC) and subsequent compaction—and this has led to speculation that a particularly narrow distribution of grain sizes in such materials might create a metastable arrangement of grain boundaries with an energy barrier toward coarsening [32]. It should be noted, however, that IGC samples of Ag, Cu and Pd [33–35] were found to undergo a certain amount of coarsening even at room temperature, and, in any event, the grain-size distribution of such materials as revealed by transmission electron microscopy [36,37] shares the rather broad, roughly lognormal shape characteristic of grain-size distributions in conventional polycrystalline materials [38].

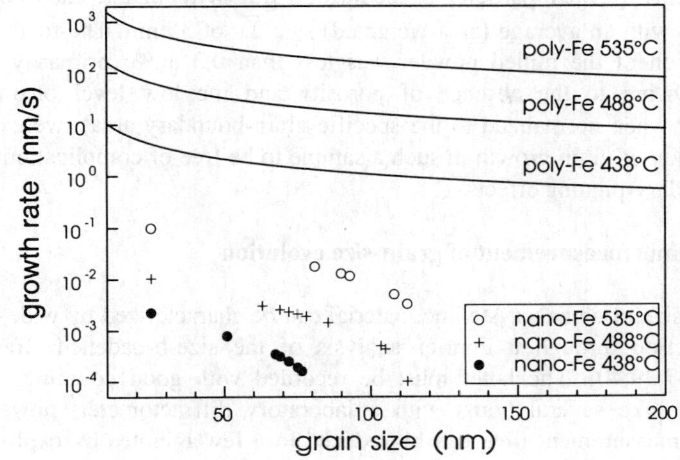

Fig. 1. Rate of grain growth in nanocrystalline Fe at 438°C, 488°C and 535°C [21] compared to that of coarse-grained Fe [30]. Curves are extrapolations for coarse-grained Fe using Eq. 3, calculated assuming an Arrhenius dependence of m on temperature.

Until recently, grain-growth inhibition like that shown in Fig. 1 was thought to be caused primarily by the same factors known to contribute to grain-boundary pinning in conventional materials, such as second-phase precipitates, pores or impurity atoms segregated to the boundaries. All of these have been shown to improve the stability of nanocrystalline materials with respect to grain growth (Sec. 4) [16,17]. Since the incorporation of a certain concentration of impurities or porosity is practically unavoidable when preparing nanocrystalline samples, the observed growth-rate suppression has generally been judged to be a byproduct of the synthesis route rather than an intrinsic feature of submicron-sized grains. However, the nanocrystalline Fe samples plotted in Fig. 1 are devoid of second phases, and the total impurity concentration normalized to the specific grain-boundary area is actually far lower in the nanocrystalline specimens than in the coarse-grained ones. Thus, it is tempting to consider the possibility that a grain-size-dependent factor may be affecting the migration rate of grain boundaries at small grain sizes, at least in this material.

3 Grain-Size-Dependent Growth Kinetics in Nanocrystalline Fe

In order to gain a deeper understanding of the mechanism for grain growth in nanocrystalline Fe and its possible grain-size dependence, we have studied the kinetics of grain growth in this material using a novel in situ x-ray diffraction technique employing synchrotron radiation [12,29]. Sample preparation was carried out by ball milling Fe powder in steel vials under high-purity Ar, resulting in fully dense powder particles exceeding 50 μm in diameter, each containing crystallites with an average (area-weighted) size $\overline{D}$ of 35 nm. The total impurity concentration of the milled powder was less than 0.3 at.% (primarily oxygen) [21,29]. Owing to the absence of porosity and the low level of impurities (especially when normalized to the specific grain-boundary area), we expect the interpretation of grain growth in such a sample to be free of complications arising from boundary-pinning effects.

3.1 Real-time measurement of grain-size evolution

The grain size of a nanocrystalline material can be characterized by wide-angle x-ray diffraction through a Fourier analysis of the size-broadened Bragg peak profiles [37,39,40]. The latter must be recorded with good counting statistics, which can take several hours with a laboratory diffractometer; however, the necessary measurement time can be reduced to a few minutes by exploiting the much higher intensity of a synchrotron source, in which case the isothermal evolution of the grain-size can be studied in situ with nearly real-time resolution [12,29]. The results of such measurements performed on ball-milled Fe powder at the high-resolution powder-diffraction beamline BM16 of the European Synchrotron Radiation Facility (ESRF) are illustrated in Fig. 2. The lattice expansion upon heating the sample is evident in the shift of the Bragg peaks to lower angles, and the subsequent grain growth manifests itself in a characteristic sharpening of the peak profiles (Fig. 2a). A Fourier analysis of the (110) and (220) peak profiles according to the Warren-Averbach method [39] yields the isothermal growth curves plotted in Fig. 2b. Following a jump in the grain size as the sample

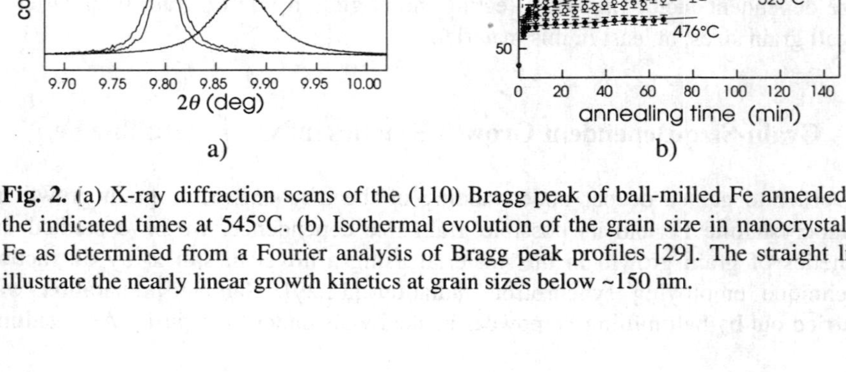

Fig. 2. (a) X-ray diffraction scans of the (110) Bragg peak of ball-milled Fe annealed for the indicated times at 545°C. (b) Isothermal evolution of the grain size in nanocrystalline Fe as determined from a Fourier analysis of Bragg peak profiles [29]. The straight lines illustrate the nearly linear growth kinetics at grain sizes below ~150 nm.

temperature equilibrated, the grains appear to grow linearly with time up to a grain size of about 150 nm, at which point the growth curves develop a nonlinearity reminiscent of the power-law behavior expected for curvature-driven boundary migration. The growth rates in the linear regions are much slower than those extrapolated from grain-growth studies performed on high-purity Fe of conventional grain size [30], and the linearity itself is inconsistent with all of the usual grain-growth laws, including those that take impurity effects into account.

3.2 Vacancy-drag model

A grain-size-dependent transition between linear and nonlinear growth is predicted by a recently developed model that considers the kinetics of excess-volume redistribution during grain growth in a nanocrystalline material. Inspired by atomistic simulations of grain growth showing the generation of lattice vacancies behind moving grain boundaries [6], Estrin and colleagues postulated that the excess volume localized in the boundary cores is transported away from shrinking grain boundaries via the generation and diffusion of vacancies through the bulk [10,11]. For this case they demonstrated that the kinetics of vacancy diffusion place an upper bound on the speed with which a grain boundary can migrate. For nanocrystalline materials their expression for the growth rate takes the form

$$\frac{d\overline{D}}{dt} = \frac{\gamma D_{SD}}{12 N Z k_B T (\Delta V/A)^2} \tag{4}$$

at temperature T, with Boltzmann constant k_B and the following material parameters: grain-boundary energy γ, lattice self-diffusion coefficient D_{SD}, number of atoms per unit volume N, atomic coordination number Z and excess volume per unit grain-boundary area, $(\Delta V)/A$. Equation 4 can be integrated immediately to yield a linear growth law for $\overline{D}$. According to Eq. 3, in the absence of vacancy effects, the boundary migration speed for curvature-driven growth diverges as $\overline{D}^{-1}$, eventually exceeding that of Eq. 4 for grains smaller than some critical grain size $\overline{D}_c$. Estrin et al. estimated $\overline{D}_c \approx 200$ nm for aluminum [10], suggesting that the initial rate-controlling step for grain growth in nanocrystalline materials is the vacancy-mediated redistribution of excess volume, with a transition to curvature-driven growth kinetics occurring at some larger grain size.

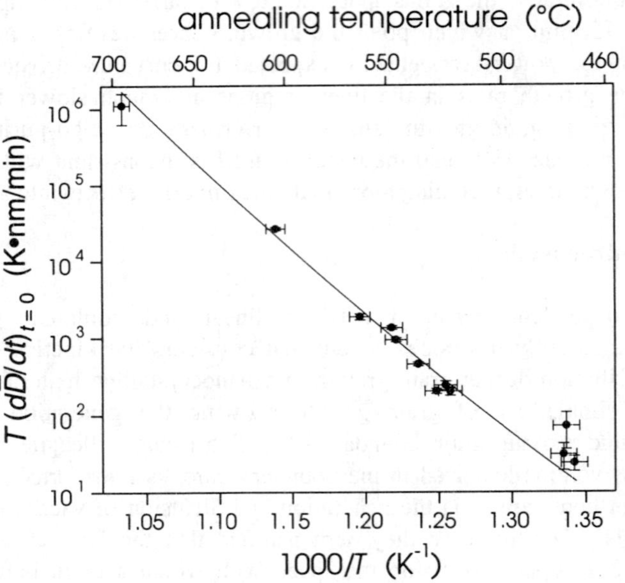

Fig. 3. Arrhenius plot of the initial rate of grain growth in nanocrystalline Fe for the data of Fig. 2b and additional annealing temperatures [29]. The solid curve is a fit of Eq. 4 with adjustable parameter (ΔV)/A.

We can test this prediction by comparing the measured $d\bar{D}/dt$ values for linear growth in nanocrystalline Fe at various temperatures with those calculated according to Eq. 4, the temperature dependence of which is dominated by that of the self-diffusion term,

$$D_{SD}(T) = D_0 \exp\left(-Q_{SD}/RT\right), \tag{5}$$

where D_0 is a constant, and Q_{SD} denotes the activation energy for lattice self-diffusion. According to Eqs. 4 and 5, an Arrhenius plot of $T\left(d\bar{D}/dt\right)_{t\to 0}$ should yield an activation energy comparable to Q_{SD} in Fe. The data points in Fig. 3 were derived from the initial slopes of the growth curves in Fig. 2b, and the solid curve is the prediction of Eq. 4 for the known values of N, Z, γ and $D_{SD}(T)$ for Fe [29]. The quantity (ΔV)/A was treated as a refinable parameter in a least-squares fit of Eq. 4 to the measured points, yielding an optimum value of 0.019 ± 0.001 nm, which lies within the range of 0.01 to 0.04 nm found in experimental studies of special boundaries in bicrystals [41]. Thus, the theory developed by Estrin et al. [10,11] is able to account for the growth-rate discrepancy illustrated in Fig. 1 between nanocrystalline and microcrystalline Fe without recourse to impurity effects. The extent to which vacancy drag can be held responsible for the sluggish grain growth observed in other nanocrystalline materials is a focus of ongoing investigations.

4 Stabilizing Nanocrystalline Materials Against Grain Growth

As discussed in Sec. 1, taking advantage of the novel properties of nanocrystalline materials in real-world applications generally necessitates stabilizing the microstructure against grain growth. To the extent that the rate of coarsening in a nanocrystalline sample is governed by Eq. 3, we can devise two strategies for improving its thermal stability: a *kinetic* one, in which the grain-boundary mobility m is reduced, or a *thermodynamic* one, in which the driving force $\gamma/\overline{D}$ for boundary motion is suppressed. Standard metallurgical techniques for realizing the former approach typically involve introducing impurity atoms or phases to pin the grain boundaries, the effectiveness of which has been demonstrated for nanocrystalline materials in numerous experiments [16,42–44]. However, owing to the Arrhenius dependence of *m* on temperature, even a drastic reduction in the grain-boundary mobility will eventually be overcome at some increased value of the temperature, at which point grain growth will commence.

Potentially more effective would be a reduction in the grain-boundary energy γ, which manifests only a weak temperature dependence (actually decreasing with increasing *T* [45]). According to the Gibbs adsorption theorem, γ can be reduced by the addition of a solute species that segregates to the grain boundaries [46]. Experiment has shown that the resulting effect on γ can be substantial [46], and theoretical considerations predict that it is even possible to drive γ to zero in alloy systems with a high enthalpy of segregation [44,47–49]. This can be understood in terms of the energy balance associated with grain growth in a strongly segregating system. Assume that the grain boundaries of such a material contain n_S^{GB} solute atoms per unit grain-boundary area and that this value is the saturation level for the system in question. When a portion dA of the total grain-boundary area is removed by grain growth, the solute atoms segregated to dA must either be redissolved in the crystalline phase, or they must precipitate out as a bulk second phase. If the latter process is kinetically hindered, such as by unfavorable conditions for nucleation, the change dG_{tot} in the total free energy caused by grain growth will be (ignoring entropic effects)

$$dG_{tot} = \left(\gamma_t - \Delta H_{seg} n_S^{GB}\right) dA, \tag{6}$$

where $dA < 0$, ΔH_{seg} is the segregation enthalpy per solute atom, and γ_t denotes the topological part of the grain-boundary energy arising from the excess volume caused by the non-ideal atomic packing in the boundary core [49]. Therefore,

$$\gamma \equiv \left.\frac{\partial G_{tot}}{\partial A}\right|_{T,P} = \gamma_t - \Delta H_{seg} n_S^{GB} . \qquad (7)$$

Substituting typical values of $\gamma_t = 0.7$ J/m^2 and $n_S^{GB} = 2.5 \times 10^{-5}$ mol/m^2 into Eq. 7, we see that γ vanishes when $\Delta H_{seg} \approx 30$ kJ/mol, a segregation enthalpy that is exceeded in many binary metallic systems [50].

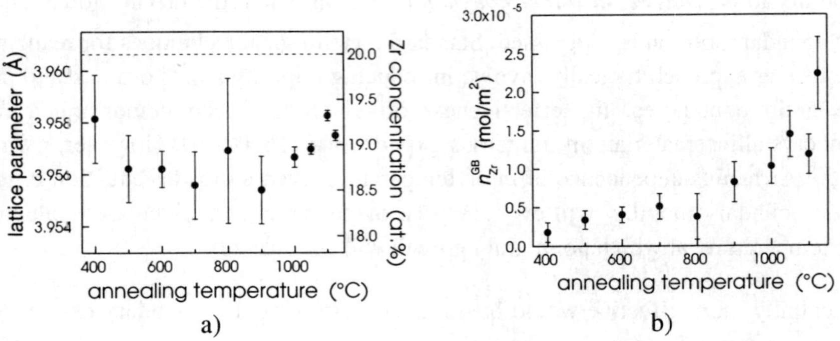

Fig. 4. (a) Lattice parameter of nanocrystalline $Pd_{80}Zr_{20}$ following 24 h anneals at the indicated temperatures [51]. (b) Areal density n_{Zr}^{GB} of Zr atoms segregated to the grain boundaries of $Pd_{80}Zr_{20}$ as a function of temperature [51], calculated from the data in (a) and Fig. 5a.

We have investigated the effectiveness of this thermodynamic strategy for suppressing grain growth in the system Pd–Zr [44,49,51], for which ΔH_{seg} equals 31 kJ/mol [50]. Using the technique of ball milling, we prepared single-phase nanocrystalline solid solutions of $Pd_{100-x}Zr_x$ containing up to 20 at.% Zr and then annealed the samples to induce segregation of Zr atoms to the grain boundaries. Owing to the large size mismatch between Pd and Zr atoms (~17%), we could determine the change in Zr concentration in the crystallites by measuring the lattice parameter through x-ray diffraction (Fig. 4a). Combining this information with the grain size (also determined from diffraction scans), we calculated n_{Zr}^{GB} as a function of annealing time (Fig. 4b). The grain-boundary energy γ could then be determined by substituting n_{Zr}^{GB} into Eq. 7 and assuming a value of 0.7 J/m^2 for γ_t of Pd [51].

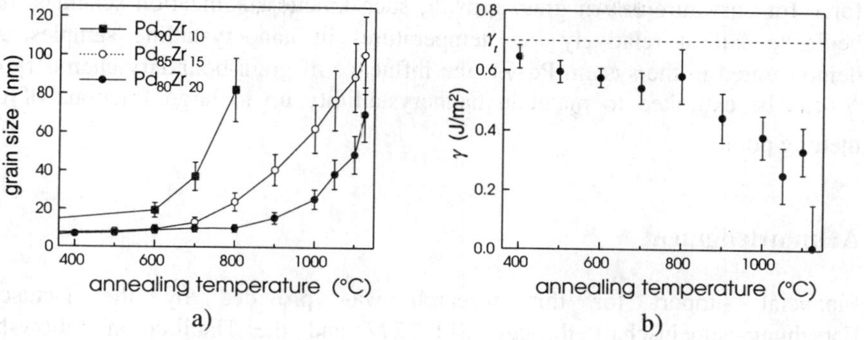

Fig. 5. (a) Grain size $\overline{D}$ in nanocrystalline Pd–Zr following 24 h anneals at the indicated temperatures [51]. (b) Grain-boundary energy γ in $Pd_{80}Zr_{20}$ [51], calculated according to Eq. 7.

The strong stabilizing influence of Zr in Pd is evident in Fig. 5a, in which the grain size is plotted for $Pd_{90}Zr_{10}$, $Pd_{85}Zr_{15}$ and $Pd_{80}Zr_{20}$ as a function of annealing temperature. Clearly, the addition of Zr increases the temperature at which significant grain growth begins, and it suppresses the rate at which it progresses. For two reasons we believe that this growth behavior has a primarily thermodynamic, rather than kinetic, origin: on the one hand, the stabilization persists to temperatures well beyond the range at which Zr-induced solute drag would be expected to be effective and, on the other hand, the plot of γ as a function of temperature indicates that the driving force for grain growth essentially vanishes in the samples manifesting the greatest degree of Zr segregation (Fig. 5b).

5 Conclusions

In many respects, grain growth in nanocrystalline samples resembles grain growth in coarser-grained polycrystalline samples: in the absence of boundary pinning factors, power-law growth with an exponent of 2 is observed, and the activation energy for growth is close to that for grain-boundary self-diffusion. In nanocrystalline Fe prepared by ball milling, on the other hand, evidence for grain-size-dependent growth kinetics was found, with an activation energy close to that of lattice self-diffusion. This behavior is consistent with the predictions of a model for grain growth postulating that the redistribution of excess volume localized in the boundary cores is the rate-controlling step for growth of nanometer-sized grains. The rate of grain growth in nanocrystalline materials can be suppressed using the same strategies for grain-boundary pinning that are successful in conventional materials. However, owing to the $\overline{D}^{-1}$ dependence of the driving force for curvature-driven grain growth, such kinetic stabilization schemes may begin to fail at relatively low temperatures in nanocrystalline samples. As demonstrated in the system Pd–Zr, the influence of grain-boundary chemistry on γ can be exploited to maintain nanocrystallinity up to large fractions of the melting point.

Acknowledgment

Financial support for this research was provided by the Deutsche Forschungsgemeinschaft through SFB 277 and the Habilitation fellowship program (CEK).

References

[1] H. Gleiter, *Prog. Mater. Sci.,* 33, 1989, pp. 223–315.
[2] H. Gleiter, in "Physical Metallurgy," eds. R. W. Cahn, P. Haasen, North-Holland, Amsterdam, 1996, 4th ed., Vol. 1, Chap. 9, pp. 908–942.
[3] N. P. Louat, *Acta Metall.,* 22, 1974, pp. 721–724.
[4] P. A. Mulheran, J. H. Harding, *Mater. Sci. Forum,* 94-96, 1992, pp. 367–372.
[5] G. Palumbo, K. T. Aust, U. Erb, *Mater. Sci. Forum,* 225-227, 1996, pp. 281–286.
[6] M. Upmanyu, D. J. Srolovitz, L. S. Shvindlerman, G. Gottstein, *Interface Science,* 6, 1998, pp. 287–298.
[7] D. T. Wu, *Mat. Res. Soc. Symp. Proc.,* 343, 1994, pp. 55–59.
[8] W. W. Mullins, *Acta Mater.,* 46, 1998, pp. 6219–6226.
[9] U. Czubayko, V. G. Sursaeva, G. Gottstein, L. S. Shvindlerman, *Acta Mater.,* 46, 1998, pp. 5863–5871.
[10] Y. Estrin, G. Gottstein, L. S. Shvindlerman, *Scripta Mater.,* 41, 1999, pp. 385–390.
[11] Y. Estrin, G. Gottstein, E. Rabkin, L. S. Shvindlerman, *Scripta Mater.,* 43, 2000, pp. 141–147.
[12] C. E. Krill III, L. Helfen, D. Michels, H. Natter, A. Fitch, O. Masson, R. Birringer, in "ESRF Highlights 2000," ed. D. Cornuéjols, ESRF, Grenoble, 2001, pp. 72–73.
[13] L. C. Chen, F. Spaepen, *J. Appl. Phys.,* 69, 1991, pp. 679–688.
[14] J. L. McCrea, K. T. Aust, G. Palumbo, U. Erb, *Mat. Res. Soc. Symp. Proc.,* 581, 2000, pp. 461–466.
[15] H. V. Atkinson, *Acta Metall.,* 36, 1988, pp. 469–491.
[16] T. R. Malow, C. C. Koch, in "Synthesis and Processing of Nanocrystalline Powder," ed. D. L. Bourell, The Minerals, Metals & Materials Society, Warrendale, PA, 1996, pp. 33–44.
[17] H. Gleiter, *Acta Mater.,* 48, 2000, pp. 1–29.

[18] H. Hu, B. B. Rath, *Metall. Trans.*, 1, 1970, pp. 3181–3184.
[19] E. A. Grey, G. T. Higgins, *Acta Metall.*, 21, 1973, pp. 309–321.
[20] T. R. Malow, C. C. Koch, *Acta Metall.*, 45, 1997, pp. 2177–2186.
[21] A. Michels, C. E. Krill, H. Natter, R. Birringer, in "Grain Growth in Polycrystalline Materials III," eds. H. Weiland, B. L. Adams, A. D. Rollett, The Minerals, Metals & Materials Society, Warrendale, PA, 1998, pp. 449–454.
[22] O. Senkov, N. Srisukhumbowornchai, M. L. Öveçoglu, F. H. Froes, *J. Mater. Res.*, 13, 1998 pp. 3399–3410.
[23] C. H. Shek, J. K. L. Lai, G. M. Lin, *Nanostruct. Mater.*, 11, 1999, pp. 887–893.
[24] H. Natter, M. Schmelzer, M.-S. Löffler, C. E. Krill, A. Fitch, R. Hempelmann, *J. Phys. Chem. B,* 104, 2000, pp. 2467–2476.
[25] K. W. Liu, F. Mücklich, *Acta Mater.*, 49, 2001, pp. 395–403.
[26] H. G. Jiang, M. L. Lau, E. J. Lavernia, *Nanostruct. Mater.*, 10, 1998, pp. 169–178.
[27] L. Lu, L. B. Wang, B. Z. Ding, K. Lu, *Mater. Sci. Eng. A,* A286, 2000, pp. 125–129.
[28] G. Hibbard, K. T. Aust, G. Palumbo, U. Erb, *Scripta Mater.*, 44, 2001, pp. 513–518.
[29] C. E. Krill III, L. Helfen, D. Michels, H. Natter, A. Fitch, O. Masson, R. Birringer, *Phys. Rev. Lett.*, 86, 2001, pp. 842–845.
[30] R. A. Vandermeer, H. Hu, *Acta Metall. Mater.*, 42, 1994, pp. 3071–3075.
[31] M. Hillert, *Acta Metall.*, 13, 1965, pp. 227–238.
[32] R. W. Cahn, in "Physical Metallurgy," eds. R. W. Cahn, P. Haasen, North-Holland, Amsterdam, 1996, 4th ed., Vol. 3, Chap. 28, pp. 2399–2500.
[33] B. Günther, A. Kumpmann, H.-D. Kunze, *Scripta Metall.*, 27, 1992 pp. 833–838.
[34] V. Y. Gertsman, R. Birringer, *Scripta Metall. Mater.*, 30, 1994, pp. 577–581.
[35] J. Weissmüller, J. Löffler, M. Kleber, *Nanostruct. Mater.*, 6, 1995, pp. 105–114.
[36] C. E. Krill, R. Birringer, *Mater. Sci. Forum,* 225-227, 1996, pp. 263–274.
[37] C. E. Krill, R. Birringer, *Phil. Mag. A,* 77, 1998, pp. 621–640.
[38] C. S. Pande, *Acta Metall.*, 35, 1987, pp. 2671–2678.
[39] B. E. Warren, "X-Ray Diffraction," Addison-Wesley, Reading, MA, 1969, Chap. 13.
[40] C. E. Krill, R. Haberkorn, R. Birringer, in "Handbook of Nanostructured Materials and Nanotechnology," ed. H. S. Nalwa, Academic Press, San Diego, 2000, Vol. 2 "Spectroscopy and Theory," Chap. 3, pp. 155–211.
[41] A. P. Sutton, R. W. Balluffi, "Interfaces in Crystalline Materials," Clarendon Press, Oxford, 1995, p. 353.
[42] C. D. Terwilliger, Y.-M. Chiang, *Nanostruct. Mater.*, 4, 1994, pp. 651–661.
[43] K. J. Bryden, J. Y. Ying, *Acta Mater.*, 44, 1996, pp. 3847–3854.

[44] C. E. Krill, H. Ehrhardt, R. Birringer, in "Chemistry and Physics of Nanostructures and Related Non-Equilibrium Materials," eds. E. Ma, B. Fultz, R. Shull, J. Morral, P. Nash, The Minerals, Metals & Materials Society, Warrendale, PA, 1997, pp. 115–124.

[45] L. E. Murr, "Interfacial Phenomena in Metals and Alloys," Addison-Wesley, Reading, MA, 1975, pp. 130–133.

[46] E. D. Hondros, M. P. Seah, S. Hofmann, P. Lejček, in "Physical Metallurgy," eds. R. W. Cahn, P. Haasen, North-Holland, Amsterdam, 1996, 4th ed., Vol. 2, Chap. 13, pp. 1201–1289.

[47] J. Weissmüller, *Nanostruct. Mater.*, 3, 1993, pp. 261–272.

[48] J. Weissmüller, *J. Mater. Res.*, 9, 1994, pp. 4–7.

[49] C. E. Krill, R. Klein, S. Janes, R. Birringer, *Mater. Sci. Forum,* 179-181, 1995, pp. 443–448.

[50] F. R. de Boer, R. Boom, W. C. M. Mattens, A. R. Miedema, A. K. Niessen, "Cohesion in Metals: Transition Metal Alloys," North-Holland, Amsterdam, 1988.

[51] H. Ehrhardt, Ph.D. thesis, Universität des Saarlandes, 1998.

Producing Ultra Fine Grain Structure in Large Scale Billets

O.A. Kaibyshev

Keywords: grain refinement, dynamic recrystallization, grain boundary sliding, phase transformation, titanium alloys, stainless steels, nickel based alloys.

Abstract. The The results of investigations of grain refinement methods are presented in this paper. Titanium alloys, stainless steels and superalloys are the materials that were studied. The most significant effect of microstructure refinement was observed during severe plastic deformation in alloys with an initial lamellar type microstructure due to dynamic recrystallization and phase transformation. The combination of dynamic recrystallization and phase transformation makes it possible to produce micro- and sub-microcrystalline structures in massive billets. Hydrogen alloying of Ti-base alloys decreases the temperature of ($\beta \rightarrow \alpha$) transition, and it becomes possible to generate a highly dispersed microstructure upon quenching. After that during subsequent severe plastic deformation a nanocrystalline state can be formed. Conditions for sub-microcrystalline structure formation in austenitic stainless steels and Ni based alloys are considered, too. On the basis of the results of the investigations the general principles of grain refinement methods in commercial alloys are discussed.

1 Introduction

The methods of grain refinement in metals and alloys are among the most advanced directions in physical metallurgy. The resultant fine grain structure provides a significant increase in low temperature strength and high temperature ductility [1-2]. One of the possible mechanisms of grain refinement is dynamic recrystallization (DRX) [3-6] occurring during severe plastic deformation (SPD) [7]. The possibility of microstructure refinement to micro-crystalline (MC), sub-microcrystalline (SMC) and nanocrystalline (NC) states by means of SPD was widely reported [8-10]. However, it is rather difficult to utilize the described methods for obtaining a MC structure in large-scale billets. The aim of the present work is to analyze the possibility of producing fine grain microstructures in large-scale billets out of (α+β)-titanium alloys and to investigate the influence of temperature, strain-rate conditions, modes of deformation and combinations of DRX with phase transformation on the formation of SMC and NC structures in different commercial alloys.

2 Experimental Procedure

Three typical commercial alloys having phase transformation - VT9, VT6, 18Cr-10Ni, and the dispersion-hardening EP962 alloy were investigated. The titanium alloy BT9 was examined in the form of hot deformed rods 20.0 mm in diameter, and the VT6 alloy in the form of an ingot. The chemical compositions are shown in Table1. The temperatures of $(\alpha + \beta \rightarrow \beta)$ - transition in VT9 and VT6 alloys were 1000°C, and 980°C, respectively.

Table 1. Compositions of the titanium alloys used in the present study.

Alloys	Amount of chemical elements,% wt.					
	Ti	Al	V	Mo	Zr	Si
VT9	Base	6.5	-	3.5	1.6	0.27
VT6	Base	6.5	4.5	-	-	-

Hydrogen alloying of the VT9 alloy was conducted according to the technique [11]. In order to investigate the process of SMC structure formation in an austenitic steel, the 18Cr-10Ni alloy was used. Its chemical composition is shown in Table 2. Steel has a matrix type structure and is not strengthened by thermal treatment.

Table 2. Composition of the 18Cr-10Ni stainless steel.

Alloy	Amount of chemical elements, % wt.						
18Cr-10Ni stainless steel	Fe	C	Cr	Ni	Ti	Mn	Si
	Base	0.1	13.1	10.7	2.8	4.6	0.36

The EP962 alloy was examined in the form of hot rolled billets of 200.0 mm in diameter. In the initial state the billets have coarse-grain microstructure. The temperature of dissolution of γ'-phase is 1150°C. The chemical composition is shown in Table 3.

Table 3. Chemical composition of nickel based EP962 alloy.

Alloy	Amount of chemical elements,%.									
EP962	Ni	C	Cr	Co	W	Mo	V	Al	Ti	Nb
	Base	0.1	13.1	10.7	2.8	4.6	0.4	3.2	2.6	3.4

Mechanical tests were conducted over the temperature range of 600-950°C using a special testing machine for complex loading [12] and universal testing machine Shenk-PSA. Hydraulic presses of 100 and 1600 ton-force power and a rolling mill, model SRD-800, were also used for deformation of large-scale billets. The scanning electron microscope JSM840 and transmission electron microscope JEM-2000EX were used in this study. An optical microscope Axiotech 100 and PC-software program KS300 were used to perform the quantitative microstructure analysis.

3 Results

3.1 Titanium alloy VT9

Uniaxial tension. In Fig.1 a diagram of tension of VT9 alloy with lamellar (curve 1) and micro-crystalline (curve 2) structures is presented. The diagram of the alloy with the coarse grain lamellar type microstructure has a dropping character. The flow stress has its maximum at a strain of 0.05 and after that it sharply drops. The flow stress σ weakly depends on a strain value in a fine-grained alloy during deformation under similar temperature-strain rate conditions. The systematic studies [3] showed that during plastic deformation of alloy with the lamellar type microstructure (Fig. 2a) a gradual fragmentation of α-plates and formation of equiaxed microstructure were observed. After small e a sub-structure is formed in α- plates. While the strain value is increased the mis-orientation of sub-grain boundaries is increased too, and at e > 0.4 the microduplex type microstructure is finally formed. However, during tension the process of grain refinement occurs non-uniformly over the sample's length due to a non-uniform character of the strain state developed in samples with initial lamellar type microstructure. The most significant effect of grain refinement was observed in a neck zone. Away from the neck α-particles retain their plate-like shapes.

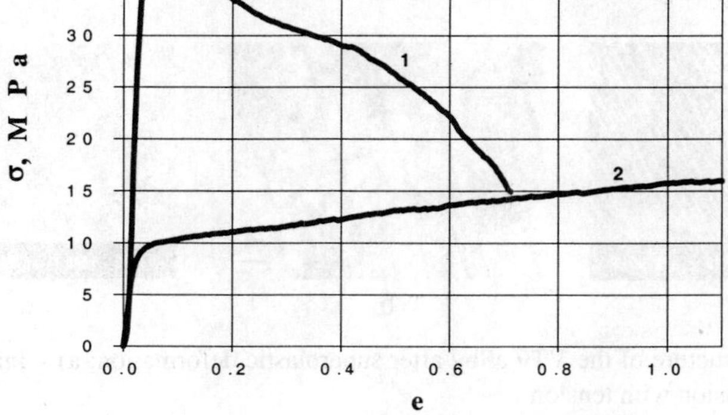

Fig. 1. σ-f(e) dependence for the VT9 alloy. 1 – coarse grain microstructure, D=340.0μm, 2 – fine grain microstructure, d=3.0 μm.

It is obvious that stress-strain states influence significantly the development of fine grain microstructure during severe plastic deformation. Magnetically forced grain boundary motion In order to check this assumption the samples were deformed under different loading modes. The special testing machine was used for this purpose.

Torsion. During torsion tests as well as during tension tests a significant softening is observed. The flow stress σ decreases by almost a factor of 3 when stress is increased up to 1.0. However, the localization of deformation in the gauge zone of samples was not observed. The microstructure changes uniformly over the sample's length and non-uniformly over its radius. After torsion some of the α-plates were bent and other were divided into fragments. Almost all fragments were shifted relative to each other. In the β-phase the areas with different orientation related to the initial matrix were observed in the peripheral zone (max. deformation) of the deformed samples. The most significant feature of torsion is that the initial microstructure, temperature and strain-rate regimes do not depend on changing strain state conditions. Therefore, material flows without localization and neck is not formed in the working zone [13]. During high temperature torsion the grain refinement occurs in accordance with the mode of deformation. It is uniform over the sample's length and non-uniform in the cross section. However, the efficiency of this process is low. The lamellar type microstructure does not completely transform to the globular one even after deformation by torsion up to e=1.1.

Torsion combined with tension. Torsion with simultaneous tension has high stability to severe deformation taken from torsion (Fig. 1, curve 2) and uniformity of stress conditions derived from tension. Such combination increases a resource of plasticity in deformed materials. As a result, the process of grain refinement becomes more efficient (Fig. 2c). Moreover, uniform plastic deformation under superplastic conditions provides more uniform transformation of the lamellar type microstructure to the micro-duplex one. This has been confirmed by deformation of large-scale billets out of the VT6 alloy using two different schemes of high temperature loading.

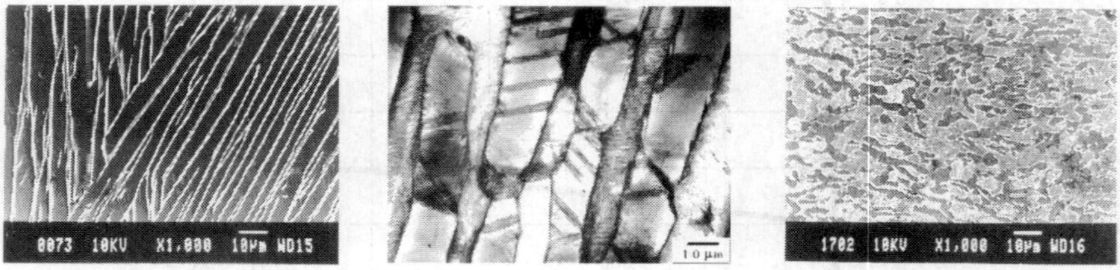

a b c

Fig. 2. Microstructure of the VT9 alloy after superplastic deformation: a) – initial state; b) – tension, c)- torsion with tension.

The billets, 250 mm in diameter and 300 mm in length (Fig. 3a) were treated under conditions of hot multi-axial forging [14]. The billets (Fig. 3b), 130 mm in diameter and 800 mm in length were deformed by torsion combined with tension under isothermal conditions. The alloy was in the form of ingots. The mean size of β- transformed grains was D=10-15 mm. The ingots were preliminarily subjected to homogenizing annealing and subsequent hot deformation in β-region. After such processing the ingots had a lamellar type microstructure with a size of β-transformed grains equal to 1-2 mm. Then ingots were subjected to deformation according to different loading schemes: 1 – multi-axial forging on a hydraulic press; 2 - complex loading by torsion with concurrent tension on a rolling mill SRD-800. In both cases the deformation temperature was 800-950°C. After thermomechanical processing in both cases the lamellar type microstructure completely transforms to the MC one. The billets had a microstructure with globular grains uniformly distributed within their volume. The mean grain size of finally transformed microstructure was 8.0-9.0 μm (Fig. 4) in both cases.

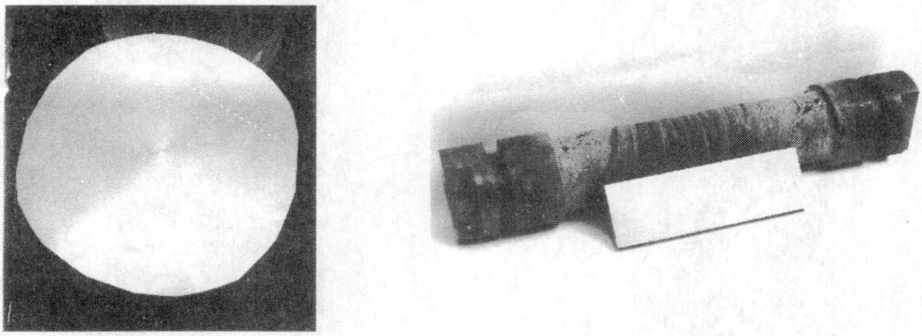

a b

Fig. 3. Microstructure of forged billets out of VT6 alloy: a) – after multi-axial forging, b) – after torsion combined with tension on a rolling mill SRD800.

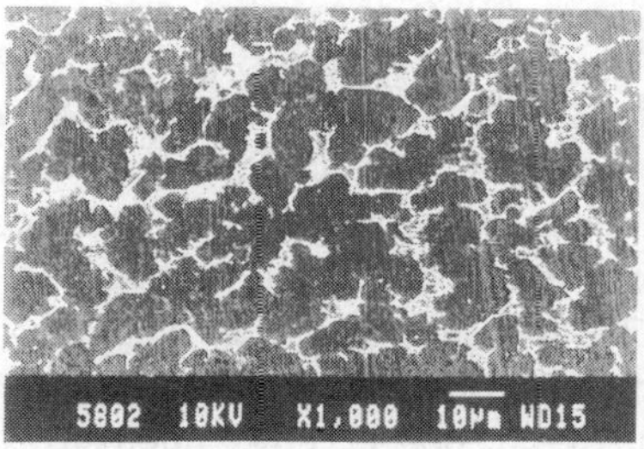

Fig. 4. Microstructure of VT6 alloy after thermo-mechanical processing.

The results of investigation show that during deformation under isothermal conditions as strain is increased the final size of microstructure constituents significantly decreases. Therefore, in order to form a microstructure with mean grain size of a nanometric scale the initial coarse grain lamellar type microstructure should be highly dispersed in order to provide high plasticity at low temperatures.

Hydrogen softening [11] was selected as a method to promote a high dispersion of microstructure. Being an efficient β-stabilizer for titanium alloys, the hydrogen decreases temperature A_{c3} of $(\alpha+\beta)\rightarrow\beta$ transition, critical cooling rate and diffusion rate of alloying elements in β-phase. All these contribute to the formation of highly dispersed martensite during quenching. The VT9 was hydrogen alloyed up to 14% at. (0.34% mass.). This leads to decreasing a temperature of allotropic transformation by more than 150°C as compared to the base alloy. Quenching from temperature exceeding A_{c3} by 10°C promotes the formation of the meta-stable β-transformed martensite type microstructure (Fig. 5a). The mean β-grain size in VT9 and VT9+13.7H alloys was 270 ± 30 and 30 ± 3 μm, respectively. Heating of samples prior to deformation led to partial decomposition of martensite, and precipitation of α-plates was observed.

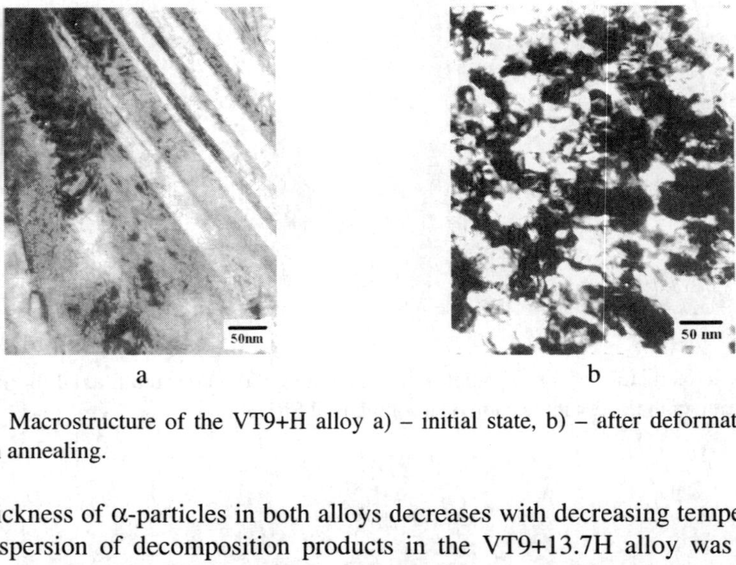

a b

Fig. 5. Macrostructure of the VT9+H alloy a) – initial state, b) – after deformation and vacuum annealing.

The thickness of α-particles in both alloys decreases with decreasing temperature. The dispersion of decomposition products in the VT9+13.7H alloy was higher than in the base alloy after its processing under similar conditions. For example, after both alloys were annealed at 600°C the thickness of precipitated particles in the VT9 and VT9+13.7H was less than 0.15 and 0.05 μm, respectively. The microstructure in preliminary quenched samples completely transforms to micro-duplex one after deformation at 600°C. Finer microstructure was formed in this case. The mean size of grains in the VT9 was 0.1 μm, but in the hydrogen saturated alloy it was less by almost twice (d=0.06 μm). Moreover, it was possible to deform hydrogen alloyed samples up to ε=80% at temperature 525°C, whereas VT9 alloy samples failed already after ε=20% at temperature 580°C.The hydrogen was removed to safe concentration (0.004% mass.) from alloy with NC structure during subsequent annealing at 550-570°C. It was confirmed by the results of spectrum analysis.

3.2 18Cr-10Ni stainless steel

Let us consider the grain refinement method in an austenitic steel. The steel has a small amount of second phases, that is why for titanium alloys the grain refinement processing by the above described routes is quite difficult. However, due to the combination of dynamic recrystallization and phase transformation a significant effect of grain refinement can be attained. The possibility of grain refinement due to DRX occurring during deformation over a wide range of temperature and strain-rate conditions was systematically studied. Table 4 summarizes data on a mean size of recrystallized grains (d) and its volume fraction (Vr) obtained under compression tests to strains up to ε=70%. Dynamic recrystallization occurs within a whole volume during plastic deformation at sufficiently high temperatures. At low deformation temperatures, over a range of 400-600°C, DRX is retarded. In this case the microstructure with a crystallite size of 0.2 μm can be formed only by severe plastic deformation via special technological routes. It is necessary that a definite level of strain should be reached during low temperature deformation. So that, a dynamically recrystallized microstructure with a mean grain size of 4-6 μm and recrystallized volume of 100% was formed at the first stage of processing. This microstructure was obtained by utilizing the following regime: temperature - 900 C, total strain - 150%. The martensite transformation usually occurs during deformation of austenite at temperatures below Md. The influenceof the initial grain size on thickness of martensite plates after martensitic transformation was studied on samples with initial coarse and fine grain (d=4-6 μm) microstructure formed during DRX. The samples were deformed at 196°C (temperature of liquid nitrogen) up to strains 20-50%. The investigation of microstructure evolution during deformation shows that the size of initial austenite grains significantly influences the dispersion of martensite plates. So, after deformation at the cryogenic temperature the samples with initially coarse grained structure had a mean grain size of 0.5-1 μm., whereas in fine-grained samples the mean grain size was 0.05-0.1 μm. In this case a strain value of 30% provides martensitic transformation within the 95% of the sample's volume. The lamellar type SMC structure was formed there (Fig. 6a). However, unlike titanium alloys, in steels a globular type microstructure can be formed by means of annealing at 650°C. Annealing causes a reverse $\alpha' \rightarrow \gamma$ transformation and recrystallization. It was determined that the grain size of newly formed austenite was limited by thickness of martensite plates. Due to that a SMC austenite structure with a mean grain size of 0.1-0.2 μm (Fig. 6b) was formed in fine-grained samples. At the same time the coarse-grained samples had a microstructure with a mean grain size of austenite d=0.5-3 μm. Moreover microstructure was not uniform there. Thus, microstructure formed during DRX under preliminary hot deformation favorably influences the martensite transformation, i.e. provides formation of highly dispersed martensite structure without crystallographic texture. In case of reverse $\alpha' \rightarrow \gamma$ transformation the thickness of martensite plates determines the size of newly formed austenitic grains.

Table 4. The mean grain size and volume fractions of recrystallized grains in a stainless steel. Compression at temperatures 0.6-0.7 T_m, ε=70%.

$\dot{\varepsilon}\cdot 10^{-4} s^{-1}$	700°•		800 °C		900 °C		1000 °C	
	D, μm	Vr, %	d, μm	Vr, %	D, μm	Vr, %	D, μm	Vr ,%
5.5	0,5	<10	2	34	4	80	12	100
550	-	-	1	12	2	55	5	85

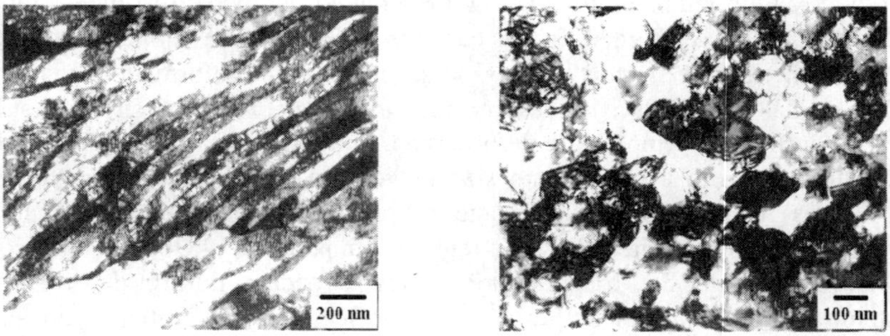

a b

Fig. 6. Microstructure of the 18Cr-10Ni: a) - after deformation, b) - after deformation and annealing.

3.4 Superalloy EP962

The alloy under study was subjected to thermo-mechanical processing. Isothermal deformation, intermediate annealing and gradual, step-by-step decreasing of deformation temperature are main steps of processing. In the initial state the alloy had a coarse grain microstructure with a mean grain size of 80 μm (Fig. 7a). Coherent dispersed γ-phase precipitates were uniformly distributed within the grains. The microstructure had grain boundary precipitates of γ'-phase, too. These particles were coarser by 2-3 times than intergranular ones and they had partially coherent interphase boundaries. It is known [1,2] that formation of the MC structure in nickel based alloys is a very complicated process. It is conditioned by mutual influences of both recrystallization and solid solution decomposition. Therefore, it is determined not only by the heterogenization but also by microstructure evolution during hot deformation. Prior to hot deformation the

EP962 alloy was subjected to homogenizing annealing. The samples were heated to 1180°C. Then they were held at this temperature for an hour. After that they were cooled down to the temperature 800°C with a rate of 25°C/h. During this processing the matrix grains did not change significantly. At the same time the intragranular γ'-phase became coarser (up to 0.5 μm). Moreover, the coherency at γ/γ' interphase boundaries was partially violated, that was confirmed by the presence of dislocations discrepancy at the given boundaries. Preliminary heterogenization accelerates the formation of a microduplex structure. The results of investigation of micro- structure evolution

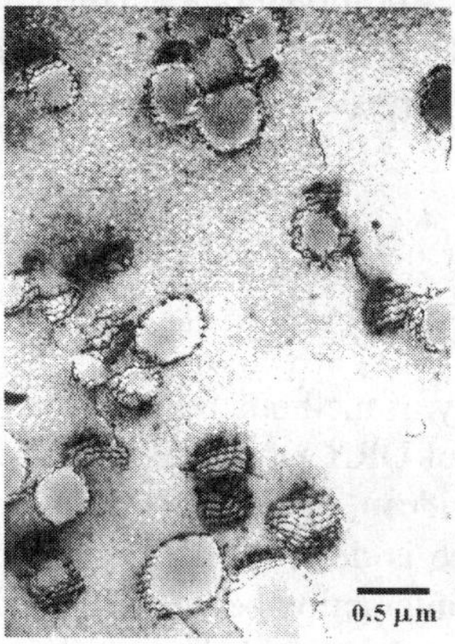

a

b

Fig. 7. Microstructure of the EP962 alloy: a) - initial state, b) - after hot deformation and annealing.

during deformation at 1100°C and over a strain-rate interval of 10^{-5}-10^{-3} s^{-1} allowed us to establish the following. At the initial stage of plastic flow (ε=5-10%) a sub-granular structure was formed within the initial coarse grains. The mean size of cells was comparable with a distance between γ'-phase particles. At the same time chains of fine recrystallized grains were formed at the grain boundaries of initial grains. These chains were fringed with non-coherent γ'-phase particle grains having a globular shape. With increasing a strain value up to 50-75% the recrystallized volume increased up to 33-53%. The following post deformation annealing at deformation temperature for 2-4 hours resulted in recrystallization and formation of a microduplex structure. Step-by-step deformation at 1100-1075°C to the total strain value ε=80% also led to transformation of the initial coarse grain matrix structure (d=80μm) to the microduplex one. The sizes of γ and γ' particles had a comparable size and were equal to 5.5 and 2.0 μm, respectively in the transformed microstructure. When temperature of processing was further decreased (from 975°C to 875-850°C) the dispersed γ'-phase particles with size 0.1-0.2 μm, within the γ-phase grains (with mean size 2.5 μm) were precipitated. The total amount of γ'-phase particles of both dispersed and coagulated ones was

45%. After deformation of the samples with the MC structure (d=2.5 μm) at 850°C, with the strain rate 10^{-3} s^{-1} and ε=65%, the SMC structure with the mean size of γ and γ'-phase grains d=0.1-0.4 μm and the recrystallized volume 20% was formed. This fact as well as the effect of refinement of γ and γ'-phases testify that recrystallization evidently develops not only in the matrix but also in the γ'-phase. Significant elastic stresses are observed in recrystallized grains. After post-deformation heat treatment (annealing at 850 °C during 8 hours) the volume of recrystallized structure was increased up to 70%. The mean grain size of γ and γ'-grains increased up to d=0.25 μm, too. However, the elastic stresses were not eliminated completely after such heat treatment. Additional deformation at 700°C up to ε=65% refines the initial SMC structure. The NC structure with a mean size of γ and γ'- grains d=0.05 μm (Fig. 7b) forms after additional plastic deformation.

4 Discussion

The data given above demonstrate a wide variety of methods for refining grains in commercial alloys. In general, the combination of DRX with phase transformation is most efficient. In reality, DRX provides significant refinement of microstructure. However, a minimum grain size is determined by a size of sub-grains formed during deformation as a result of interaction between dislocations of different slip systems. When a value of strain is increased, low angle sub-grain boundaries transform to high angle boundaries of a common type. Grain boundary sliding can be realized after that. During further deformation a microduplex structure with a grain size of 3-5 μm can be formed. Both the phase transformation and the dispersion hardening processes are accompanied by formation of interphase boundaries that provides a more significant refinement effect on microstructure. For example, it has been shown that martensite transformation occurring in steel where the initial microstructure was preliminary refined during DRX, provides formation of microstructure with grain size of several sub-microns. The same effect can be attained during processing of superalloys. However, there are some small differences. Firstly, dynamic recrystallization refines a matrix microstructure. Secondly, during heterogenization the coherent γ-γ' boundaries transform to the usual interphase boundaries. As a result, a microduplex microstructure is formed during high temperature deformation. A SMC structure can be formed if deformation temperature is decreased. A quite similar process occurs during hot deformation of titanium alloys. Quenching from the β-region results in the formation of a lamellar microstructure. This structure can be transformed to a microduplex one by means of hot deformation under temperature and strain-rate conditions corresponding to superplastic deformation. The process of microstructure refinement during deformation under superplastic conditions can be described as follows. Plastic deformation is realized by movement of dislocations in different slip systems. As a result, sub-grains are formed. Plates of α-phase are divided into fragments. When the strain value is increased the low angle boundaries are transformed to high angle boundaries of a common type. After that, grain boundary sliding can proceed at these boundaries. As a result, during further deformation a microduplex structure is formed. The mode of loading influences the microstructure refinement process. It is not only due to uniform stress distribution within the deformed volume but also to more favorable conditions for operation of multiple slip systems that facilitates sub-grains formation. A significant effect of microstructure refinement can also be attained by decreasing the temperature of deformation as well as the temperature of $\beta \rightarrow \alpha$ transition, for example, due to hydrogen alloying. When microstructure is refined to a very fine level the effect of optimum superplasticity shifts to a lower temperature (low temperature superplasticity). That provides an opportunity to refine a lamellar microstructure to a microduplex one, and even to sub-microcrystalline and nanocrystalline scales.

5 Conclusions

1. The systematic investigations of grain refinement processes during hot plastic deformation have been conducted both in VT9 and VT6 titanium alloys with initial lamellar type structure. The results show that the lower the temperature of deformation the finer the microstructure. Hydrogen alloying decreasing the deformation temperature to 560-600°C provides an opportunity to form SMC and NC structures.
2. It is shown that application of hot multi-axial forging and torsion combined with tension under isothermal conditions provides an opportunity to process large-scale billets out of VT6 titanium alloy.
3. Severe hot plastic deformation contributes to grain refinement due to dynamic recrystallization. Application of martensite transformation and subsequent tempering provides formation of a MC structure in austenitic stainless steel.
4. Severe plastic deformation of nickel based superalloys over the temperature range of 0.6-0.9 Tm and at strain rates $10^{-4} - 10^{-2}$ s^{-1} combined with homogenizing annealing provides the formation of MC, SMC and NC structures due to combination of DRX and coagulation of γ'-phase processes during deformation.

References

[1] O.A. Kaibyshev, "Plasticity and Superplasticity of Metals", Metallurgy, Moscow, 1975.
[2] O.A. Kaibyshev, "Superplasticity of Alloys, Intermetallides and Ceramics", Springer-Verlag, Berlin, 1992.
[3] O.A. Kaibyshev, R.Ya. Lutfullin, G.A. Salishchev, *FMM*, 66, 1988, pp. 1163-1171.

[4] T. Sakai, J.J. Jonas, *Acta Metallurgica*, 32, 1984, pp. 189-209.

[3] R.O. Kaibyshev, O.Sh. Sitdikov, *DAN SSSR*, 321, 1981, pp. 306-310.

[6] O.A. Kaibyshev, G.A. Salishchev, *FMM*, 4, 1992, pp. 132-142.

[7] V.M. Segal, V.I. Reznikov, A.E. Drobyshevskiy and V.I. Kopylov, *Russ.Metall.*, 1, 1981, p. 99.

[8] G.A. Salishchev, R.G. Zaripova, A.A.Zakirova, H.J.McQueen, in "Submicrocrystalline Materials: Structure Formation and Properties: Hot Workability of Steels ans Light Alloys-Composites – Proc.Int.Symp." (eds. H.J. McQueen, E.V. Konopleva, N.D. Ryan), CIM, Montreal, 1996, pp. 217-226

[9] G.A. Salishchev, R.G. Zaripova, in "Structure formation in stainless steels during hot deformation: Strip Casting, Hot and Cold Working of Stainless Steels", CIM, Quebec, 1993, pp. 79-91.

[10] M.I. Mazursky, M. A. Murzinova, G.A. Salishchev, D.D. Afonichev, *Metalls* 6, 1995, pp. 83-88.

[11] V.K. Nosov, B.A. Kolachev, "Hydrogen plasticization at hot deformation of titanium alloys", Metallurgy, Moscow, 1986.

[12] O.A. Kaibyshev, R.A. Vasin, V.K. Berdin, R.M. Kashaev, *Zav.Lab.*, 4, 2000, pp. 50-53.

[13] O.A. Kaibyshev, V.K. Berdin, M.V. Karavaeva, R.M. Kashaev,L.F. Syutina, in "The 9th world Conf. On Titanium, 7-11 June 1999, Abstract Booclet", CRISM "Pometey", St. Petersburg, Russia, p. 53-2.

[14] G.A. Salishchev, R.M. Galeev, O.R. Valiakhmetov, S.Z. Zherebtsov, V.F.X. Gigliotti, B.P. Dewlay, in "The 9th world Conf. On Titanium, 7-11 June 1999, Abstract Booclet", CRISM "Pometey", St. Petersburg, Russia, p. 59-3.

Fundamentals, Modeling and Simulation of Grain Growth

Computer Simulation of Grain Growth Based on Orientation Data of Pure Aluminum

M. Kobayashi[1], Y. Takayama[2], H. Kato[2], and T. Shibayanagi[3]

[1] Satellite Venture Business Laboratory, Utsunomiya University, 7-1-2 Yoto, Utsunomiya, Tochigi, Japan

[2] Department of Mechanical Systems Engineering, Utsunomiya University, 7-1-2 Yoto, Utsunomiya, Tochigi, Japan

[3] Joining and welding Research Institute, Osaka University, 11-1 Mihogaoka, Ibaraki, Osaka, Japan

Keywords: grain growth, computer simulation, SEM/EBSP analysis.

Abstract. The modified Monte Carlo (MC) simulation of grain growth has been performed in the model applying practical orientation data by SEM/EBSP analysis in a pure aluminum to an initial microstructure. Changes in microstructure feature and textures have been compared between practical and simulated results. The SEM/EBSP analysis reveals that the pure aluminum has a residual rolling texture after annealing. The texture becomes sharper with the increase of annealing time. Microstructural and textural developments in the simulation were in good agreement with those in the experiment.

1 Introduction

The scanning electron microscopy / electron backscatter diffraction pattern (SEM/EBSP) method is an effective technique to get information of crystallographic orientation [1]. The data format of SEM/EBSP analyzing system is similar to that of the Potts model in MC simulation [2]. In the present study, in order to predict microstructural evolution, the modified MC simulation of grain growth is performed in the model applying practical orientation data to an initial microstructure. Changes in microstructure feature and texture are compared between experimental and simulated results, and moreover, what factor plays an important role during grain growth is discussed.

2 Experimental Procedure and Simulation Method

An experimental sample used in this study was an annealed sheet of a 99.99% pure aluminum (A1N99) after rolling. Small specimens cut from this sheet were isothermally annealed in salt bath at 673K for 60 s~57.6 ks. After annealing the specimens were polished mechanically and electrochemically in the RD-TD section. Orientation mapping data of them were investigated by SEM/EBSP method in area 2×2 mm^2 by a step size of 10 μm.

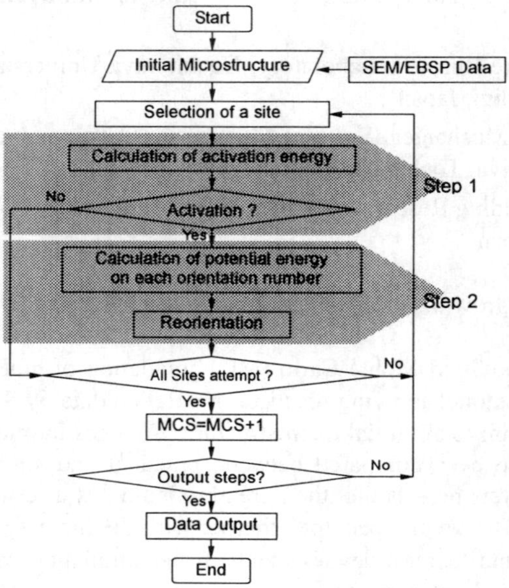

Fig. 1. Flow chart for simulation of grain growth.

The MC simulation using the Potts model has been already performed in consideration of reorientation accompanied with activated state for grain boundary (GB) migration [3]. In the present study, the simulation model with the activated state was also carried out using the practical orientation data not only to predict the microstructural evolution but also to clarify effects of GB energy and mobility. The flow chart of the simulation is shown in Fig. 1. A site is selected at random, and then a reorientation trial of the site is done by two steps. The first step is the activation step of the site, in which whether site is activated is determined with the following equation.

$$W_a \propto \exp\left(-E_a/\alpha^3 k_B T\right), \tag{1}$$

where W_a is activation probability in the selected site, E_a is activation energy of the site, α is a constant concerned with the ratio of site to atom, k_B is Boltzman constant, and T is absolute temperature. Furthermore the activation energy E_a is given by

$$E_a = E_{act} - (U - E_m), \tag{2}$$

where E_{act} is potential energy of activated state of the site, which is the maximum value calculated as GB energy for the neighbors shells, U is GB energy of the selected site, and E_m is reduction of the energy for the presence of impurities in boundary, which is defined as sum of interaction factor μ in each site and represents mobility of GB migration. If the selected site is activated according to the probability (1), then it goes on to the next. The second step is a selection of next orientation corresponding to the GB energy on the site. The orientation is changed with the following probability

$$W_i = f_i \Big/ \sum_{n=q} f_n , \qquad (3)$$

$$f_i \propto \exp\left(-U_i / \alpha^3 k_B T\right). \qquad (4)$$

In practice, the equations (1) and (4) are normalized by probabilities at melting point T_m of a material. The simulation proceeds to repeat the above trial. The unit of time is defined as 1 Monte Carlo steps (MCS) per site, which corresponds to trials of a number of site in the model. In this study, the governing parameter of simulations $J/\alpha^3 k_B T_m$ was appropriately given 0.8 for a pure Al and the temperature of simulations, $T=0.72T_m$ was selected to be consistent with the experiments, where J is an unit GB energy.

3 Results and Discussion

3.1 Experimental results

Isothermal grain growth in the pure aluminum was investigated by SEM/EBSP. An initial mean grain size was 79.9 μm in equivalent area diameter. Figure 2 shows an inverse pole figure map in the normal direction (IPF Map [ND]) and pole figures in the specimen annealed at 673K for 57.6 ks. Grains grow with the increase of annealing time to reach the grain size of 175.4 μm for 57.6 ks (Fig. 2a) It is found that the sample has a residual rolling texture after annealing or recrystallization as shown in Fig. 2b. The texture became sharper than that before annealing. Further, normal grain growth was confirmed by the scaling rule observed in grain size distributions during annealing.

3.2 Effects of grain boundary energy and mobility in simulation

To clarify effects of grain boundary energy and mobility, the microstructural evolutions in the twelve hexagonal grain model were simulated for various energy J and interaction factor μ of GBs surrounding a central grain (CG) and rest grains (RG). Figure 3a is a result set up only GB energy with $(J_{CG}, J_{RG}) = (0.8J, 1.0J)$ for a central and rest grains, respectively. A central grain surrounded with low energy boundaries has grown into neighbor grains. Figure 3b shows the microstructure set as $(\mu_{CG}, \mu_{RG}) = (0.6J, 0J)$. The boundaries of the central grain migrate slowly owing to the low mobility reflected by a large interaction factor μ. However, the low energy boundaries elongate from center to periphery due to annihilation of the high energy boundaries connected with them. These are apparently common results from balance of GB energy [4]. On the other hand, the central grain shrinks in Fig. 3c because of its boundaries with higher mobility, nevertheless the boundaries have a relatively low energy. It is, however, found in Fig. 3d that the lower energy of J_{CG} leads to a predictable growth of the central grain in spite of being surrounded with high mobility boundaries, as well as in Fig. 3a. These results reveal that the direction of migration at a triple junction is dependent not only on the GB energy but also on the interaction factor. Consequently, equilibrium at the triple junction of GBs ought to be understood from *GB energy compensated by the mobility*.

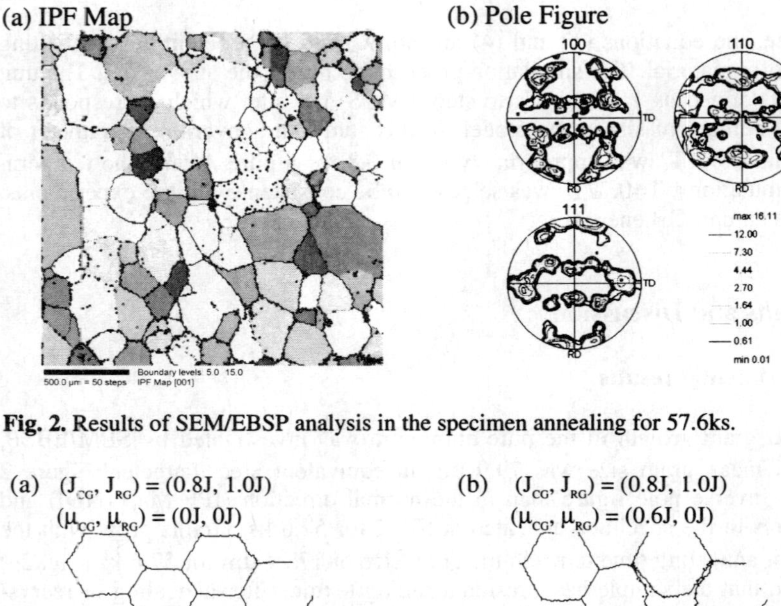

Fig. 2. Results of SEM/EBSP analysis in the specimen annealing for 57.6ks.

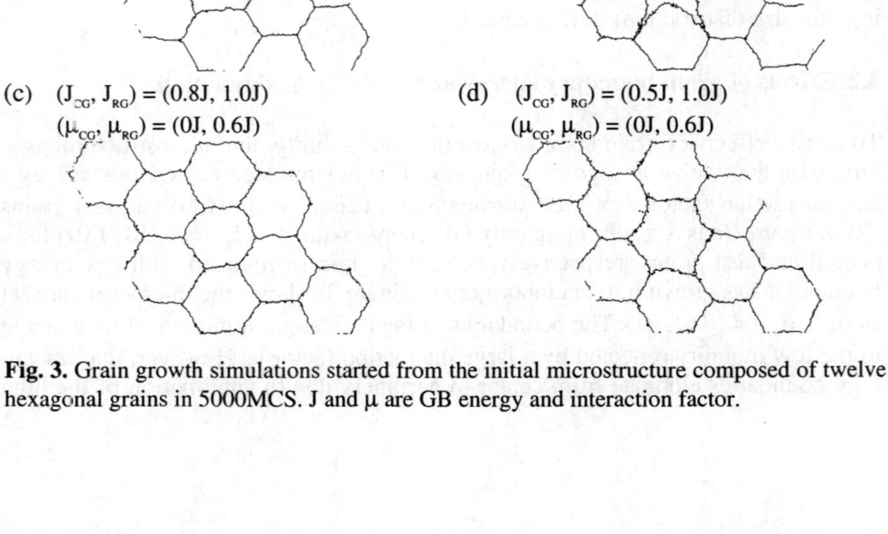

Fig. 3. Grain growth simulations started from the initial microstructure composed of twelve hexagonal grains in 5000MCS. J and μ are GB energy and interaction factor.

3.3 Simulations based on orientations data of a pure Aluminum

The simulations based on orientation mapping data obtained in the specimen annealed for 60s have been performed to compare changes in microstructure feature and texture between practical and simulated results. Following four types of simulations were carried out:

1) basic model, which has a constant energy for all of the GBs.
2) LAGB/HAGB model, in which dependence of GB energies on crystallographic misorientation is taken into account.
3) LAGB/HAGB & CSL model, in which coincidence site lattice (CSL) boundaries is considered to extend 2).
4) GBE + Mobility model, in which GB mobility is considered, in addition to 3).

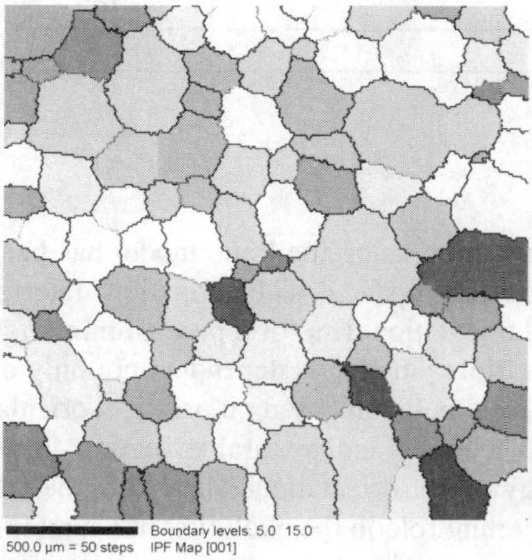

Fig. 4. IPF Map [ND] simulated in a basic model (1000MCS).

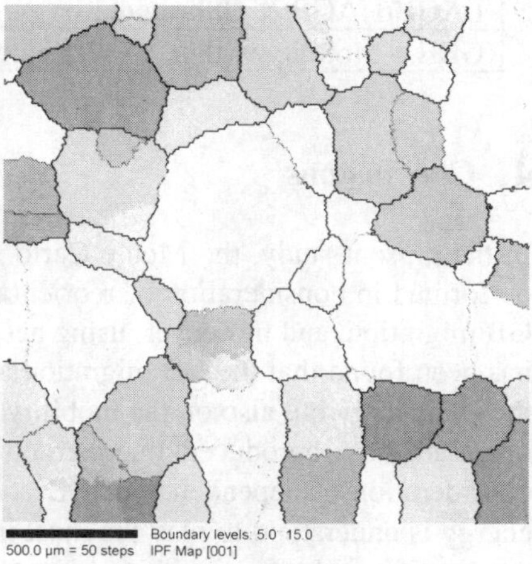

Fig. 5. IPF Map [ND] simulated in a LAGB/HAGB model (5000MCS).

The first simulation is a basic model, which has been often seen in early studies of the MC simulation. The grains have grown commonly in the simulation. The grains after simulation are equiaxed and have the same size as shown in Fig. 4. It was found in the model that the texture was randomized.

In actual microstructure, GBs have various energies depending on crystallographic misorientation [5]. Fig. 5 shows the microstructure evolved in the LAGB/HAGB model assuming low and high angle GBs with low and high energies, respectively. Much-curved and broken-off boundaries are observed in the microstructure. Some grains are colored or oriented with a little difference, which are regard as consisting of subgrains. The simulated microstructure is analogous to experimental one. It was also found the texture was sharpened as corresponding to the experiment.

In the LAGB/HAGB & CSL model, energy cusps are considered for coincidence site lattice (CSL) boundaries of $\Sigma \leq 21$. It was confirmed that fraction of CSL boundaries in the microstructure was almost equivalent to that of experiment.

Slow microstructural evolution was observed owing to low mobility set for HAGB.

Grain growth exponents, n in the experiment and the simulations performed in the present study are listed in Table 1. The exponent in the experiment is 0.119 and then that of the three simulations without mobility are more than 0.3. The grain growth exponent, n of 0.128 in the GBE + Mobility model with μ_{HAGB}=0.5 is in the best agreement with the experimental value.

Table 1. Grain growth exponents, *n* in the experiment and the simulations.

Experimental (673K)	0.119
Basic Simulation (a constant GB energy)	0.327
LAGB/HAGB model	0.320
LAGB/HAGB & CSL model	0.362
GBE + Mobility with μ_{HAGB}=0.5 model	0.128

4 Conclusions

In the present study, the Monte Carlo simulation using the Potts model has been performed in consideration of reorientation accompanied with activation state for GB migration, and moreover, using actual orientation data for a pure aluminum. It has been found that the GB migration at a triple junction is dependent not only on the GB energy but also on the mobility. The simulation based on practical orientation data has reproduced microstructural features and textural evolution under consideration of dependence of GB energy on misorientation. Therefore, the GB energy is understood to play the most important role in the evolution of actual microstructure and texture, although the GB mobility should be also noted.

References

[1] Ph. Bocher and J.J. Jonas, in "Proc. 4th Inter. Conf. on Recrystallization and Related Phenomena", 1999, pp. 25-35.

[2] M.P. Anderson, D.J. Srolovitz, G.S. Grest and P.S. Sahni, *Acta Metall.*, 32, 1984, pp. 783-791.

[3] M. Kobayashi, Y. Takayama, H. Kato, H. Nakajima, *J. Japan Inst. Metals*, 65, 2001, pp. 85-92.

[4] C.S. Smith, *Trans. AIME*, 175, 1948, pp. 15-51.

[5] G.C. Hasson and C. Goux, *Scripta Metall.*, 5, 1971, p. 889.

Recrystallization and Grain Growth
Proceedings of the First Joint International Conference
Eds. G. Gottstein and D.A. Molodov, Springer-Verlag, 2001

Dimensional Effects on Anisotropic Grain Growth

Elizabeth A. Holm[1], Gregory N. Hassold[2], and Mark A. Miodownik[3]

[1] Sandia National Laboratories, Albuquerque, NM 87185-1411 USA
[2] Dept. of Sci. and Math., Kettering University, Flint, MI 48504 USA
[3] Mech. Eng. Dept., Kings College London, Strand, London, WC2R 2LS, UK

Keywords: grain growth, anisotropy, CSL boundaries, molecular dynamics, mobility.

Abstract. Monte Carlo simulations were used to explore grain growth in polycrystals with anisotropic boundary energy and mobility. In crystallographically constrained systems (i.e. strongly textured or two-dimensional systems) low energy boundaries are found to proliferate during grain growth, altering the microstructural morphology and evolution dramatically; however, low mobility boundaries have much less effect. For randomly textured, three-dimensional polycrystals, grain size distribution, topology, and evolution kinetics are relatively insensitive to the details of the boundary property functions. These results emphasize the need to incorporate realistic crystallography in grain growth models.

1 Introduction

Grain boundary properties influence the bulk behavior of polycrystals. For example, increasing the frequency of coincident site lattice (CSL) grain boundaries in a randomly textured polycrystal (i.e. grain boundary engineering) has been shown to improve corrosion resistance and ductility in several commercial alloys [1].
The mechanisms by which CSL boundaries increase in frequency during grain boundary engineering are not understood. Because CSL boundaries have a high degree of lattice coincidence between grains, they have more fulfilled atomic bonds and tend to possess a lower free energy and mobility than general boundaries. Hypotheses have been proposed that their lower energy [2], lower mobility [3], or ease of formation during strain annealing [4] bias evolution toward CSL boundaries. While some theories suggest CSL boundaries increase during grain growth, others assert that mechanical deformation is required. Recent computer simulations of Ono *et al.* [2] seem to show the frequency of some CSL boundaries increases during grain growth.
In this paper, we use a Monte Carlo model to study grain growth in systems where boundary properties depend on crystallography. We examine a two-dimensional (2D) system of tilt boundaries, in which CSL-like boundary properties were dis-

covered by molecular dynamics simulations. We contrast this to a three-dimensional (3D) system of true CSL boundaries, with properties derived from classical theory.

2 Simulation Method

The Monte Carlo Potts model used here has been described in theoretical and computational detail [5]. A continuum microstructure is bitmapped onto a discrete lattice. Each lattice site is allocated a discrete crystallographic orientation $\mathbf{O}_i$ using a method that allows both the initial texture and the boundary misorientation distribution of the ensemble to be defined [6], and grain boundaries are represented by interfaces between neighboring sites of unlike orientation. The misorientation angle between grains i and j, θ_{ij}, is the minimum misorientation angle between orientations $\mathbf{O}_i$ and $\mathbf{O}_j$. The grain boundaries have a misorientation-dependent excess energy $\gamma(\theta_{ij})$. This allows us to specify the total system energy by the Hamiltonian:

$$H = \sum_{i=1}^{N} \sum_{j=1}^{n} \gamma(\theta_{ij}) \tag{1}$$

where the sums are taken over the n sites within the neighbor shell of site i and for all N lattice sites.

Grain growth is simulated by selecting a site i at random and choosing an orientation $\mathbf{O}_j$ at random from the set of neighbor orientations. (Note that this scheme prohibits nucleation of new grains.) The orientation of site i is changed to $\mathbf{O}_j$ with probability

$$P(\Delta E) = \begin{cases} M_{ij} & \Delta E \le 0 \\ M_{ij} \exp(-\Delta E / kT) & \Delta E > 0 \end{cases} \tag{2}$$

where ΔE is the change in energy, $0 \le M_{ij} \le 1$ is the relative mobility of the boundary between grains i and j, and kT is an energy that determines the amount of noise present in the system. After each attempted growth step, the simulation time t is increased by 1/N Monte Carlo steps (MCS).

Simulations were performed on a 250,000 site, fully periodic, triangular lattice with first and second neighbor interactions. To avoid pinning due to simulation lattice anisotropy, kT was set to $0.5\gamma_0$, where γ_0 is the minimum grain boundary energy in the system. To simulate a continuum crystallographic texture, Q = 999 different orientations were allowed. Numeric data points represent the average of 10 independent simulation runs.

3 Results and Discussion

3.1 Two-dimensional crystallography

Upmanyu and Srolovitz performed molecular dynamics simulations of the motion of 2D tilt boundaries in hairpin bicrystals [7]. The atomic lattice is triangular, with a Lennard-Jones interatomic pair potential. Because all boundaries were tilt boundaries, boundary properties are a function of a scalar misorientation angle $0 \leq \theta \leq 30°$. Fig. 1 shows boundary free energy and mobility at half the melting temperature. Because the simulation loses resolution at low tilt angles, a Read-Shockley boundary energy [8] was used for $\theta < 10°$, and mobility was linearly extrapolated to zero for $\theta < 5°$.

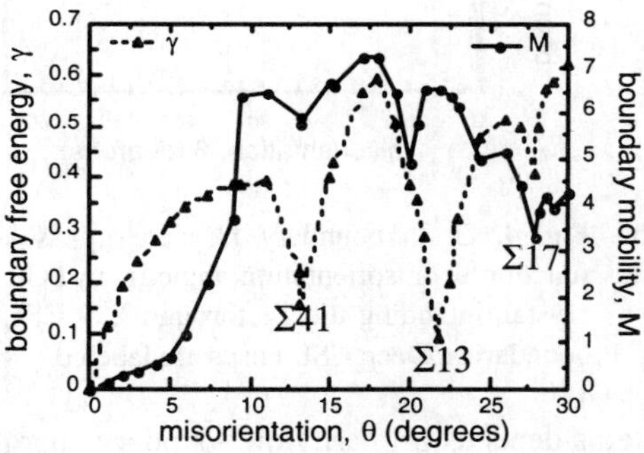

Fig. 1. 2D tilt boundary free energy and mobility as a function of misorientation, as determined by molecular dynamics simulations [7]. Cusps in properties occur for the Σ41, Σ13, and Σ17 CSL misorientations.

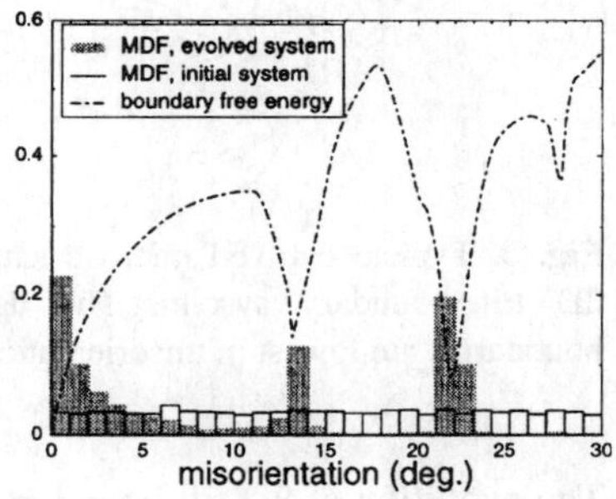

Fig. 2. Misorientation distribution function (MDF) for initial and evolved 2D tilt boundary systems. The boundary free energy function (dashed line) is included for comparison.

The grain boundary energy and mobility data in Fig. 1 were used in Monte Carlo grain growth simulations [Eqs. (1) and (2) respectively]. To initialize the simulations, scalar orientations (i.e. tilt angle relative to a global axis) were assigned at random to each lattice site. This resulted in a randomly-textured polycrystal with a uniform distribution of tilt boundaries, as shown in Fig. 2.

In these systems, grain growth follows power-law kinetics, with average grain area $\langle A \rangle \propto t^{0.5}$. Kinetics are depressed from the linear relationship $\langle A \rangle \propto t$ observed in isotropic grain growth because the average boundary mobility, which scales boundary velocity, continually decreases [9]. Evolved microstructures in these systems also differ from the isotropic case, as shown in Fig. 3. Grains are not equiaxed, and many four-grain and higher order junctions occur. The microstructure is dominated by low energy boundaries.

This increase in low energy boundaries is documented in Fig. 2, which shows the misorientation distribution function (MDF) for a typical 2D tilt boundary microstructure. Low angle, Σ41, and Σ13 boundaries are far more prevalent than other boundary types; the Σ17 boundary, which is not as low in energy as the others, is not enhanced.

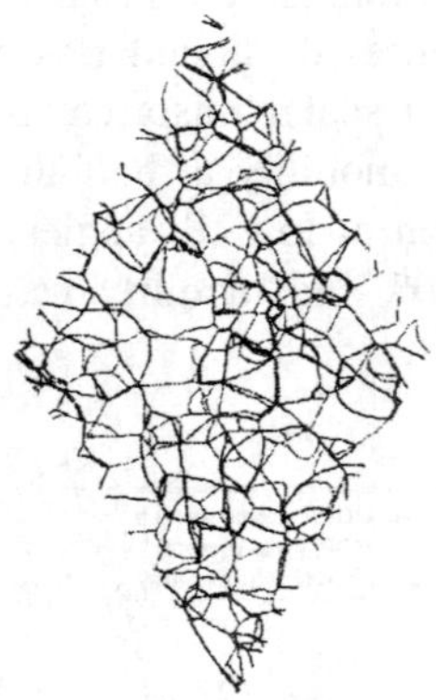

Fig. 3. Typical evolved microstructure in 2D tilt boundary system. The darkest boundaries are lowest in misorientation.

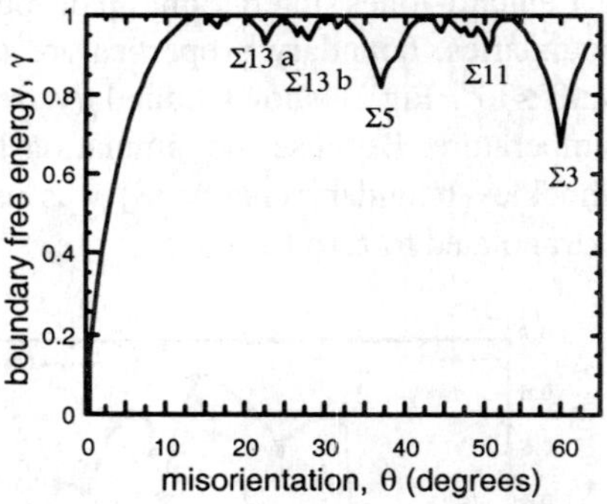

Fig. 4. Grain boundary free energy as a function of misorientation angle in an FCC crystal, including the Σ1 through Σ29 CSL boundaries. Deep CSL cusps are labeled.

The evolution of 2D tilt-boundary systems depends primarily on boundary energy and not on mobility. With anisotropic energy (Fig. 1) but isotropic mobility $M = 1$, the system evolves exactly as described above. In contrast, with anisotropic boundary mobility (Fig. 1), but isotropic energy $\gamma = 1$, the evolution is normal: microstructures are equiaxed and tricoordinated, the MDF remains uniform, and $\langle A \rangle \propto t$.

3.2 Three-dimensional crystallography

For certain 3D misorientation rotations, a fraction $1/\Sigma$ of atomic lattice sites at the grain boundary are coincident [8]. Such a boundary is termed a CSL boundary and is characterized by Σ. According to the Brandon criterion [10], misorientations with angular deviations of less than $\delta\theta_\Sigma = 15°/\Sigma^{1/2}$ from the true CSL misorientation are within the Σ CSL. Using a Read-Shockley model [8], the normalized CSL boundary energy is given by

$$\gamma = 1 - \frac{1}{\Sigma}\left\{1 - \frac{\delta\theta}{\delta\theta_\Sigma}\left[1 - \ln\left(\frac{\delta\theta}{\delta\theta_\Sigma}\right)\right]\right\} \qquad (3)$$

where $\delta\theta$ is the angular deviation from the true CSL rotation. Fig. 4 shows boundary energy as a function of misorientation angle for the Σ1 through Σ29 CSL boundaries in an FCC crystal. Note that the Σ1 boundary type is special; this is the low angle boundary (LAB) and is characterized by misorientation angle alone.

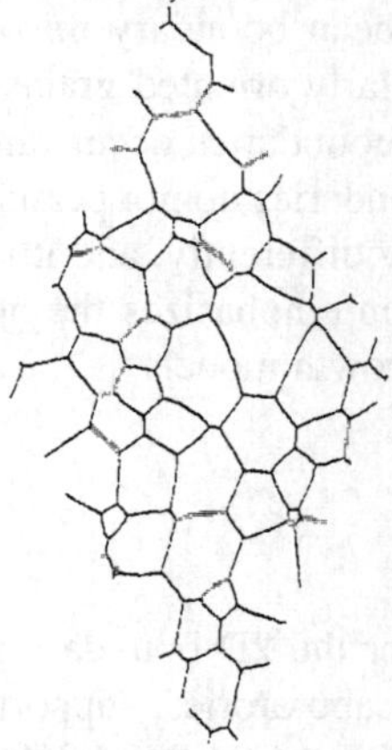

Fig. 5. Typical evolved microstructure in 3D CSL boundary system. The thick, light boundaries have CSL misorientations.

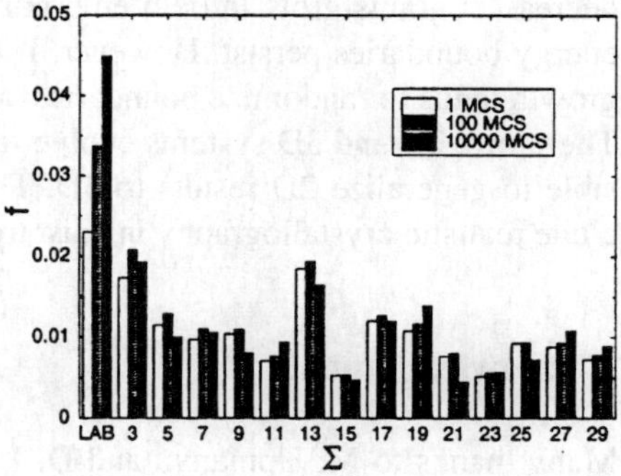

Fig. 6. Frequency of CSL boundaries during microstructural evolution at times of 1, 100, and 10,000 MCS.

Equation (3) defined the boundary free energy in Monte Carlo grain growth simulations [equation (1)], and the boundary mobility M = 1. To initialize the simulations, random 3D orientations were assigned to each lattice site. This resulted in a randomly-textured polycrystal with a Mackenzie MDF [11].
In these systems, grain growth is normal, with $\langle A \rangle \propto t$ as for isotropic grain growth. Evolved microstructures in these systems also appear normal, with equiaxed grains, trijunction angles near 120°, and no higher order junctions, as shown in Fig. 5.
During evolution, except for the LAB, CSL boundaries do not increase in frequency, as shown in Fig. 6. These results contradict those of Ono [2]; we believe that his observed changes in CSL frequency were statistical fluctuations due to small system size. As in 2D, changing the CSL boundary mobility had no effect on these results.

4 Discussion and Conclusions

While low energy boundaries increase during grain growth in systems with 2D crystallography (i.e. tilt boundaries or strongly textured structures), they do not in systems with 3D crystallography. The exception to this is LABs, which increase in both 2D and 3D. The creation of a tilt boundary or a LAB requires that a scalar misorientation angle fall within a certain range. In contrast, the creation of a low-energy 3D CSL boundary requires three variables – the misorientation angle and the rotation axis unit vector – to have a particular relationship. Thus, it is geomet-

rically more difficult to create a 3D CSL boundary than a 2D CSL boundary or a LAB.
If low energy boundaries are readily created, the mean boundary misorientation decreases; grains grow into an environment of similarly oriented grains; and low energy boundaries persist. However, if low energy boundaries occur rarely, grain growth tends to randomize boundaries, and such boundaries do not persist.
Therefore, 2D and 3D systems evolve fundamentally differently, and it is not possible to generalize 2D results to 3D. This conclusion emphasizes the need to include realistic crystallography in anisotropic grain growth models.

Acknowledgements

Many thanks to M. Upmanyu and D. J. Srolovitz for the 2D boundary properties data. This work was performed at Sandia National Laboratories, supported by the U.S. Department of Energy under contract number DE-AC04-94AL85000 and by the Office of Basic Energy Sciences New Initiative program.

References

[1] G. Palumbo, E. M. Lehockey, and P. Lin, *JOM,* 50, 1998, pp. 40-43.
[2] N. Ono, K. Kimura, and T. Watanabe, *Acta Mater.,* 47, 1999, pp. 1007-1017.
[3] G. Palumbo and K. T. Aust, in "Grain Growth in Polycrystalline Materials III", eds. H. Weiland et al., TMS, Warrendale, PA, 1998, pp. 311-320.
[4] V. Randle, *Acta Mater.,* 47, 1999, pp. 4187-4196.
[5] M. P. Anderson, G. S. Grest and D. J. Srolovitz, *Phil. Mag. B,* 59, 1989, pp. 293-329.
[6] M. Miodownik *et al.*, *Acta Mater.,* 47, 1999, pp. 2661-2668.
[7] M. Upmanyu, D. J. Srolovitz, L. S. Shvindlerman, and G. Gottstein, *Acta Mater.,* 47, 1999, pp. 3901-3914.
[8] A. P. Sutton and R. W. Balluffi, "Interfaces in Crystalline Materials", Oxford Science Publications, Oxford, U.K., 1995.
[9] Y. Huang and F. J. Humphreys, *Acta Mater.,* 48, 2000, pp. 2017-2030.
[10] D. G. Brandon, *Acta Metall.,* 14, 1966, pp.1479-1484.
[11] J. K. Mackenzie, *Biometrika,* 45, 1958, pp. 229-240.

Simulation of Grain Boundary Migration in Well Topologically Defined Grain Configurations

J. Kroc[1] and V. Paidar[2]

[1] Research Center, University of West Bohemia, Univerzitni 8, 306 14 Plzen, Czech Republic

[2] Institute of Physics, Academy of Sciences, Na Slovance 2, 182 21 Praha 8, Czech Republic

Keywords: grain boundary migration, computer simulation, cellular automata, tricrystals and quadricrystals.

Abstract. The grain boundary migration of well defined grain configurations is studied using the method of cellular automata. The attention is focused on the quantitative description of special configurations with relatively small number of grains (tri- and quadricrystals). It is tested how a local transition rule, that defines behaviour of a cellular automaton, describes the grain boundary migration and grain growth. The model provides a tool to investigate development of complex microstructure as it takes properly into account changes of the topology of boundary shapes.

1 Introduction

Current knowledge of the grain boundary migration has been recently reviewed in the book [1]. A great attention is paid to the results obtained on fully defined samples such as bi- and tricrystals [2-7]. Grain boundary migration can be theoretically described using models on different scales. The analytical models consider the crystals as homogeneous continua contrary to the discrete atomistic models such as molecular dynamics or Monte Carlo. The results of two-dimensional simulations using molecular dynamics were reported in [8,9]. Recently, the simulations for a single triple junction [10] have been conducted for the same geometry as that used in the experiments [7]. A strong variation of the triple junction mobility with deep minima at the grain misorientations of coincidence boundaries with low Σ values was found. A new approach to the three-dimensional simulation of grain boundary migration using the method of molecular dynamics was proposed in [11]. A constant driving force is generated by imposing an anisotropic elastic strain on a bicrystal such that the elastic energy densities in two grains are different. Under these conditions even a planar grain boundary can move.
Migration of grain boundaries is a fundamental mechanism controlling grain growth in polycrystals. A general characteristic of this process, so called Von

Neumann-Mullins relation for the grain area change during the growth process in the quasi-two-dimensional model, was derived assuming that the triple junctions do not affect the grain boundary motion. When it is not the case, a triple junction drag can have important consequences for the grain growth as it has been recently discussed in [12]. Nevertheless, we will study as the first approximation only changes of the microstructure when no particular properties are attributed to the triple junctions.

In this paper, a model based on cellular automaton that can be considered as a theoretical tool on the mesoscopic level will be applied. Two simple configurations will be considered, a tricrystal containing only one triple junction where a perpendicular boundary segment meets a straight boundary (T configuration), and a quadricrystal containing two triple junctions (H configuration). Development of these configurations as a function of annealing time will be studied and the motion of grain boundary segments will be analysed.

2 Cellular Automaton Model

In order to better understand the physical principles of such fundamental mechanism of microstructure formation as grain growth, the experiments and theoretical modelling have to be performed on well-defined simple samples containing a small number of microstructure elements. The properties of grain boundaries are measured on bicrystals and the effects of triple junctions are examined on specially prepared tricrystals. To study theoretically the effect of triple junctions we have chosen tri- and quadricrystal of simple shapes.

Our objective is to perform modelling by a cellular automaton and to inspect physical features of the obtained results. This method allows us to study details of grain shape changes in addition to kinetics of the grain boundary movement. The method has a probabilistic nature and is governed by the local rules; the global quantities such as total energy given by the total grain boundary area are not explicitly considered. Since the driving force of grain boundary migration is a reduction of the total boundary area it should be tested how a model based on the local rule only describes the system development.

A few essential characteristics of the used cellular automaton have to be mentioned. The sample is divided into cells, for simplicity square cells of the same size were chosen, and each cell is assigned to one of considered differently oriented grains. Since no other parameters characterizing the crystals are taken into account, the number of states that a cell can possess is given by the number of grains. The neighbourhood of a cell that affects its transition to the next temporal state is composed of 8 nearest neighbouring cells (Moore neighbourhood). The rule of transition of a cell - its reorientation into the neighbouring crystal - requires two conditions to be fulfilled:

a) the cell is situated at the grain boundary
b) the number of contiguous cells (going one after the other without interrupts in the Moore neighbourhood) belonging to the same grain is equal to 4 or higher, or in the restricted neighbourhood on the free surface it is equal to 2 or higher.

No special rule attributed to the behaviour of triple junctions (e.g. an enhanced drag effect) is considered.
The reorientation is occurring with the probability of 50% in order to prevent non-physical anisotropic growth reflecting the symmetry of the square grid. The grid size is 532x256 cells. In every time step the entire grid of automata cells is updated simultaneously according to the local transition rule. Therefore, the time is also a discrete quantity similarly to the space divided into discrete cells. The calculations proceed in a parallel way and are different at this point from the Monte Carlo methods. Since a straight grain boundary segment represents a stable configuration, the process of grain boundary migration is triggered by the fluctuations of cell reorientations at the ends of the straight segments meeting at the triple junctions. For more general information on cellular automata see for example [13-16].
The only driving force for the grain boundary migration in this model is local curvature of the grain boundary shape. It leads to gradual lowering of the system energy owing to the reduction of grain boundary area. We will study how the initial configurations of tri- and quadricrystals are developing towards the final configurations formed by a single crystal and a bicrystal with the single straight grain boundary perpendicular to the side free surfaces, respectively.

3 Results

Our initial configurations contain only straight grain boundary segments as depicted in Fig. 1 and 2. Those are T and H configurations. The triple junctions with the dihedral angles of 90°-90°-180° are not in equilibrium and thus the system starts to develop by their displacements. In equilibrium, the dihedral angles should be 120° as no anisotropy of grain boundary energy is taken into account. Notice that there is no force acting on the grain boundary segments perpendicular to the free surfaces.
Local rotation of the grain boundary about the triple junction leads to curvature of the boundary surfaces. When the curved segments are established they move towards their centre of curvature. Due to the symmetry of the initial T configuration, the horizontal segment in Fig. 1 starts to move downwards accidentally. When the lower grain disappears in the lower-left corner, the intersection S3 of the grain boundary with the left-hand vertical free surface starts to move upwards. Finally, the upper grain disappears in the upper-left corner and the tricrystal is transformed into a single crystal.
In the case of H configuration, the central-upper grain first disappears upwards at about the same time when the two triple junctions meet. A new triple junction formed at the apex of the central-lower grain then starts to move downwards. Notice that a new grain boundary between the grains 3 and 4 is created during this stage. This is an elementary topological event that may happen during the grain growth in polycrystals. Finally, the central-lower grain disappears and the quadricrystal is transformed into a stable bicrystal.

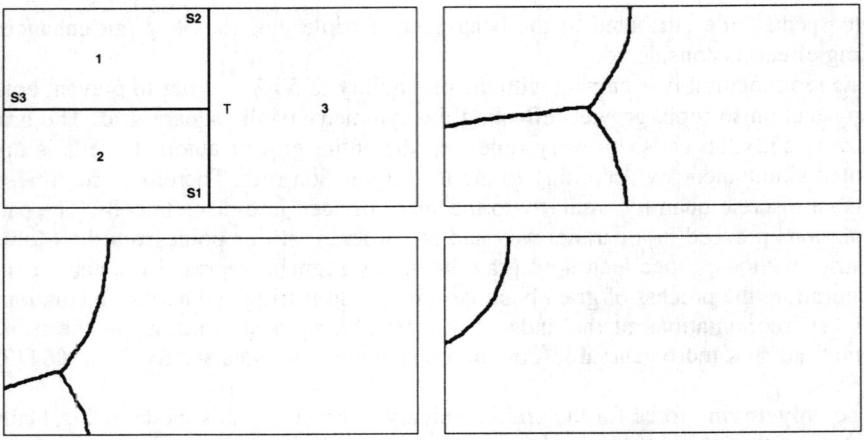

Fig. 1. Development of T configuration. (a) t = 0 (initial configuration), (b) t = 15k, (c) t = 55k, (d) t = 83k.

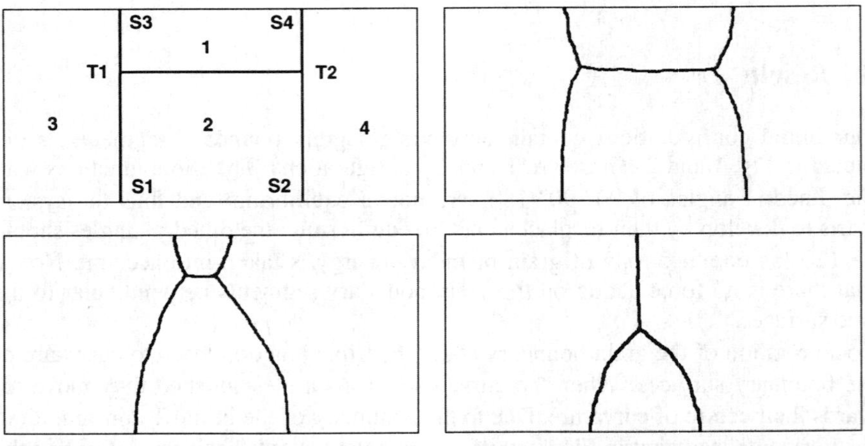

Fig. 2. Development of H configuration. (a) t = 0 (initial configuration), (b) t = 5k, (c) t = 25k, (d) t = 42k.

Time development of the two studied configurations T and H is shown in Fig. 3 and 4, respectively. The mean migration rate of the triple junction in the x-direction is about 0.0028 in the units of the cell size per time step for the T configuration. Linear parts of the dependences presented in Fig. 4 for the H configuration correspond to the rate of 0.0038 and 0.0027. Therefore, the migration rate in the horizontal direction is in average about 0.003. Notice that it is the direction of triple junction motion towards the equilibrium of dihedral angles between the meeting boundaries. The vertical y-coordinate of the triple junctions is first constant for both studied configurations and then deviates towards the side free surfaces where smaller grains shrink (downwards and upwards for the T and H configurations, respectively).

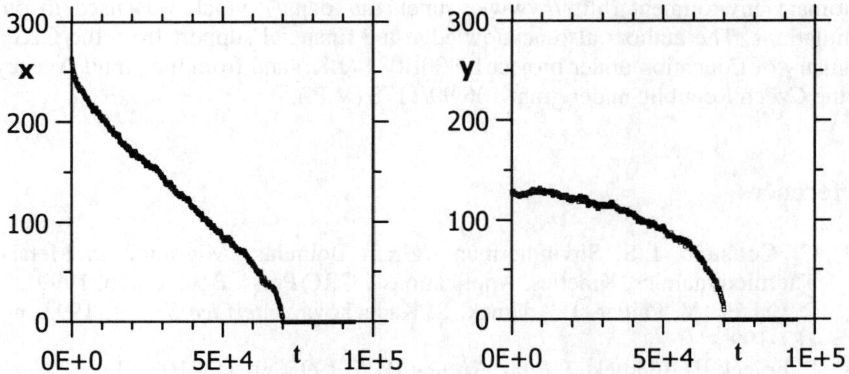

Fig. 3. Migration of the triple junction for T configuration. x- and y-coordinates.

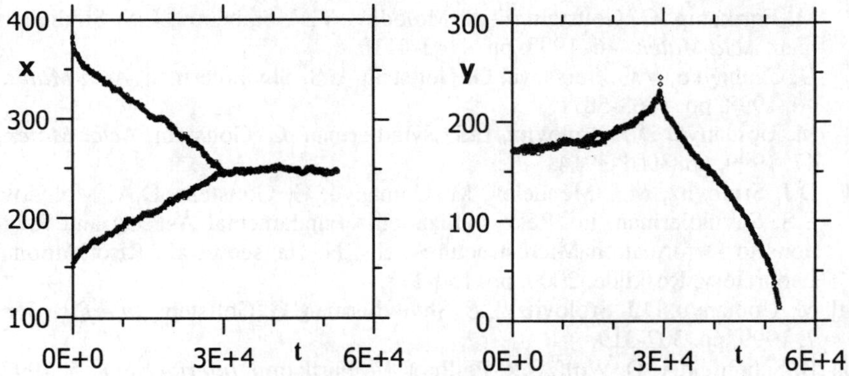

Fig. 4. Migration of the triple junctions for H configuration. x- and y-coordinates.

4 Conclusions

Since a wide range of technologically important properties of polycrystalline materials is controlled by the microstructure, the study of processes leading to the changes of microstructure deserves necessary attention. Contrary to the analytical methods where the changes of grain boundary topology must be carefully treated, they come out naturally from the cellular automata simulations.

Acknowledgement

We are greatly indebted to Prof. D.J. Eckart for his publicly available cellular automata environment (http://www.cs.runet.edu/~dana/) which was used in our simulations. The authors also acknowledge the financial support from the Czech Ministry of Education under project LN00B084 (J.K.) and from the Grant Agency of the Czech Republic under grant 106/99/1178 (V.P.).

References

[1] G. Gottstein, L.S. Shvindlerman, "Grain Boundary Migration in Metals: Thermodynamics, Kinetics, Applications", CRC Press, Boca Raton, 1999.

[2] P. Lejcek, V. Paidar, J. Adámek, S. Kadecková, *Interface Sci.,* 1, 1993, pp. 187-199.

[3] P. Lejcek, J. Adámek, *J. Phys. France IV,* 5, 1995, pp. C3_107-114.

[4] S. Tsurekawa, T. Ueda, K. Ichikawa, H. Nakashima, Y. Yoshitomi, H. Yoshinaga, *Mater. Sci. Forum,* 204-206, 1996, pp. 221-226.

[5] M. Furtkamp, P. Lejcek, S. Tsurekawa, *Interface Sci.,* 6, 1998, pp. 59-66.

[6] M. Furtkamp, G. Gottstein, D.A. Molodov, V.N. Semenov, L.S. Shvindlerman, *Acta Mater.,* 46, 1998, pp. 4103-4110.

[7] U. Czubayko, V.G. Sursaeva, G. Gottstein, L.S. Shvindlerman, *Acta Mater.,* 46, 1998, pp. 5863-5871.

[8] M. Upmanyu, D.J. Srolovitz, L.S. Svindlerman, G. Gottstein, *Acta. Mater.,* 47, 1999, pp. 3901-3914.

[9] D.J. Srolovitz, M.I. Mendelev, M. Upmanyu, G. Gottstein, D.A. Molodov, L.S. Shvindlerman, in "Rekrystallization - Fundamental Aspects and Relations to Deformation Microstructure", eds. N. Hansen et al., Riso National Laboratory, Roskilde, 2000, pp. 157-178.

[10] M. Upmanyu, D.J. Srolovitz, L.S. Shvindlerman, G. Gottstein, *Interface Sci.,* 7, 1999, pp. 307-319.

[11] B. Schönfelder, D. Wolf, S.R. Phillpot, M. Furtkamp, *Interface Sci.,* 5, 1997, pp. 245-262.

[12] G. Gottstein, A.H. King, L.S. Shvindlerman, *Acta Mater.,* 48, 2000, pp. 397-403.

[13] G.Y. Vichniac, *Physica,* 10D, 1984, pp. 96-116.

[14] T. Toffoli, *Physica,* 10D, 1984, pp. 117-127.

[15] T. Toffoli, N. Margolus, "Cellular Automata Theory", MIT Press, Cambridge, 1987.

[16] J.R. Weimar, "Simulation with Cellular Automata", Logos Verlag, Berlin, 1997.

Modelling Subgrain Growth during Annealing of Deformed Aluminium Alloys

S. Abtahi, K. Sjølstad, K. Marthinsen, and E. Nes

Norwegian University of Science and Technology, Department of Materials Technology and Electro-Chemistry, N7491, Trondheim, Norway

Keywords : aluminium, recovery, subgrain growth, modelling.

Abstract. Subgrain growth, like normal grain growth, is commonly found to follow an empirical growth law in the form of $\delta^n - \delta_0^n$ = const.×t, where the exponent, n, varies with the annealing temperature. The value of n, in aluminium, being typically around 2 at temperatures above 350 °C and as high as 5-7 at lower temperatures. In the present work, on the basis of subgrain growth observations in an Al-0.5Si alloy, this growth behaviour has been analyzed in terms of two different modelling approaches. The NTNU-model [1,2] interprets the change in growth exponent with temperature as a result of solute pinning or clustering, while in the UMIST-model [3] this exponent-effect is attributed to mobility changes associated with changes in sub-grain boundary misorientation.

1 Introduction

In pure metals and stable solid-solution alloys, the kinetics of normal grain growth is in general well described by the following empirical equation, first proposed by Beck and co-workers [4]:

$$D^n - D_0^n = Kt \quad (1)$$

Where D and D_0 are the instantaneous and the initial average grain sizes, respectively, t is the isothermal annealing time, K a parameter depending on material and temperature, and the exponent n, which also varies with material and temperature, is usually referred to as the grain growth exponent. It follows from the classic treatment by Burke and Turnbull [5] that n = 2. However, in high purity metals and solid solution alloys n is usually found to be larger than 2 as illustrated by the celebrated results shown in Fig. 1a, taken from Beck et al. [4]. As can be seen from this diagram n varies from as low as 3 to as high as 16. No accepted theory exists which accounts for n being larger than 2, but a common assumption is that non-linear dependence of velocity on driving pressure due to solute effects is the main cause for this phenomenon, see e.g. Humphreys and Hatherly [6].

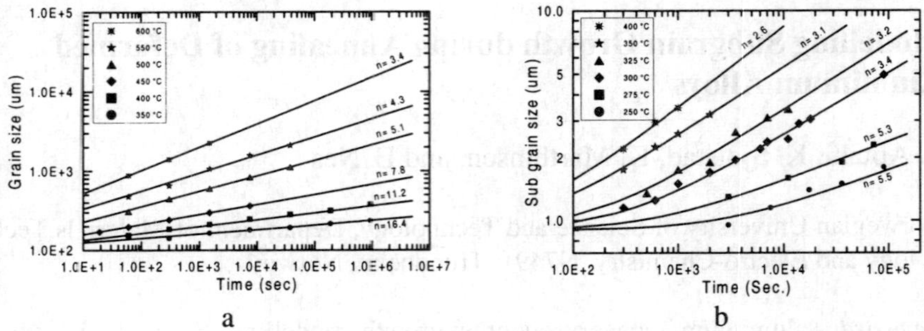

Fig. 1. Grain growth a) in high purity aluminium (99.99) [5] and b) in an Al-0.05Si-alloy [3].

Observations of subgrain growth [1-3] demonstrate that also this growth reaction is well covered by an empirical relationship in the same form as Eq. 1. This is illustrated in Fig. 1b, where the experimental data are from an investigation of Humphreys and Huang [3] on single crystals of an *Al-0.5Si* alloy. Why subgrain growth results in growth exponents larger than 2 has been interpreted in various ways by Humphreys and coworkers [3] and Nes and coworkers [1,2]. While the UMIST-group focused on the observed change in sub-boundary misorientation as the reason for a larger than 2 growth exponent, the NTNU-group has interpreted this exponent-effect as caused by a non-linear effect of solute on sub-boundary mobility. The objective of this work is to compare and contrast these two different modelling approaches, using the Humphreys and Huang data as a case study.

2 Models

2.1 UMIST-model

This model takes a standard approach in defining the growth rate of a subgrain structure (average size δ) as follows:

$$\frac{d\delta}{dt} = M\frac{\gamma_{SB}}{\delta} \tag{2}$$

where M is the mobility of an average boundary of energy γ_{SB}, an energy which is assumed to be well represented by the Read-Shockley equation, i.e. $\gamma_{SB}=\gamma_{GB}(\varphi/\varphi_c)\ln(e\varphi_c/\varphi))$, where γ_{GB} is the energy of a high angle boundary and φ_c is the minimum angle of such a boundary, typically 15°. According to Humphreys and Huang [3], an important consideration in analyzing the subgrain growth rates is that the mean misorientation of the sub-boundaries is not a constant but decreases during subgrain growth. It follows from their detailed investigation that this misorientation varied during subgrain growth according to an empirical relationship in the form $\varphi=\varphi_0(\delta/\delta_0)^{-0.13}$, where φ_0 and δ_0 refer to initial values. How this change in misorientation affects sub-boundary mobility has been modeled using two alternative empirical relationships:

$$M = M_m\left[1-\exp(-B(\varphi/\varphi_c)^A)\right] \tag{3a}$$

$$M = M_0^{SB}\varphi^C \exp(-U_{SB}/kT) \tag{3b}$$

where $M_m = M_{GB} \exp(-U_{GB}/kT)$ is the mobility of a high angle boundary, A = 4, B = 5, and C = 5.18. The pre-exponential parameters, M_0^{SB} and M_0^{GB}, and the corresponding activation energies U_{SB} and U_{GB} need to be determined experimentally, as part of the model tuning procedure.

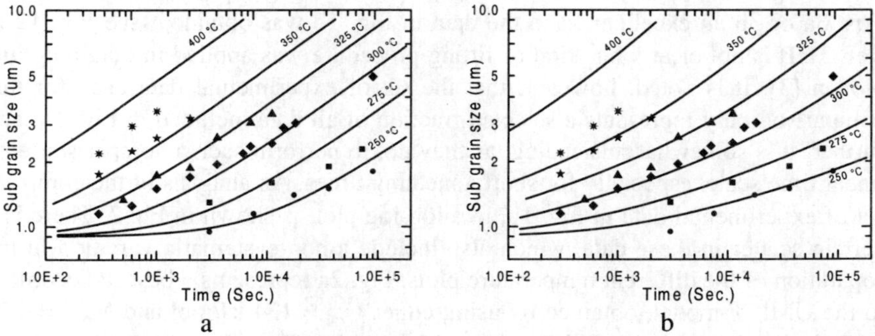

Fig. 2. Subgrain size vs. annealing time at the temperatures given (following PSC of 70% at RT). Model predictions: a) UMIST-model, b) NTNU-model.

2.2 NTNU-model

This treatment [1,2] is similar to the one above in the sense that subgrain growth is analysed analogously to that of normal grain growth, with the migration of sub-boundary dislocations being the rate controlling reaction. The growth rate takes the form

$$\frac{d\delta}{dt} = v(P) \tag{4}$$

where v(P) is the migration rate of a sub-boundary acted upon by a pressure $P \equiv 2\gamma_{SB}/\delta$ making the same assumption as above, namely that the sub-boundary energy can be approximated by using the Read-Shockley equation. Equation 4 is a more general version of Eq. 2. The rate controlling reaction in sub-boundary migration is assumed to be dislocation climb, with the climb rate controlled by solute pinning (or clustering). The boundary migration rate can then be written:

$$v(P) = b\nu_D B \exp\left(-\frac{U_s + \Delta U}{kT}\right) 2\sinh\left(\frac{PV_a}{kT}\right) \tag{5}$$

where ν_D is the Debye frequency, b Burgers vector, B a constant, U_s the activation energy for solute diffusion, ΔU the interaction energy between the solute atom and the sub-boundary and V_a is an activation volume. In the case of solute pinning; $V_a=\omega_b b^3 c^{-1}$ where c is the solute concentration and ω_b a geometrical constant (for more details see [2]). In the case $PV_a \ll kT$ (which is the condition assumed to prevail in the UMIST-model) Eq. 4 takes the same form as Eq. 2.

3 Application of Models and Discussion

According to Humphreys and Huang [3], by combining Eq. 2 with either one of Eqs. 3a or 3b an excellent fit to the data in Fig. 1b was obtained (see Fig. 12 in Ref. 3). It is not clear what kind of fitting procedure was applied in obtaining this plot in [3]. It is noted, however, that the set of experimental data used for this comparison only represents a selected fraction of all data included in Fig. 1b, and further, it is somewhat misleading, as they do, to perform such a comparison on a linear time scale, especially for short annealing times. An analysis of the complete set of experimental data in Fig. 1b, in a log-log plot, is shown in Fig 2. There is a certain scatter in these data, which also include a non-systematic variation in the separation of the different temperature plots. Fig. 2a represents a best fit according to the UMIST-model, obtained by using either U_{GB} = 134 kJ/mol and M_0^{GB} = 0.22 m^4/Js in Eq. 3a or U_{SB} = 134 kJ/mol and $M_0^{SB} = 4.2\times10^{-6}$ m^4/Js (with φ given in degrees) in Eq. 3b. A fit which is less impressive than the impression given by Fig. 12 in [3]. Within the time-range covered in Fig. 2a, the difference in predictions due to either one of the two expressions (Eqs. 3a and 3b) was very small, indeed (not possible to separate in this figure). Fig. 2b shows the predictions due to the NTNU-model, using ω_b/c = 1900, B_δ = 300 and $U_s + \Delta U$ = 190 kJ/mol. A cursory inspection of the two diagrams in Fig. 2, leaves the impression that the difference in the predictions due to the two models is not that big, with both models failing the 400 °C observations. There is, however, an important difference in the prediction-power of the two models which pertain to the temperature dependency of the growth-exponent, n. In the UMIST-model, the form of the growth curve is temperature invariant (constant n-value of 2.8) in conflict with experimental observations, see Fig. 1. This important aspect is well covered by the NTNU-model. It should be noted that because of the work-term PV_a in Eq. 5, the effective activation energy ($U_s + \Delta U - PV_a$) in the NTNU-model becomes of comparable size to that in the UMIST-model, for the conditions modeled in Fig. 2. If these conditions change, however, the apparent activation energy will change, such a situation is shown in Fig. 4. It is also noted that the size of the activation volume is far too large to be caused by Si in solid solution ($V_a=\omega_b b^3 c^{-1}$ = 1900b^3 or $\omega_b \approx 95$). This large activation volume is assumed to reflect interactions between sub-boundary dislocations and iron in solid solution (or, alternatively, some sort of low-temperature clustering phenomenon).

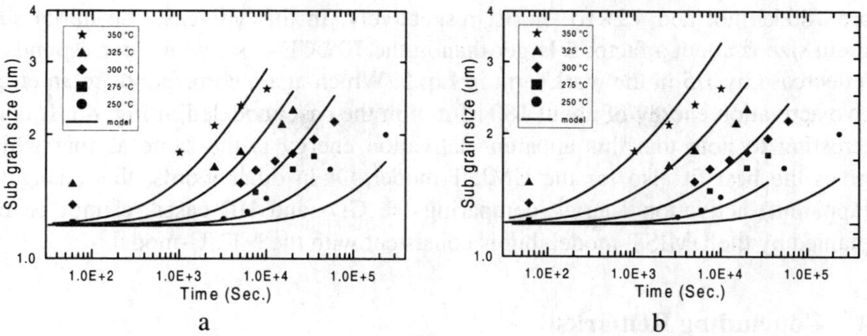

Fig. 3. Subgrain size vs. annealing time at the temperatures given (PSC of 30% at RT). Model predictions: a) UMIST-model, b) NTNU-model.

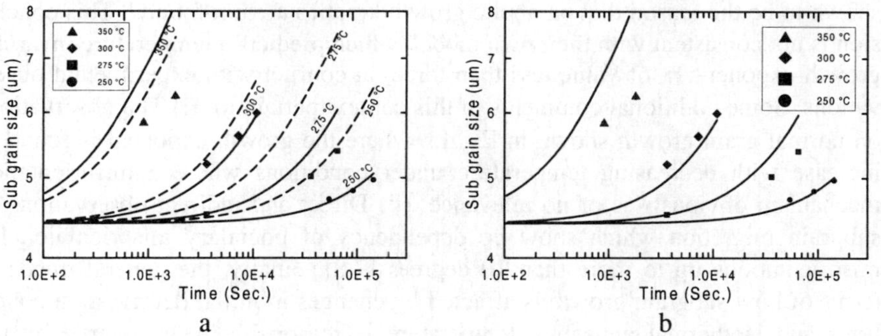

Fig. 4. Subgrain size vs. annealing time at the temperatures given (PSC of 70% at 350 °C). Model predictions: a) UMIST-model, b) NTNU-model.

A further testing of the models is now possible by considering the additional data sets in Figs. 3 and 4 (also from [3]) showing subgrain growth during isothermal annealing after 30% cold deformation (30%CD) and 70% hot deformation (70%HD), respectively. If the models are to be trusted, they should predict these results by only changing the initial conditions represented by δ_0 and φ_0 (values to be found in [3]) in the two models. As can be seen from Fig. 3a the UMIST-model fails in predicting the 30%CD data. The growth curves are too steep, and the model under-predicts the low-temperature observations. The NTNU-model is only slightly better, providing a more realistic shape of the growth curves, Fig. 3b (in this figure the pre-exponential parameter B_δ has been reduced by a factor 0.3 compared to the 70%CD-condition). The HD-data in Fig. 4b represents an especially interesting situation. In this case a reasonable fit is obtained by the NTNU-model by increasing the B_δ -parameter by a factor 10 compared to the 70%CD-condition. The UMIST-model, however, requires a more extensive re-tuning as illustrated by the broken lines in Fig. 4a, reflecting a change both in activation energy and pre-exponential factor compared to that used for the 70%CD-case, fully drawn lines. New values used for the activation energy and pre-exponential factor are 180 kJ/mol and 4.2×10^{-3} m^4/Js, respectively. In this HD-case the initial subgrain size is about a factor 5 larger than in the 70%CD-case, which corresponds to a decrease by 1/5 in the work term in Eq. 5. Which again corresponds to an effective activation energy of about 180 kJ/mol in the case modeled in Fig. 4b. It is interesting to note that this apparent activation energy is the same as that which gives the best fit also for the UMIST-model. Or in other words, this change in (apparent) activation energy, comparing the CD- and HD-cases, cannot be explained by the UMIST-model, but is consistent with the NTNU-model.

4 Concluding Remarks

Regarding the mechanism controlling subgrain growth, Humphreys and Huang write in the conclusion of their paper [3]: "The decrease in mean misorientation is shown to be the reason that parabolic growth kinetics are not found." This conclusion is not consistent with their own model, which predicts a temperature invariant growth-exponent, n, of value less than three, in conflict with experimental observations. Some additional comments in this context pertain to: (i) The observations on normal grain growth shown in Fig 1a, where the growth-exponent is found to increase with decreasing temperature under conditions where a misorientation mechanism obviously is of no relevance. (ii) Direct and indirect observations of subgrain migration which show no dependency of boundary misorientaion for misorientations up to more than 10 degrees [7,8]. Finally, the general trends in terms of how subgrain growth is affected by changes in initial deformation conditions and isothermal annealing temperature is reasonably well covered by the NTNU-model, based on solute- or cluster-pinning.

References

[1] T. Furu, R. Ørsund, and E. Nes, *Acta Metall. Mater.* 43, 1995, pp. 2209-2232.

[2] J.A. Sæter, B. Forbord, H.E. Vatne and E. Nes, in "Proceedings of the 6th International Conference on Aluminum Alloys, ICAA-6" eds. T. Sato et al., JILM, Japan, 1998, pp. 113-126.

[3] F.J. Humphreys and Y. Huang, in „Proceedings of the 21st Risø International Symposium on Material Science", 2000, pp. 71-89.

[4] P.A. Beck, J.C. Kremer, L.J. Demer, and M.L. Holtzworth, *Trans AIME*, 175, 1948, pp. 372-400.

[5] J.E. Burke and D. Turnbull, *Progress in Metal Physics*, 3, 1952, pp. 220-293.

[6] F.J. Humphryes and M. Hatherly, "Recrystallization and Related Annealing Phenomena", Pergamon, Oxford, 1995.

[7] M. Winning, G. Gottstein, and L.S. Shvindlerman, in „Proceedings of the 21st Risø International Symposium on Material Science", 2000, pp. 645-652.

[8] C.C. Yang, A.D. Rollett and W.W. Mullins, in „Proceedings of the 21st Risø International Symposium on Material Science", 2000, pp. 659-664.

Analytic Model of Grain Growth Based on a Generalized LS Stability Argument and Topological Relationships

P. Streitenberger

Institut für Experimentelle Physik, Abteilung Materialphysik, Otto-von-Guericke-Universität Magdeburg, PF 4120, D-39016 Magdeburg, Germany

Keywords: grain growth, size distribution function, topological relationships

Abstract. In the present paper an analytic theory of 2D grain growth is presented, which combines a generalization of the Lifshitz-Slyozov (LS) stability argument for the existence of a stationary self-similar size distribution function (SDF) of grains with the von Neumann-Mullins topological requirements for a 2D grain network of uniform boundaries. It is shown that the LS stability theory generalized for an arbitrary upper cutoff grain radius as presented earlier corresponds to the assumption of a non-linear (quadratic) topological relationship for the average number of sides for a fixed scaled grain size. By a suitable choice of the cutoff parameter the present model yields an excellent analytic description of the SDF of 2D grain growth as it results from recent simulations.

1 Introduction

In the theory of normal grain growth an assembly of grains is described by a size distribution function (SDF) $F(R,t)$, where $F(R,t)dR$ gives the number of grains per unit volume with radius between R and $R+dR$ at time t. $F(R,t)$ obeys the continuity equation in R-space

$$\frac{\partial F(R,t)}{\partial t}+\frac{\partial}{\partial R}\left(\dot{R}F(R,t)\right)=0, \tag{1}$$

where $\dot{R}=\dot{R}(R,t)$ is the growth law, i.e. the average growth rate of grains with radius R. In the classical theory of coarsening due to Lifshitz, Slyozov (LS) [1], Wagner [2] and Hillert [3] the asymptotic stable state of the system is characterized by the self-similar scaling form

$$F(R,t)=\frac{A}{R_c^{D+1}}f(\rho),\ \rho=\frac{R}{R_c(t)}, \tag{2}$$

which explicitely fulfils the requirement of volume preservation $a_D\int R^D F(R,t)dR=\text{const}$. $D=1,2,3$ is the spacial dimension of the system and

a_D the volume of the D-dimensional unit sphere. $R_c = R_c(t)$ is the critical grain radius defined by $\dot{R}(R_c, t) = 0$. Equation 2 allows one to transform the continuity equation (1) into the ordinary differential equation

$$\frac{df(\rho)}{d\rho} + \frac{1}{U(\rho)}\left(\frac{dU(\rho)}{d\rho} - D\right)f(\rho) = 0 \tag{3}$$

with $U(\rho) = (R_c / \dot{R}_c)\dot{\rho}$ as the growth law in the scaled variable ρ. In the theory of Hillert [3] $U(\rho)$ takes the form

$$U(\rho) = -\frac{\gamma}{\rho} + \gamma - \rho, \quad \gamma = \frac{k}{R_c \dot{R}_c}, \tag{4}$$

which follows from the growth law $\dot{R} = k(1/R_c - 1/R)$ for curvature driven grain growth. The form of $U(\rho)$ for different values of γ is sketched in Fig. 1a.

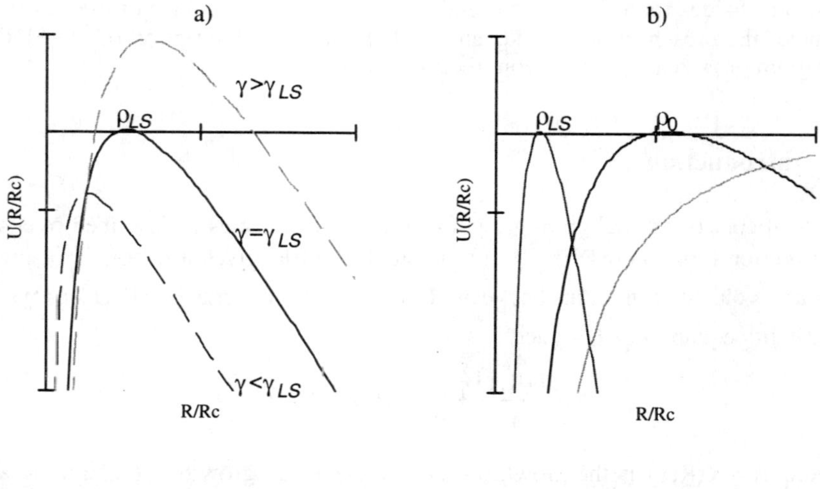

Fig. 1. (a) The scaled growth law (4) for different values of γ. (b) The parametrized growth law (8) for different cutoff parameters ρ_0 (schematically).

Due to an argument of LS [1] based on the requirement of volume preservation, the only asymptotic stable state which has physical significance is given by the curve in Fig. 1a which is tangent to the R/R_c axis leading to the requirement that $U(\rho)$ must have a double root at $\rho = \rho_{LS}$:

$$U(\rho = \rho_{LS}) = 0, \quad \left(\frac{dU(\rho)}{d\rho}\right)_{\rho=\rho_{LS}} = 0. \tag{5}$$

The scaled SDF $f(\rho)$ is then defined only in the range $0 \le \rho \le \rho_{LS}$, whereas $f(\rho)$ and all its higher derivatives must be zero at the cutoff parameter ρ_{LS} [1]. Both conditions (5) fix the value of γ and ρ_{LS} to $\gamma = \gamma_{LS} = 4$ and $\rho_{LS} = 2$, respectively, yielding by integration of Eq. 3 the well known SDF $f_H(\rho)$ derived by Hillert [3]. The Hillert-distribution disagrees, however, with those observed experimentally and found by simulations based on curvature driven growth.

The Hillert theory combines scaling with the LS stability argument Eq. 5 for the special growth law Eq. 4. However, as demonstrated by computer simulations of 2D normal grain growth [4 to 6] the average growth law Eq. 4 is not correct because it does not take into account spatial grain size correlations observed in the microstructure of the grain ensembles. Modifications of the Hillert theory for 2D grain growth incorporating the size correlation effect have been presented in [4, 6, 7]. But differently to the original Hillert approach these modifications do not employ the LS stability conditions (5). Quite the reverse, Mullins [7] has pointed out that the LS stability argument one cannot use because that argument is based on a fixed growth law, while the effective growth law changes with changing local environment as the system evolves toward the self-similar state. On the other hand, we have demonstrated in [8] that the fulfilment of the LS stability conditions is essential to bring the analytical size distribution function of a modified Hillert theory in agreement with the 2D simulation results found in [4]. In the present paper we generalize our approach and show that the LS stability requirement can be made compatible with an effective growth law based on topological and microstructural relationships found in the recent simulations [5, 6] yielding a simple and accurate analytic description of the SDF.

2 LS-Stable Size Distribution for an Arbitrary Cutoff Parameter

As pointed out in [7, 9], for grain growth velocities to be bounded the scaled SDF $f(\rho)$ must have a finite cutoff ρ_0, even if larger than ρ_{LS} of the Hillert distribution. In order to fulfil the LS stability conditions for an arbitrary cutoff ρ_0 we generalize the scaled growth law Eq. 4 by the following parametrization:

$$U(\rho) = -\frac{a_{-1}}{\rho} + a_0 + a_1\rho . \tag{6}$$

The three parameters a_{-1}, a_0, a_1 are determined by the LS stability conditions after which the expression (6) must have a double root at $\rho = \rho_0$,

$$U(\rho = \rho_0) = 0, \quad \left(\frac{dU(\rho)}{d\rho}\right)_{\rho=\rho_0} = 0, \tag{7}$$

and by the requirement that, per definition, $\dot{R} = \dot{R}_c(U(\rho)+\rho) = 0$ for $\rho = 1$, i.e. $U(\rho = 1) = -1$. This yields after a little algebra

$$a_{-1} = \frac{\rho_0^2}{(\rho_0 - 1)^2}, \quad a_0 = \frac{2\rho_0}{(\rho_0 - 1)^2}, \quad a_1 = -\frac{1}{(\rho_0 - 1)^2}$$

and

$$U(\rho) = -\frac{(\rho - \rho_0)^2}{(\rho_0 - 1)^2 \rho}. \tag{8}$$

The generalized scaled growth law Eq. 8 is illustrated in Fig. 1b for different cut-off parameters ρ_0. For $\rho_0 = \rho_{LS}$ Eq. 8 reduces to the Hillert form Eq. 4. With Eq. 8 we obtain by integration of Eq. 3 the following one-parameter family of scaled and normalized SDFs which all obey the LS stability constraints:

$$f(\rho) = a\rho_0^a e^a \frac{\rho}{(\rho_0 - \rho)^{a+2}} \exp\left(-\frac{a\rho_0}{\rho_0 - \rho}\right), \quad \rho \le \rho_0$$
$$= 0, \qquad \rho > \rho_0, \tag{9}$$
$$a = D(\rho_0 - 1)^2.$$

For $\rho_0 = \rho_{LS} = 2$ Eq. 9 yields the Hillert distribution [3], for $\rho_0 \to \infty$ the scaled distribution proposed by Louat [10], $f_L(\rho) = D\rho\exp(-D\rho^2/2)$, results. Therefore, for a finite cutoff radius in the range $\rho_{LS} \le \rho_0 < \infty$ our scheme interpolates the LS scaling state between the two limiting distributions of Hillert and Louat, respectively.

In order to use ρ_0 as an adjustable parameter we have to transpose the scaled size variable ρ to the measurable scaled unit $x = R/\langle R \rangle = x_c\rho$, where $\langle R \rangle$ is the average grain radius and

$$x_c = 1 / \int_0^\infty \rho f(\rho) d\rho \tag{10}$$

is the scaling factor. The corresponding cutoff parameter x_0 is then related to ρ_0 by $x_0 = x_c\rho_0$. Eq. 10 is equivalent to the relationship $\int f(x)dx = \int xf(x)dx = 1$. The above SDF Eq. 9 obeys apart from volume conservation and the LS stability requirements all the general scaling equations considered in [7, 9]. Especially, in terms of the scaled variable x the average growth law takes the form

$$\dot{R} = \langle \dot{R} \rangle (\tilde{U}(x) + x), \tag{11}$$

where

$$\tilde{U}(x) = -\frac{(x_0 - x)^2}{\left(\frac{x_0}{x_c} - 1\right)^2 x}.$$

3 Size Correlation Effect in 2D Grain Growth

In the following we use the size correlation effect observed in the micostructure of simulated 2D grain ensembles to determine the cutoff parameter ρ_0 and x_0, respectively, in the above phenomenological description of grain growth. For this purpose we rewrite the scaled growth law Eq. 11 in the von Neumann-Mullins form (cf. [6, 7])

$$\dot{R} = \frac{k}{6 < R > x}[s(x) - 6], \tag{12}$$

where $s(x)$ is a topological correlation function representing the average number of sides of grains of relative size $x = R / < R >$ [6, 7]. By comparison of Eq. 12 with Eq. 11 we obtain

$$s(x) = c_0 + c_1 x + c_2 x^2, \tag{13}$$

$$c_0 = 6\left(1 - \frac{x_0^2}{(x_0 / x_c - 1)^2 \tilde{\gamma}}\right), c_1 = \frac{12 x_0}{(x_0 / x_c - 1)^2 \tilde{\gamma}}, c_2 = \frac{6}{\tilde{\gamma}}\left(1 - \frac{1}{(x_0 / x_c - 1)^2}\right),$$

where $\tilde{\gamma} = k / < R >< \dot{R} >$ is the modified kinetic growth parameter analogue to γ in Eq. 4. Because of the volume conservation during grain growth, for any value of the two remaining free parameters x_0 and $\tilde{\gamma}$ Eq. 13 fulfils the average von Neumann-Mullins law $\int s(x) f(x) dx = 6$, provided the scaling factor x_c in the coefficients of Eq. 13 is determined self-consistently from Eqs. 9 and 10 using the relationship $\rho_0 = x_0 / x_c$.

The topological function Eq. 13 is precisely of the quadratic form as found by recent computer simulations of 2D grain growth [5, 6]. Brandt et. al. [6] e.g. found from their simulation results of 2D grain growth the empirical topological relationship $s(x) = 4.0 + 1.26x + 0.627x^2$. Equating the first two coefficients to the corresponding coefficients in Eq. 13, i.e. $c_0 = 4.0$ and $c_1 = 1.26$, we find from the expressions of the first two coefficients in Eq. 13 in combination with Eqs. 9, 10 and $\rho_0 = x_0 / x_c$ for the cutoff parameter the value $x_0 = 3.175$, for the kinetic growth parameter $\tilde{\gamma} = 7.20$ and for the scaling factor $x_c = 1.04$. This yields for the third coefficient in Eq. 13 $c_2 = 0.635$, which is in good agreement with the numerical value of c_2 in the above empirical expression for $s(x)$. As shown in Fig. 2 the corresponding analytic LS-stable SDF Eq. 9 is in excellent agreement with the simulation results.

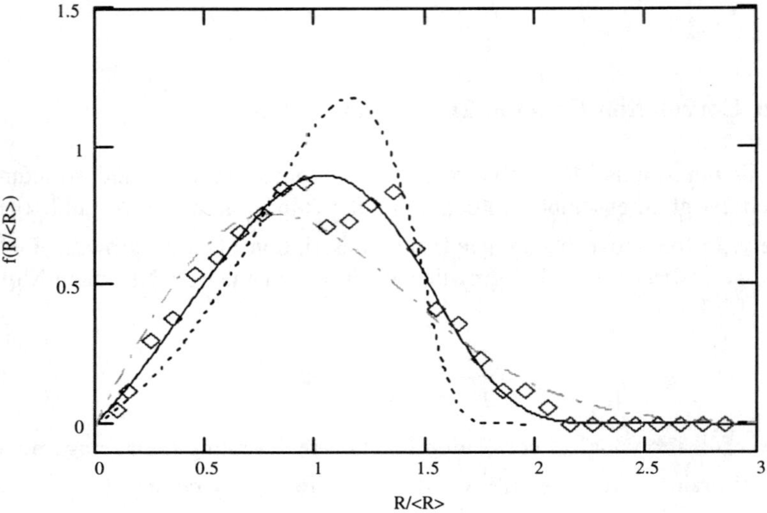

Fig. 2. Grain size distribution function $f(R/<R>)$ versus $R/<R>$ for $D = 2$. Full line: present (Eq. 9 for $\rho_0 = 3.05$ and $x_0 = \rho_0 / x_c = 3.175$, respectively); ◊ simulation results [6]; dotted line: Hillert ($x_0 = x_{LS} = 2$); dot-dashed line: Louat ($x_0 \to \infty$).

References

[1] I. M. Lifshitz, V. V. Slyozov, *J. Phys. Chem. Solids,* 19, 1961, pp. 35-50.
[2] C. Wagner, *Z. Elektrochem.,* 65, 1961, pp. 581-591.
[3] M. Hillert, *Acta Metall.*, 13, 1965, pp. 227-238.
[4] K. Marthinsen, O. Hunderi, and N. Ruym, *Acta Mater.*, 44, 1996, pp. 1681-1689.
[5] D. Weygand, Y. Brechet, *Phil. Mag. B*, 78, 1998, pp. 329-352.
[6] R. Brandt, J. Svoboda, and K. Lücke, in „Grain Growth in Polycrystalline Materials III“, eds. H. Weiland et. al., TMS, Warrendale, 1998, pp. 131-136.
[7] W. W. Mullins, *Acta Mater.*, 46, 1998, pp. 6219-6226.
[8] P. Streitenberger, *Scripta Mater.*, 39, 1998, pp. 1719-1724.
[9] O. Hunderi, N. Ruym, *J. Mater. Sci.*, 15, 1980, pp. 1104-1108.
[10] N. P. Louat, *Acta Metall.*, 22, 1974, pp. 721-724.

Re-Examination and Extension of Hillert Theory of Two Dimensional Normal Grain Growth

G. Liu[1], H. Yu[1,2], X. Song[1], X. Qin[1], C. Wang[1], and S. Liu[1]

[1] School of Materials Science and Engineering, University of Science and Technology Beijing, Beijing 100083, China

[2] Department of Applied Physics, Beijing Institute of Technology, Beijing 100081, China

Keywords: grain growth theory, grain size distribution, computer simulation.

Abstract. Hillert first undertook the simultaneous calculation of the grain size distribution and the rate law for grain growth, which resulted in a classical 2D grain growth analytical model predicting a single quasi-stationary grain size distribution function (GSDF) reached from arbitrary initial state. In this paper, this theory is re-examined based upon theoretical analysis, computer numerical simulation as well as Potts-type Monte Carlo simulation. It is found that the theory is well obeyed by the 2D grain growth processes starting from gamma and lognormal initial GSDFs. However, the same basic equation may also lead to quasi-stationary GSDFs different from Hillert's, as demonstrated by our computer simulation started from theoretical and experimental initial GSDFs.

1 Introduction

Hillert first undertook the simultaneous calculation of the grain size distribution and the rate law for grain growth [1,2], the classical two-dimension grain growth (2DGG) theory so resulted has been accepted widely. The original Hillert theory predicted a single quasi-stationary GSDF that can be reached from any arbitrary initial state. By a systematic re-examination based upon theoretical analysis, computer numerical and Potts Monte Carlo simulations starting from different types of initial structures, the form of this quasi-stationary GSDF has been reconfirmed in this paper. On the other hand, we do find that other quasi-stationary GSDFs may also be reached in 2DGG processes under certain conditions.

2 Theoretical Re-Analysis

2.1 Hillert's original analysis

For the 2D ideal grain growth, Hillert [1,2] proposed following equations for the grain with a radius R:

$$dR/dt = m\sigma(R/R_{cr}-1)/2R = m\sigma(u-1)/2R \qquad (1)$$

$$dR^2/dt = m\sigma(R/R_{cr}-1) = m\sigma(u-1) \qquad (2)$$

$$du^2/d(\ln R_{cr}^2) = m\sigma(u-1)(dt/dR_{cr}^2)-u^2 \qquad (3)$$

where $u = R/R_{cr}$ is the relative grain size, m is the mobility of and σ the specific energy of the grain boundary, respectively. The critical grain radius R_{cr} is intimately related to the average grain size of the material:

$$R_{cr} = \langle R^2\rangle/\langle R\rangle$$

where < > denotes the value averaged over all the grains in the system. For 2DGG and 3DGG, $R_{cr} = \langle R\rangle$ and $(8/9)\langle R\rangle$, respectively.
Let $\tau = \ln R_{cr}^2$, $\nu = m\sigma(dt/dR_{cr}^2)$, Hillert rewrote Eq. 3 as

$$du^2/d\tau = \nu(u-1)-u^2 \qquad (4)$$

and predicted that only when $\nu = 4$, a quasi-stationary grain size distribution

$$f(u) = [(2e)^{\beta}\beta u/(2-u)^{2+\beta}]\exp[-2\beta/(2-u)] \qquad (5)$$

can be achieved, where $\beta = 2$ and 3 for 2DGG and 3DGG respectively.

2.2 Further analysis of the problem

From the above analysis, one also has that

$$dR_{cr}^2/dt = m\sigma/\nu, \text{ or } dR_{cr}/dt = m\sigma/(2\nu R_{cr}).$$

Then the relative grain growth rate, w, can be written as

$$w = (dR/dt)/(dR_{cr}/dt) = \nu\,(u-1)/u \qquad (6)$$

Besides Hillert's plot of the growth rate, $du^2/d\tau$, as function of relative size [1,2], we also examined the relative growth rate w, as function of u (Fig. 1).

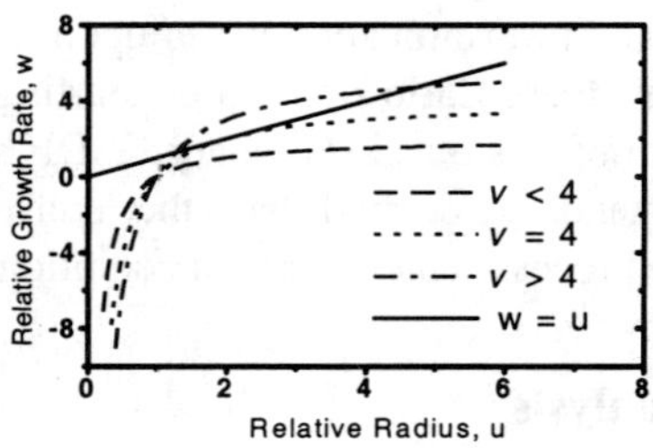

Fig. 1. The relative grain growth rate as function of relative grain size at various ν values.

From Eq. 6 one can obtain

$$du/dt = d(R/R_{cr})/dt = (w-u)(dR_{cr}/dt)/R_{cr} \qquad (7a)$$

or

$$du/dt = (dR_{cr}/dt)[\nu(u-1)-u^2]/(uR_{cr}) \tag{7b}$$

From Fig. 1, Eqs. 6 and 7, the following conclusions can be drawn for the 2D idea grain growth.

(1) When $\nu = 4$, $w \le u$ for all grains, $w = u$ and $du/dt = 0$ for grains with $u = 2$. This is the case described by Hillert in [1,2].

(2) When $u = 1$, $w \equiv 0$ for all ν values.

(3) When $w = u$, $du/dt \equiv 0$.

(4) By setting $\nu > 4$ and $w = u$, one can obtain the two roots (u_1 and u_2) of Eq. 6, shown as the two intersections of the curve $\nu > 4$ with the line $w = u$ shown in Fig. 1. We found that the relative frequency of grain size distribution, f(u), is always negative in the range of $u_1 < u < u_2$. It means that $u \le u_1$ for all the grains, "all the grains larger than u_1" should not be included in Hillert's analysis. When $u = u_1$, $du/dt = 0$, the relative size of grains with u_1 will keep constant during grain growth process. On the other hand, $w \le u$ for all grains with $u \le u_1$.

(5) When $\nu < 4$, one always has $w < u$. For grains with limited sizes, u should always decrease. This is true for a system with limited number of grains. However, if the maximum of u approaches ∞, $f(u) \to 0$, quasi-stationary grain size distributions would again be possible.

Following the procedure described in [3], the whole distributions of the individual grain sizes can be calculated as follows.

For $\nu = 4$, the quasi-stationary GSDF is the same as shown in Eq. 5, in which $0 \le u \le 2$;

For $\nu > 4$, let $\Delta = \nu^2 - 4\nu$ and $H_0 = (\nu/2)(f/u)_{u=0}$, f(u,t) is the relative grain size distribution function, then the quasi-stationary GSDF is taking the form:

$$f(u)=\frac{2H_0\ u\ \nu^{H_0}[(2u-\nu+\sqrt{\Delta})(-\nu-\sqrt{\Delta})]^{\frac{H_0\nu}{\sqrt{\Delta}}}}{[u^2-\nu(u-1)]^{1+H_0}[(2u-\nu-\sqrt{\Delta})(-\nu+\sqrt{\Delta})]^{\frac{H_0\nu}{\sqrt{\Delta}}}} \tag{8}$$

where $0 \le u \le u_1$, u_1 is the smaller one of the two roots

$$u_{1,2} = \frac{\nu m\sqrt{\nu^2-4\nu}}{2} \tag{9}$$

of the equation $u^2 - \nu(u-1) = 0$.

For $\nu < 4$, the quasi-stationary GSDF is taking the form:

$$f(u)=\frac{2uH_0\nu^{H_0}}{\left[u^2-\nu(u-1)\right]^{1+H_0}\exp\left\{\frac{2H_0\nu}{\sqrt{-\Delta}}\left[\operatorname{arctg}\frac{2u-\nu}{\sqrt{-\Delta}}-\operatorname{arctg}\frac{-\nu}{\sqrt{-\Delta}}\right]\right\}} \tag{10}$$

where $u = (0, \infty)$. When $u \to \infty$, $f(u) \to 0$ with $u^{-(1+2H_0)} \to 0$.

For a real material system with $\nu < 4$, the case of $u \to \infty$ never happens, such a GSDF is not approachable theoretically. However, the GSDF in such a system changes very very slowly at late stages, what one feels is that the 2DGG process is already in the quasi-stationary stage. Thus we call such kind of GSDF as near quasi-stationary functions.

3 Computer Simulation Results

3.1 Monte Carlo simulation

The Monte Carlo algorithm used here was developed and modified by Song [4,5] based on that of Anderson et al. [6]. The initial grain size distribution formed in the simulation can be well described by a gamma distribution (Fig. 1a). After the grain growth exponent n had become larger than 0.49, the initial gamma distribution was developed into that predicted by Hillert, as shown in Fig. 2b.

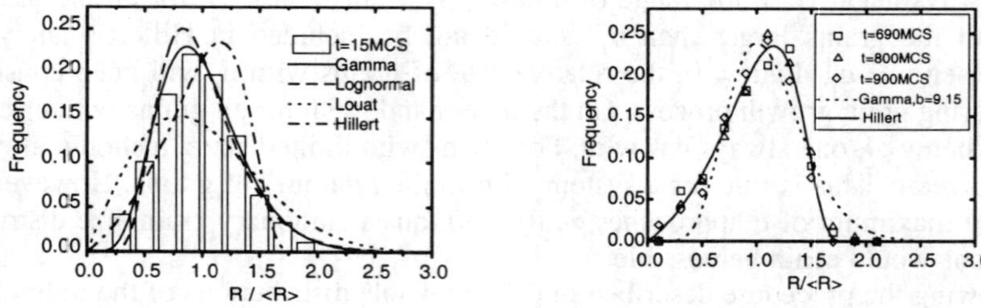

(a) (b)

Fig. 2. Evolution of GSDF during 2D grain growth simulated with a modified Monte Carlo algorithm. From an initial GSDF of gamma type (a), the Hillert's quasi-stationary DSDF (b) is obtained. The grain growth exponent of the power growth law, n, becomes larger than 0.49 and reaches 0.5 when the simulation time t≥600 and ≥800 MCS (Monte Carlo Steps), respectively. Here R and <R> denote the grain size of individual grain and the average grain size at time t, respectively.

It can be seen from Fig. 3 that for the clustering of grains larger than the average size, du/dt<0, other than du/dt≥0 for many grains with u≥1 in the idea case predicted in section 2.2. This means the growth rate of individual grains is also dependent on its local environment in the system. However, it seems that such fluctuations do not affect much the behavior of whole system shown in Fig. 2.

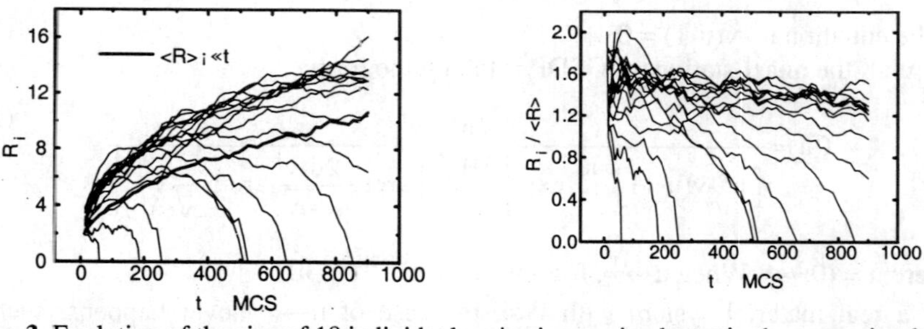

Fig. 3. Evolution of the size of 18 individual grains in a grain cluster in the same simulation described in Fig. 2. R_i and <R> denote the grain size of grain i and the average grain size at time t, respectively.

3.2 Numerical simulation

Such simulation is based on Eq. 1. When the initial GSDFs for simulation are gamma type with maximum relative grain radius $u_{max} = 3$ and with different distribution width, we finally obtain Hillert's quasi-stationary GSDF ($\nu = 4$), as shown in Figs. 4 and 5. The results are almost the same when the initial GSDF is lognormal type with $u_{max}>2$ (such as 3.0 and 2.4).

While starting from an initial non--stationary GSDF with $\nu<4$ and $u_{max}<2$ (such as 1.9) using the same method and simulation time as used in Figs. 4 and 5, one always finally obtains a quasi-stationary GSDF different from Hillert's. Fig. 6 gives such an example based on the data of high-purity Fe [7]. Similar results have been obtained based on the experimental data of Al [8] as an initial GSDF. However, Fig. 7 demonstrates that an initial quasi-stationary GSDF with $\nu<4$ (here is 2.0) as described by Eq. 10 does not change during the simulation process.

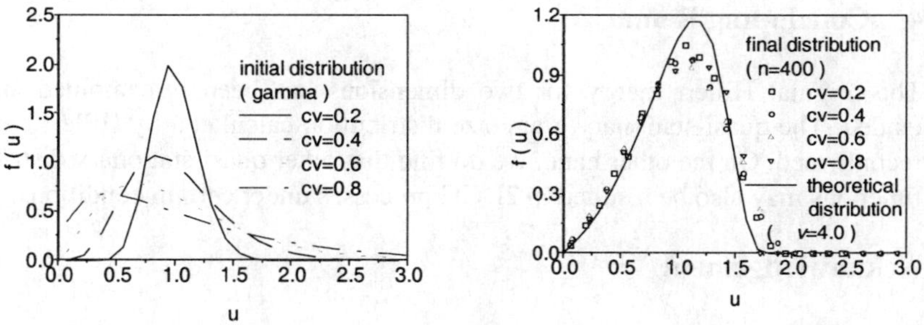

Fig. 4. Evolution of the GSDF in 2DGG with a initial gamma distribution. There existed 4000 grains (i.e., n = 4000) with $u_{max} = 3$ in the initial microstructure.

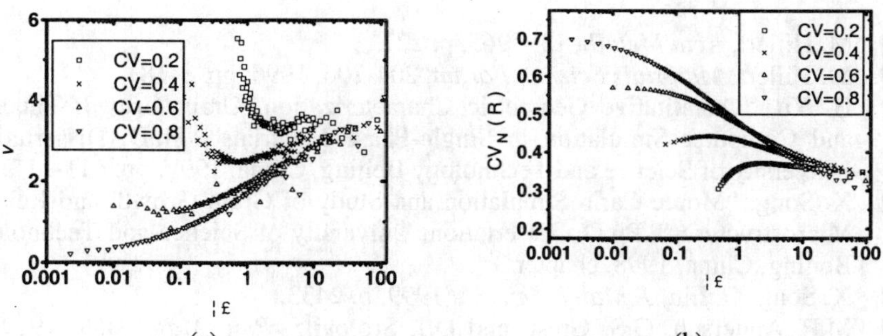

Fig. 5. Evolution of the GSDF in the same 2DGG process. The scale Γ is related to time t as $\Gamma = m\sigma t$. Whether the initial distribution with its width CV = 0.2, 0.4, 0.6, or 0.8, the final values of ν(a) and distribution width (b) are approaching to those corresponding to Hillert's quasi-stationary grain size distribution function [1,2]. When the initial distribution is of lognormal type, the corresponding plots are very similar to these two.

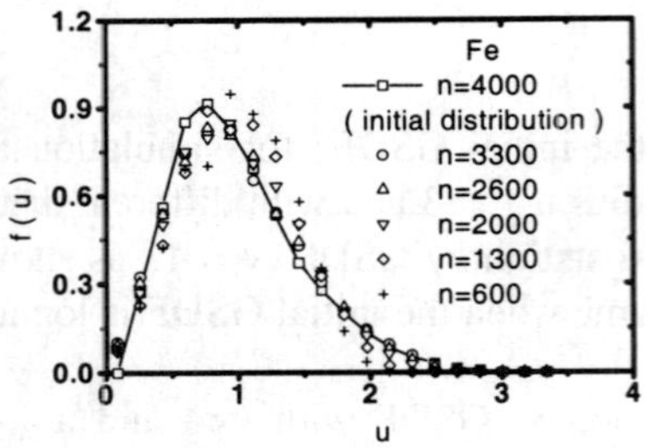

Fig. 6. Evolution of GSDF in a 2DGG simulation, using the experimental GSDF in high-purity Fe [7] as its initial GSDF (ν≈1.5). The simulation results are consistent well with the experimetal observation, other than Hillert's function.

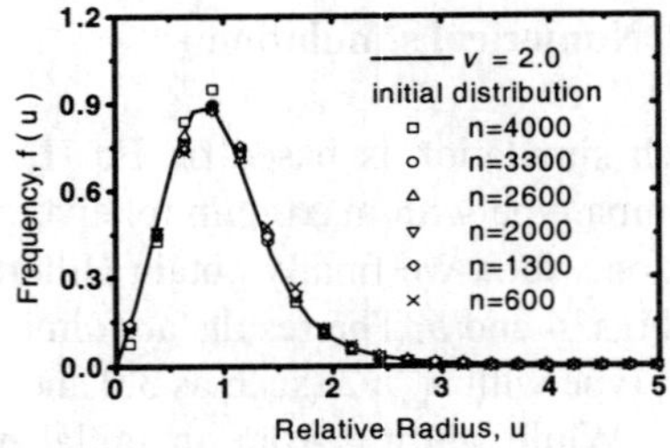

Fig. 7. Evolution of GSDF in a 2DGG simulation, using a quasi-stationary GSDF described by Eq. 10 with ν = 2.0 as the initial GSDF. It shows that such a GSDF does not change during 2DGG, thus it is a quasi-stationary GSDF different from Hillert's.

4 Concluding Remarks

The original Hillert theory for two dimensions has been re-examined and extended. The quasi-stationary grain size distribution calculated by Hillert has been reconfirmed. On the other hand, we do find that other quasi-stationary distribution functions may also be resulted in 2DGG processes under certain conditions.

Acknowledgement

The authors are grateful to the National Natural Science Foundation of China for financial support of the projects No. 59872003 and No. 59771070.

References

[1] M. Hillert, *Acta Metall.*, 13, 1965, p. 227.
[2] M. Hillert, *Materials Science Forum*, 204-206, 1996, pp. 3-18.
[3] H. Yu, "Quantitative Geometric Characterization, Grain Growth Theories and Computer Simulation in Single-Phase Materials", Ph.D. Dissertation, University of Science and Technology Beijing, China, 1997, pp. 111-117.
[4] X. Song, "Monte Carlo Simulation and Study of Grain Growth and Related Microstructures", Ph.D. Dissertation, University of Science and Technology Beijing, China, 1998, chapter 3.
[5] X. Song, G. Liu, *J. Mater. Sci.*, 34, 1999, p. 2433.
[6] M.P. Anderson, G.S. Grest, and D.J. Srolovitz, *Phil. Mag.*, B59, 1989, p. 293.
[7] C. Antonione, L. Battezzati, A. Lucci et al., *J. Mater. Sci.*, 15, 1980, pp. 1730-1735.
[8] V.E. Fradkov, A.S. Kravchenko, L.S. Shvindleman, *Scripta Metall.*, 19, 1985, p. 1291.

Dynamic Grain Growth in Superplastic AA5083

Pete Bate and Luke Mackenzie

Manchester Materials Science Centre, Manchester M1 7HS, England

Keywords: grain growth, Zener pinning, superplasticity.

Abstract. During 'superplastic' deformation of AA5083+Cr, the grain size increases and grain elongation occurs. Several theories for this behaviour have been proposed, but it can be rationalised simply by considering the geometric effect of deformation on a Zener- pinned grain structure. This effect has been modelled by both 'Network' and 'Monte Carlo -Potts' techniques, both of which give results roughly similar to the experimental observations. The effect is complicated by coarsening of the particulate at high temperatures, which leads to static grain growth and an enhanced rate of dynamic grain growth, but this can be included, approximately, in the numerical models.

1 Introduction

During slow plastic deformation of many materials at elevated temperatures, an enhanced grain growth rate is observed. This dynamic grain growth is most apparent in superplastic deformation, where the grain size of the materials involved is small and static growth is slow. There has been a significant amount of both experimental and theoretical work on this phenomenon [1-7]. Generally the growth is effectively strain controlled: the relative growth rate is proportional to the strain rate. Most theories for the phenomenon have been based on the effects of grain boundary sliding or diffusional creep; commonly assumed to be important in superplastic deformation. That is not, however, necessary. Nes *et al* [8] considered several possible reasons for dynamic grain growth, including the simple geometric effect of deformation but concluded that this was unlikely to have a significant effect. Rabinovich and Trifonov [9] suggested that the geometric effect would be significant, but in common with most analyses and in contrast to Nes *et al* ignored the pertinent fact that superplastic materials are invariably Zener-pinned systems. That may not be obvious in duplex alloys such as Ti-6Al-4V, but it remains true; those systems are 'vertex pinned'. It is more obvious in superplastic aluminium alloys. An analysis by Bate [10], relevant to cases where the Zener pinning particles are so finely distributed that a uniform drag field can be assumed, showed that the simple geometric effect of deformation perturbing the balance of boundary tensions and Zener drag will lead to significant dynamic growth.

A further feature observed in superplastic aluminium alloys is grain elongation. Li *et al* [11] and Martin *et al* [12] have observed and analysed this in Al-Mg alloys. This indicates that there is some contribution from intragranular dislocation creep to the deformation and that boundary or near-boundary processes are not wholly responsible for the deformation. Irrespective of such issues, superplastic Al-Mg provides a very useful system to study dynamic growth, and some results using that material are presented here.

2 Experiments

Tensile tests were carried out on 1.6 mm thick sheet of Al-4.4%Mg–0.7%Mn–0.1%Cr, supplied in the recrystallised condition with small equiaxed grains. These tests were carried out in air with strain rates of 10^{-4} s^{-1} and 10^{-3} s^{-1}, at temperatures from 450°C to 550°C, and revealed typical superplastic behaviour, with $m \approx 0.38$ at 450°C rising to 0.54 at 500°C and 0.59 at 550°C. An estimate of the activation energy of deformation is 150 kJ mol^{-1}; about the value for lattice self-diffusion of aluminium.
An ageing treatment to give magnesium segregation to boundaries prior to etching was used for optical metallography, and some specimens were examined using EBSD to give orientation maps. The EBSD grains sizes were consistently smaller than the optical measurements, and did not become coincident until an unrealistically high cut-off level of boundary misorientation was used. Aspect ratio values were very similar for both techniques.
There was a rapid initial increase in grain size during static annealing, indicating that the initial microstructure was not at the static Zener limit. At 450°C and 500°C very little static grain growth occurred after that transient, and a realistic 'initial' grain size was 12 μm. At 550 °C, static growth was noticeable, and the grain size increased to above 20 μm after 10^4 s.
The dynamic grain growth is shown in fig. 1. A useful approximation for dynamic growth is that the relative rate of mean grain diameter, D, is proportional to the strain rate:

$$\dot{D}/D = \lambda\dot{\varepsilon} \quad \text{i.e.} \quad \partial \ln D/\partial\varepsilon = \lambda \tag{1}$$

The values of λ are indicated on the graphs of Fig. 1, and range from 0.36 at 450°C (similar for both strain rates) to about 1 at 550°C with a rate of 10^{-3} s^{-1}. Although a higher value resulted at 550°C with a rate of 10^{-4} s^{-1}, particle coarsening contributes to that figure- this will be discussed below. The data of Martin *et al* [12] gives $\lambda \approx 0.5$ at 525°C with a rate of 10^{-4} s^{-1}. The data from Li *et al* [11] shows λ decreasing with strain, but up to a strain of about unity, $\lambda \approx 0.6$ with a strain rate of 4 x 10^{-4} s^{-1} at 550°C, at which temperature the static growth in their alloy was negligible. At the lower temperatures, the aspect ratio approximately doubled after a strain of 0.92, and at 550°C it increased by about 50% at that strain. Although both Martin *et al* and Li *et al* reported broadly similar increases in aspect ratio, they did not start from a near-equiaxed structure.

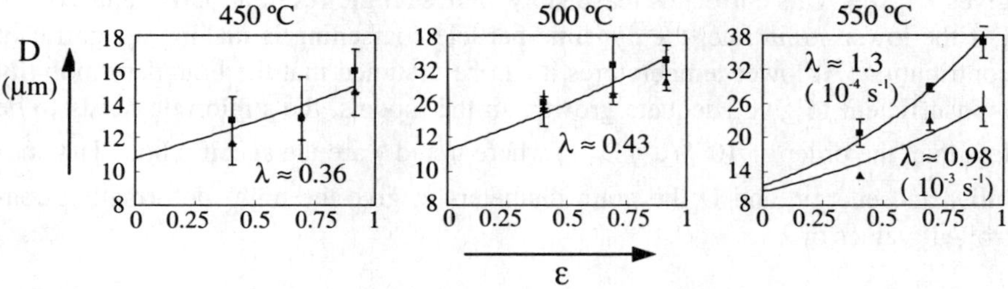

Fig. 1. Dynamic growth behaviour of the alloy at the temperatures indicated. Note the very different scale for 550 ºC.

The measured volume fraction of the $Al_6(Mn,Cr)$ pinning particles in the temperature range used was close to 0.02, and the particles were plates or discs, with an aspect ratio typically 3:1. These were aligned to some extent with the rolling plane, and their spatial distribution (assessed by autocorrelation functions) was reasonably homogeneous. The mean particle section diameter was 0.15 μm. Following extended times at 550 ºC, however, there was particle coarsening and this was noticeably more advanced in deformed material. After about 10^4 s at that temperature, the particle size had increased to 0.16 μm without deformation; with deformation at a strain rate of 10^{-4} s^{-1} it had reached 0.22 μm. This coarsening was only significant for the material deformed at the highest temperature and slowest strain rate.

3 Modelling

The analysis of dynamic grain growth mentioned above [10] used a vertex, or network, model of grain growth [e.g. 13-15] with a spatially uniform Zener drag. In two dimensions, with plane strain, this gives $\lambda \approx 0.4$ with homogeneous deformation and $\lambda \approx 0.5$ if deformation inhomogeneity of a type expected with slip deformation is included. Although these are comparable with experimental results for the lower two temperatures, they are too low to account for the behaviour at 550ºC and particle coarsening is probably only significant at the slower strain rate in that case. It was noted in [10] that the expected growth rate will increase if discrete, rather than averaged, particle pinning were involved. This is true in reality, of course, and in situations like the one reported here, relatively few particles per grain are involved.

An approximation to discrete particles can be made in the linear boundary network model by introducing a spatial field of Zener pressure, which is composed of Cauchy distributions of drag potential [16]. To approximate the situation of interest here, modelling with an average of about five Cauchy distribution centres per grain at the static limit was used (Fig. 2). With plane strain deformation at a rate comparable with that used for the homogeneous drag results noted above, this gives $\lambda \approx 0.9$. This correlates reasonably well with the result at 550°C and 10^{-3} s^{-1}. For the lower strain rate, the dynamic particle coarsening is making a significant contribution. At lower temperatures it can be assumed that the boundary mobility is insufficient to give adequate growth. In the models, the strain rate needs to be less than the order of $10^{-2}(\mu\gamma/D^2)$, where μ and γ are the specific boundary mobility and energy, and D the grain diameter- to give the high 'deformation controlled' values of λ.

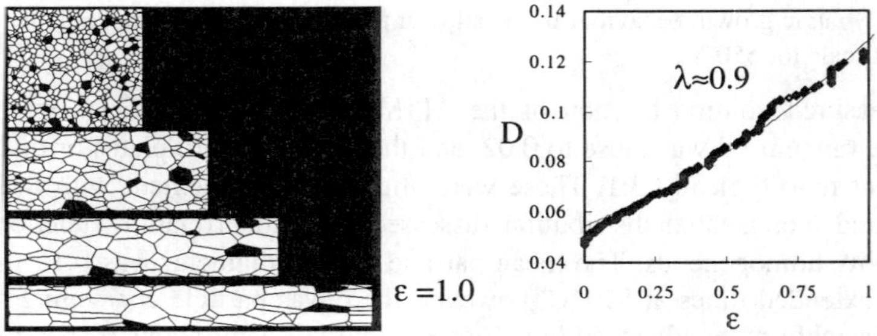

Fig. 2. Dynamic growth behaviour predicted by the vertex model using an inhomogeneous field of Zener drag potential. Grain structures are shown on the left, beginning at the static limit, and the growth characteristics on the right.

The validity of the vertex model used here- with its straight boundaries- to this intermediate discrete particle case is questionable. A better network model for discrete particle pinning has been used by Weygand *et al* [17], but in the present work the Monte Carlo- Potts (MCP) method [18,19] was employed. Introducing realistic particle pinning into MCP models requires some care: the particles must occupy several grid sites and a relatively high 'entropy' factor must be used [20]. Even in two dimensions, this means that models need to be large, in terms of computer memory. The only simple way to introduce deformation into a MCP simulation is by discrete shearing of the grid: a row is selected from a predefined distributed sequence and all rows above that one are effectively shifted one location. The translational symmetry, and the fact that particles should not be sheared, need to be considered. This simple shear deformation is intrinsically spatially and temporally inhomogeneous, and is even further from uniform uniaxial deformation than the plane strain used in the network model, but for relatively small deformations the deviation from a proportional plane strain- at about 45° to the shear- is quite small.

Even in two dimensions, simulations with a reasonable number of grains for low particle volume fractions means single processor computation times of several weeks. Because of this, larger volume fractions were used which allowed smaller ... Using $f=0.1$ with an equivalent plane strain rate of 10^{-7} MCS^{-1} gave $\lambda \approx 0.7$. This is in line with the network model prediction. Some example results are shown in Fig. 3. It remains to be seen if a lower particle volume fraction would affect the dynamic growth rate. There is slightly less increase in aspect ratio with strain in this type of simulation than with the network model. The greater freedom of boundary geometry even allows occasional grain 'splitting' or 'pinching'. Additional factors which may affect aspect ratios in practice were mentioned in [10].

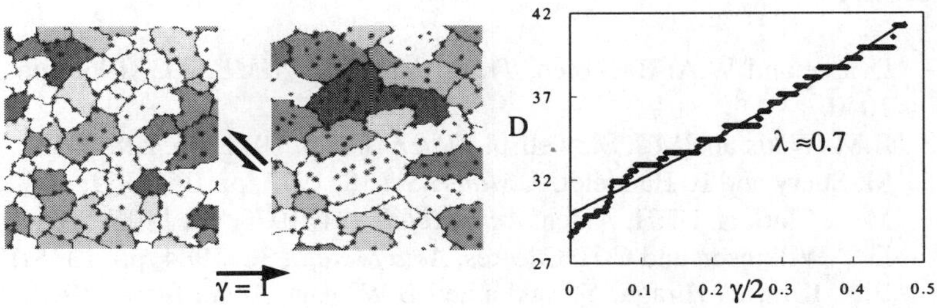

Fig. 3. Dynamic growth behaviour predicted by the 'sheared' MCP model using discrete particles with $f_v = 0.1$. Grain structures are shown on the left, starting at the static limit, and the growth characteristics on the right.

Dynamic particle coarsening has been reported several times in superplastic deformation, and was likely to be significant in the behaviour at 550°C with the slow strain rate. An approximation to this behaviour was presented in [10] using the homogeneous drag field vertex model. Using that gives an increase in λ from 0.5 to 0.65. If that ratio were to scale to the higher values characteristic of inhomogeneous drag, then it would give a predicted value very close to that observed. Although the total growth is reasonably close to a simple addition of the 'geometric' and 'particle coarsening' contributions, subtracting a static growth to give a 'true' dynamic growth should be treated with caution.

4 Summary

The dynamic grain growth in superplastic Al-4.5%Mg is consistent with that predicted by both vertex and Monte Carlo- Potts models of the geometric effect of deformation on a Zener- pinned microstructure, even though the models were only 2-d. It is clear that the discrete nature of the particle pinning needs to be considered. There was a significant contribution from dynamic coarsening of the pinning particles at the highest temperature and slowest strain rate used. An implicit assumption in the modelling is that the deformation is fully intragranular, and that boundary and near- boundary deformation mechanisms can be neglected. There is evidence from various sources that support this view [21], and the fact that dynamic grain growth may also be explained on that basis can be taken as further indication for a dominant role of intragranular deformation in superplasticity. There seems to be no need to invoke any special 'superplastic' mechanism to give dynamic grain growth, which is fortunate, as the phenomenon occurs in non-superplastic materials such as Al-6%Ni [16], and it may therefore be possible to decouple the phenomena of dynamic grain growth and superplasticity.

References

[1]. D. Lee and W.A. Backofen, *Trans. Met. Soc. AIME*, 293, 1966, pp. 1034-1040.

[2]. B.M. Watts and M.J. Stowell, *J. Mater. Sci.*, 6, 1971, pp. 228-237.

[3]. M. Suery and B. Baudelet, *J. Mater. Sci.*, 8, 1973, pp. 363-369.

[4]. M.A. Clark and T.H. Alden, *Acta Metall.*, 21, 1973, pp. 1195-1206.

[5]. D.S. Wilkinson and C.H. Cáceres, *Acta Metall.*, 32, 1984, pp. 1335-1345.

[6]. B.N. Kim, K. Hiraga, Y. Sakka and B.W. Ahn, *Acta Mater.*, 1999, 47, pp. 3433-3439.

[7]. J.R. Seidensticker, and M.J. Mayo, *Acta Mater.*, 46, 1998, pp. 4883-4893.

[8]. E. Nes, N. Ryum and J.K. Solberg, in „Proc. 4th Risø Int. Symposium on Metallurgy and Materials Science", eds. J. B. Bilde-Sørensen *et al*, Risø National Laboratory, Roskilde, Denmark, 1983, pp. 459-465.

[9]. M.Kh. Rabinovich and V.G. Trifonov, *Acta Mater.*, 44, 1996, pp. 2073-2078.

[10]. P. Bate , *Acta Mater.*, 49, 2001, pp. 1453-1461.

[11]. F. Li, D. H. Bae and A. K. Ghosh, *Acta Mater.* , 45, 1997, pp. 3887-3895.

[12]. C.F. Martin, J.J. Blandin, L. Salvo, *Mat. Sci. Eng.*, A297, 2001 pp. 212-222.

[13]. H.J. Frost, C.V. Thompson, C.L. Howe and J.H. Wang., *Scripta Metall.*, 22, 1988, pp. 65-70.

[14]. F.J. Humphreys, *Materials Science Forum*, 113-115, 1993, pp. 329-334.

[15]. D. Weygand, Y. Bréchet and J. Lépinoux, *Acta Mater.*, 47, 1999, pp. 961-970.

[16]. F.J. Humphreys and P.S. Bate, *Materials Science Forum*, 357-359, 2001, pp. 477-488.

[17]. D. Weygand, Y. Bréchet and J. Lépinoux, *Mat. Sci. Eng.*, A292, 2000, pp. 34-39

[18] M.P. Anderson, D.J. Srolovitz, G.S. Grest, and P.S. Sahni, *Acta Metall.*, 32, 1984, pp. 783-791.

[19]. M.P. Anderson, G.S. Grest, and D.J. Srolovitz, *Phil. Mag. B.*, 59, 1989, pp. 293-329.

[20]. M. Miodownik, E.A. Holm, and G.N. Hassold, *Scripta Mater.*,42, 2000, pp. 1173-1177.

[21]. P.L. Blackwell and P.S. Bate, *Materials Science Forum*, 304-306, 1999, pp. 189-194.

Multi-Phase-Field Model for the Motion of Multiple Interfaces

Britta Nestler[1] and Harald Garcke[2]

[1]Giesserei Institut, Intzestr. 5, D-52072 Aachen, Germany

[2]Institut für Angewandte Mathematik, Wegelerstr. 6, D-53115 Bonn, Germany

Keywords: phase-field model, grain boundary motion, grain growth, grain structure formation, numerical simulation, sharp interface model.

Abstract. A general class of anisotropic multi-phase-field models is proposed to describe the motion of multiple interfaces that may intersect at multiple junctions. The model reduces to classical sharp interface models in the limit of vanishing interfacial thickness. Numerical simulations show that our model is capable to recover generic features of grain growth such as crystalline curvature flow, an anisotropic as well as a crystalline version of Young's law at triple junctions, wetting phenomena and the (in-)stability of quadruple junctions.

1 Introduction

The phase-field methodology has recently emerged as a powerful mathematical and computational modelling technique to capture a wide variety of crystal growth and solidification structures. In this paper, we present the formulation of a general anisotropic and crystalline multi-phase-field model which has the flexibility to be applied either to simulate multi-phase alloy systems or to model growth phenomena in polycrystalline microstructures. The proposed model uses N order parameters to describe different phases in alloy systems or different orientational variants in a polycrystal. To design the multi-phase-field model, an approach by Steinbach et al. [1] has been extended to incorporate effects of smooth and crystalline time relaxation and surface energy anisotropy [2]. In phase field models the interface has a positive but small thickness and it can be shown by formal asymptotic expansions that classical physical laws can be recovered in the sharp interface limit of vanishing interfacial thickness. In [2], the relation of the diffuse interface model to a sharp interface model in the asymptotic limit is shown for smooth anisotropies. In particular, an anisotropic form of the Gibbs-Thomson equation at interfaces and an anisotropic equilibrium condition at triple junctions, known as Young's law, with additional shear forces due to surface energy anisotropy is derived. By introducing generalized Cahn-Hoffman ξ-vectors and a stress tensor formalism for the multi-phase-field model, an alternative way to obtain the sharp interface equations has been introduced in [3]. For the case of crystalline, non-differentiable surface energy anisotropies, the local force balance at triple junctions can be obtained using the notion of subdifferentials of convex functions [4].

The analysis in [2,3] ensures the validity and reliability of the new multi-phase-field model and allows for numerical simulations e.g. of grain growth phenomena. First numerical results in [2,4] illustrated a high agreement with the sharp interface analysis and demonstrated that the multi-phase-field model is capable to describe the influence of shear forces acting on interfaces and junctions when smooth or crystalline surface energy anisotropy is taken into consideration. The aim of this Contribution is to show the extensive applicability of the multi-phase-field model as a computational tool for microstructure simulations describing general features observed in real recrystallisation experiments.The application of the multi-phase-field model to multi-phase systems such as eutectic, peritectic and monotectic alloy systems can be found in [3,5].

2 Phase-Field Model

The vectorial multi-phase-field model is presented in terms of a generalized Ginzburg-Landau free energy F(u) which is a functional of the multi-phase-field vector $u=(u_1,\ldots,u_N) : (0,T) \times V \to \Sigma$, with $V \subset R^d$, $d = 1,2,3$. In the simplest case (not taking into account the dependence on e.g. temperature and concentration) the functional is of the form

$$F(u) = \int_V (\varepsilon f(u,\nabla u) + 1/\varepsilon\ \psi(u))dV. \tag{1}$$

Here, ε is a generally small length scale parameter which is related to the thickness of the interface. A vector of phase fields u is necessary in order to model the motion of N different grains or phases. Each component u_i, $i=1,\ldots,N$, describes the fraction of a specific grain or phase, so that $u \in G$ where $G := \{u \in R^N : u_i \geq 0, \Sigma_{i=1}^N u_i = 1\}$ is the Gibbs simplex in $\mathbf{R}^N$. The association with fractions motivates the following definitions of the space $\Sigma := \{u \in R^N :\ {}^{N}\, u_i = 1\}$ and of the projection T onto the tangent plane $T\Sigma := \{u \in R^N : \Sigma_{i=1}^N u_i = 0\}$, given by $(Tu)_i := u_i - 1/N\ \Sigma_{j=1}^N u_j$. For the free energy density $\psi(u)$, a higher order form of a multi-obstacle potential is used. Precisely we take

$$\psi(u) := 16/\pi^2\ \Sigma_{i<j}\ \mu_{ij}\sigma_{ij}u_iu_j + \Sigma_{i,j,k}\ \sigma_{ijk}u_iu_ju_k, \tag{2}$$

if u lies on the Gibbs simplex G and ∞ for all other u. In the above formula σ_{ij} and μ_{ij} are positive, symmetric (in i and j) surface energy and kinetic parameters and the σ_{ijk} are positive constants. The global minima for ψ on G are $e^1,\ldots,e^N$ and they correspond to the individual grains or phases. Here, e^i denotes the i-th standard unit vector in R^N. Choosing an obstacle potential gives that the interfacial regions are localized, so that in principle, the phase-field equations (Eq. 4) need to be solved only in the vicinity of the interfaces. The higher order terms $\Sigma_{i,j,k}\ \sigma_{ijk}u_iu_ju_k$ allow, by suitably choosing the parameters σ_{ijk}, to increase the energy in the interior of the Gibbs simplex. In systems with more than three order parameters these terms ensure that a stationary wave connecting two minima of ψ is forced to lie close to the edges of the Gibbs simplex G. As a result a good calibration for the parameters σ_{ij} and μ_{ij} of the phase field system with the surface tension σ_{ij} and the kinetic coefficient μ_{ij} of the sharp interface model is achieved. The surface energy contributions of all interfaces in the system are realized by the anisotropic gradient energy potential

$$f(u,\nabla u) := \Sigma_{i<j} (\sigma_{ij}/\mu_{ij}) (A_{ij})^2 |u_i \nabla u_j - u_j \nabla u_i|^2 \tag{3}$$

Anisotropy of the surface energies is incorporated into the model by the dependence of A_{ij} on $(u_i\nabla u_j - u_j\nabla u_i)$. As smooth anisotropies, in two dimensions, one can use e.g. ${}_sA_{ij}(\nu) = (1 + \delta_{ij}\cos(M_{ij}(\theta-\alpha_{ij})))$, which is a natural extension of an expression for solid/liquid phase systems by Kobayashi, [6]. Here, $\theta = \arg(\nu)$ is the angle which the vector ν forms with the x-axis, $M_{ij} \in N$ realizes an M_{ij}-fold anisotropy, $\alpha_{ij} \in [0,2\pi)$ describes the relative orientation of the anisotropy with respect to the x-axis and $\delta_{ij} \in (-1,1)$ measures the strength of the anisotropy. The surface potential f and the bulk potential ψ are chosen in such a way that the energy of the two phase stationary wave solution connecting e^i and e^j in direction of the unit normal ν to the i/j interface is exactly $\sigma_{ij}(\nu) = \sigma_{ij} \cdot {}_sA_{ij}(\nu)$. For crystalline anisotropies, we choose non-differentiable gradient energy coefficients of the form ${}_{cr}A_{ij}(\nu) := \max \{(\nu/|\nu|) \cdot \eta^k_{ij}; k=1,\ldots,M_{ij}\}$ with given $\eta^k_{ij} \in R^d$ defined by $\eta^k_{ij} := (\cos(2\pi k/M_{ij} - \alpha_{ij}), \sin(2\pi k/M_{ij} - \alpha_{ij}))$. The Wulff shape associated to ${}_sA_{ij}$ is the convex hull of the η^k_{ij}. The governing equations of motion for u are derived as the L^2-gradient flow of the Ginzburg-Landau free energy F(u). In the case of a smooth ψ we obtain

$$\varepsilon\beta u_t = \varepsilon \operatorname{div} Tf_{,x}(u,\nabla u) - \varepsilon Tf_{,u}(u,\nabla u) - 1/\varepsilon\, T\, \psi_{,u}(u) \tag{4}$$

together with the natural boundary condition $Tf_{,x}(u,\nabla u)\cdot\nu\partial v = 0$ on ∂V. We remark that the obstacle potential $\psi(u)$ leads to a parabolic variational inequality instead of a partial differential equation (see [2,4] for details).The notation $f_{,u}$ denotes the derivative with respect to u, and $f_{,x}$ denotes the derivative with respect to the gradient variable ∇u. The scalar kinetic coefficient β can be of anisotropic character by letting β depend on $(u,\nabla u)$.

3 Sharp Interface Limit

For smooth anisotropic free energies, the phase field system in Eq. 2 aims to approximate a sharp interface model in the limit when the interfacial thickness parameter ε tends to zero. The full derivation of the sharp interface model is presented in [2].

In summary: The limiting motion of sharp interfaces, separating two grains or phases i and j, is given by the kinetic Gibbs-Thomson law

$$\mu_{ij}(\theta)V = -(\sigma_{ij}(\theta) + \sigma_{ij}''(\theta))\kappa \tag{5}$$

where V and κ are the normal velocity and the mean curvature of the interface. For anisotropic systems, both, the limiting kinetic coefficient μ_{ij} and the surface energy σ_{ij}, may depend on the orientation of the individual interface. At triple junctions we obtain the classical force balance law (Young's law) including shear forces (Herring torque terms) which are due to surface energy anisotropy

$$(\sigma_{ij}(\theta_1)\nu_1 + \sigma_{ij}'(\theta_1)\tau_1) + (\sigma_{jk}(\theta_2)\nu_2 + \sigma_{jk}'(\theta_2)\tau_2) + (\sigma_{ki}(\theta_3)\nu_3 + \sigma_{ki}'(\theta_3)) \quad (6)$$

Here, ν_i and τ_i are the normal and tangent vector to the interface. At external boundary points, where an interface and the domain boundary intersect, the natural boundary condition

$$(\sigma_{ij}(\theta)\nu + \sigma_{ij}'(\theta)\tau)\nu d\nu = 0 \quad (7)$$

has to hold. For crystalline, non-differentiable energies, a crystalline motion law for faceted interfaces can be formulated (see [7,8]). The notion of subdifferentials $D\sigma_i(\nu_i)$, i = 1,2,3 for convex functions σ_i is introduced to derive a crystalline version of Young's law at triple junctions, [4]. Herewith, the force balance law reads $\xi_1 + \xi_2 + \xi_3 = 0$, where $\xi_i \in D\sigma_i(\nu_i)$. The vector $\xi_i^\perp$ is the force acting on an interface labelled i. The tangential component of $\xi_i^\perp$ represents surface tension effects, while the normal component of $\xi_i^\perp$ is the surface shear. Furthermore, a crystalline version of the condition at external boundary points $\xi \cdot \nu_{\partial V} = 0$ is fulfilled, where $\xi \in D\sigma(\nu)$ and $\nu_{\partial V}$ is the outer normal to V.

4 Numerical Simulations

The numerical simulations for the phase-field model (1) are performed by implementing a discretization with finite differences. In Fig. 1 the evolution of a horizontally aligned planar interface with isotropic, anisotropic and crystalline surface energies in situations where the interface intersects the external domain boundary is reproduced. The simulations are performed with an initial 90° angle configuration between the interface and the domain boundary. For non-isotropic interfaces, shear forces acting in normal direction to the interface occur which rotate the boundaries in a direction of lower anisotropic or crystalline surface energy. The respective angles at the domain boundary fulfill the anisotropic or crystalline form of the equilibrium force balance law. The angle φ measured in the simulation Fig. 1b compares favourably with the theoretically predicted value which is a solution of the equation

$$(1 + \delta_{ij}\cos(M_{ij}(\varphi - \alpha_{ij})))\cos\varphi + \delta_{ij}M_{ij}\sin(M_{ij}(\varphi - \alpha_{ij}))\sin\varphi = 0 \quad (8)$$

Due to crystalline surface energy anisotropy in Fig. 1c, strong facets are formed. In accordance with the definition of flow by crystalline curvature, the facet in the middle which is with respect to its neighbours neither convex nor concave has zero velocity.

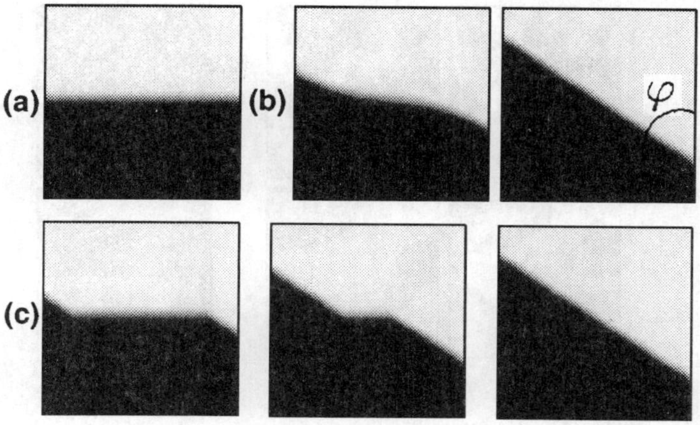

Fig. 1. Evolution of a horizontally aligned planar interface for (a) isotropic surface energies (the solution is stationary), (b) smooth anisotropic surface energies with $M_{ij} = 2$, $\delta_{ij} = 0.3$, $\alpha_{ij} = 120°$ and (c) crystalline surface energies with $M_{ij} = 4$, $\alpha_{ij} = 10°$ and $i, j = 1,2$, $i \neq j$.

For numerical computations of triple junctions we used a T-shaped initial configuration, Fig. 2a. We investigate the influence of surface energy anisotropy on the angle condition at the triple junction. In accordance with the analytical expression, Eq. 8, angle conditions differing from 120° are obtained as a result of additional torque terms in this anisotropic situation.
In isotropic systems, wetting occurs if the surface energies violate the stability condition $\sigma_{ij} < \sigma_{ik} + \sigma_{kj}$. An example of recrystallisation and wetting is illustrated in Fig. 3 for a system of four different order parameters.

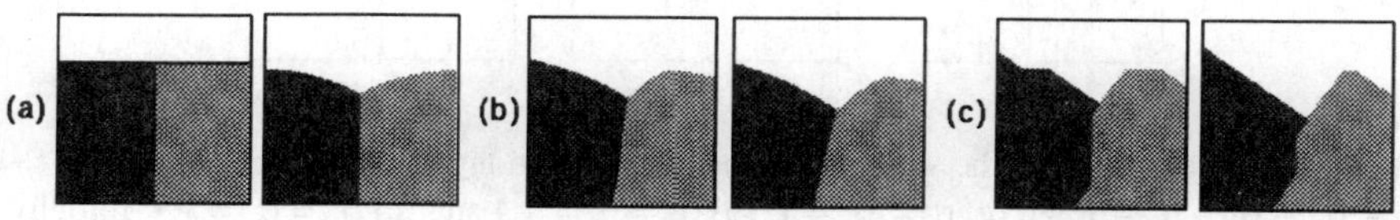

Fig. 2. Phase-field simulations of triple junctions for (a) isotropic surface energies $\sigma_{ij} = 1$, (b) anisotropic surface energies with $\sigma_{ij} = 1$, $M_{ij} = 4$, $\delta_{ij} = 0.065$, $\alpha_{ij} = 10°$ and (c) crystalline surface energies with $\sigma_{ij} = 10°$ and $i, j = 1,\ldots,3$, $i \neq j$.

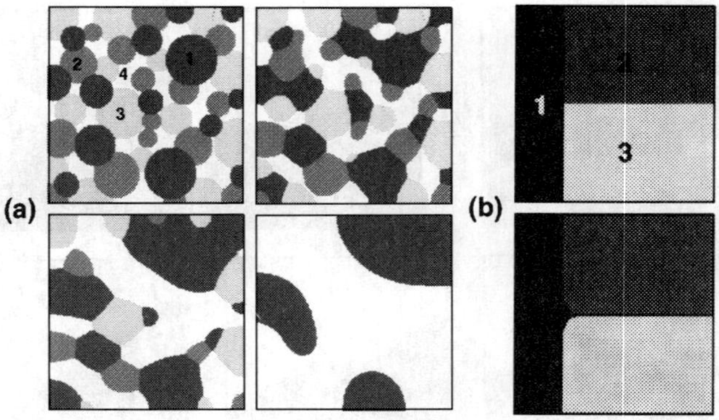

Fig. 3. Recrystallisation and wetting in an isotropic system with surface energies $\sigma_{12} = \sigma_{13} = \sigma_{14} = \sigma_{24} = \sigma_{34} = 0.7$ and $\sigma_{23} = 1.8$.

To investigate the stability of quadruple junctions, we use different sets of surface energy parameters σ_{ij}, Fig. 4. As expected from theory [2,9], the simulations prove that a quadruple point is unstable if $\sigma_{23},\sigma_{14} < \sqrt{2}\,\sigma$ (Fig. 4a and b) and stable if $\sigma_{23},\sigma_{14} > \sqrt{2}\,\sigma$ (Fig. 4c). According to this condition, a quadruple point splits into two triple junctions if the outer angles at the triple junctions are larger than the corresponding angles in the quadruple junction.

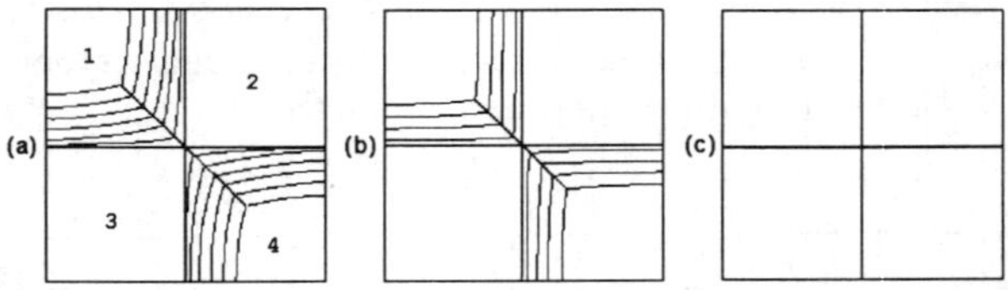

Fig. 4. Numerical study of the stability of quadruple junctions using surface energies $\sigma := \sigma_{12} = \sigma_{13} = \sigma_{24} = \sigma_{34} = 1$ and (a) $\sigma_{23} = \sigma_{14} = 1$, (b) $\sigma_{23} = \sigma_{14} = 1.4$ and (c) $\sigma_{23} = \sigma_{14} = 1.9$. Initially we chose a cross-shaped configuration. In all three cases the solution is plotted for different times and it is seen that the quadruple junction remains stable only for the last set of parameters.

In Fig. 5, the multi-phase-field model is applied to describe the symmetry properties of adjacent grain boundary triple junctions in hetero-epitaxially growing tri-crystalline microstructures in thin metallic aluminum films. In such films the number of different grain orientations can be controlled via epitaxial constraints (see [10,11]). For a system with three orientational variants we constructed, on the basis of geometrical arguments of grain boundary symmetries, an eight-fold crystalline surface energy anisotropy to model the experimentally observed grain structures. This energy formulation corresponds to a regular octagonal Wulff shape. In accordance with experimental observations, the phase-field-simulations show that two adjacent triple junctions always belong to different symmetry classes. It can be verified that all triple junctions in the system fulfill Young's force balance. Starting with the initial grain configuration of Fig. 5a, the framed region in Fig. 5b (the solution at a later time) illustrates that the simulated system relaxes into a grain configuration that can be found in thin film experiments such as in Fig. 5c.

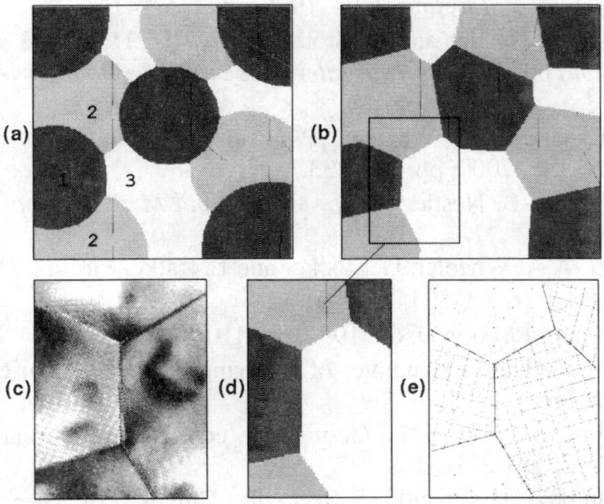

Fig. 5. (a) and (b) Phase-field-simulation of the stability and symmetry properties of grain boundary triple junctions in a microstructure of three orientational variants, (c) experimental grain structure of a thin aluminum sample (taken from [10]) in comparison with (d) the magnification of the simulation and (e) schematical drawing of the cubic crystal axes which are rotated by 30° with respect to each other.

5 Conclusion

In conclusion, we have shown that phase-field-models are able to describe a wide variety of phenomena in the case of multiple interfaces that may intersect at junctions. We incorporated anisotropy in the model and got anisotropic versions of equilibrium conditions at junctions. In particular we were able to quantitatively recover features observed in experiments (e.g. in the case of the polycrystalline thin film). In further studies it is planned to incorporate heat and mass diffusion as well as convection into the model and to perform numerical computations in three dimensions.

Acknowledgment

Britta Nestler gratefully acknowledges the support provided by the German Research Foundation (DFG) under Grant No. SA 335/25-3.

References

[1] I. Steinbach et al., *Physica D*, 94, 1996, pp. 135-147.

[2] H. Garcke, B. Nestler and B. Stoth, *Physica D*, 115, 1998, p. 87; SIAM J. *Appl. Math.*, 60, 1999, p. 295; *Interfaces and Free Boundaries*, 1, 1999, pp. 175-198.

[3] B. Nestler and A.A. Wheeler, *Phys. Rev. E*, 57, 1998, pp. 2602-2609; *Physica D*, 138, 2000, pp. 114-133.

[4] H. Garcke and B. Nestler, *Math. Models and Meth. in Appl. Sc.*, 10, 2000, pp. 895-921.

[5] B. Nestler, A.A. Wheeler, C. Stöcker and L. Ratke, *Physica D*, 141, 2000, pp. 133-154.

[6] R. Kobayashi, *Physica D*, 63, 1993, pp. 410-423.

[7] M.E. Gurtin, "Thermodynamics of Evolving Phase Boundaries in the Plane", Clarendon Press, 1993.

[8] J.E. Taylor, in "Differential Geometry", eds. B. Lawson and K. Tenenblat, Pitman, 1991, pp. 321-336.

[9] J.W. Cahn, E.A. Holm and D.J. Srolovitz, *Mat. Sc. For.*, 94-96, 1992, pp. 141-158.

[10] N. Thangaraj and U. Dahmen, *MRS Proc.*, 238, 1992, pp. 171-178.

[11] J.W. Cahn and G. Kalonji, *J. Phys. Chem. Solids*, 55, 1994, pp. 1017-1022.

Simulation of Grain-Growth Using the Multi-Phase Field Method

I. Steinbach, M. Apel, and P. Schaffnit

ACCESS e.V., Intzestraße 5, 52072 Aachen, Germany

Keywords: phase field method, grain growth, impurity drag.

Abstract. The multi-phase field method is a modern and flexible numerical tool to calculate the structural development during phase transitions. Primarily being developed for solidification problems, it can be effectively applied to solid state transformations driven by diffusion and/or reduction of interfacial area. Curvature effects and topological rearrangement of grain boundary networks are incorporated in a comprehensive way. The paper presents simulations of grain growth scenarios in 2D and 3D: isotropic growth and the effect of impurity drag. Impurity drag is described by a simple model, where the "loaded" and "free" part of the grain boundary is distinguished depending on the local velocity of the grain boundary. This treatment allows for the consideration of hysteresis effects. The model is calibrated by comparison with bicrystal experiments taken from the literature.

1 Introduction

Reduction of total free energy is the basic driving force of any phase transformation process. Based on this principle, the phase-field method has emerged as a method of choice to simulate microstructural evolution during materials processing [1–5]. The applications reach from the investigation of morphological instabilities e.g. dendritic growth [7-10] to structural phase transitions in solid-solid systems [11-13].The multi phase field method [14,15] being a natural extension of the basic phase field concept to the interaction of more than two individual phases, is especially suited to simulate ripening effects and grain growth in solids. A serious demand to technical applications, however, is the incorporation of realistic models for grain boundary mobilities into the general framework of the multi-phase field method. These models may relay on various material constants and process conditions on one hand as well as on the specific solution of the grain growth problem like local curvature and velocity of grain boundaries and triple junctions.

The paper summarizes shortly the underlying multi-phase field model and the selected model for grain boundary mobility. The examples concentrate on the simulation of a bicrystal experiment and on grain growth in 2 and 3 dimensions.

2 The Multi-Phase Field Model

We consider a system of N field variables $\phi_\alpha(\vec{x},t)$, $\alpha=1\ldots N$, $0\leq\phi_\alpha\leq 1$. These variables may be identified with the local phase state (liquid, solid matrix, precipitate, ...) or with individual grains of a solid phase identified by their orientation. In the bulk of one grain α $\phi_\alpha=1$ holds, on grain boundaries between grain α and grain β $\phi_\alpha+\phi_\beta=1$ and $\phi_\alpha<1$, $\phi_\beta<1$. In general, the constraint

$$\sum_{\alpha=1}^{N}\phi_\alpha = 1 \tag{1}$$

must be fulfilled everywhere. Using the approach of [14] the free energy functional F for this system is expanded in a sum over all pairwise energy differences between two phases α and β $f_{\alpha\beta}$

$$F(\phi,\nabla\phi) = \int_v \sum_{\alpha=1}^{N}\sum_{\beta=1}^{N} f_{\alpha\beta}(\phi,\nabla\phi)\,dV \tag{2}$$

where the integral has to be taken over the whole system volume v.
We use the explicite formulation of the local free energy differences $f_{\alpha\beta}$, which is

$$f_{\alpha\beta} = \sigma_{\alpha\beta}\,\eta_{\alpha\beta}\left[\frac{(\phi_\alpha\nabla\phi_\beta-\phi_\beta\nabla\phi_\alpha)^2}{2}+\frac{\pi^2\phi_\alpha\phi_\beta}{2\eta_{\alpha\beta}^2}\right] + G_{\alpha\beta}\,h(\phi_\alpha,\phi_\beta) \text{ with} \tag{3}$$

$$h(\phi_\alpha,\phi_\beta) = \frac{2}{\pi}\left[(\phi_\alpha-\phi_\beta)\sqrt{\phi_\alpha\phi_\beta}+\frac{1}{2}\arcsin(\phi_\alpha-\phi_\beta)\right]. \tag{4}$$

The constants are $\sigma_{\alpha\beta}$: boundary energy [J/cm²]
$\eta_{\alpha\beta}$: boundary thickness [cm]
$G_{\alpha\beta}$: enthalpy difference between the bulk phases [J/cm³]

Minimizing the total free energy and using a relaxation ansatz (for details see [14]) leads to the set of phase field equations for each field variable ϕ_α

$$\dot{\phi}_\alpha = \sum_{\beta\neq\alpha}\mu_{\alpha\beta}\left\{\sigma_{\alpha\beta}\left[\left(\phi_\beta\nabla^2\phi_\alpha-\phi_\alpha\nabla^2\phi_\beta\right)+\frac{\pi^2}{2\eta^2}\left(\phi_\alpha-\phi_\beta\right)\right]+\frac{\pi}{\eta}\sqrt{\phi_\alpha\phi_\beta}\quad\Delta G_{\alpha\beta}\right\} \tag{5}$$

and $\mu_{\alpha\beta}$: mobility of boundary [cm⁴/Js]

In (5) we have set all grain boundaries to the same thickness $\eta_{\alpha\beta}=\eta$ and neglected higher order triple point contributions [14,15]. It can easily be checked, that for dual phase boundaries $\phi=\phi_\alpha$, $1-\phi=\phi_\beta$, Eq. 5 reduces to the standard formulation

$$\dot{\phi}=\mu\left\{\sigma\left[\nabla^2\phi+\frac{\pi^2}{2\eta^2}(2\phi-1)\right]+\frac{\pi}{\eta}\sqrt{\phi(1-\phi)}\;\Delta G\right\}. \tag{6}$$

For the application of grain growth in multicrystalline materials it is important to notice, that the contribution in square brackets [ ] in (5) and (6) represent the curvature of the grain boundaries. In the multicrystalline case (5) this contribution as well accounts for the interfacial stress balance at triple junctions.

3 The Impurity Drag Model

Despite of the complexity of the underlying physics the effect of the drag of impurities on the motion of a grain boundary can in first approximation be described by a "two step" model of grain boundary mobility μ (Fig. 1). This postulates the existence of a critical velocity v* above which the grain boundary breaks free from the impurities and moves like a free boundary with a high mobility that is almost independent of temperature (between 1000°C and 1100°C for Fe3%Si alloy [16]). Below the critical velocity, the boundary is considered to be "loaded" and the temperature dependence of its mobility can be described by an Arrhenius law.

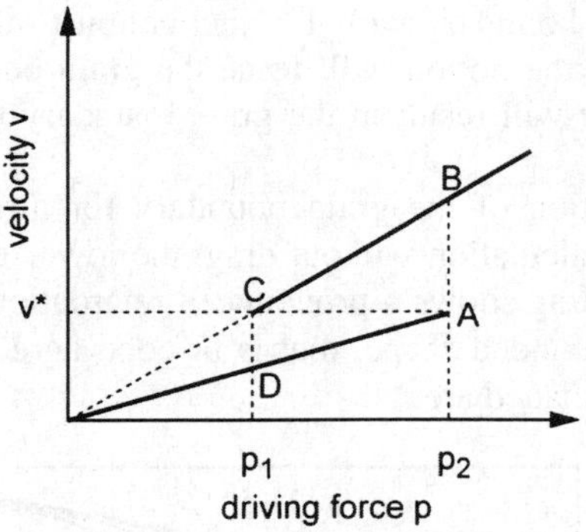

Fig. 1. Schematic dependence of grain boundary velocity on driving force, after [17].

This model includes the hysteresis effect, that the grain boundary keeps its high mobility while decreasing from the regime with high driving forces p ($p>p_2$, see Fig. 1) down to the point p_1, while it keeps low mobility up to p_2 when increasing from a state below p_1.
This hysteresis effect is considered naturally by treating the mobility of the grain boundary, μ, i.e. the slope of the lines in Fig. 1, as a function of velocity

$$\mu=\mu(v)=\begin{cases}\mu_f & v>v^*\\ \mu_l & v<v^*\end{cases}$$

For our calculations we use the following parameters.

μ_f	: mobility of free boundary	$5\cdot10^{-3}$ cm^4/Js
μ_l	: mobility of loaded boundary	$5\cdot10^{-4}$ cm^4/Js
v*	: critical velocity	0.018 μm/s

The values are for Fe3%Si at 940°C taken from the analysis of a reversed capillary experiment [16].
The impurity drag model uses an explicit time discretization for the dependence of mobility on velocity to resolve the hysteresis problem. The local velocity of the grain boundary, calculated at each time step, is used for the definition of the local mobility at the next time step.

4 Results

4.1 Benchmark calculation for the drag model

The benchmark test is based on the paper "The Shape of the Moving Grain Boundary in the Reversed-Capillary Technique under Consideration of the Drag Effect" by M. Furthkamp, G. Gottstein, L.S. Shvindlerman [16].
The calculation domain is 2-D, 6000x1200 μm divided into two grains with an initially straight grain boundary with 12° inclination to the x-axis. The adiabatic boundary condition at the bottom will force the grain boundary to a 90° inclination. This curved edge will result in the grain boundary motion driven by curvature.
Fig. 2 shows the position of the grain boundary for different times. The upper graph represents the calculation without drag, the lower graph the one with drag. The calculation with drag shows a pronounced retardation of motion after 4000 s as well as a slightly rounded shape, that is in good agreement to the experimentally observed shape (shaded area).

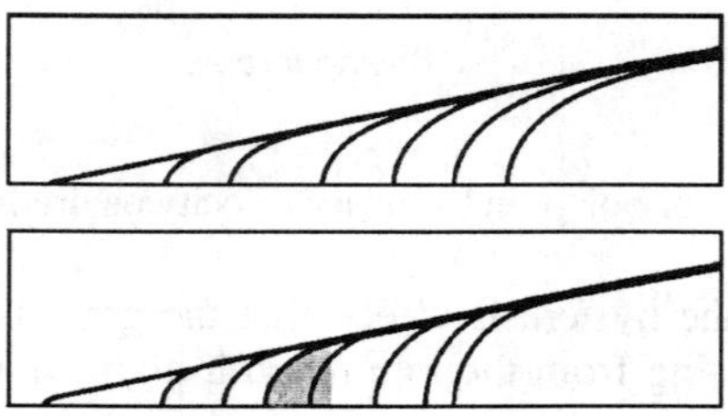

Fig. 2. Calculated grain boundary profiles without (upper case) and with drag model (lower case). Times are from left to right 100s, 2000s, 4000s, 8000s, 12000s, 16000s, 20000s. The shaded area represents the experimental observation, compared to the 6000s grain boundary profile.

4.2 Grain growth with solute drag

The impurity drag model described in Chapter 1 shall be applied to grain growth. The material data are taken same as in section 3.1. The grain boundaries were taken to be isotropic with respect to orientation in surface tension and mobility. We present the results for 2D simulation with and without impurity drag and a 3D calculation without drag. The 2D and 3D calculation domains are 600 x 600 μm² and 200 x 200 x 200 μm³, respectively. In the simulation spherical seeds were set randomly in the calculation domain.

For initialization, the seeds grow with constant driving force into a virtual matrix in order to produce the initial solid-solid grain structure. The initial 2D grain structure is shown in Fig. 3. It consist of 117 grains with an average radius of 31,3 μm and is the same for both 2D calculations. The initial 3D structure consists of 64 grains with the same average radius. After the initialization procedure, the grains grow competitively according to local curvature and interfacial stress at the junctions.

Fig. 3 shows the 2D grain structures of the simulations with and without impurity drag model. The average grain radius as a function of time is shown in Fig. 4. As expected, grain growth is slowed down in the case of impurity drag.

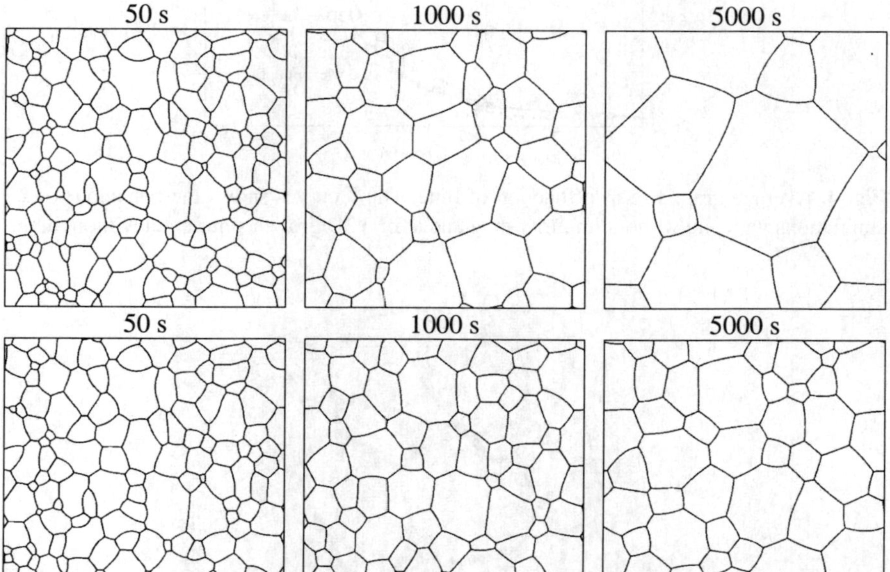

Fig. 3. Snapshots of the grain structures for 0s, 1000s and 5000s. Upper row: Grain growth without impurity drag. Lower row: with impurity drag model.

The "no drag" simulation tends to a growth law $R^n \sim t$ with an exponent n = 2.6 for longer times. The "drag" simulation tends to a mean exponent n = 8.2. Compared to the "no drag" case the curve shows distinct steps. For this specific grain structure the steps can be attributed to changes in the grain structure when configurations of low curved grain boundaries with low mobilities switch over to structures with a number of high curved boundaries with high mobilities. As a consequence the local growth exponent n oscillates between 4 and 12. Further simulations are necessary to clarify this effect.
Compared to the 2D simulations the 3D grain growth (Figs. 4, 5) seems to have a slightly slower growth kinetics. But one has to consider that the initial grain structures differ in their specific arrangements and numbers although the initial average radius is the same. Therefore a number of simulations with various initial structures are necessary to figure out the geometrical effect.

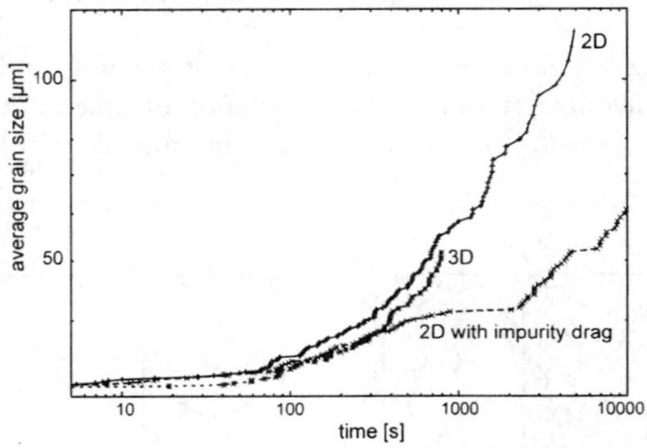

Fig. 4. Average grain size as a function of time. The 3 curves shows the behaviour of 2D simulations with and without impurity drag and a for a 3D growth simulation without drag.

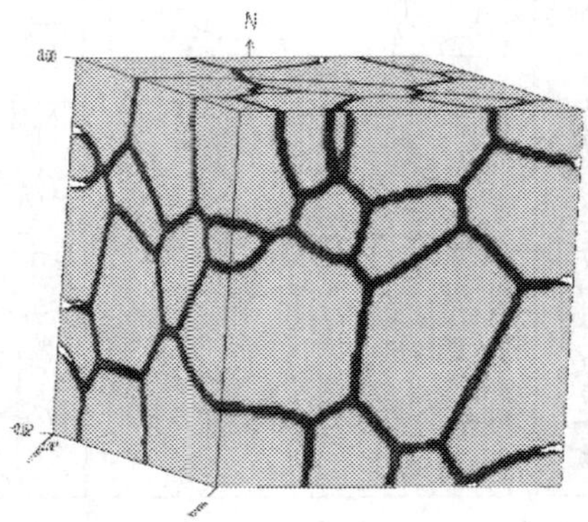

Fig. 5. Simulated 3D structure after 100s of grain growth. The cube is 200 x 200 x 200 μm^3 in volume.

5 Conclusion

The multiphase field method is applied to grain growth in 2D and 3D. The mobility of the grain boundary is taken to be a nonlinear function of the local grain boundary velocity including hytenesis effects. This "impurity drag model" was successfully validated against a bicrystal benchmark case. The application of the model to a 2D-grain growth simulation predicts a growth exponent n of 8.2 compared to the exponent of 2.6 for unrestricted growth. 3D calculations do not show the expected enhancement of the kinetic growth factor. These results are however not well settled due to the low statistics of the 3D calculations performed.

Acknowledgment

The work was sponsored by Corus Staal BV, Thyssen Krupp AG and Irsid-Usinor.

References

[1] G.J. Fix, in "Free Boundary Problems: Theory and Applications", eds. A. Fasano and M. Primicerio, Piman, Boston, 1983 (2), p. 580.
[2] A.A. Wheeler, W.J. Boettinger and G.B. Mc Fadden, *Phys. Rev.*, E47, 1993, p. 1893.
[3] J. Tiaden, B. Nestler, H.J. Diepers, I. Steinbach, *Physica D*, 15, 1998, p. 73.
[4] H.J. Diepers, C. Beckermann and I. Steinbach, *Acta Mat.*, 47, 1999, p. 3663.
[5] C. Beckermann, H.J. Diepers, I. Steinbach, A. Karma and X. Tong, *J. of Comp. Phys.*, 154, 1999, p. 468.
[6] S.G. Kim, W.T. Kim and T. Suzuki, *Phys. Rev. E*, 60, 1999, p. 7186.
[7] R. Kobayashi, *Physica D*, 63, 1993, p. 410.
[8] J.A. Warren and W.J. Boettinger, *Acta Metall. Mater.*, 43 (2), 1995, pp. 689-703.
[9] A. Karma and W.-J. Rappel, *Phys. Rev.*, E53, 1996, p. R3017; *Phys. Rev.*, E57, 1998, p. 4323.
[10] N. Provatas, N. Goldenfeld and J. Dantzig, *Phys. Rev. Lett.*, 80, 1998, p. 3308; *J. Comp. Phys.*, 148, 1999, p. 265.
[11] D.Y. Li and L.Q. Chen, *Acta Materialica*, 45, 1997, p. 2435.
[12] Y. Wang, D. Banerjee, C.C. Su and A.G. Khachaturyan, *Acta Met. Mater.*, 46 (9), 1998, pp. 2983-3001.
[13] U. Grafe, B. Böttger, J. Tiaden and S.G. Fries, *Scripta Mat.*, 42, 2000, p. 1179.
[14] I. Steinbach, F. Pezzolla, B. Nestler, M. Seeßelberg, R. Prieler, G.J. Schmitz and J.L.L. Rezende, *Physica D*, 94, 1996, pp. 135-147.
[15] I. Steinbach and F. Pezzolla, *Physica D*, 1999, p. 385.
[16] M. Furthkamp, G. Gottstein and L.S. Shvindlerman, *Interface Science*, 6, 1998, p. 279.
[17] G. Gottstein and L.S. Shvindlerman, "Grain boundary Migration in Metals", CRC Press, 1999.

Recrystallization and Grain Growth
Proceedings of the First Joint International Conference
Eds. G. Gottstein and D.A. Molodov, Springer-Verlag, 2001

The Effects of Surface Boundaries and Triple Junctions on Grain Growth in Polycrystalline Films

Amanda L. Giermann[1], Elizabeth A. Holm[2], and Alexander H. King[3]

[1]Dept. Mat. Sci. Eng., University of Illinois, Urbana, IL 61801 USA
[2]Sandia National Laboratories, Albuquerque, NM 87185-1411 USA
[3]School of Mat. Eng., Purdue University, West Lafayette, IN 47907-1289 USA

Keywords: grain growth, triple junctions, mobility, thin films, pinning

Abstract. Three-dimensional Monte Carlo grain growth simulations were used to explore mechanisms for stagnation of grain growth in polycrystalline films. When grain boundaries at the film surface have a low mobility due to grooving, grain growth slows but does not stop, nor do the catenoidal grains predicted by theory occur. If low trijunction mobility causes trijunction drag, grain vertex angles deviate from equilibrium, but the total curvature per grain boundary, thus the driving force for growth, remains constant, so stagnation does not occur.

1 Introduction

Polycrystalline thin films are important in the microelectronics industry. The microstructure of these films is particularly critical, as it determines many film properties. It is widely observed that in these films, grain growth stagnates when grains become columnar, having an average diameter approximately equal to the film thickness.

When a grain boundary intersects a free surface, a groove may form in order to reduce the system energy [1]. When grooving slows surface boundary motion, isolated, cylindrical grains shrinking by curvature attain a catenoidal shape, as shown in Fig. 1a. Mullins suggests that similarly impeded boundary motion in a columnar polycrystal will result in grain boundaries that are catenoidal sections [2]. Since catenoidal surfaces have zero geometric curvature (i.e. the two orthogonal curvature components are equal in magnitude and opposite in sign), such surfaces will cease to move by curvature. This stagnation mechanism becomes operative when the microstructure becomes columnar, that is, when the average grain diameter equals the film thickness. However, it is not possible to connect three catenoidal surfaces so as to maintain zero curvature and trijunction equilibrium simultaneously; thus, this curvature removal mechanism is not well understood for polycrystals. Furthermore, experimental studies of gold thin films have not observed catenoidal grain boundaries, but rather indicate that boundaries in stagnated structures are flat [3].

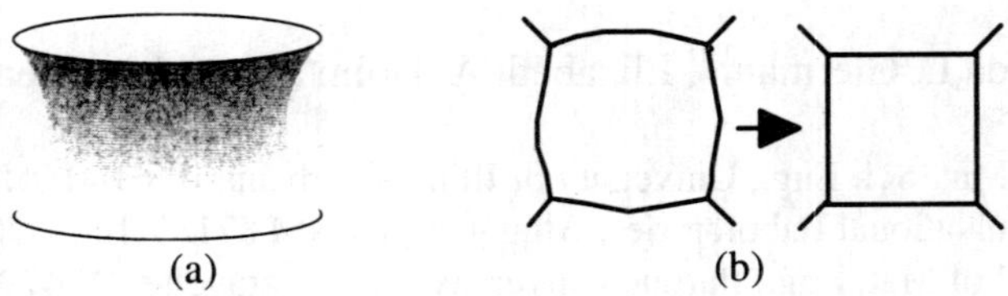

Fig. 1. Mechanisms for curvature removal in polycrystalline films. (a) Formation of catenoidal grains with zero geometric curvature. (b) Flattening of grain boundaries due to trijunction drag.

Gottstein, King and Shvindlerman have proposed an alternate curvature removal mechanism for columnar polycrystals [4]. They suggest that trijunctions may possess a lower mobility than their constituent grain boundaries. Slow trijunctions must be dragged by moving boundaries, causing non-equilibrium trijunction angles, which remove curvature from grain boundaries and cause stagnation, as shown in Fig. 1b. While this curvature removal mechanism is quite reasonable, the theory considers only isolated grains, not a network of boundaries.
In this paper, we utilize a Monte Carlo Potts model for grain growth to examine and compare these grain stagnation hypotheses.

2 Simulation Methods

The Monte Carlo Potts model used here has been described in theoretical and computational detail [5]. A microstructure is mapped onto a discrete lattice. Each lattice site is allocated a spin number so that all sites within a grain have the same spin, and grain boundaries are interfaces between neighboring sites of unlike spin. The energy associated with a site equals the Boltzmann constant k times the number of neighbors of unlike spin, thus is proportional to the amount of boundary present at that site. Grain growth is simulated by selecting a site at random and choosing a candidate spin at random from the set of neighbor spins. (Note that this scheme prohibits nucleation of new grains.) The spin of the selected site is then changed to the new spin with probability

$$P(\Delta E) = \begin{cases} M & \Delta E \leq 0 \\ M\exp(-\Delta E / kT) & \Delta E > 0 \end{cases} \tag{1}$$

where ΔE is the change in energy, $0 \leq M \leq 1$ is the relative mobility of the site, and kT is an energy that determines the amount of noise present in the system. After each attempted growth step, the simulation time is increased by 1/N Monte Carlo steps (MCS) where N is the total number of lattice sites.

Three-dimensional simulations were performed on a 50x50x10 simple cubic lattice with first, second and third neighbor interactions and with periodic boundary conditions in the *x* and *y* directions. Two-dimensional simulations were performed on a 500x500 fully periodic square lattice with first and second neighbor interactions. To avoid pinning due to simulation lattice anisotropy, simulation temperatures were T=0.5 for 2D simulations and T=1.0 for 3D simulations.

3 Results and Discussion

3.1 Grain Boundary Mobility

To simulate changes in grain boundary mobility due to surface grooving, the mobility of grain boundary sites on both free surfaces of the film was reduced to M_s, while the interior site mobility M_b remained unity. Since stagnation is observed primarily in columnar films, a fine-grained columnar microstructure was used as the initial structure. This structure was formed by seeding the lattice with random spins and performing normal grain growth with uniform boundary mobility until the grain structure became fully columnar.

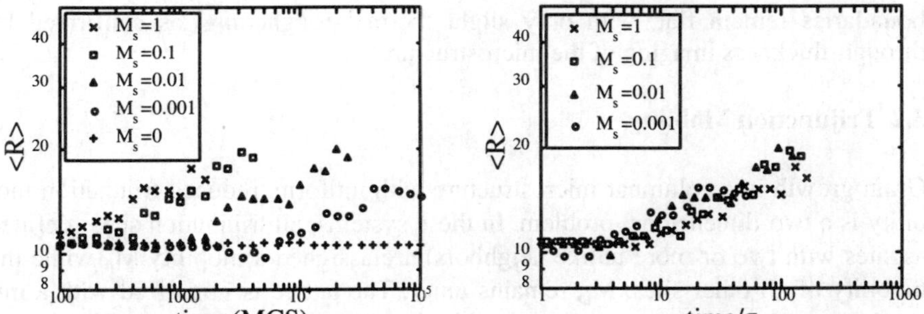

(a) (b)

Fig. 2. Average grain radius versus (a) time and (b) scaled time for a 3D columnar polycrystal with reduced surface boundary mobility.

The simulation was validated using a single cylindrical grain in a uniform matrix. With M_s=1, the grain shrinks uniformly with the radius proportional to the square root of time [6]. When M_s=0, the cylindrical grain collapses into a catenoid. This catenoid remains stable for a base radius to height ratio of 1 and pinches off into two separate grains at a ratio of 0.5. Since these are the results predicted by theory [7], the simulation technique was deemed appropriate.

In a polycrystalline microstructure with uniform boundary mobility (i.e. M_s=1), grains grow normally. In fact, grains grow continuously without stagnation for all M_s>0, although growth slows as M_s decreases, as shown in Fig. 2a.

For $M_s>0$, shrinking grains disappear by necking. The formation of a neck, which gives the grain a catenoidal appearance, is actually an early stage in the Rayleigh instability that ultimately leads to pinching off. However, except for these rare instabilities, grain boundaries remain flat (defined as possessing zero curvature normal to the film surface) during growth, and trijunction angles are 120°.

The unit process for grain growth in a columnar structure is the addition or removal of a column of grain boundary sites. The time required for this process is proportional to a time constant given by

$$\tau = \frac{2}{M_S} + \frac{H-2}{M_b} \tag{2}$$

where the film thickness is H sites, so there are two surface sites and H-2 interior sites per column. When the observed growth kinetics are scaled by τ, the data collapse onto a single curve, as shown in Fig. 2b. Since the kinetics scale with the speed at which the unit process occurs, it appears that the type and number of unit processes are the same for every M_S. That is, grain growth is self-similar over the range of M_S, and no stagnation mechanisms such as curvature removal become operative when $M_S<1$.

When $M_s=0$, no grain growth occurs. In this case, a few boundaries assume curved shapes in cross section. It is unclear whether these are stable catenoid formations or thermal fluctuations about a mean position. However, the vast majority of boundaries remain flat, with only slight thermal roughening, as confirmed by through-thickness imaging of the microstructures.

3.2 Trijunction Mobility

Grain growth in a columnar microstructure with uniform, reduced trijunction mobility is a two-dimensional problem. In these systems, all trijunction sites (defined as sites with two or more unlike neighbors) are assigned a mobility M_t, while the mobility of all other sites, M_b, remains unity. The lattice is initialized with a microstructure that had been pre-coarsened in the absence of mobility modifications.

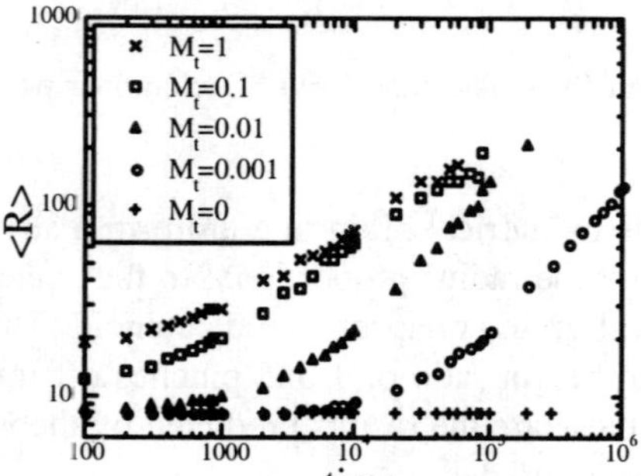

Fig. 3. Average grain radius versus time for 2D systems with reduced trijunction mobility.

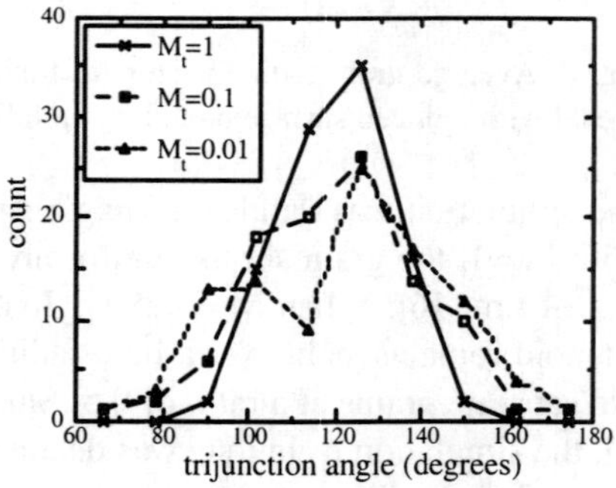

Fig. 4. Trijunction angle distribution for various trijunction mobilities. 99 angles were measured in each system.

As shown in Fig. 3, grains grow continuously and without stagnation for all $M_t>0$, and growth slows as M_t decreases. When $M_t=0$, no grain growth occurs. If grain growth is limited by trijunction motion, the average grain radius should increase linearly with time [8]. In these simulations, the late-time, asymptotic grain growth exponent increases from n=0.5 to n=1 as M_t decreases.

As suggested by theory [4], as M_t decreases, trijunctions deviate from their equilibrium angles. Figure 4 shows the distribution of trijunction angles for various M_t. When $M_t=1$, there is a symmetric distribution of trijunction angles centered about the equilibrium angle of 120°. (The distribution is not a delta function, because in this model, the time scales for trijunction equilibration and boundary motion are similar.) As M_t decreases, the distribution of trijunction angles spreads. This is consistent with the hypothesis that trijunction drag alters trijunction angles.

Figure 5 shows the evolution of boundary curvature during grain growth in these systems. While the total boundary curvature in the system decreases as grain size increases, the curvature per grain boundary remains essentially constant for all simulation times and systems. Evidently, trijunction drag does not decrease the average curvature per grain boundary, and therefore does not decrease the driving force for boundary motion.

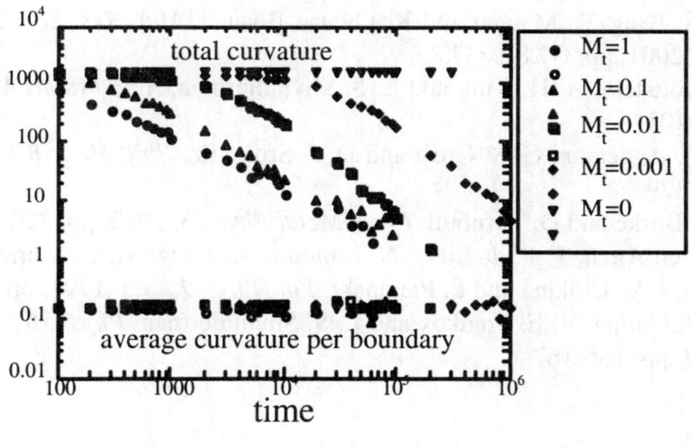

Fig. 5. Total curvature (filled symbols) and average curvature per grain boundary (open symbols) versus time for systems with varying trijunction mobility M_t.

The microstructure is an interconnected network of grain boundaries. While trijunction drag may decrease curvature in one grain, as shown in Fig. 1b, the integrated boundary curvature of the neighboring grains increases to offset this. The net effect is to conserve curvature per boundary. It is interesting to consider whether the average curvature per boundary is a universal grain growth constant, for various materials systems, textures, and morphologies.

4 Conclusions

Neither grain boundary grooving nor low trijunction mobility provides the curvature removal mechanism necessary to stagnate grain growth in polycrystalline thin films. Future studies should focus on alternate stagnation hypotheses, including crystallographic pinning, solute drag, and long-range trijunction interactions.

Acknowledgements

This work was performed at Sandia National Laboratories, supported by the U.S. Department of Energy under contract number DE-AC04-94AL85000 and by the Office of Basic Energy Sciences New Initiative program.

References

[1] W. W. Mullins, *J. Appl. Phys.*, 28, 1957, pp. 333-339.
[2] W. W. Mullins, *Acta Metall.* 6, 1958, p. 414-427.
[3] A. H. King, R. Mangat and K. Owusu-Boahen, *Mat. Res. Soc. Symp. Proc.* 615, 2001, pp. G7.8.1-G7.8.6.
[4] G. Gottstein, A. H. King and L. S. Shvindlerman, *Acta Mater.* 48, 2000, pp. 397-403.
[5] M. P. Anderson, G. S. Grest and D. J. Srolovitz, *Phil. Mag. B* 59, 1989, pp. 293-329.
[6] J. E. Burke and D. Turnbull, *Prog. Metal Phys.*, 3, 1952, pp. 220-292.
[7] M. Ben Amar, P. P. da Silva, N. Limodin, A. Langlois, C. Mbrazovskaia, C. Even, I. V. Chikina and P. Pieranski, *Eur. Phys. J., B* 3, 1998, pp. 197-202.
[8] A. V. Galina, V. E. Fradkov and L. S. Shvindlerman, *Phys. Met. Metall.*, 63, 1987, pp. 165-167.

Recrystallization and Grain Growth
Proceedings of the First Joint International Conference
Eds. G. Gottstein and D.A. Molodov, Springer-Verlag, 2001

Experimental Verification of Finite Element Simulation of Curvature-Driven Grain Growth in Al-Foil

M. C. Demirel[1], A. P. Kuprat[2], D. C. George[2], B. S. El-Dasher[1], G. K. Straub[2], and A. D. Rollett[1]

[1] Carnegie Mellon University, Materials Science & Engineering, PA, USA
[2] Theoretical Division, T-1, Los Alamos National Laboratory, NM, USA

Keywords: interfaces, microevolution, grain growth, mobility, finite element, simulation.

Abstract. Experimental results on grain boundary properties and grain growth that are obtained from Electron Backscattered Diffraction (EBSD) technique are compared with the Finite Element simulation results of an Al-foil with a columnar grain structure. The starting microstructure and grain boundary properties are implemented as an input for the three-dimensional grain growth simulation. In the computational model, minimization of the interface energy is the driving force for the grain boundary motion. The computed evolved microstructure is compared with the final experimental microstructure, after annealing at 550 °C. Good agreement is observed between the experimentally obtained microstructure and the simulated microstructure. The constitutive description of the grain boundary properties was based on a 1-parameter characterization of the variation in mobility with misorientation angle.

1 Introduction

Computational modeling and experimental characterization can be linked to study the interface properties and microstructural evolution of materials. However, new experimental data is needed for the comprehensive understanding of the kinetics of grain evolution. Grain boundary energy and mobility can be extracted with the measurement of the geometry of triple junctions between grain boundaries [1]. These experimental results on grain boundary properties that are obtained from EBSD technique can be used to simulate the topological changes in grain boundary motion.

Simulation [2, 3], theory [4], and the experimental observation [5] of grain boundary evolution of three-dimensional microstructures have been studied by several authors in the literature. A new technique for three-dimensional grain growth simulations was introduced by Kuprat [6]; this method utilizes gradient-weighted moving finite elements (GWMFE) [7] combined with algorithms for performing topological reconnections on the evolving mesh [8]. In this model, minimization of the interface energy is the main driving force for the grain boundary motion. Interface motion is assumed to obey a linear equation ($v=\mu\kappa$) where μ is reduced mobility, v is the velocity, and κ is the curvature of the grain boundary. An important verification of the model is that the expected power law dependence of growth kinetics is obtained [9]. The gradient in mobility has a major effect on the growth process. In this paper, the orientation dependence of the boundary mobility is introduced in order to break the symmetry in GWMFE simulations. With the same simulation technique, both normal and abnormal grain growth in three-dimensions can be studied.

In the following section, a brief summary of experimental details is presented. This is followed by results from GWMFE simulation, then comparison with annealing experiment, and ending with a discussion and conclusion.

2 Experimental Procedure

The samples, which have $1x1cm^2$ lateral dimensions and 120 μm thickness, were cut from 99.98% pure Al foil. The starting microstructure was polycrystalline and annealed conventionally in a horizontal tube furnace at a temperature of 550 °C for 9 hours in a N_2 environment. Under these conditions, a thin oxide layer is always present on the foil surfaces which prevents thermal grooves from forming along grain boundaries. After annealing, the specimen was quickly removed from the furnace and quenched under water, then mounted on a glass slide. The sample was then electro-polished at room temperature for 60 seconds with a solution of 730 ml ethanol, 100 ml ethylene glycol monobutylether, 78 ml perchloric acid, and 90 ml distilled water. A strong cube texture, {100}<001>, was observed after the annealing, with a small number of grains deviating from cube texture. In order to locate the scanning area, microhardness indents were used. A second annealing for 20 minutes at 550° C, under the same conditions, was performed to generate the final microstructure.

A secondary electron image was recorded from the surface of the crystal and the grain morphology obtained by applying the Gabor Wavelet [10] image-processing algorithm, which detects the boundaries. The orientation information was extracted from EBSD patterns using standard EBSD methods. EBSD scans were performed in beam control mode on the Al foil on an XL-40 Philips FEG Scanning Electron Microscope. A 52 mm diameter phosphor scintillation screen was used to observe the EBSD patterns, and the working distance (sample surface to gun tip) was fixed throughout the data acquisition process. Scan areas of 800x800 μm^2 with a three-μm step size were used. It is important to note that the scan size was chosen so as to minimize the possible errors and accurately measure orientation in the scan plane [11].

3 Simulation Details

To simulate the grain structure evolution, an area of 400x400 μm^2 was chosen from the scanned experimental microstructure. It is assumed that the grains are columnar in nature and as such, their boundaries are perpendicular to the surface [12]. Hence, the area is extruded in the third direction to create the three dimensional mesh. In the initial configuration, the microstructure consists of 26 grains which were used to construct a three-dimensional finite element mesh of 4328 tetrahedral elements in which the mesh lines conformed to the grain boundaries. Each grain is meshed independently and combined with LAGRIT [13] mesh generation software. The initial lattice orientation for each grain is then assigned based on the EBSD measurements. The total volume of the structure is kept constant during the simulation and exterior boundaries are assumed to be quasi periodic. The GWMFE method minimizes the function, $\int(v-\mu\kappa)^2 dA$ over all possible values of dx/dt values, where A is the surface area, x is the vector containing coordinates of all interface nodes, and t is the explicit time. All other details related to GWMFE method can be found in reference [6]. The GWMFE method solves an ordinary differential equation, where computational time scales with the number of tetrahedral elements in the simulation box. In an SGI-O2/R10000, our simulations took approximately 10 minutes of CPU time.
Anisotropy of the model is introduced by a single parameter dependence in the grain boundary mobility. In the calculations, the relative mobility in columnar Al-foil depends on misorientation angle and is low for small misorientations but undergoes a sharp transition to high mobilities between ten and fifteen degrees misorientation. [12]. It is assumed that capillarity is the only driving force for boundary migration. We should also note that relative mobility values are calculated based on the low angle boundaries and all high angle boundaries are considered to behave similarly. The local equilibrium condition at triple junctions is determined by the grain boundary energies. Relative boundary energies and mobilities were extracted through a statistical/multiscale analysis [14] and are given in Table 1 for the pre-annealed microstructure. The variation in energy was not considered in these simulations, however. The first column in Table 1 indicates the relative grain boundary mobility. The second column denotes the grains that have an interface with the relative mobility value in the initial microstructure.

4 Results

The simulated results of microstructure evolution show large differences between isotropic and anistropic dependence in terms of grain morphology. It was found that when starting with identical initial microstructures, varying the mobility dependence caused significant differences in the final microstructure. Isotropic (i.e., mobility is constant and equal at all boundaries) and anisotropic simulation results are contrasted in Fig 1. In the isotropic case, Fig 1c, the final microstructure has preserved the larger grains, but the smaller grains have disappeared, as would be expected from having fewer than six sides. In contrast to the simulation with isotropic properties, examination of the experimental microstructure, Fig 1b, reveals that four sided grain number 16 did not disappear after annealing. Additionally, some of the small grains (number 13 and 20) shrank but preserved their identities. In addition to these grains, there are also some important topological changes in the final annealed structure. Grains numbered 8, 10, 19 vanished completely, and numbers 21, 22, 25 coalesced into grain number 21. As shown in Fig 1d, for the case of anisotropic simulation [15], grain topology is correctly predicted except for the grain number 8.

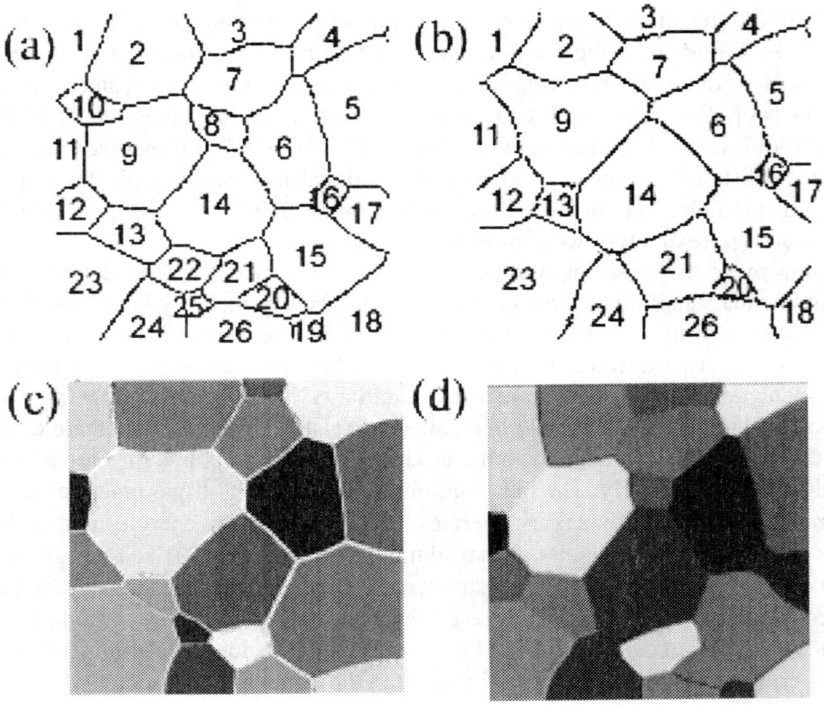

Fig. 1. Comparison of experiment and predicted grain morphology: Experimentally obtained microstructure (a) before annealing (500 μm by 500 μm), (b) after annealing, and simulated final microstructure (top view) with (c) isotropic properties (d) anisotropic properties.

Table 1. Experimentally measured relative mobility values of each grain interface.

Relative mobility values	Grain numbers for each grain interface
0.03239	3-4 4-7 5-17 7-4 9-8 11-9 25-21
0.00355	2-9 5-6 5-16 7-9 13-23 15-21 19-20 19-26
0.037287	2-7 4-6 6-16 9-13 12-13 21-26 25-26
0.0239	2-10 3-7 6-14 15-16 15-19 25-24
0.029	6-8 7-8 12-9 13-14 15-18 16-17 19-18
0.005728	6-15
0.00934	6-7
0.0107	9-10 15-20 23-24
0.05033	24-26
0.254698	14-21
0.282448	14-9
0.404045	1-2 1-10 1-11 8-14 13-22 14-15 14-22 22-21
0.8279	22-25 23-22 24-22

5 Discussions and Conclusion

The comparison between the experiment and computed results provides important details related to grain evolution. A strong similarity is observed between growth experiments and anisotropic three-dimensional GWMFE simulations. It is clear from the simulation results that anisotropy in the grain boundary energy and mobility has a major effect on growth process and the growth is driven by these parameters. Combining the three dimensional grain growth simulations results with experimental outputs will provide a computational tool for the discovery of some new methods to design meso-structures.

Absolute grain boundary mobility values were measured by Gottstein et. al. in Aluminum bi-crystals and reduced mobilities of order 10^{-8} to 10^{-7} $m^2.s^{-1}$ for two similar types of high angle boundary and various purities were obtained [16]. For curvatures about 100 μm^{-1} (i.e., boundary interface between grain 1 and 2) suggest migration rates of approximately $v=A\kappa=5.10^{-8}$ 100 = 5 $\mu m.s^{-1}$. These migration rates are high when compared to the estimated migration rates in the experiment (0.1 $\mu m.s^{-1}$). We note, however, that the purity of the material used in the bi-crystal experiments was significantly higher than that in our experiments. For the comparison of simulation times with experiment, these values can be used to estimate the migration distance during the annealing at various temperatures.

We conclude that the present simulation method verified the experimentally determined microstructure. The simulation model was an idealized grain growth case, and future extensions may be made to systems that are more complex and these further studies should foster a more comprehensive understanding of grain growth. Extension of this work to a five-parameter dependence of grain boundary energy and mobility is in preparation.

Acknowledgement

This work was supported by the MRSEC program of NSF under award number DMR-0079996 and Department of Energy under Contract W-7405-ENG-36 and completed during visits supported by the Computational Materials Science Network, US Department of Energy.

References

[1] B. L. Adams, D. Kinderlehrer, W. W. Mullins, A. D. Rollett, S. Ta'asan, *Scripta Materialia*, 38, 1998, pp. 531- 536.
[2] M. P. Anderson, G. S. Grest, D. J. Srolovitz, *Philosophical Magazine B*, 59, 1998, pp. 293-329.
[3] F. Wakai, N. Enomoto, H. Ogawa, *Acta Materialia*, 48, 2000, pp. 1297-1311.
[4] W. W. Mullins, *Acta Metallurgica*, 37, 1998, pp. 2979-2984.
[5] D. J. Jensen, H. F. Poulsen, T. Lorentzen, "Plastic Deformation, Recrystallization and Internal Stresses studied by a new 3D X-Ray Microscope", MRS Fall'99 Meeting, Boston, MA, USA, 1999.
[6] A. Kuprat, *Siam J. Scientific Computing*, 22, 2000, pp. 535 - 560.
[7] N.N. Carlson, K. Miller, *Siam J. Scientific Computing*, 19, 1998, pp. 766 - 798.
[8] A. Kuprat, *Los Alamos National Laboratory Report LA-UR-00-3475*, 2000.
[9] G. K. Straub, D. G. George, A. P. Kuprat, *Los Alamos National Laboratory Report LA-UR-00-1*, 2000.
[10] C. E. Shannon, *Proc. IRE*, 10, 1949.
[11] M.C. Demirel, A.P. Kuprat, D.C. George, B.S. El-Dasher, N.N. Carlson, G. K. Straub, A.D. Rollett, in "Proceedings of Materials Research Society Fall Meeting, 2000".
[12] C.-C. Yang, A. D. Rollett, W. W. Mullins, *Scripta Materialia*, 2000, submitted.
[13] D. George, *User Manual*, http://www.t12.lanl.gov/~lagrit/, 1995.
[14] D. Kinderlehrer, I. Livshits, S. Ta'asan, D. E. Mason, in "Twelfth International Conference on Textures of Materials", Montréal, Canada, 1999.
[15] For movies of the finite element computer simulation in Fig. 1, see http://mimp.mems.cmu.edu/links.html, 2000.
[16] G. Gottstein, D. A. Molodov, L.S. Shvindlerman, *Interface Science*, 6, 1998, pp. 7 - 22.

Grain Growth in Thin Films With Columnar Microstructure

V.Yu. Novikov

Treptower Str. 74d, D-22147 Hamburg, Germany

Keywords: thin films, microstructure, texture, grain growth, modelling.

Abstract. Grain growth in thin films is modelled on the assumptions that all grains are columnar, the energy and mobility of their boundaries are identical, and the boundary migration is inhibited by thermal grooves. It is found for the first time that the character of grain growth depends on the ratio of the most probable initial grain size to the film thickness. If this ratio is close to the experimentally observed, only abnormal grain growth (AG) is detected. The AG cessation typical of thin films on substrates is found to result from the insufficient number of the matrix grains which can be consumed by the abnormal grains. It has been shown that the fibre texture of the film can enhance owing to the size advantage of some grains.

1 Introduction

It is common knowledge that the properties and relevance of metallic and semiconductor thin films are controlled by their grain size and texture which form during film deposition and can change during the subsequent grain growth. The available experimental data on grain growth are contradictory, and systematic studies of the growth process in thin films of various thickness are lacking. Grain growth in films with columnar structure under the influence of both the grain boundary curvature and thermal grooves has recently been modelled [1], but in a very limited range of driving and pinning forces. Results of a more complete and systematic study of the microstructure and texture evolution during grain growth in films [2] are presented in this report.

2 Method

To study the growth process in films with columnar microstructure, a statistical model [3] was slightly modified. The growth of all the grains was supposed to be inhibited by a constant drag force F* exerted by thermal grooves [4]:

$$F^* = \gamma / 3\delta \tag{1}$$

where γ is the grain boundary energy and δ the film thickness. In some cases, an additional driving force, G*, was assigned to some grains. The net driving force, ΔG_i, for growth of a grain of size D_i at the expense of a grain of size D_j can be described as

$$\Delta G_i = k\gamma(1/D_j - 1/D_i) - F^* \tag{2a}$$

when neither of the grains have the additional driving force or both have it, as

$$\Delta G_i = k\gamma(1/D_j - 1/D_i) + G^* - F^* \tag{2b}$$

when the growth of the grain of size D_i is promoted by G^*, whereas the adjacent grain has no additional driving force, and as

$$\Delta G_i = k\gamma(1/D_j - 1/D_i) - G^* - F^* \tag{2c}$$

in the case opposite to the preceding one. The first term in the right-hand part of Eq. 2 describes a capillarity driving force, G_{gb}, and a factor $k \cong 2$. The grain of size D_i can grow if $\Delta G_i > 0$.

Changes of the grain size distribution with time were used to trace both microstructure and texture evolution during grain growth (the initial grain size distribution was taken from [5]). The following microstructure parameters were calculated: the apparent average grain diameter, D (it was found as if the measurements were conducted on the film surface by the random intercept method), the most probable grain diameter, M, and the maximum grain diameter, D_m. The character of the growth process was determined from the scatter of grain sizes characterized by the ratio D_m/M (initially D_m/M was equal to 3.69). Normal grain growth (NG) was assumed to proceed at $D_m/M \leq 5$, and AG at $D_m/M > 5$. The latter condition makes it possible to find the volume fraction of abnormal grains, V.

Texture was characterized by the volume fraction, V_T, of the grains of a certain type. Since the grain boundary energy as well as their mobility were assumed independent of time and identical for all the grains, the modelling data relate to a fibre texture with its axis normal to the film surface.

3 Results and Discussion

3.1 Character of grain growth process

The first series of simulations was performed on film models with various thickness and various initial microstructure. The latter was characterized by the same grain size distribution but with different M_0 = 28.4, 56.8 or 113.7 nm; the ratio M_0/δ was varied between 0.2 and 2.84. The growth inhibition was shown to increase with a reduction of δ and, thus, with an increase of F*(see Eq. 1). NG was observed at $M_0/\delta < 0.5$, AG at $0.5 \leq M_0/\delta < 2.84$ its rate reducing with an increase of M_0, and the growth process did not practically evolve at $M_0/\delta = 2.84$. The most intensive AG was observed at $M_0/\delta \cong 1$ (see Fig. 1a).

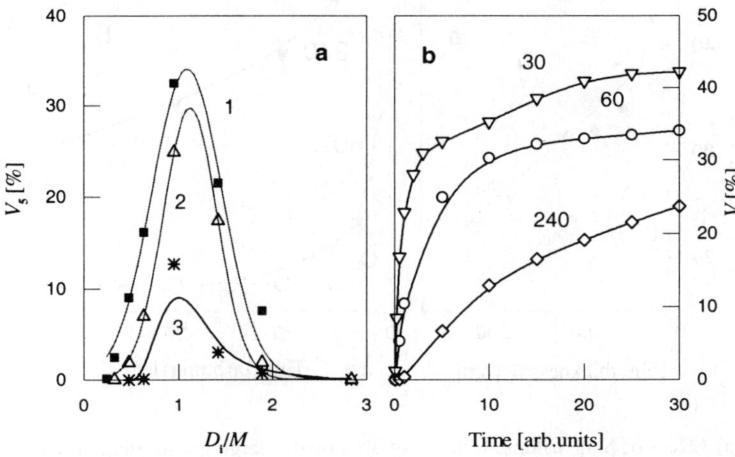

Fig. 1. (a) Effect of M_0/δ on the volume fraction of abnormally large grains attained by time 5 arb. units, V_5, in film models with M_0=28.4 (1), 56.8 (2) and 113.7 nm (3) and various δ. (b) AG kinetics in film models with $M_0/\delta \cong 1$(numbers show δ values).

Since in as-deposited films with columnar structure M_0/δ is usually close to 1 [5-7], the growth process was investigated further on film models with the same $M_0/\delta \cong 0.95$ but various δ in the range from 30 to 600 nm. In a good accordance with Fig. 1a, AG evolves in all models independent of *δ*, although an increase of *δ* results in a reduction of the AG rate (Fig. 1b). A distinct AG cessation is observed at δ =30–120 nm after the growth duration of 30 arb.units, whereas at a greater δ the cessation is hardly noticeable even for a longer growth period, obviously owing to a very slow AG evolution. The AG cessation characterized by the volume fraction of abnormal grains at t=30 arb.units, V_{30}, is shown in Fig. 2a; the dependence $V_{30}(\delta)$ agrees qualitatively with the experimental observations [8]. Since grain growth in film models develops as AG in a wide range of δ (on condition $M/\delta \cong 1$) this permits the supposition that the majority of the experimental data on grain growth in thin films relate to the AG evolution.
The next simulation series concerned the effect of G* on the growth kinetics in film models with $M_0/\delta \cong 0.95$ and δ from 30 to 600 nm; the initial size distribution of grains with G* > 0 was assumed identical to the overall distribution. The initial fraction of these grains was 0.02 which is typical of non-restricted fibre texture [9]. G* was assumed associated with a decreased surface energy owing to which $G^* = 0.09\gamma/\delta$ [10]. The introduction of G* does not affect the character of the growth process however it causes a more complete AG development accompanied by noticeable changes in the grain size: e.g. D increases to ~20δ in the film model with δ = 60 nm.

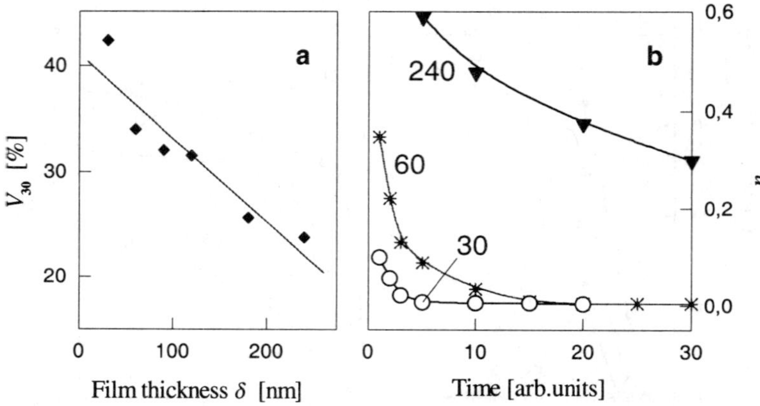

Fig. 2. (a) Effect of δ on volume fraction of abnormally large grains attained by t = 30 arb. units, V_{30}, at $M_0/\delta \cong 1$. (b) Number fraction of grains which can be consumed by abnormally large crystallites (numbers show δ values).

Since G* is always smaller than F*, in grains with G* > 0 it compensates to some extent the drag force hindering the movement of their boundaries and increases thereby their net driving force (see Eq. 2). So G* accelerates the growth of large grains for which $G_{gb} > F^*$ and promotes the growth of grains with $G_{gb} < F^*$. At the same time G* acts as an additional drag force upon the boundaries of grains with G* = 0 adjacent to grains with G* > 0 and enhances thereby the stabilization of the fine-grained matrix. So we may suggest that the term "surface-energy assisted AG" should be used in the case under consideration, instead of commonly accepted "surface-energy induced AG".

3.2 Cessation of abnormal grain growth

The simulation results conform qualitatively with the experimental data on thin films with columnar microstructure in the following major aspects: (i) grain growth is found to be retarded the retardation decaying with an increase of the film thickness; (ii) AG ceases before the fine-grained matrix is wholly consumed; and (iii) the maximum AG fraction attained decreases with the increasing film thickness. The most interesting features, namely the AG cessation and its dependence on the film thickness, can be explained as follows.

AG is known to evolve provided the fine-grained matrix is stabilized and some of the grains are able to grow rapidly [11]. The matrix in our experiments is fairly stable. Simultaneously, both the AG commencement and evolution prove the ability of certain grains to grow rapidly. So neither the matrix instability nor the absence of growing crystallites can affect the AG cessation. This effect may apparently be attributed to the exhaustion of the material consumed by abnormally large grains which should be accompanied by the reduction of their growth rate. The maximum size of the consumed grains, d_c, can be derived from Eqs. 1 and 2a as:

$$d_c = 1/(1/3\delta + 1/\langle D_a \rangle) \qquad (3)$$

where $\langle D_a \rangle$ is the true average size of abnormally large grains. Both the number fraction of the grains, n_c, whose size is smaller than d_c, and $\langle D_a \rangle$ can be found from the grain size distribution at any point of the growth process. As can be seen from Figs. 1b and 2b the AG retardation commences when n_c does become quite low. It is also possible that impingements of abnormally large grains contribute to their growth rate reduction.

3.3 Texture development

Since grains with $G^*>0$ can grow faster than the others of the same size, AG in all the models studied leads to an increase of V_T (Fig. 3). It may be supposed that texture can also amplify only under the influence of G_{gb} when crystallites of a certain orientation have an increased size in the initial microstructure, owing to their increased growth rate during the film deposition [12].
To study the effect of the size advantage, the initial size distribution of a group of grains was displaced toward large sizes owing to which their most probable size increased to ~$2M_0$. The initial number fraction of these grains was 0.02 and the volume fraction ~0.07. The properties of their boundaries were the same as those of the other grains. V_T was observed to increase significantly (Fig. 3), especially in models with an increased tendency to the AG development. These data allow the suggestion that an enhancement of {111} fibre texture in Au, Ge and Si thin films during annealing [5,8] does not necessarily be a result of a decreased surface energy of {111} plane.

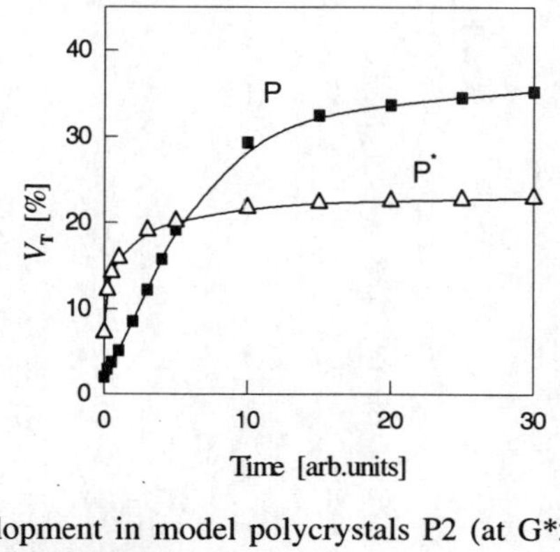

Fig. 3. Texture development in model polycrystals P2 (at $G^*>0$) and P2*(at $G^*=0$) with $\Theta_{AA}= 2°$, $M_0/\delta \cong 1$ and $\delta = 60$ nm.

References

[1] H.J. Frost, C.V. Thompson, D.T. Walton, *Acta Metall.Mater.*, 38, 1990, pp. 1445-1462.
[2] V.Yu. Novikov, *Acta Mater.*, 47, 1999, pp. 4507-4514.
[3] V.Yu. Novikov, *Acta Metall.*, 26, 1978, pp. 1739-1744.
[4] W.W. Mullins, *Acta Metall.*, 6, 1958, pp. 414-427.
[5] H.-J. Kim and C.V. Thompson, *J.Appl.Phys.*, 67, 1990, pp. 757-767.
[6] A. Gangulee, *J.Appl.Phys.*, 45, 1974, pp. 3749-3756.
[7] A. Gangulee and S. Krongelb, *Thin Solid Films*, 24, 1974, pp. 273-277.
[8] J.E. Palmer, C.V. Thompson, H.I. Smith, *J.Appl.Phys.*, 62, 1987, pp. 2492-2497.
[9] J.K. Mackenzie, *Acta Metall.*, 12, 1964, pp. 223-235.
[10] K. Detert, *Acta Metall.*, 7, 1959, pp. 589-598.
[11] V.Yu. Novikov, “Grain Growth and Control of Microstructure and Texture in Polycrystalline Materials”, CRC Press, 1996.
[12] D.B. Knorr, *Mater.Sci.Forum*, 157–162, 1994, pp. 1327-1336.

Computer Simulations of Abnormal Subgrain Growth

M. A. Miodownik[1] and E. A. Holm[2]

[1] Mechanical Engineering Department, Kings College London, Strand, London, WC2R 2LS, UK
[2] Materials and Process Modeling, Sandia National Laboratories, Albuquerque, NM 87185-1411, USA

Keywords: abnormal growth, subgrain, dislocation cell, cellular automaton

Abstract. The annealing behaviour of subgrain structures is explored using a cellular automaton that simulates motion by curvature. The initial microstructures are statistically equivalent to experimentally measured deformation structures in terms of grain boundary structure, texture and misorientation distribution. The subgrain boundary energies and mobilities are taken from experimental measurements of low-angle dislocation boundaries. Abnormal subgrain growth naturally occurs in these systems, and the rate of growth of these abnormal subgrains is in good agreement with analytical theory.

1 Introduction

Recently there has been considerable interest in the recovery of subgrain structures and their subsequent microstructural evolution [1-3]. Huang and Humphreys [2] have studied the subgrains formed through cold deformation of high purity Al-0.05% Si single-phase alloys. A strong correlation between subgrain growth and average misorientation was observed and used by the authors to measure boundary mobility on the assumption that boundary migration was occurring simply through curvature driven growth [2,3]. This hypothesis, that the driving force for subgrain boundar motion is the reduction of interfacial energy, is still controversial. Internal stress fields from dislocations are also may affect microstructural evolution [4], and migrating boundaries may be rate-limited by the triple points [5]. Nevertheless, Humphreys [6,7] has analysed of curvature driven subgrain assemblies and concluded that under certain conditions they could undergo discontinuous subgrain growth. He found that the likelihood depended on the sharpness of texture and the presence of orientation gradients. Perhaps his most interesting result is that low-angle subgrain structures are intrinsically unstable whilst medium to high angle structures may be quite stable.

2 Simulation Methodology

Subgrain growth is modeled by a cellular automaton with a probabilistic transition function. The details of the model are very similar to that of the Monte Carlo Potts model for grain growth [8]. A continuum microstructure is mapped onto a discrete lattice. Each lattice site is given a discrete Euler orientation using an allocation method developed by Miodownik *et al.* [9], which allows both the texture and misorientation distribution function (MDF) of the ensemble to be defined to match experimental measurements. The grain boundaries have an excess energy, given by a normalized Read-Shockley formulation:

$$\gamma = \gamma_o \frac{\theta}{\theta_o}\left(1 - \ln\frac{\theta}{\theta_o}\right) \tag{1}$$

where γ_o is the maximum boundary energy, and θ_0 is a constant observed to lie between 10° and 30°. Grain growth occurs via a Monte Carlo technique. For each site i, the orientation of one of its neighbour orientations is chosen at random, and the change in boundary energy, ΔE, associated with a switch to that orientation is computed. The switch is performed with probability

$$P(\Delta E) = \begin{cases} m_{ij} & (\Delta E \leq 0) \\ m_{ij}\exp(-\Delta E/\Gamma) & (\Delta E > 0) \end{cases} \tag{2}$$

where m_{ij} is the reduced mobility of the boundary between site i and its neighbour j. Γ is a variable that imparts thermal energy (random noise) to the system. Time is related to the number of attempted switches (TS).

In this study we used a 250,000 site, fully periodic, 2D square lattice with first and second nearest neighbour interactions. To minimize lattice pinning, the noise term Γ was set to 0.5 γ_0/k, where γ_0 is the minimum grain boundary energy encountered. The deformation texture was simulated using Q = 999 different possible orientations.

3 Results and Discussion

The initial subgrain structure texture and MDF were constructed to be statistically equivalent to the experimental incidental dislocation boundary (IDB) structures reported by Hughes *et al.* [10] for deformed aluminium single crystals. Initially simulations were carried out with Read-Shockley boundary energy and uniform boundary mobility. Normal subgrain growth was observed in these cases, with the average misorientation of the ensemble decreasing with increasing subgrain size. However when a sharp, exponential mobility function of the form measured by Huang and Humphreys [2] was included in the model, abnormal sub-grain growth was observed to occur. Fig. 1 shows an example of abnormal subgrain growth; the gray scale of the boundaries is proportional to their misorientation, thus the darker boundaries have higher misorientation. At first normal subgrain growth occurs, and average misorientation decreases. Then a subgrain begins to grow abnormally. This abnormal subgrain emerges from the distribution; *it is not seeded in any way.* Its average misorientation is larger than the average, as indicated by the dark colour of its boundaries

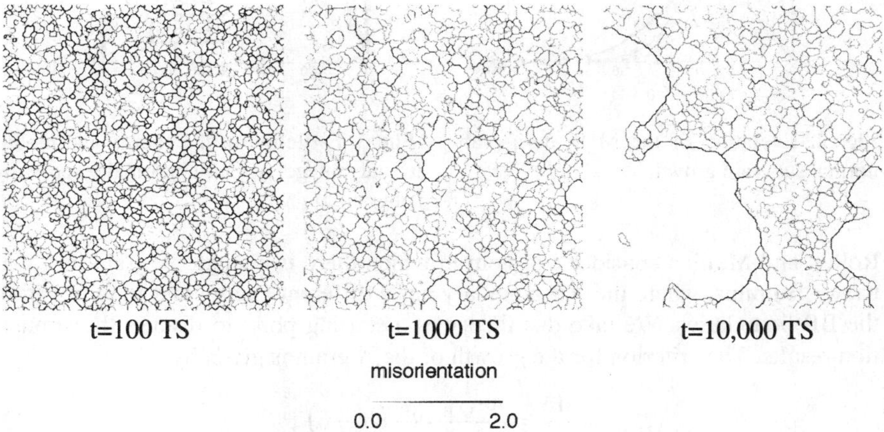

Fig. 1. Snapshots of microstructures during subgrain growth. The boundary grey scale is proportional to misorientation.

Figure 2 shows the MDF of the simulated microstructures. The distribution of misorientations narrows and the average misorientation decreases as normal grain growth occurs. At later times the high-angle boundaries of the abnormal grain start to dominate the MDF. The average mobility of the boundaries surrounding the abnormal grain normalised by the average boundary mobility of the system, $m/\langle m \rangle$, is defined as the 'mobility advantage' of the abnormal grain. The energy advantage is defined similarly, $\gamma/\langle\gamma\rangle$. Fig. 3 shows the evolution of these two quantities in an abnormal grain. Note that the abnormal grain has an initial energy advantage of 2. In the early stages of growth its energy advantage and mobility advantage decrease. This occurs because the average misorientation of the system, including the local environment of the embryo, is decreasing. When t=1000 TS, this effect is reversed locally, and the energy and mobility advantages of the embryo start increasing. This is the critical point of the simulation and represents the triggering event. The embryo's environment changes such that the misorientation of its boundaries increases. Because the embryo has higher boundary energy, it is more likely to be consumed by its neighbours. However the increase in misorientation also gives the embryo high boundary mobility. This trade-off between energy and mobility determines whether a grain will grow abnormally, as pointed out by Rollett and Mullins [11] and also by Humphreys [6,7].

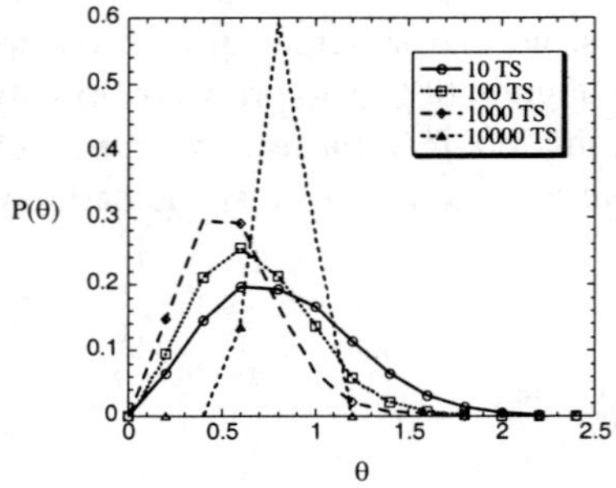

Fig. 2. Evolution of the MDF during abnormal subgrain growth.

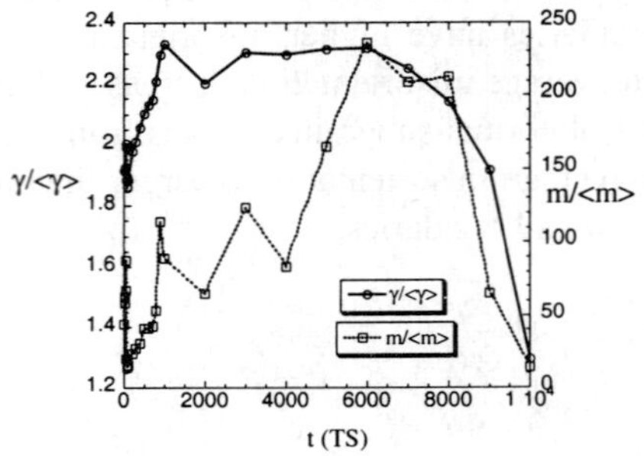

Fig. 3. Evolution of the energy and mobility advantage of an abnormally growing subgrain.

Rollett and Mullins consider a system in which there is a single A grain in a matrix of B grains, where the AB boundary has a different energy and mobility from the BB boundaries. We take this theory as a starting point to discuss the simulation results. The criterion for the growth of the A grain is given by:

$$\frac{dA}{dt} = \frac{\pi m_{AB} \gamma_{AB}}{3} (an - 6) \tag{3}$$

where m_{AB} is the mobility of the AB boundaries, γ_{AB} is the energy of the AB boundaries, n is the number of sides of the A grain, and a is defined as:

$$a = \left(\frac{6}{\pi} \right) \sin^{-1} \left(1 - \frac{2\gamma_{AB}}{\gamma_{BB}} \right) \tag{4}$$

where γ_{BB} is the boundary energy of the BB boundaries. This reduces to the von Neuman law for isotropic grain growth. The presence of anisotropic boundaries can also lead to abnormal behaviour, the criterion being given by the equation:

$$\frac{d\rho}{dt} = \frac{m_{BB}\gamma_{BB}}{2R_B^2} \left\{ \frac{m_{AB}\gamma_{AB}}{m_{BB}\gamma_{BB}} \left(a + (a-2)\frac{1}{\rho} \right) - \frac{\rho}{4} \right\} \tag{5}$$

where m_{BB} is the boundary mobility of the BB boundaries, R_B is the average grain size of the matrix and ρ is the size advantage of the A grain, R_A/R_B. If dρ/dt is zero, the A grain remains in the distribution and its growth is normal. If dρ/dt is greater than zero the A grain grows abnormally (i.e. relatively faster than the matrix grains).

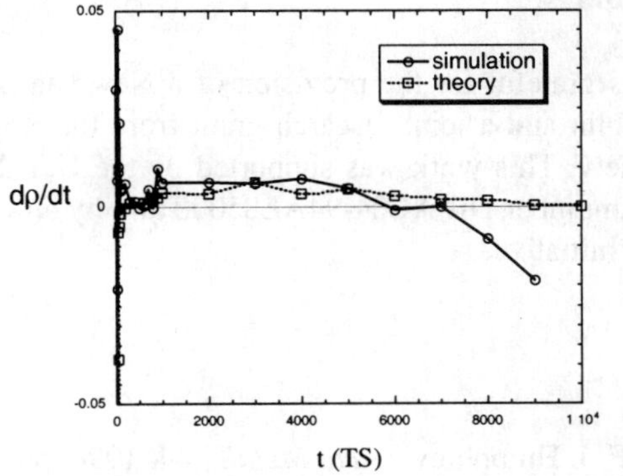

Fig. 4. A comparison between the predictions of the Rollett and Mullins theory and the simulation results.

Substituting the simulation parameters γ for γ_{AB}, $\langle\gamma\rangle$ for γ_{BB}, m for m_{AB} and $\langle m\rangle$ for m_{BB} into equation (5) allows us to compare the predictions of the theory to the behaviour of the simulation. Fig. 4 shows this comparison for an abnormal event; $d\rho/dt$ of the abnormal grain is measured from the simulation and plotted along with that calculated from equation (5). The agreement is good although there is quite a lot of scatter at early times due to the stochastic nature of the simulation. At late times there is also some deviation due to impingement. The important point to note is that the abnormal behaviour of this high-angle grain is predicted by theory. Of course, the term 'high-angle' is relative, as the mean misorientation of the abnormal grain was about 1°. However as Humphreys [6,7] points out, one expects low-grain structures, $\langle\theta\rangle < 5°$, to be intrinsically unstable precisely because of the shape of the Read-Shockley function. Grain growth in these systems can only be observed in the absence of recrystallising grains sweeping through the structure and destroying the evidence. Thus experimentally it difficult to observe low-angle abnormal instabilities; however a simulation such as the present one does not have this disadvantage.

4 Conclusions

When boundary mobility is uniform, normal subgrain growth, in which the average boundary misorientation decreases with time, is observed. When an exponential mobility function, measured experimentally, is used, abnormal subgrain growth is triggered after an initial incubation period of normal subgrain growth. The boundary surrounding the abnormal subgrain is always high-angle in character with respect to the average misorientation in the system. The rate of growth of these abnormal subgrains is in good agreement with analytical theory.

Acknowledgements

Mark Miodownik is grateful for the provision of a Newman Scholarship at University College Dublin and a joint research grant from the Royal Irish Academy and the Royal Society. This work was supported by the U.S. Department of Energy under contract number DE-AC04-94AL85000 and by an Office of Basic Energy Sciences New Initiative.

References

[1] M. Ferry and F. J. Humphreys, *Acta Metall.,* 44, 1996, pp. 1293-1308.
[2] Y. Huang and F. J. Humphreys, *Acta Metall.,* 48, 2000, pp. 2017-2030.
[3] Y. Huang, F. J. Humphreys and M. Ferry, *Acta Metall.,* 48, 2000, pp. 2543-2556.
[4] M. Winning, G. Gottstein and L. S. Shvindlerman, *Acta Metall.,* 49, 2001, pp. 211-219.
[5] G. Gottstein, A. H. King, L. S. Shvindlerman, *Acta Metall.,* 48, 2000, pp. 397-410.
[6] F. J. Humphreys, *Acta Metall.,* 45, 1997, pp. 4321-4240.
[7] F. J. Humphreys, *Acta Metall.,* 45, 1997, pp. 5031-5039.
[8] D. J. Srolovitz, M. P. Anderson, G. S. Grest, P. S. Sahni, *Acta Metall.,* 32, 1984, pp. 1429-1438.
[9] M. A. Miodownik, A. Godfrey, E. A. Holm, and D. A. Hughes, *Acta Mat.*, 47, 1999, pp. 2661-2668.
[10] D. A. Hughes, D. C. Chrzan, Q. Liu, & N. Hansen, *Phys. Rev Lett.,* 81, 1999, pp. 4664-4669.
[11] A. D. Rollett and W. W. Mullins, *Scripta Mat.*, 36, 1997, pp. 975-980.

Recrystallization and Grain Growth
Proceedings of the First Joint International Conference
Eds. G. Gottstein and D.A. Molodov, Springer-Verlag, 2001

Grain Growth and Texture Evolution in Polycrystals With Strong Initial Texture

V.Yu. Novikov

Treptower Str. 74d, D-22147 Hamburg, Germany

Keywords: grain growth, microstructure, texture, modelling.

Abstract. Microstructure and texture alterations during grain growth in textured materials are simulated by a new statistical model taking into account the disorientation dependence of the grain boundary energy and mobility. Boundaries between textured grains A and randomly oriented grains B were assumed high-angle; the A/A boundaries had disorientation of 2, 5, or 10°. The character of the growth process, displayed in a novel map, is governed by the intensity and scatter of the initial texture. Both normal and abnormal grain growth lead to the vanishing of the main texture component. However, this component amplifies at the incubation period of abnormal grain growth, especially when the initial size distribution of randomly oriented grains is displaced toward small sizes.

1 Introduction

Up to now there have been no systematic experimental studies into the effect of the initial texture on the grain growth process and texture evolution [1]. The first computer modelling of the texture influence on grain growth [2] concerned materials with grain boundaries of the same energy and slightly different mobility i.e. materials with a scattered texture. The same relates to the model [3] which cannot be applied to strongly textured materials either because it does not take into account the difference in the energy of boundaries between grains of different texture components [4] typical of materials of this kind. In the present study the effect of both the intensity and scatter of a strong initial texture on the character of the growth process and accompanying texture evolution was investigated by means of a new model.

2 The Model and Objects Studied

A model used in the present simulations is a modification of the model [2] and is described in [5]. In contrast to [2,3], different grain boundaries are characterized here by various energy and mobility, they both depending on the boundary disorientation. Thus the new model can be applied to grain growth in materials with a single-component texture. It was assumed in the model that textured grains A are slightly disoriented in respect to each other, and the average angle of their disorientation, Θ_{AA}, may be considered as a measure of the scatter of the main texture component. The other grains, B, belong to the texture background and are of random orientation owing to which the boundaries B/B and A/B are assumed high-angle.

In the present work the model was used to study the influence of the scatter, Θ_{AA}, and the initial intensity of the main texture component on the microstructure and texture development in materials with a strong single-component texture. Relative magnitudes of the energy, γ, and mobility, m, of different boundaries taken from [6] are listed below:

Θ_{AA}	γ_{AA}	m_{AA}	$\gamma_{BB} = \gamma_{AB}$	$m_{BB} = m_{AB}$
2°	0.4	0.15	1.0	10.0
5°	0.7	0.55	1.0	10.0
10°	0.9	6.25	1.0	10.0

The evolution of microstructure and texture was traced by variation of the grain size distribution with time, the latter being expressed in arbitrary units. Texture intensity was defined as the volume fraction of grains A, V_A. Microstructure was described by the apparent average grain diameter, D (it was calculated as if the measurements were conducted by the random intercept method), the most probable grain diameter, M, and the maximum grain diameter, D_m. The character of the grain growth process was judged from the value of D_m/M: it was assumed that the ratio is ≤5 when normal grain growth (NG) takes place and >5 when abnormal grain growth (AG) evolves. AG development was characterized by the time dependence of the volume fraction of abnormal grains, V_s.

Grain growth was simulated on the following 3-D models

polycrystal	D_0, μm	M_0, μm	D_m/M_0
P1	26.5	23.0	3.15
P2	22.5	13.9	3.05
P3	7.4	6.1	3.09

having various scatter ($2° \leq \Theta_{AA} \leq 10°$) and different initial intensity of the main texture component ($0.30 \leq V_{A0} \leq 0.99$). The initial grain size distribution was of nearly log-normal shape in all the models; grains A and B had identical initial size distributions.

3 Results

3.1 Character of the growth process

Grain growth evolves either as NG or as AG depending on Θ_{AA} and V_{A0}. The AG evolution at Θ_{AA}=2° is caused by the stabilization of the fine-grained matrix (known as texture inhibition) due to a decreased boundary mobility of the majority of grain boundaries. An increase of the mobility (due to an increase of Θ_{AA} to 5°) or a decrease of V_{A0} reduces texture inhibition. This reveals in the AG interruption when V_s stops increasing before approaching 100% (Fig. 1a), and the NG process evolves afterwards. A further increase of the mobility (due to an increase of Θ_{AA} to 10°) results in the NG evolution at any V_{A0}. Fig. 1b summarizes effects of V_{A0} and Θ_{AA} on the character of the growth process and may be considered as a map of the grain growth modes in strongly textured materials. Since the data for all the models studied coincide (see Fig. 1) it may be supposed that the character of the growth process depends mainly on the initial texture.

Although the initial microstructure affects the character of the growth process only moderately, its effect on the growth kinetics and grain size obtained is quite significant. For instance, AG in polycrystal P3 evolves much faster and the grain size at the point of the AG completion is much smaller than in polycrystal P1 with a greater initial grain size (see Fig. 2). The initial grain size as well as texture parameters also affect the length of the AG incubation period.

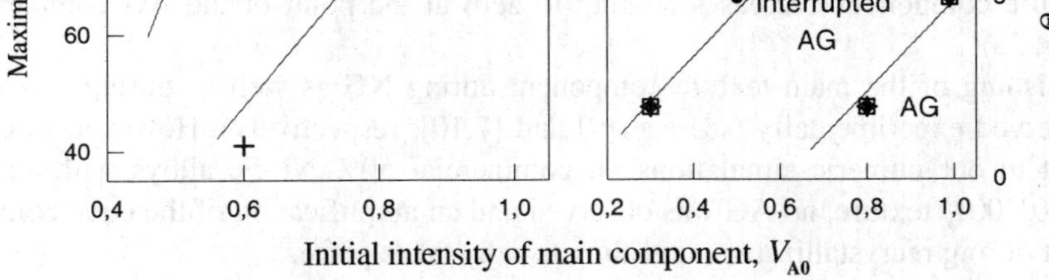

Fig. 1. (a) Influence of V_{A0} on maximum V_s at Θ_{AA} = 2° (curve 1) and 5° (curve 2), and (b) map of grain growth character in polycrystals P1 (squares), P2 (crosses), and P3 (circles); the fields in the map are separated by lines corresponding to $V_s \geq 0.97$ and $V_s \leq 0.005$.

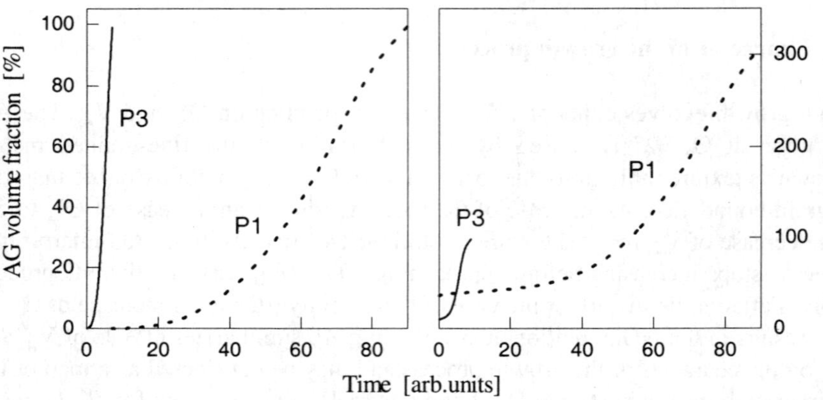

Fig. 2. AG evolution (left) and grain growth kinetics (right) in polycrystals P1 and P3 at $\Theta_{AA} = 2°$ and $V_{A0} = 0.90$.

The above results conform qualitatively with the experimental observations: AG development under texture inhibition is known for a long time [7], NG was recently observed in Al with a significantly scattered cube texture [8].

3.2 Texture evolution

Texture development during grain growth is dependent on its character. A gradual weakening of the main texture component is observed during NG in model polycrystals where AG does not occur. In contrast, the main texture component amplifies at the AG incubation period. An extension of this period by introducing a small pinning force makes this effect much more obvious. During AG the main texture component decreases tending to zero at the point of the AG completion (Fig. 3a).

Vanishing of the main texture component during NG as well as during AG was observed experimentally (see e.g. [9] and [7,10], respectively). However, in contrast to our numeric simulations, in commercial 50% Ni–Fe alloys with strong {100}⟨001⟩ texture, no AG was observed and an amplification of the cube component during recrystallization annealing usually takes place.

This contradiction deserves a more detailed consideration and may be explained as follows. First, the annealing duration may be shorter than that of the AG incubation period which is quite possible in alloys of commercial purity where impurities inhibit grain growth and increase thereby the length of the AG incubation period (see Fig. 3a). Second, cube grains in primary recrystallized alloy can be larger than the other grains, just as in Al [11]. The size advantage of cube grains makes the consumption of the non-cube grains easier and should result in the extending of the AG incubation period. This situation was modelled on polycrystal P2 by shifting the initial size distribution of grains B to smaller sizes so that $M^A{}_0/M^B{}_0 \cong 1.2$; at the same time the overall microstructural parameters in a new polycrystal P2* were remained as in P2. Fig. 3b shows that this increases not only the length of the AG incubation period but also the magnitude of V_A, especially in the presence of a small pinning force.

4 Discussion

The observed features of the growth process in textured polycrystals may be explained as follows. The rates of growth and consumption of any grain should depend on the effective mobility [12] of its boundaries, along with the driving force. So at high V_A the behaviour of grains A and B with the same driving force should be quite different: grains B with boundaries of a higher effective mobility should be consumed faster than grains A, whereas the survived grains B should grow faster than grains A. As a result, the highest size classes in the overall grain size distribution should be occupied by grains B which was observed in the simulations. As a consequence, all the crystallites growing during AG and most of those growing during NG are grains B.

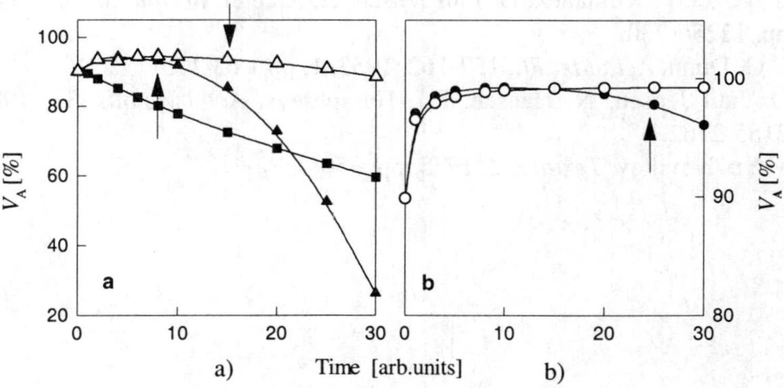

Fig. 3. Texture evolution (a) in polycrystal P2 during NG at Θ_{AA}= 10° (squares) and AG at Θ_{AA}=2° (triangles) and (b) in polycrystal P2* during AG at Θ_{AA}= 2° (circles). Arrows show the point of AG commencement. Open symbols correspond to the case of grain growth affected by small pinning force.

However, in some cases the grain growth process results in an enhancement of the main texture component A. So, the question arises why the volume fraction of grains B may increase in some cases and decrease in others. The answer is obviously the following. The behaviour of V_B does not only depend on an increased effective boundary mobility of grains B, but also on the number of the survived grains B. If this number is great enough, the growth process results in an increase of V_B; in the opposite case it can temporarily lead to the V_B reduction. This gives an insight into why V_A increases at the AG incubation period, especially in the presence of a small pinning force.

References

[1] V.Yu. Novikov, "Grain Growth and Control of Microstructure and Texture in Polycrystalline Materials", CRC Press, 1996.
[2] V.Yu. Novikov, *Acta Metall.*, 27, 1979, pp. 1461-1466.
[3] G. Abbruzzese and K. Lücke, *Acta Metall.*, 34, 1986, pp. 905-914.
[4] V.Yu. Novikov, *Scripta Metall.*, 23, 1989, pp. 1235-1236.
[5] V.Yu. Novikov, *Acta Mater.*, 47, 1999, pp. 1935-1943.
[6] F.J. Humphreys, *Acta Mater.*, 45, 1997, pp. 4231-4220.
[7] M.L. Kronberg and F.G. Wilson, *Trans. AIME*, 185, 1949, pp. 501-514.
[8] L.S. Shvindlerman, V.G. Sursaeva, V.Yu. Novikov, R.G. Faulkner, *Mat. Sci. Forum*, 157–162, 1994, pp. 1063-1068.
[9] S. Vogel, P. Klimanek, D. Juul Jensen, H. Richter, *Scripta Mater.*, 34, 1996, pp. 1225-1230.
[10] C.G. Dunn, *Acta Metall.*, 157-162, 1953, 1, pp.1 63-175.
[11] D. Juul Jensen, N. Hansen, F.J. Humphreys, *Acta Metall.*, 33, 1985, pp. 2155-2162.
[12] V.Yu. Novikov, *Texture*, 2, 1975, pp. 35-44.

Energy Barriers for the Morphological Evolution and Coarsening of Faceted Crystals

G. S. Rohrer and W. W. Mullins

Department of Materials Science and Engineering, Carnegie Mellon University, Pittsburgh, PA, 15213-3890, USA

Keywords: nucleation, faceted crystals, coarsening, morphological evolution.

Abstract. Crystals completely bounded by flat facets change shape, grow, or shrink only through the lateral motion of steps across the facets. If the crystal is free of step producing defects such as screw dislocations, then such steps must be created by a two dimensional nucleation process. Here, we quantify the energy barrier for the nucleation of steps during the morphological evolution and coarsening of faceted crystals. Our analysis indicates that for defect free, faceted crystals larger than a few nanometers, the nucleation energy barrier will prevent crystals from attaining their equilibrium shape. Furthermore, this energy barrier also limits the growth and dissolution of crystals during coarsening.

1 Introduction

The nucleation (free) energy barrier (NEB) present during the growth of a faceted crystal is well known [1]. We have recently considered how the NEB influences the morphological evolution of isolated, defect free crystals with equilibrium and non-equilibrium shapes [2,3]. The principal results were that if the facets are larger than a few nm, fluctuations around the equilibrium shape will not occur. Furthermore, the NEB will prevent defect free crystals larger than a few nm from evolving to their equilibrium shape [3]. In the present paper, we review these calculations for the specific case of a crystal whose equilibrium shape is a cube. We calculate the NEB for fluctuations about, and morphological changes toward, the equilibrium shape. Using the same approach, we then consider the role of the NEB on the coarsening of faceted crystals by examining its influence on growth and dissolution when the crystal can exchange matter with a reservoir at a fixed chemical potential.

2 NEB for Intraparticle Transport on a Cube Shaped Crystal

We begin by considering an isolated crystal with its equilibrium cube shape, an edge length of L, and a surface energy of σ. We wish to calculate the NEB that must be overcome to transport material from one facet to another. As illustrated in Fig. 1, material is removed from the top facet to form a square island with area s_2^2 on the side facet. Since atoms will be removed from the periphery, a square island with area s_1^2 is left behind on the top facet. To conserve matter, it must be true that $L^2 = s_1^2 + s_2^2$. Obviously, the only energy change in this process is associated with the change in the edge energy. Assuming that the step edges have the same energy per area as the facets (σ), then the energy for the transfer is:

$$\varepsilon_b^e = 4a\sigma s_1 + 4a\sigma s_2 - 4a\sigma L \tag{1}$$

In Eq. 1, the first two terms on the right hand side represent the added perimeter energy while the last one represents the energy eliminated. It is clear from Eq. 1 that there is no net energy change to transfer a layer in either direction and this point is illustrated graphically in Fig. 1. Figure 1 also demonstrates that to transfer the layer, the crystal must pass through an intermediate state with a higher energy. The maximum is reached when the area of the partial layer on each facet is the equal ($s_1=s_2= L/\sqrt{2}$). If we return these numbers to Eq. 1, we see that the barrier to the transfer of a layer from one facet to another is 1.66 aσL.

We have previously shown nucleation on a facet can be considered negligible if the barrier energy is ≥ 40kT [2]. This estimate is based on a classical expression for the steady-state nucleation rate on an infinite singular facet and defines nucleation rates ≤ $10^{-4}nm^{-2}s^{-1}$, equivalent to a rate of less than one atomic layer per hour on a facet of 1 nm^2, as negligible [4]. If we take a to be 2.5 Å, σ to be 1 J/m^2, and T= 1000 °C, then the barrier will only be surmounted for particles where L < 1.7 nm. In other words, for defect free particles of sizes greater than about 2 nm, the NEB will prevent fluctuations about the equilibrium shape.

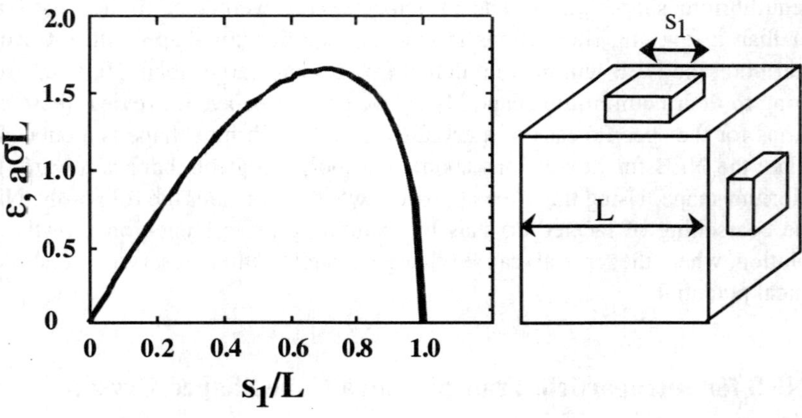

Fig. 1. The NEB for transferring a layer from the top facet of a cube to side (Eq. 1) and a schematic of the process.

3 NEB for Intraparticle Transport on a Crystal with a Nonequilibrium Shape

We now consider non-equilibrium shapes with volumes $L_1(L_2)^2=L^3$, square facets perpendicular to z (L_2 by L_2), and length along z, L_1, such that $L_1>L$. Here, we consider only the NEB as the crystal evolves toward its equilibrium shape. The barriers are the free energies of the critical state representing the maximum free energy of the lowest free energy path for material transfer. In calculating the NEB for this shape, we note that the removal of a complete layer from the face perpendicular to z results in only a partial layer on the other participating face. A subsequent transfer will encounter a different barrier, in general, because of the different degree of coverage of the participating face. We confine the calculation to the maximum of the sequence of barrier free energies. If this is prohibitive, changes will not occur. If it is not prohibitive, neither will the other barriers be prohibitive.

For the long, thin crystal (prolate shape) to evolve toward equilibrium, the facets perpendicular to z must move toward the center. In this case, the critical state of maximum free energy is the unstable equilibrium in which the material removed from the z face leaves a square nucleus (sxs) and also forms a square nucleus of the same size on one of the side faces. Mass balance requires $2s^2=(L_2)^2$. The free energy change to form the configuration and hence the NEB is

$$\varepsilon_b(-) = 8a\sigma s - 4a\sigma L_2 = 4(\sqrt{2}-1)a\sigma L_2. \qquad (2)$$

Since Eq. 1 and 2 are identical when $s_1=s_2=s$ and $L_2=L$, we see that $\varepsilon_b(-)/\varepsilon^e_b = L_2/L$. Using the constant volume condition, we obtain

$$\varepsilon_b(-)/\varepsilon^e_b = (L/L_1)^{1/2}; \qquad (3)$$

as L_1/L increases, the barrier slowly decreases. It is interesting to note that even when the crystal is far from equilibrium, the barrier can be significant. For example, consider a crystal for which L = 10 nm. Using the values for the physical parameters cited in section 2, the barrier for a 5 nm x 5 nm x 40 nm crystal to move material from a small face to a large one is about 2×10^{-18} J or more than 100 kT at 1000°C. In other words, if the crystal had no step generating defects, it would be kinetically frozen in a stationary configuration far from its equilibrium shape. Elsewhere, we have shown that there are similar barriers for oblate ($L_1<L$) shapes [3]. The influence of the NEB on morphological changes has been observed experimentally. The erratic morphological evolution of cavities in sapphire annealed for up to 16 h at 1900°C [5] and the stationary states of 50 nm zirconia crystals annealed on a hot stage in the TEM for several hours at 850 °C [6] were both explained in terms of the NEB.

4 Effect of the NEB on Coarsening

We now consider the NEB for the addition or subtraction of layers when the cube is able to exchange matter with a reservoir at a fixed chemical potential. This is intended to represent the processes that must occur during the coarsening of faceted particles. As before, the crystal has its equilibrium shape and an edge length of L. We assume that the crystal is situated in a medium in which it is slightly soluble so that it can exchange material with a reservoir of the same substance, at a fixed chemical potential of μ_∞. We can imagine this reservoir to be a large number of identical cubes of the same material where the distance from the center of each cube to any of its six facets is R* so that $\mu_\infty=2\sigma/R^*$. Now consider the transfer of material from the reservoir to the crystal. To do so, a square nucleus of area s^2 and height a must be placed on the facet. The energy for this transfer is:

$$\varepsilon(s) = 4as\sigma - as^2 \frac{2\sigma}{R^*}, \qquad (4)$$

for $0 \le s \le L$. In Eq. 4, the first term on the right hand side is the energy of the perimeter of the nucleus and the second term is the change in the free energy on moving a volume of material (as^2) from the reservoir to the flat surface. Note that if the crystal is in equilibrium with a reservoir of identical particles ($R^*=L/2$), then there should be no energy change when a single layer of atoms is transferred from the reservoir. Equation 4 shows that this is the case when s=L and $R^*=L/2$. However, to add even a single layer, the crystal must past through a relatively higher energy state that contains a partial layer. By differentiation of Eq. 4, we see that the maximum occurs at $s = R^*$ and that the barrier to addition, ε_+, is:

$$\varepsilon_+ = 2a\sigma R^* \qquad (5)$$

where R* is between 0 and L. In this case, the NEB is independent of the crystal size and depends only of R*, which characterizes the size of the particles in the reservoir. If we assume the same physical quantities as before, then the crystals will grow only when R* is less than a critical value of about 1.4 nm. In other words, only crystals with $L < 3$ nm can provide supersaturation that is high enough to drive nucleation on a flat facet. If we take R* as a representation of a mean crystal size in a distribution, then as soon as R* rises above the critical value, the NEB will halt growth.

The energy to remove a layer from the crystal and place it in the reservoir is:

$$\varepsilon(s) = 4as\sigma - 4aL\sigma + (aL^2 - as^2)\frac{2\sigma}{R^*} \qquad (6)$$

Where the first two terms on the right hand side represent the energy difference between a complete layer and a partial layer; this is ≤ 0 for all valid values of s. The final term on the right hand side is the energy increase associated with moving a volume of material from the flat surface to the reservoir. The maximum of Eq. 6 is at s = R* and the barrier to removing a layer, ε, is:

$$\varepsilon_{-} = 2a\sigma\left[R^{*} + L\left(\frac{L}{R^{*}} - 2\right)\right] \quad (7)$$

where R* is between L/2 and L. Note that for a crystal in contact with a reservoir of identical particles (R* = L/2), $\varepsilon_{+} = \varepsilon_{-} = 2a\sigma R^{*}$, as we would expect for the equilibrium case. The barrier to remove a layer is smaller if the reservoir is made up of crystals that are somewhat larger (L/2 < R*< L) and vanishes completely if the particles in the reservoir are twice as large as the crystal (R* ≥L).

According to these results, illustrated in Fig. 2, the coarsening of defect free faceted crystals will be limited by the NEB; when the mean field chemical potential (μ_{∞}) increases to the point where corresponding values of R* are beyond a few nm, growth will stop. The results are in sharp contrast to conventional coarsening theory, in which atoms are driven to or away from surfaces by chemical potential gradients where they attach or detach without a nucleation barrier [7,8]. In real populations of crystals, some will contain defects that produced steps and these crystals will continue to grow after the perfect crystals become immobilized. We believe that this might explain abnormal coarsening in systems of faceted crystals.

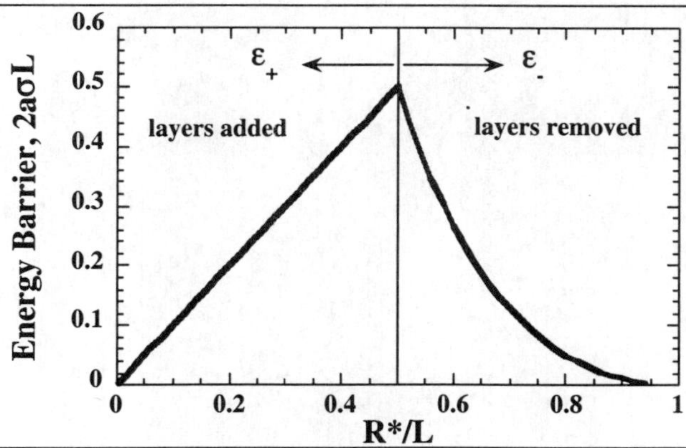

Fig. 2. The barriers for addition of layers to the crystal (from a reservoir containing smaller crystals) and removal of layers (from a reservoir containing larger crystals).

Acknowledgement

This work was supported primarily by NASA under grant number 8-1674 and partially by the MRSEC program of the NSF under award number DMR-0079996.

References

[1] W.K. Burton, N. Cabrera, and F.C. Frank, *Philos. Trans. R. Soc. London, Ser. A*, 243, 1951, pp. 300-358.

[2] W.W. Mullins and G.S. Rohrer, *J. Amer. Ceram. Soc.*, 83, 2000, pp. 214-216.

[3] G.S. Rohrer, C.L. Rohrer and W.W. Mullins, *J. Amer. Ceram. Soc.*, 2001, in press.

[4] J. Feder, K.C. Russell, J. Lothe and G.M. Pound, *Adv. Phys.*, 15, 1966, pp. 111-178.

[5] J. Rankin, *J. Am. Ceram. Soc.*, 82, 1999, pp. 1560-1564.

[6] M. Kitayama, T. Narushima, and A.M. Glaeser, *J. Am. Ceram. Soc.*, 83, 2000, pp. 2572-2582.

[7] I.M. Lifshitz and V.V. Slyozov, *J. Phys. Chem. Solids*, 19, 1961, pp. 35-50.

[8] C. Wagner, *Z. Electrochem.*, 65, 1961, pp. 581-591.

Recrystallization and Grain Growth
Proceedings of the First Joint International Conference
Eds. G. Gottstein and D.A. Molodov, Springer-Verlag, 2001

Investigation of Two-dimensional Grain Growth in Al

V.G. Sursaeva and S.G. Protasova

Institute of Solid State Physics, Russian Academy of Sciences, Chernogolovka, Moscow Distr., 142432 Russia

Keywords: single crystals, grain boundary, aluminium, normal grain growth.

Abstract. Microstructure evolution in 2–D aluminium strips during normal grain growth has been examined. The strips were produced by rolling deformation of Al polycrystals and Al (100), (110) and (111) single crystals. Time dependences of mean grain area are presented. The data on mean grain boundary mobility obtained in the experiments gives us possibility to suppose, that final grain boundary microstructure is defined by "orientation nucleation". But the subsequent grain growth is determined by low angle grain boundaries kinetics.

1 Introduction

In spite of a great amount of papers concerning grain growth in polycrystals the relationship between kinetics of the process and the final structure is poorly understood [1-10]. The question discussed in these papers is: how different factors influence the grain growth kinetics and the resulting microstructure. These factors are the initial state of material and heat and mechanical treatment. In this paper we try to clarify how the initial state of material influence the recrystallization process in aluminium. We investigate this phenomenon in 2–D polycristals at the stage of normal grain growth.

2 Experimental Procedure

Three types of flat single crystals and bicrystals with [100] tilt grain boundary were grown by directional crystallization, using high purity (99.999 at%) aluminium. The orientation of the crystallographic axis of the single crystals seeds and produced single crystals and bicrystals was measured with an accuracy of 1°. Strips from the rolled (ε=75%) single crystals ((100), (110) and (111) – rolling plane), bicrystals and polycrystals were annealed for 0.5 h at 823 K to obtain the 2–D "columnar" structure (Fig. 1). The term 2–D "columnar" structure labels a

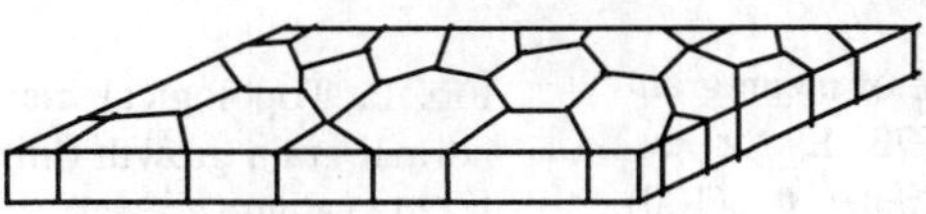

Fig. 1. The scheme of 2–D "columnar" structure.

system where grain boundaries are perpendicular to the sample plane and where sample thickness is of the order of the mean grain size. The third dimension remained homogeneous. Strips thickness was 1 mm. Thus, the initial material (single crystals, bicrystals and polycrystals) received the same preliminary mechanical and heat treatment. Finally the strips were annealed at the same conditions at 773 K for 0.5, 2.5, 4, 6, 8 and 10 h for grains growth. To study the kinetic parameters of the grain growth (mean grain area $\overline{S}$, individual grain area S and a number of neighbours n) were measured for each annealing time. The orientations of the grains in the "columnar" structure were determined by etch pits technique and optical microscope.

3 Results and Discussion

Self-similar (scaling) behaviour is often observed experimentally in polycrystals and called normal grain growth. The two natural characteristics of 2–D polycrystals are mean grain area $\overline{S}$ and topological class n (the number of its neighbours). One of the most important indication of the normal grain growth is the linearity of mean grain area $\overline{S}$ vs. annealing time t [11,12] (Fig. 2). An additional confirmation of the normal grain growth is the time independent topological class distribution (Fig. 3) and linearity of the mean topological class $\overline{n}(S)$ vs. relative grain area $S/\overline{S}$ (Fig. 4) [11-14]. The data on normal grain growth in "columnar" structure are presented in Table 1.

It is obvious from Fig. 2 that the normal grain growth velocity depends on the initial state of material. In the single crystals after rolling deformation there is blockwise slip on definite slip systems, especially for certain symmetric initial orientation of the rolling plane and direction. Nuclei then appear at the border of

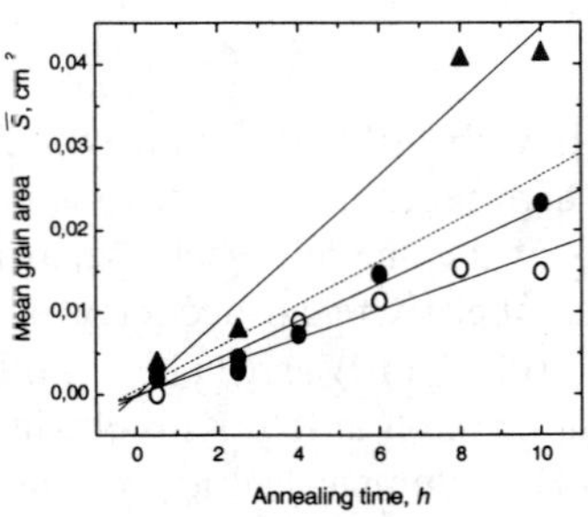

Fig. 2. Time dependence of mean grain area increment at T=773 K for deformed Al: ▲- (110) single; ●- (100) single and bi-; ○- (111) single crystal;

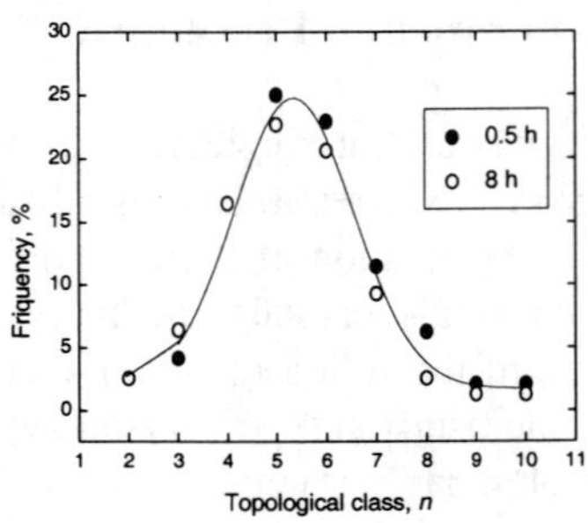

Fig. 3. Topological class distribution at normal grain growth (initial state of strips is (111) single crystal).

Table 1. Kinetic parameters of grain boundaries ensemble in Al.

Initial state of samples	Grain boundary mobility A, cm^2/sec	The velocity of the change of grain area V, cm^2/sec	Mean topological class $\bar{n}\ (0)$
(100) single crystal	$5.11 \cdot 10^{-7}$	$6.34 \cdot 10^{-7}$	3.7
(110) single crystal	$5.70 \cdot 10^{-7}$	$1.23 \cdot 10^{-6}$	3.2
(111) single crystal	$3.01 \cdot 10^{-7}$	$4.47 \cdot 10^{-7}$	4.0
polycrystal	$3.00 \cdot 10^{-7}$	$7.46 \cdot 10^{-7}$	3.3

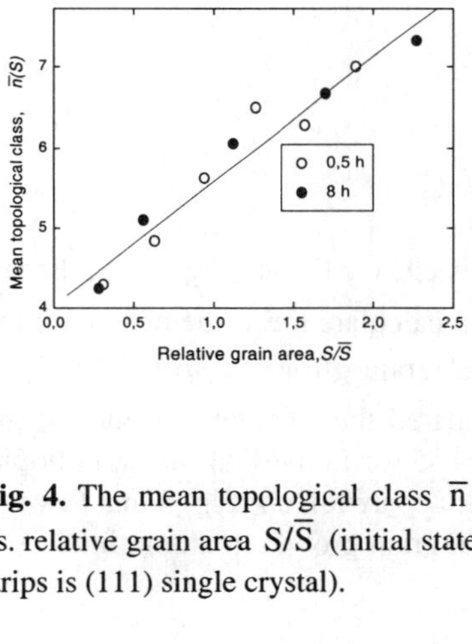

Fig. 4. The mean topological class $\bar{n}(S)$ vs. relative grain area $S/\bar{S}$ (initial state of strips is (111) single crystal).

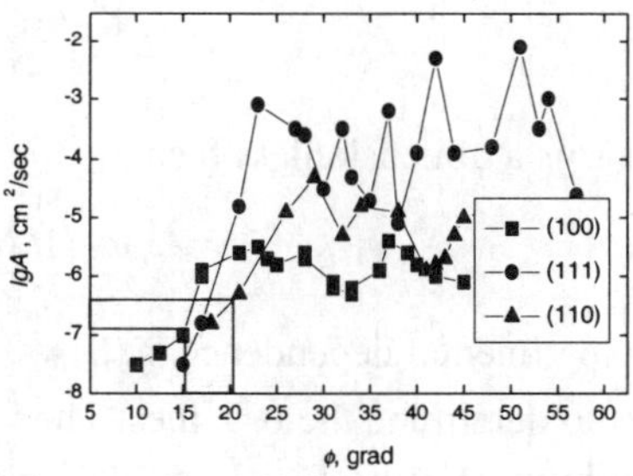

Fig. 5. Misorientation dependencies of reduced grain boundary mobility A for <100>, <110> and <111> tilt grain boundaries in aluminium (T=773 K) [17].

two slipped blocks which takes the form of sharp deformation – produced high angle grain boundaries. In the polycrystals we obtain an average deformation structure.

We observed that the structure in annealed Al strips was rather different. In the strips from (110) single crystal there are grains with twinning orientation. Small grains with n=2, 3 ,4 are not detected. In the samples from (100) single crystal there are small grains with n=2, 3, 4. The same microstructure was observed for (100) bicrystals with tilt grain boundary. Typical picture are shown in Fig. 5. In 2–D polycrystals and in the strips from (111) single crystals grains of topological class *n=2* are rather rare (about 0.1%).

The experimental results $\bar{n}(0)_{100}=3.7\pm0.2$, $\bar{n}(0)_{110}=3.2\pm0.2$, $\bar{n}(0)_{111}=4.0\pm0.2$, $\bar{n}(0)_{polycrystal}=3.3\pm0.2$ were obtained by extrapolation of the experimental data concerning grains with finite observable area to $S\rightarrow0$. But it doesn't mean that 4 –

and 5 – sided, and even 3 – sided two – dimensional grains disappear intact. On the contrary, they eventually do lose sides and become lenses just before vanishing. Apparently, in many cases these topological changes do not occur before the grains are much smaller then their neighbours, or do not occur at all due to the effect of the finite strips thickness (Fig. 6).

Von Neumann and Mullins [15,16] showed that the velocity of change of individual grain area S does not depend on the size and geometrical shape of the grain but depends only on its topological class n:

$$\frac{dS}{dt}=\frac{A\pi}{3}(n\text{-}6) \tag{1}$$

From this equation Fradkov and Udler [11,12] have derived an expression for mean topological class $\overline{n}_s$ (for definite grain area S):

$$\overline{n}_s\text{-}6=\frac{3\alpha}{\pi}\left(\frac{S\text{-}\overline{S}}{S}\right) \tag{2}$$

where α is a dimensionless factor:

$$\alpha=(1/A)\left(d\overline{S}/dt\right) \tag{3}$$

The experimental dependence $\overline{n}(S)$ vs. $S/\overline{S}$ is clearly linear (Fig. 4), so the slope allows to determine the α values. Then we can calculate the value of grain boundary mobility A from Eq. 3 and the velocity of grain growth $d\overline{S}/dt$ (Fig. 2). Results are presented in Table 1. We have compared the calculated values of grain boundary mobility A in 2–D structure with those for individual tilt grain boundaries in Al bicrystals at 773 K (Fig. 5) [17]. It is apparent from Fig. 6 that low angle grain boundaries are responsible for the normal grain growth in aluminium.

4 Conclusion

The reported experiments throw light on the old controversy of whether the new orientations developing during primary recrystallization are already present in the deformation microstructure ("oriented nucleation" [18]) or are kinetically selected from random nuclei by later growth ("growth competition" [19]). It was shown that after strong rolling deformation during primary recrystallization single crystals of different orientation have the different grain and grain boundary microstructure and consequently different final grain and grain boundary microstructure during normal grain growth. We suppose that mean grain growth area velocity $d\overline{S}/dt$ (Fig. 2) is defined by the motion of low angle grain boundaries (Fig. 5, Table 1).

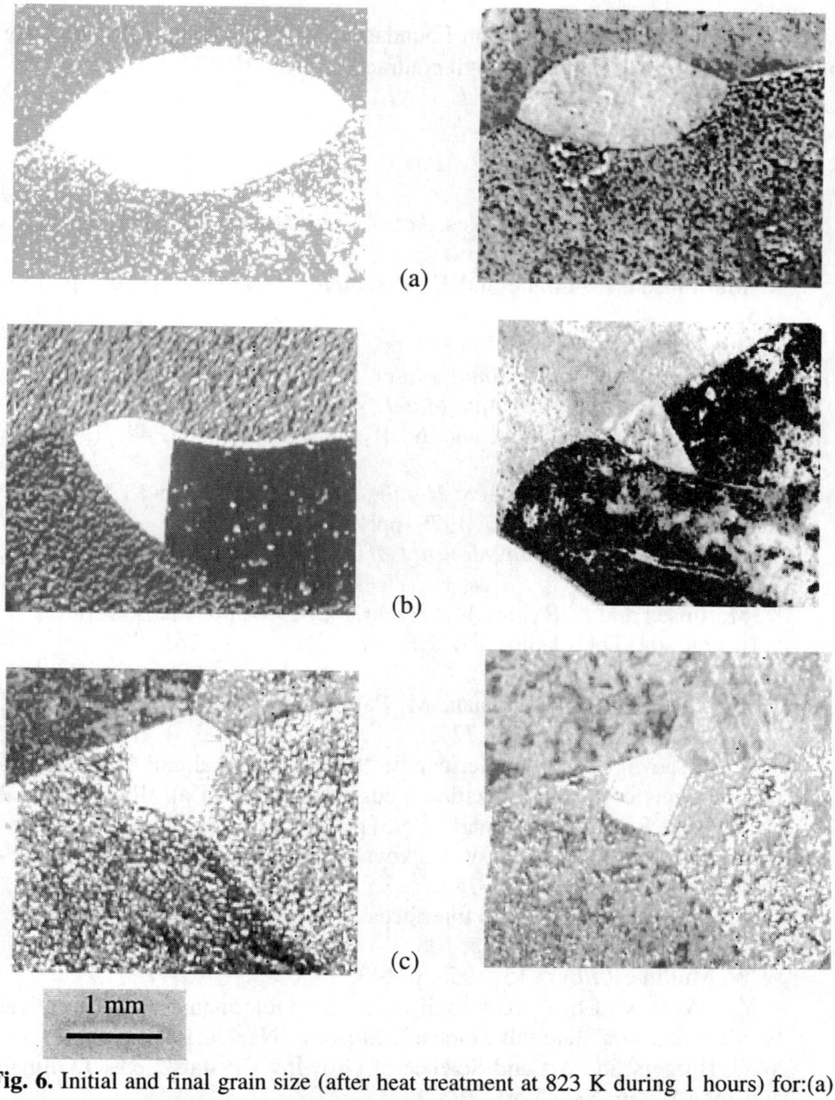

Fig. 6. Initial and final grain size (after heat treatment at 823 K during 1 hours) for:(a) -2-sides grains; (b) -3-sides grains; (c) -5-sides grains.

Acknowledgment

The authors are grateful to the Russian Foundation for Fundamental Research for financial support of the program through contracts 99–02–04017.

References

[1] J. Hjelen, R. Orsund and E. Nes, *Acta Metall. Mater.*, 39, 1991, pp. 1377-1404.
[2] K. Marthinsen, O. Lohne and E. Nes, *Acta Metall.*, 37, 1989, pp. 135-145.
[3] O. Hunderi, *Acta Metall.*, 27, 1979, pp. 167-169.
[4] K. Martinsen, *Modelling Simul. Mater. Sci. Eng.*, 4, 1996, pp. 87-100.
[5] O. Hunderi and N. Ryum, *Acta Mater.*, 44, 1996, pp. 1673-1680.
[6] K. Martinsen, O. Hunderi and N. Ryum, *Acta Mater.*, 44, 1996, pp. 1681– 1689.
[7] N. Ryum and O. Hunderi, *Acta Metall.*, 37, 1989, pp. 1375-1379.
[8] O. Hunderi, *Acta Metall.*, 27, 1979, pp. 167-169.
[9] O. Hunderr and N. Ryum, *Journal of Materials Science*, 15, 1980, pp. 1104-1108.
[10] K. Martinsen and N. Ryum, *Acta Mater.*, 45, 1997, pp. 1127-1136.
[11] V.E. Fradkov, D.G. Udler and R.E. Kris, *Phil Mag.*, 58, 1998, pp. 277-283.
[12] V.E. Fradkov, M.E. Gliksman, M. Palmer and K. Rajan, *Acta Metall. Mater.*, 42, 1994, pp. 2719-2727.
[13] V.G. Sursaeva, L.S. Shvindlerman, in "Microstructural and Crystallographic Aspects of Recrystallization", eds. N. Hansen et al., Risø National Laboratory, Roskilde, Denmark, 1995, pp. 559-563.
[14] V. Sursaeva, S. Protasova, W. Lojkovski and J. Jun, *Texture and Microstructure*, 32, 1999, pp. 175-185.
[15] J. Von Neumann, in "Metals Interfaces", American Society for Testing Materials, Cleveland, 1952, p. 108.
[16] W.W. Mullins, *J.Phys.*, 15, 1956, p. 143.
[17] V.Yu. Aristov, Ch.V. Kopetskii, L.S. Shvindlerman, in "Theoretical Fundamentals of Materials Science", Moscow, Nauka, 1981, p. 84.
[18] W.G. Burgers, in "Art and Science of Growing Crystals", eds. J.Gilman and John Wiley, New York, 1957.
[19] P.A. Beck, *Adv. Phys.*, 3, 1954, p. 245.

Analysis of Grain Growth Kinetics Using the Cellular Automaton Method

W. Yu[1], S.P. Banks[2], and E.J. Palmiere[1]

[1]Department of Engineering Materials
[2]Department of Automatic Control and Systems Engineering, The University of Sheffield, Sir Robert Hadfield Building, Mappin Street, Sheffield S1 3JD, UK

Keywords: grain growth, kinetics, computer simulation.

Abstract. The exponent (n) of grain growth is an important parameter in quantitatively describing the grain growth process. In the past, there have been some discrepancies between the experimentally determined results compared to those predicted theoretically. A two-dimensional Cellular Automaton (CA) model has been developed for the description the process of grain coarsening. In this paper, the analysis of the grain growth exponent based on this model. Because temperature and experimental results can be taken into account using this model, it shows the grain growth exponent increases with temperature, and to be in reasonable agreement with experimental data.

1 Introduction

It is well known that grain growth is a thermally activated process. High temperatures provide atoms with sufficient mobility to overcome the boundary energy barrier, and to become one part in the neighbouring grain. The driving force for grain growth is the reduction in the grain boundary energy, which is the function of the curvature of a boundary. For a long time, many investigations have been carried out, and a variety of theories and models have been put forward in describing the grain coarsening process. To a great extent, this is due to the importance of reheating practice as setting the initial microstructure prior to thermomechanical processing. However, even now, no empirical model and equation can predict the grain size for any given condition with sufficient accuracy. The reason for this is that grain growth is a dynamic process which occurs at elevated temperatures, and many factors such as structure, particles, texture, and heating rate may affect the final grain size and its distribution. It is an elusive and challenging target to predict the grain coarsening process accurately. Therefore, computer simulation becomes a necessary tool in predicting microstructure evolution.
Burke and Turnbull [1] deduced a parabolic relationship for grain growth kinetics in the 1950s. It was assumed that: (1) the boundary energy γ is independent of grain size and time, and is the same for all boundaries, (2) the radius of curvature (R) is proportional to the mean radius of an individual grain, and (3) the boundary velocity is proportional to the driving pressure (P). Based on these assumptions,

the following equation for grain growth can be obtained for isothermal annealing conditions:

$$R_t^2 - R_0^2 = kt \quad (1)$$

where R_t is the mean grain size at time t, R_0 the initial mean grain size and k a constant. In the limit where $R_t >> R_0$,

$$R_t = k't^n \quad (2)$$

Here, n is often called the grain growth exponent and n=0.5 is referred as the theoretical value for an ideal, single phase material.

Considerable theoretical and experimental effort has been devoted to both the prediction and explanation of the grain growth exponent [2]. Computer simulations have also been carried out to model the grain growth kinetics. Among these, the Exxon group, using a Monte Carlo method [3], predicted the grain growth kinetics to be 0.41 ± 0.03. Fan and Chen, using a continuum field model [4] determined the n value to be 0.5. Liu [5], using the CA method, arrived at an n value of 0.45. A common feature of all these models, however, is that each one describes the grain growth kinetics as a single value or as a range of values.

However, experimental studies of grain growth in metals often show grain growth exponents that are temperature dependent. Hsun Hu [6] presented a detailed study of isothermal grain growth in zone-refined iron in the temperature range 550-850°C, which showed that the n value increases with temperature from approximately 0.25 to 0.5. There exists differences among grain growth exponents for a series of isothermal grain growth experiments on zone refined metals [2]. How to reflect this fact is vital to the accurate and quantitative simulation of grain growth.

A two dimensional CA model has been developed for the description of the normal grain coarsening process [7]. In this paper, we report our analysis of the grain growth exponent based on this CA model. Because temperature and the results from experimental microstructural characterisation can be taken into account in this CA model, it shows that indeed the grain growth exponent increases with temperature. The developed CA model can both qualitatively (i.e., graphically) and quantitatively analyse the grain coarsening process.

2 Cellular Automaton Method

Cellular automaton represents a dynamic system in which space and time are discrete, and has been widely developed in both the physical and natural sciences with the development of computer technology. Hesselbarth et al. [8] first tried to use the CA method to simulate the recrystallisation process. Y. Liu [5] applied the CA method to simulate normal grain growth using Von Neumann's definition of neighbouring cells with deterministic transformation rules. Raabe [9] shows some consecutive two-dimensional sections from a three-dimensional grain growth simulation of a recrystallised aluminium single crystal, where all equations of mo-

tion are simultaneously integrated using a weighted stochastic sampling integration scheme. In order to predict the grain growth quantitatively, further work still needs to be done.
From the present work on the CA model for grain growth, the overall process is divided into two stages: nucleation and grain growth. A detailed explanation of the model has been given in another paper [7], and the probability for the cell to change its state at every calculation time step is defined as P.

3 Results and Discussion

3.1 Simulation result

In the following simulation, the nucleation condition is set to be the same for grain growths under different conditions. If we assume the distance between two neighboring cells is 1 μm, the model can output the average grain size, grain growth kinetics and grain size distribution automatically. Fig. 1 shows the grain size evolution with calculation time under different values of P within 20000 steps.

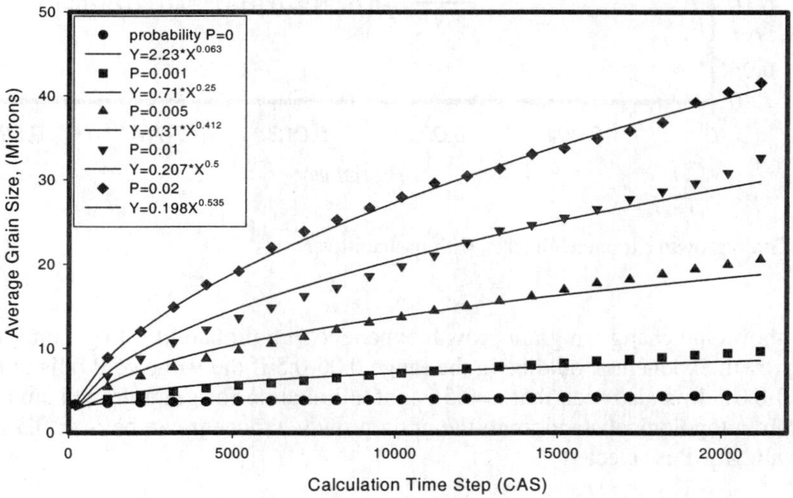

Fig. 1. Grain size evolution under different probability P at 20200 CAS.

Initially, the value of P is arbitrarily set 0, which means that the CA program becomes a deterministic method, and not a probabilistic one [7]. In Fig. 1, the curve with P=0 shows that the grain size only has a small amount of growth after nucleation. It reflects the fact that in relatively low temperature, such as iron at room temperature, not enough energy is available to make an atom jump over the boundary, and therefore the grain size can maintain stability. This completely fits the natural phenomena observed through experiments. Following this initial stage, a different value for P is selected, and the evolution of grain coarsening with calculation time is observed. It can be seen that grain growth can briefly satisfy the general relation $R_t = k t^n$, where the n value increases gradually as P increases. The explanation for this is as follows: the higher probability, the more chance that a given cell can change its state, so the faster the grain growth. This is a quite interesting and useful result, for if the value of n is known for a given material, a corresponding value of P may be selected to simulate the grain coarsening behaviour.

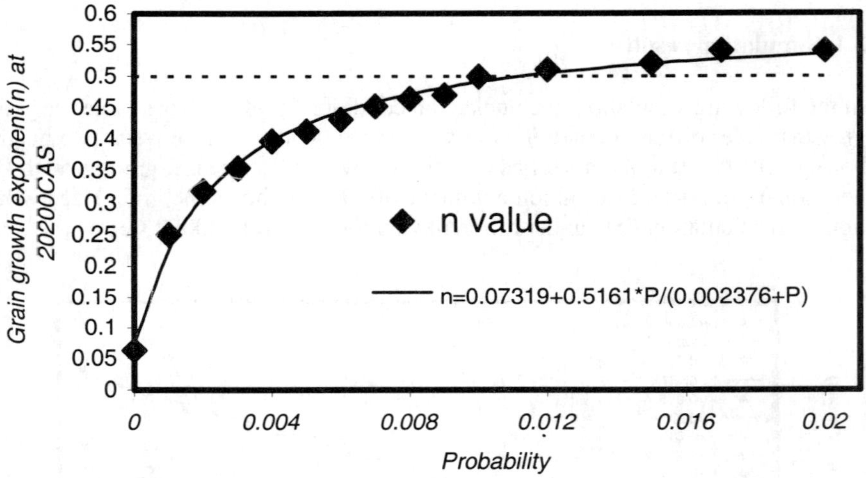

Fig. 2. Grain growth exponent changes with probability.

Fig. 2 shows the change in grain growth exponent with probability. In general, the grain growth exponent should be in the range 0.06-0.5 if the value of P falls in the range 0-0.01. It is also seen that n=0.5 is not a limitation for grain growth simulation from a topological standpoint; the grain growth exponent can be over 0.5 if a large value for P is selected.

3.2 Application of CA method to simulate the materials

The next issue for the development of the model was how to reflect the n value change with temperature. The expression which was subsequently incorporated into the model was given as:

$$P = A \cdot \exp(-Q_B / RT) \tag{3}$$

Where P is the probability for the cell to change the state, R the universal gas constant (8.3 J/mol·K), Q_B represents the activation energy for the movement of cell

across the boundary in J/mol, T is the absolute temperature (K). The constants A and Q_B can be determined from experiment results.
This research has selected Hu's [6] experimental results as a comparison with the developed model. According to his experimental results, the time exponent (n) increases with temperature from approximately 0.25 at 550°C to 0.5 at 850°C . In the present CA program, P should be 0.001 when n=0.25 and P should be 0.01 when n=0.5. Using Eq. 3, and values of A=5.54, Q_B=59.4 kJ/mol can be obtained. $P = 5.54\exp(-59.358/8.3T)$ can be input into the program to reflect the grain size evolution with temperature. Fig. 3 shows the comparison between the simulation of the temperature dependence of the time exponent during isothermal grain growth along with Hu's [6] experimental result.

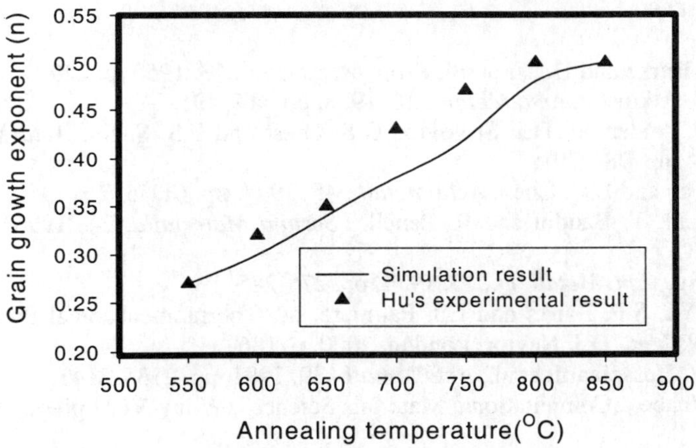

Fig. 3. Dependence of n on T for isothermal grain growth.

From the above example, it is shown that experimental result from quantitative metallography and temperature can be coupled in the present program. The current CA model can therefore accurately and quantitatively assess the process of grain coarsening within limits as determined from experimental results.

4 Conclusions

A two-dimensional CA model has been developed for the description of the normal grain coarsening process. The program shows that the grain growth kinetics gradually increase with probability, and that temperature can be coupled into the program. It is further shown that the grain growth kinetics can increase with temperature, which is close to experimental observations. Simulation results show that an ideal material with grain growth exponent n=0.5 is actually non-existent, since different materials with different grain growth exponents are sensitive to microstructure and temperature. The current model can be used to describe normal grain growth, provided that experiments related to the given material can be properly coupled.

Acknowledgements

W. Yu is grateful to The Institute for Microstructural and Mechanical Process Engineering: The University of Sheffield (IMMPETUS), and the Overseas Research Students Awards Scheme(ORS) award for financial support.

References

[1] J.E. Burke and D. Turnbull, *Prog. Metal. Phys.*, 3, 1952, p. 220.
[2] H.V. Atkinson, *Acta Metall.*, 36, 1988, pp. 469-491.
[3] M.P. Anderson, D.J. Srolovitz, G.S. Grest and P.S. Sahni, *Acta Metall*, 32, 1984, pp. 783-791.
[3] D. Fan and L.Q. Chen, *Acta Metall.*, 45, 1997, pp. 611-622.
[4] Y. Liu, T. Baudin and R. Penelle, *Scripta Materialia*, 34, 1996, pp. 1679-1683.
[5] H. Hu, *Can. Metall. Q.*, 13, 1974, pp. 275-285.
[7] W. Yu, S.P. Banks and E.J. Palmiere, in "Thermomechanical Processing of Steels", ed. D.J. Naylor, London, 2000, p. 186.
[6] H.W. Hesselbarth et al., *Acta Metall.*, 39, 1991, pp. 2135-2143.
[7] D. Raabe, „Computational Materials Science", Wiley-VCH press, 1998.

Recrystallization and Grain Growth
Proceedings of the First Joint International Conference
Eds. G. Gottstein and D.A. Molodov, Springer-Verlag, 2001

The Effect of Temperature on Grain Growth in $YBa_2Cu_3O_{7-x}$ Superconductive Ceramics

M.F. Imayev, D.B. Kabirova, H.A. Churbaeva, and G.A. Salishchev

Institute for Metals Superplasticity Problems, Russian Academy of Sciences,
Khalturina str. 39, Ufa 450001, Russia

Keywords: grain growth, superconductor, Y-Ba-Cu-O.

Abstract. Grain growth during annealing of $YBa_2Cu_3O_{7-x}$ (Y123) ceramics was studied. It was established, that grain growth begins near 900°C. The mean grain size is increased only up to 975°C while at higher temperatures, up to the basic incongruent melting point of the Y123 at m_1=1015°C, the grain growth becomes slower. The grain growth is connected with formation of a thin liquid film on grain boundaries. The first portion of liquid is formed at about 900°C as a result of eutectic reaction. With increasing temperature, several peritectic reactions take place with second solid phase Y_2BaCuO_5 (Y211) forming on grain boundaries. The grain growth inhibition is caused by accumulation of Y211 phase and pores, which quantity is increased with approaching the temperature m_1.

1 Introduction

Grain growth during annealing of fine-grained deformed Y123 ceramics was thoroughly investigated in our previous paper [1]. It has been established that grain growth of Y123 matrix phase occurs in the presence of a thin liquid film on grain boundaries. The grain growth begins near 900°C that correlates with the temperature of appearance of first portions of liquid resulted from decomposition the Y123 compound according to reactions of double (e_1) and/or threefold (e_2) eutectics. The fastest grain growth occurs above 925°C due to increasing liquid amount as a result of peritectic reaction (p_1). However, grain growth has been investigated only up to 975°C. This paper reports the investigation of grain growth phenomenon in Y123 ceramics in a wider temperature range, up to the point of basic incongruent melting (m_1).

2 Experimental Procedure

The samples of Y123 ceramics (Ø10 mm and 10 in height) were prepared by sintering in flowing oxygen at 930°C for 10 h. Then they were extruded at 875°C to

Ø5 mm. The obtained extruded rods were cut out into samples, 3 mm in height, by a wire saw. Isothermal grain growth experiments were carried out in air at temperatures 900, 925, 950, 975, 980 и 990 и 1000°C for 5 h. Phases formed in the samples were analyzed by XRD measurement using Cu K_α radiation. Microstructures were examined on the polished and etched surfaces by means of optical and scanning electron (SEM) microscopes. The growth of grains was investigated by measuring the length (l), thickness (h), and grain aspect ratio (a=(l/h)) of separate grains. The necessity in such a method of calculation was proved elsewhere [1]. A number of grains sized exceed 300. The thermal behavior of the batch was studied by DTA. The sample densities were measured by Archimedes method.

3 Results

Fig.1 shows optical micrographs of Y123 samples after annealing at different temperatures. It is seen that grain growth occurs up to 950-975°C, but with increasing temperature above 975°C the grain growth sharply slows down.

The XRD method does not reveal any changes in phase composition up to 990°C (Fig. 2). However, at 1000°C the material becomes two-phase and in addition to the Y123 peaks of the Y211 phase are observed. Traces of other phases, in particular, $BaCuO_2$, CuO and CuO_2 are not revealed.

Since the reflectivity of Y123 and Y211 phases are almost similar it is impossible to distinguish them by means of the optical microscope. That is why the numerical analysis of the microstructure was made by means of SEM in the secondary electron image regime at low accelerating voltage. This regime provides reliable distinction of Y123 and Y211 phases. Fig. 3 shows the SEM image of the sample annealed at 1000°C. A large amount of Y211 particles are seen. Moreover, an increased porosity is observed.

Analysis of the microstructure shows that with increasing temperature up to 950-975°C the mean values of the length (L) and the thickness (H) of Y123 grains grow (Table 1). However, above 975°C the growth of both L and H is very slow. The behavior of the average aspect ratio of Y123 grains is also very unusual. It increases up to 925°C, and then it decreases.

Below 950°C the quantity of Y211 particles does not exceed 0.3 vol.% and their mean size is about 0.5 μm. However, above 950°C the quantity and mean size of Y211 particles sharply increase. Relative density of the material shows the same behaviour. Below 975°C the density does not almost change, but above this temperature the decrease in density is rather prominent (Table 1). Theoretical densities of Y123 and Y211 phases differ only by 2.2%, that is why the change in density is almost completely connected with increasing porosity of the material.

Fig. 4 shows a DTA curve for the initial state. Alongside the basic incongruent melting of the Y123 compound at a temperature of about 1015°C (peak m_1), a small

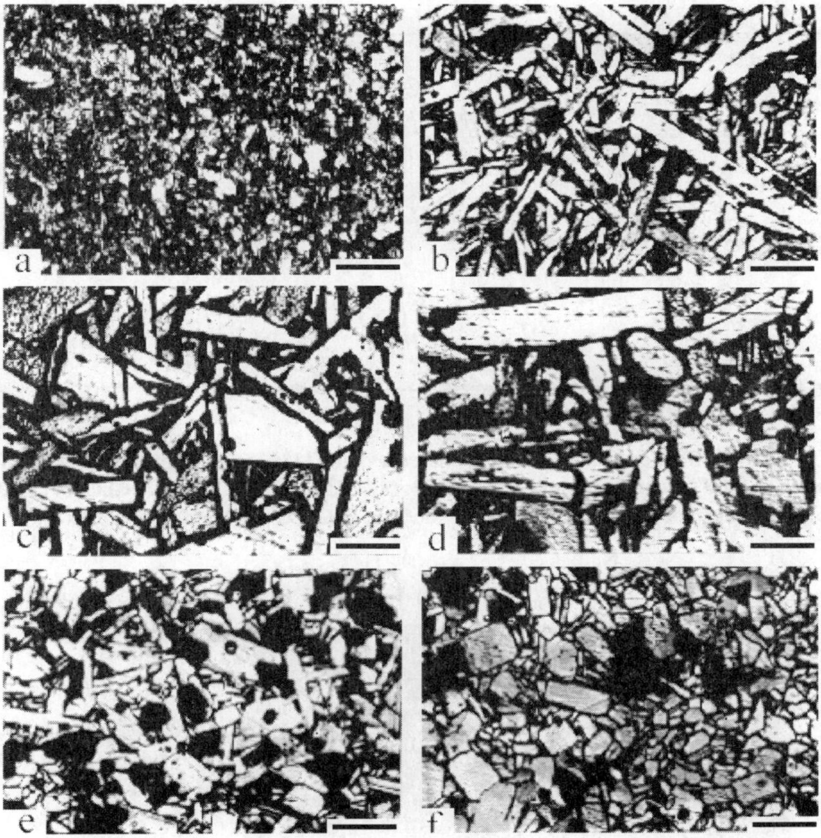

Fig. 1. Optical micrographs of the grain growth of Y123 samples (a) as-deformed at 875°C and isothermally annealed for 5 h at (b) 925°C, (c) 950°C, (d) 975°C, (e) 990°C, and (f) 1000°C. Bar = 20 μm.

endothermic peak corresponding to partial melting is observed over the temperature range 925-950°C. The possible chemical reactions near composition Y123 resulting in occurrence of liquid phases on grain boundaries [2] are given in Table 2. Liquid is formed at 925°C on grain boundaries due to p_1 reaction.

4 Discussion

The results obtained show that with increasing the temperature of annealing from 900°C up to 1000°C some fraction of the Y123 phase is decomposed due to a series of eutectic and peritectic reactions. The possibility of Y123 decomposition

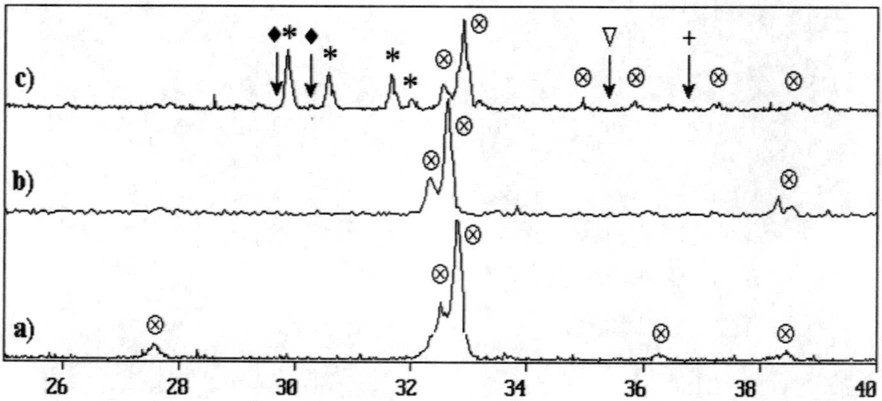

Fig. 2. X-ray diffraction patterns of (a) as-deformed at 875°C and annealed at (b) 990°C for 5 h and (c) 1000°C for 24 h samples. (⊗) Y123, (*) Y211, (♦) $BaCuO_2$, (∇) CuO, (+) Cu_2O. Arrows indicate positions of principal characteristic $BaCuO_2$, CuO and Cu_2O peaks.

Fig. 3. Secondary scanning electron image of sample annealed at 1000°C for 5 h: (1) Y123, (2) Y211, and (3) pore.

Table 1. Parameters of microstructure of annealed samples.

T, (°C)	Mean length L (μm)	Mean thick-nessH (μm)	Mean aspect ratio A=L/H	Relative density ρ (%)	Volume fraction of Y211 V_{211} vol.%	Mean size of Y211 d_{211}(μm)
875	0.5	0.5	1	97		
900	3.25	0.67	6.02	97		
925	6.61	0.98	9.12	97		
950	15.1	3.39	5.25	96	0.3	0.5
975	14.8	4.07	4.17	93	0.7	1.0
980	7.82	3.67	2.24	91	0.75	1.0
990	6.51	2.64	2.43	82	2	1.3
1000	2.88	1.57	1.95	77	8.7	1.6

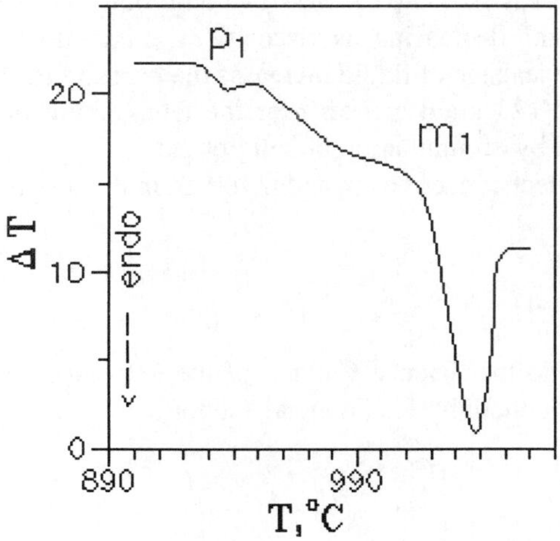

Fig. 4. DTA curve for Y123 sample as-deformed at 875°C.

at temperatures lower than an incongruent melting point by almost 100°C is attributed to the fact that Y123 phase is a strict line compound with an area of solid state solubility by cations less than 0.5 wt% [3]. Thus, small local deviations from stoichiometry can result in partial melting due to e_1, e_2 and p_1 reactions (Table 2). The preferential sites for composition deviation are grain boundaries.

Table 2. Invariant reactions in the Y-Ba-Cu-O system observed in air

Invariant points	Reactions	T (^oC)	
		DTA	Ref. [2]
e_1	$Y123+BaCuO_2+CuO \rightarrow L(e_1)+O_2$	913	898
e_2	$BaCuO_2+CuO \rightarrow L(e_2)+O_2$		902
p_1	$Y123+CuO \rightarrow Y_2BaCuO_5+L(p_1)+O_2$	929	933
m_1	$Y123 \rightarrow Y_2BaCuO_5+L(m_1)+O_2$	1015	1015

The first portions of Ba-Cu-rich liquid are formed due to eutectic reactions e_1 and e_2. Liquid "switches on" the process of grain growth. Over the temperature range 925-950°C there occurs peritectic reaction p_1 with formation of additional amount of liquid. However, the second solid phase Y211 occurs in this case. With increasing temperature above 950°C, the amount of liquid and Y211 phase increases significantly due to onset of reaction m_1 on grain boundaries. During annealing the major amount of liquid leaves the sample but Y211 particles remain completely in its interior. The XRD and SEM data testify this. Unlike phase Y211 traces of liquid phase are not revealed even after annealing at 1000°C for 24 h (Fig. 2 and 4). As seen from Table 2 all reactions of Y123 decomposition are accompanied by formation of oxygen. That is why the easiness of liquid outflow from the sample can be connected both with decreasing its viscosity as temperature increases and with oxygen release. The leakage of liquid increases the porosity of the material. Thus, the inhibition of Y123 grain growth near the temperature of macroscopic melting can be explained by accumulation of solid phase Y211 particles and pores due to occurrence of peritectic reactions p_1 and m_1 on grain boundaries.

Acknowledgement

The authors are grateful to the Science Council of the Russian Academy of Sciences on High-T_c superconductivity for financial support.

References

[1] M.F. Imayev, D.B. Kazakova, A.N. Gavro, A.P. Trukhan, *Physica C*, 329, 2000, pp. 75-87.
[2] T. Aselage and K. Keefer, *J. Mater. Res.*, 3, 1988, pp. 1279-1291.
[3] H.J. Scheel and F.Licci, *Thermochim. Acta*, 174, 1990, pp. 115-120.

Stagnation Effects in Grain Growth

3D Massively Parallel Cellular Automaton Simulations of Zener Pinning

M. A. Miodownik[1], E. A. Holm[2], and G. N. Hassold[3]

[1] Mechanical Engineering Department, Kings College London, Strand, London, WC2R 2LS, UK

[2] Materials and Process Modeling, Sandia National Laboratories, Albuquerque, NM 87185-1411, USA

[3] Department of Science and Mathematics, Kettering University, Flint, MI 48504, USA

Keywords: Zener pinning, simulation, particles, cellular automaton

Abstract. Massively parallel 3D cellular automaton simulations of polycrystalline grain growth were performed in the presence of a dispersion of incoherent pinning particles. Over the range of particle volume fractions simulated, $0.025 \leq f \leq 0.15$, grain growth stagnates in agreement with the Zener analysis, such that pinned grain radius scaled by particle radius is inversely proportional to the particle fraction. Although predicted by theory, other pinning regimes, characterised by decreased pinning exponents, are not observed. Using a frozen boundary model, we investigate the requirements for pinning a boundary network and show that pinning can occur even when a majority of the constituent boundaries are not themselves pinned.

1 Introduction

The relationship between pinned grain size and the volume fraction of particles is still a matter of controversy. Over fifty years ago Zener proposed an equation that has been argued about ever since. The Zener equation relates the maximum attainable grain size R during grain growth in a random dispersion of particles of radius r

$$R = A\frac{r}{\beta f^n} \tag{1}$$

where f is the volume fraction of particles, A and n are constants and β is a function of R/r. The original formulation due to Zener and Smith [1] gives A= 4/3, β=1 and n = 1. A number of subsequent investigators have proposed corrections to their approach [2-6]. Most notably some investigators [5,6] have suggested that the volume fraction exponent should decrease from n = 1 to n = 1/3 above a critical volume fraction of particles. They argue that above this critical volume fraction there is less than one particle per grain boundary invalidating the statistical approach of Zener and that the grains continue grow until some other pinning mechanism becomes active.

The controversy about the validity of the Zener equation has continued for two reasons. First, there are few experimental results that cover a wide enough range of volume fractions to sample both proposed regimes [7], although most experiments are consistent with the Zener model. Second, a number of computer simulations of Zener pinning have found n = 1/2 in 2D and n = 1/3 in 3D over a wide range of volume fractions.

While the dimensionality of the system is not explicitly included in the Zener equation, the analysis only applies to 3D systems. This is because an underlying assumption in the derivation of the equation is that a boundary interacting with an array of particles can be treated as a single thermodynamic entity over which pinning forces and driving forces can be summed. This is true of 3D boundaries but not of 2D boundaries, as is evident if one considers what happens when a 2D boundary meets a particle. The 2D boundary must split into two boundary segments and the points at which each segment attaches to the particle effectively become nodes. Detachment from the particle can only occur if the two boundary segments rotate by an angle of π degrees. This is the crucial difference between 2D and 3D pinning as first pointed out by Hillert [5]. In 3D, the pinning force is a frictional force due to the local increase in boundary area through the formation of a dimple during particle bypass. In 2D, pinning arises from the topological consideration that the particles act as static nodes. The upshot of all this is that one cannot use 2D computer simulations to investigate Zener Pinning.

3D Monte Carlo simulations of Zener pinning have been carried out by Anderson *et al.* [8] and Hazzledine and Oldershaw [6]. The relationship between the pinned grain sizes and the volume fraction is given by equation (1), where n=0.31. However by examining the structures formed at particle/boundary intersections, Miodownik *et al.* [9] found that these simulations suffered from artificial pinning due to lattice anisotropy, and thus are invalid for examining particle pinning. Recently Miodownik *et al.* [10] reported the results of corrected 3D simulations. In this paper we report some further results and discuss in detail the requirements to pin boundary networks.

2 Simulation Methodology

Grain growth is modeled by a 3D cellular automaton with a probabilistic transition function. The details of the model are very similar to that of the Monte Carlo Potts model for grain growth [11]. A continuum microstructure is bitmapped onto a 3D lattice by assigning each lattice site an orientation index s_i, where $0 < s_i < Q$ and

Q is the number of different orientations in the system. Inert particles are incorporated by assigning clusters of sites a unique, non-changeable index (i.e. $s_i = 0$). The boundaries are defined between neighbouring sites with unlike orientations; these boundaries are given a nominal energy of γ_0. The grain boundary and particle/matrix interface energies are equal in magnitude and the neighbourhood is defined as the 1st, 2nd, and 3rd nearest neighbours in 3D simple cubic lattice. Sites were chosen using checkerboarding to prevent local correlation effects [12]. For each site the orientation of one of its neighbours is chosen at random and the change in boundary energy, ΔE, associated with a switch to that orientation is computed. The following probabilistic switching rule is then applied:

$$P(\Delta E) = \begin{cases} m_{ij} & (\Delta E \le 0) \\ m_{ij}\exp(-\Delta E/\Gamma) & (\Delta E > 0) \end{cases} \tag{2}$$

where P(ΔE) is the probability of switching, m_{ij} is the normalised mobility of the boundary defined between orientations i and j, and Γ is a variable which imparts thermal energy (random noise) to the boundaries. Time is an independent variable and is related to the number of attempted switches.
In this study, initial microstructures were constructed by placing (3 x 3 x 3) cubic particles at random until the specified particle volume fraction was reached, then assigning each grain site a random orientation. Microstructures were evolved until grain size stopped changing, typically 10^7 time steps. System sizes ranged from 200 x 200 x 200 sites for f = 0.15 to 400 x 400 x 400 sites for f = 0.025. In all the simulations the total number of discrete orientations was Q=100. Full periodic boundary conditions were utilized. Grain size was measured by a cluster enumeration technique that measured each grain volume, and the grain size R was computed as the cube root of grain volume. Particle size r was determined similarly. The code was parallelised and the biggest simulations ran on 128 processors of an Intel Paragon supercomputer.

3 Results and Discussion

3.1 Particle Pinning

The simulated microstructures appear normal in that the grain boundaries are smoothly curved and triple junctions have interior angles near 120°. In these simulations all boundaries had uniform mobilities and energies. Figure 1 shows a cross-section through the pinned grain structure from a simulation with f = 0.025. Most of the particles are located in the grain interiors and as observed in experiments and predicted by theory, boundaries exhibit dimples where they intersect particles. Note that the particles do not remove curvature from the system.

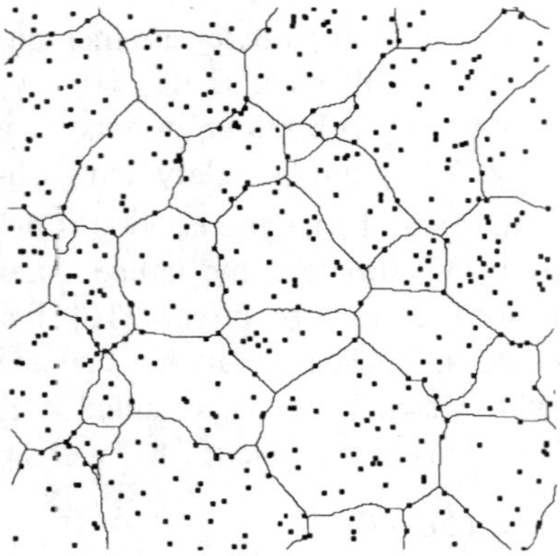

Fig. 1. Cross-section of the pinned grain structure for f=0.025.

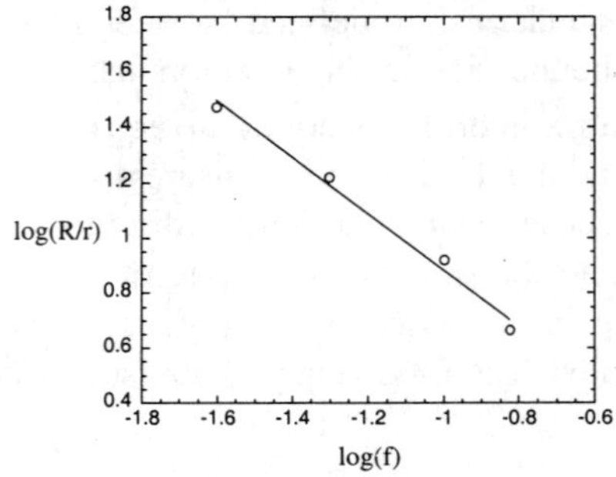

Fig. 2. Pinned grain size R scaled by the particle radius r as a function of volume fraction of particles f. The solid line is a best fit with slope -1.02.

Figure 2 shows the pinned grain size R scaled by the particle radius r as a function of volume fraction of particles f. A power-law curve fit of the form of equation (1) gives $A = 0.728$ and $n = 1.02$, in agreement with the Zener prediction.

Hazzledine and Oldershaw [6] have suggested that the volume fraction exponent should change from $n=1$ to $n = 1/3$ when $f = 0.01$. They argue that equation (1) is only applicable at values of $f < 0.01$ because for larger volume fractions, the number of particles in contact with any one boundary segment becomes small. They calculate that when $f > 0.01$ the number of particles per boundary facet is less than one. In this case a migrating boundary can no longer be thought of as a multi-dimpled interface moving through an array of particles. They further assert that for $f > 0.01$ a discrete geometric approach should be preferred. Hillert [5], using a different argument comes to a similar conclusion, expecting a transition when $f = 0.1$. The geometrical model proposed is that the structure becomes pinned when the majority of the particles reside at the grain apexes. If there are on average 24 apexes per grain each shared between four grains, then there will be six particles per grain, giving the expression identical to equation (1) but with $A=1.46$, $\beta=1$ and significantly a different volume fraction dependence of $n=1/3$. Experimental evidence for such a transition is still controversial; Manohar *et al.* [7] have summarised the data.

In these simulations, we observe no deviation from $n = 1$ over the range $0.025 \leq f \leq 0.15$. In the next section we address this question of whether a network of boundaries can be pinned even when the majority of the boundaries that comprise the network are not pinned.

3.2 Frozen Boundary Model

A set of simulations was carried out in which no particles were included in the model. Instead a fraction, g, of the boundaries was frozen by setting $m_{ij} = 0$ for these boundaries (thus switches that would cause their movement had zero probability of occurring). Grain growth was allowed to continue until either the grain structure became pinned or reached a 1/3 of the simulation size in which case the simulation was halted. Figure 3 shows a plot of the probability that the grain structure became pinned versus the initial fraction of frozen boundaries. There is a sharp transition that occurs when g=0.01 above which the grain structure always became pinned. We strongly suspect that this is a percolation phenomenon and are currently undertaking further work to investigate this possibility. Note that g is the initial fraction of frozen boundaries, not the final fraction. Figure 4 shows a cross-section through such a pinned structure; the pinned boundaries are shown with thick lines. It illustrates that a network of boundaries can become pinned despite the fact that very few of the boundaries are frozen. This result informs the theories of Hillert [5] and Hazzledine and Oldershaw [6] in that it shows that their criterion for pinning is too stringent.

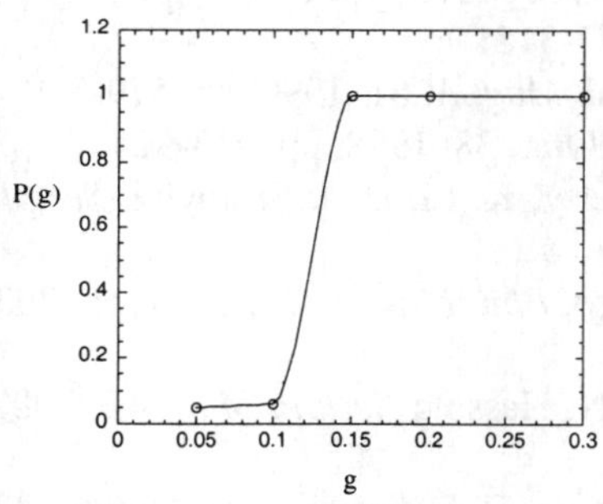

Fig. 3. A plot of the probability that the grain structure became pinned, P(g) versus the initial fraction of frozen boundaries g.

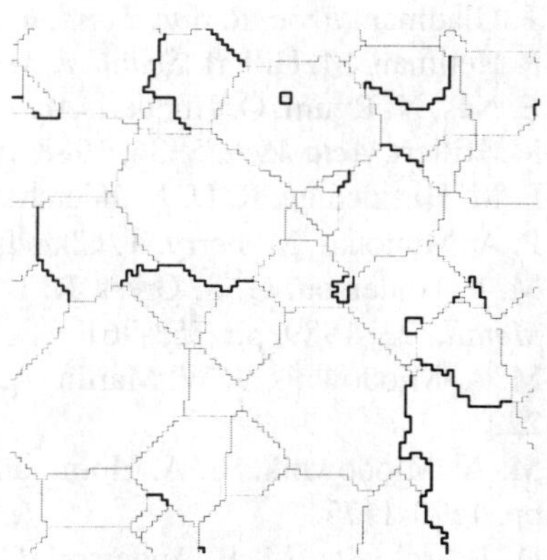

Fig. 4. A cross-section of a pinned microstructure for g = 0.12. Thick black lines show immobile boundaries.

4 Conclusions

The simulation results of the particle pinning model are in agreement with the Zener- Smith analysis and give n=1. The frozen boundary simulations show that a polycrystalline grain structure can be pinned even when the majority of its boundaries are unhindered. The results from the two models cast doubt on the theories proposed by Hillert [5] and Hazzledine and Oldershaw [6] to explain an volume fraction dependence transition. It is not obvious that the mean field theory, originally proposed by Smith and Zener, should be able to predict the pinning behaviour of the boundary networks, however fifty years on, it still seems to be the best theory we have.

Acknowledgements

Many thanks to Steve Plimpton for writing the first working version of the massively parallel grain growth code. This work was supported by the U.S. Department of Energy under contract number DE-AC04-94AL85000 and by an Office of Basic Energy Sciences New Initiative.

References

[1] C. S. Smith, *Trans. AIME,* 175, 1948, pp. 15-25.
[2] T. Gladman, *Proc. R. Soc. London, Ser. A,* 294, 1966, pp. 298-312.
[3] P. Hellman, M. Hillert, *Scand. J. Metall.,* 4, 1975, pp. 211-219.
[4] E. Nes, N. Ryum, O. Hunderi, *Acta Metall.,* 33, 1985, pp. 11-18.
[5] M. Hillert, *Acta Metall,.* 36, 1988, pp. 3177-3181.
[6] P. M. Hazzledine, R. D. J. Oldershaw, *Phil. Mag. A,* 61, 1990, pp. 579-589.
[7] P. A. Monohar, M. Ferry, T. Chandra, *ISIJ Int.,* 38, 1998, pp. 913-924.
[8] M. P. Anderson, G. S. Grest, R. D. Doherty, K. Li, D. J. Srolovitz, *Scripta Metall.,* 23, 1989, pp. 753-761.
[9] M. A. Miodownik, J. W. Martin, A. Cerezo, *Phil. Mag. A,* 79, 1999, pp. 203-222.
[10] M. A. Miodownik., E. A. Holm. and G. N. Hassold *Scripta Mat.,* 42, 2000, pp. 1173-1177.
[11] D. J. Srolovitz, M. P. Anderson, G. S. Grest, P. S. Sahni, *Acta Metall.,* 32, 1984, pp. 1429-1438.
[12] S. A. Wright, S. J. Plimpton, T. P. Swiler, R. M. Fye, M. F. Young, E. A. Holm, "Potts-model Grain Growth Simulations: Parallel Algorithms and Applications," Sandia Report SAND97-1925, August 1997.

Effect of Particle Clustering on the Grain Size Control

E. Flores[1*], J.M. Cabrera [1,2] and J.M. Prado [1,2]

[1] Departamento de Ciencia de Materiales e Ingeniería Metalúrgica, ETSEIB - Universidad Politécnica de Catalunya, Av. Diagonal 647, 08028-Barcelona, Spain
[2] Centro Tecnológico de Manresa, Av. de las Bases de Manresa 1, 08240 – Manresa, Spain
(*) Currently at Universidad Central de Panamá, Panamá

Keywords: grain size control, clustering, agglomeration

Abstract. The present work shows that the clustering of particles promotes deviation in the classical mathematical expressions describing the grain growth control by second phase particles. On the basis of experimental results and theoretical laws, a semi-empirical expression to predict the limiting grain size is presented. The latter expression takes account of the agglomeration phenomena and can be extended to large volume fractions of particles, conditions under which classical theories clearly fails. The equation remains valid as far as the nucleation of precipitates takes place at random. From a practical point of view it is shown that volume fractions larger than 0.12 cannot control significatively the grain size because the increased probability for clustering.

1 Introduction

One of the most important and frequent ways of controlling the microstructure of engineering alloys, and therefore their mechanical properties, is by refining the grain size. This in turns can be achieved by pinning the grain boundaries by second phase particles. Size and volume fraction of particles as well as the homogeneity of the nucleation sites are the factors determining the efficiency of particles in governing the grain size. Since the Zener expression [1], several equations have been proposed to relate the grain size to some characteristics of the particles (see a recent review in ref. [2]). In general, most theories maintains that the grain size is directly proportional to the particles size, and inversely proportional to the volume fraction of precipitates. The latter models fit to experimental data with high accuracy up to volume fractions of approximately 0.02. Beyond this value, strong deviations are usually observed. It is worth noting that most of the theoretical models assume that particles are distributed at random, and they do not contact themselves, i.e., they are isolated. Nevertheless there is enough experimental and theoretical evidence [3-8] showing that under certain circumstances, single particles are frequently found in contact each other, connected and forming *clusters*. The latter, on the other hand, is a usual phenomena in precipitation processes of chemical reactions [9]. Particularly the nonuniform distribution of particles seems to be a common feature in steels, especially in microalloyed steels, although most authors has not paid full attention to this phenomenon. The precipitation in clusters promotes a clear deviation of the hypothesis of homogeneous distribution assumed in many of the aforementioned theories. No account of this feature may introduce a significant error on the classical equations for the grain size control.

In many of the grain growth theories [2] the particle distribution is simply characterised by its average radius. It is implicitly assumed that particles are isolated and therefore the entire superficial area of the particle is available to pin the grain boundary. However, the particle agglomeration would decrease the above area so a larger critical grain size should be found. It seems, therefore, plausible that part of the experimental deviations of theoretical equations can be due to the clustering of precipitates. The present work is devoted to derive how the critical grain size can become affected by the latter clustering phenomena.

Finally, it must be mentioned that the clustering effect on increasing the critical grain size is expected to be negligible at low volume fraction of particles, but should be taken into account in the case of relatively large volume fractions (conditions under which classical theories fail).

2 Experimental Method and Results

Cast iron appears as an appropriate alloy for studying the clustering phenomenon at relatively large volume fractions of particles. Hence, five different nodular cast iron alloys were selected to derive the clustering ability of their graphite particles. The chemical compositions as well as the volume fractions of the graphite nodules of the cast irons are listed in Table 1.

Table 1. Chemical composition (mass %) and volume fraction f_v of graphite spheroids of the cast irons studied.

Cast iron	f_v	%C	%Si	%Mn	%Cu	%Mg
A	0,067	2,34	2,77	0,54	0,52	0,05
B	0,109	3,41	2,67	0,15	0,08	0,02
C	0,130	3,77	2,40	0,14	0,24	0,03
D	0,150	4,53	2,80	0,10	0,18	0,05
E	0,170	4,66	2,58	0,09	0,20	0,03

Samples were studied in the as-cast condition, and conventional metallography was employed in the preparation and observation of specimens. An evaluation of the number of particles grouped by pairs, trios, quartets and so on was carried out. An average of 1000 nodules (particles) were employed for that purpose. As an illustrative example of the agglomeration phenomena, the latter types of clusters, are shown in Fig. 1. The percentage of particles clustered in groups of two, three, etc., is displayed in Fig. 2.

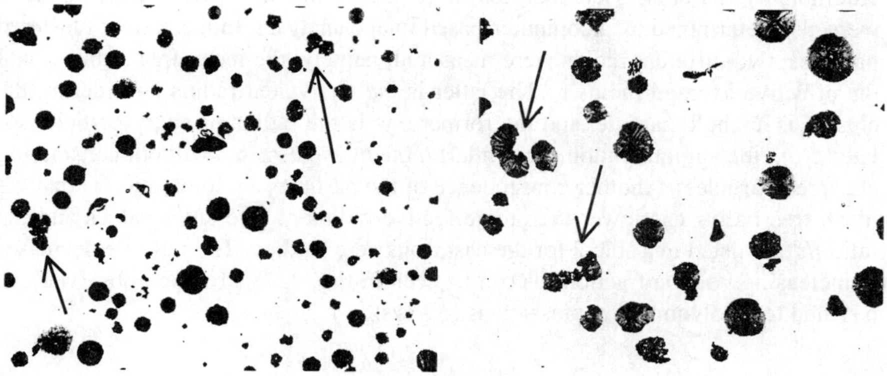

Fig. 1. Illustrative examples of the particle clustering on cast irons

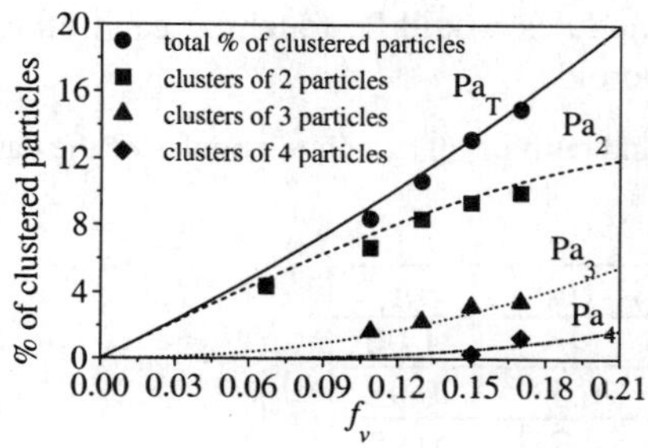

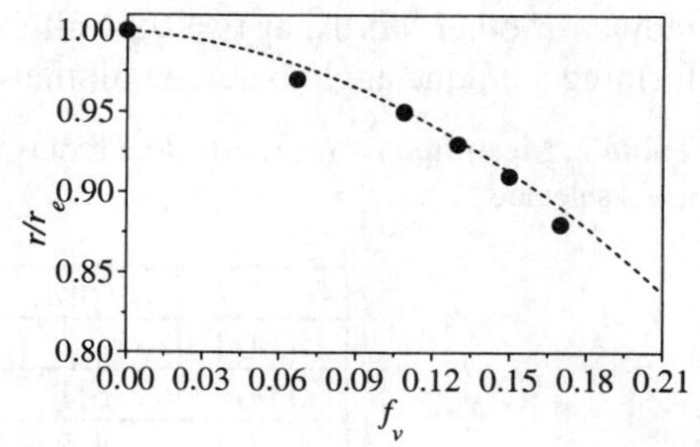

Fig. 2. % of particles clustered by groups of 2 (Pa_2), 3 (Pa_3), and so on.

Fig. 3. Dependence of the ratio r/r_e on the volume fraction f_v of particles in the current cast irons.

One can readily observe that the clustering of the graphite nodules increases with the volume fraction. Up to values of f_v ~ 0.067 only clusters consisting of a pair of spheroids could be found. Under this condition, solely a 4% of particles are agglomerated. At higher volume fractions (0.109 and 0.125) not only the percentage of spheroids forming clusters of two particles increases, but also a new type of cluster with three particles can now be detected. At even higher volume fractions (0.150 and 0.170) clusters formed by four nodules appear, and the total amount of particles forming part of clusters of any size is 14% and 16% respectively. In other words, clusters of higher order (groups of three or more) appears gradually at increasing volume fractions. It is also apparent that clusters of low order are more frequent than those of high order, at least under the experimental conditions studied.

Additionally the cross sectional area of each nodule and its equivalent diameter were also determined by a computer-based image analyser. In the case of clustered particles, two different radius were measured, namely, the mean free radius r, and the effective average radius r_e. The latter is the equivalent radius considering the cluster as a single particle, and the former one is the radius of every particle isolated from the agglomeration (or similarly, the average radius without considering clustered particles). Another consequence of the particles agglomeration is that the mean free radius r is lower than the effective radius r_e. The latter values and the ratio r/r_e are listed in Table 2 for the cast irons here studied. The ratio r/r_e decreases at increasing volume fractions as can be seen in Fig. 4. The relation observed can be fitted to a polynomial expression as follows:

$$\frac{r}{r_e} = 1 - 0{,}1 f_v - 3{,}2 f_v^2 \tag{1}$$

The positive solution of this equation has a value of $f_v = 0.54$, a volume fraction where the effective radius is very large ($r_e \to \infty$) in comparison with the mean free radius. In other words, at $f_v = 0.54$ all the nodules would be touching each other, forming a unique agglomerate of all the spheroids.

Table 2. Mean free radius r, effective radius r_e and ratio r/r_e of graphite nodules of the cast irons selected.

F_v	r (µm)	R_e (µm)	r/r_e
0,000	-	-	1,00
0,067	6,3	6,5	0,97
0,109	4,5	4,7	0,95
0,130	6,2	6,7	0,93
0,150	6,6	7,2	0,91
0,170	7,2	8,2	0,88

3 Practical Effect of Particles Clustering on Grain Growth.

In order to determine the effect of the particles agglomeration on the grain size, a theoretical model of grain growth taken from the literature will be employed. In particular, the present authors have found that the Rios model [10] predict with high accuracy experimental data of several alloys at relatively low volume fractions of particles [11-13]. The latter theory is specially useful for the purpose of the present study as it is based on the additional superficial area created at the particle-matrix interface when a grain boundary overpasses a particle. The clustering of particles promotes, in fact, a diminution of the superficial area of the precipitate available to pin the grain boundary.

Rios [10] derives the variation of the total superficial area by volume unit (S_T^N) as a function of the average grain size (R_m) as follows:

$$\frac{dS_T^N}{dR_m} = \left(\frac{4\pi}{3} R_m^3\right)\left(\frac{dS_V^M}{dR_m}\right) + 4\pi\ R_m^2 S_V^P \tag{2}$$

where S_V^M and S_V^P are the superficial area of the grains and particles by volume unit, respectively. Introducing into Eq. (2) a function F, here called "effective area function", that takes account of the diminution of the total superficial area of the particles, due to the clustering phenomenon, one obtains :

$$\frac{dS_T^N}{dR_m} = \left(\frac{4\pi}{3} R_m^3\right)\left(\frac{dS_v^M}{dR_m}\right) + F\left(4\pi\ R_m^2 S_V^P\right) \tag{3}$$

Following a similar procedure to that developed by Rios, the final expression giving the limiting grain size (R_l) promoted by particles of average radius r and volume fraction f_v, will be:

$$R_l = \left(\frac{1}{F}\right)\left(\frac{r}{6f_v}\right) \tag{4}$$

The function F in fact measures the probability of agglomeration of the second phase particles. A simple reordering of Eq. (4) gives:

$$R_l = \left(\frac{r}{F}\right)\left(\frac{1}{6f_v}\right) \tag{5}$$

One can readily see that the ratio (r/F) represents the effective average radius (r_e), and therefore

$$F = \frac{r}{r_e} \tag{6}$$

Introducing now the experimental Eq. (1) obtained for particles distributed at random into Eq. (6), one obtains the limiting grain size with account of the clustering of particles :

$$R_l = \left(\frac{r}{6f_v - 0{,}6f_v^2 - 19{,}2f_v^3}\right) \tag{7}$$

4 Validation of the present modifications.

To validate the latter equation at low volume fractions, several experimental results published in the literature [11-21] where employed as shown in Fig. 4. One can see a good agreement between these results and Eq. (7), although some points tend to diverge. It must be kept in mind that part of the scatter can be explained by the errors associated to the determination of the volume fraction and the size of the particles. It is also worth noting that, at low volume fractions, Eq. (7) is equivalent to the original one developed by Rios. Only at relatively large volume fractions ($f_v > 0{,}03$), Eq. (7) and the Rios expression tend to diverge.

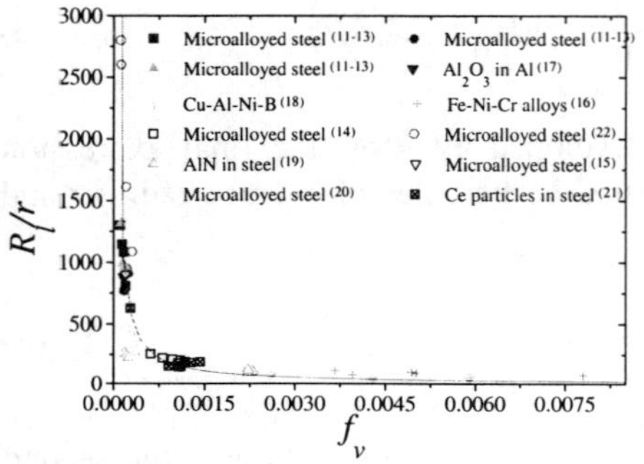

Fig. 4. Validation of Eq. (7) at low volume fractions of particles

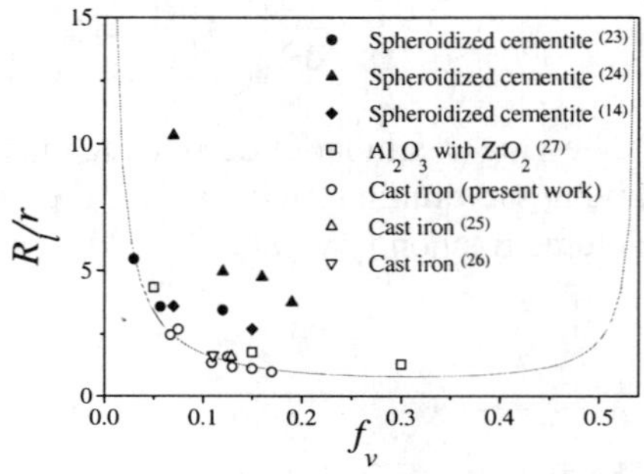

Fig. 5. Validation of Eq. (7) at high volume fraction of particles. See text.

In order to check the accuracy of Eq. (7) at large volume fractions, the present experimental results, as well as some data of additional cast irons [25,26], steels with globulized cementite [14,23,24], and a ceramic, Zirconia Toughened Alumina (ZTA), [27] will be employed. Fig. 5 displays that the experimental data of cast irons and the ceramic alloy agree with Eq. (7). However, data corresponding to globulized pearlite deviates clearly of the behaviour predicted. The latter is a clear demonstration of the effect of the randomness of the nucleation sites of particles on their ability for clustering. By comparison between both kind of materials is clear that nucleation at random promotes smaller grain sizes than localised nucleation (globulized particles are not nucleated at random). Accordingly, Eq. (7) should be modified in case that particle nucleation takes place in preferential sites. Finally, one can see that Eq. (7) shows three different zones. Continuously decreasing at $f_v < 0{,}12$, practically constant in the range $0{,}12 > f_v > 0{,}40$, and finally continuously increasing at $f_v > 0{,}40$. Therefore, and from the point of view of minimising the grain size R_l, there is no sense in employing volume fractions larger than 0.12, as the ratio R_l/r remains practically constant, provided that nucleation sites of precipitates are distributed at random.

5 Conclusions

It is proposed that the clustering of particles promotes deviation in the classical mathematical expressions describing the grain growth control by second phase particles. On the basis of results of cast irons and the Rios theory, an expression to predict the limiting grain size has been presented. This expression can be extended to large volume fractions, conditions under which classical theories fails. The equation is valid assuming randomness nucleation. From a practical point of view volume fractions larger than 0.12 (approx.) do not reduce significatively the grain size because the increased probability for clustering.

Acknowledgments

EF is grateful for a MUTIS grant provided by the Spanish AECI. Further financial support was received from the Spanish CICYT through MAT 97-0827-C02-01

References

[1] C.S. Smith. Private communication from C. Zener to C.S. Smith, quoted in page 47, *Trans. AIME*, 175, 1948, pp. 15-51.
[2] P.A. Manohar, M. Ferry and T. Chandra, *ISIJ Int.*, 38, 1998, pp. 913-924.
[3] D.W. Tian, L.P. Karjalainen, B. Qian and X. Chen, *Metall. and Materials Trans.*, 27A, 1996, pp. 4031-4038.
[4] B. Lehtinen and P. Hansson, *Scand. J. Met.*, 18, 1989, pp. 295-300.
[5] A.K. Ibraheem, R. Priestner and C. Zhou, in "Proc. 1st Int. Conf. On Modelling of Metal Rolling Process", London, 1993, pp. 309-320.
[6] N.K. Ballinger and R.K. Honeycombe, *Met. Trans.*, 11A, 1980, pp. 421-429.
[7] G. Hassold, E. Holm and D. Srolovitz, *Scripta Mater.* 24, 1990, pp.101-106.
[8] E. Underwood in "Quantitative stereology". Addislon-Wesley Publishing Company, USA, 1970.
[9] S. Wachi and A.G. Jones, *Chem. Eng. Sci.*, 47, 1992, pp. 3145-3148.
[10] P. R. Rios, *Acta metall.*, 35, 1987, pp. 2805-2814.
[11] J.M. Cabrera, A. Al Omar and J.M. Prado, *J. Mat. Sci.*, 31, 1996, pp. 1303-1309.
[12] J.M. Cabrera, A. Al Omar and J.M. Prado, *Zeitschrift Metallkd.*, 89, 1998, pp. 47-58.
[13] J.M. Cabrera, E. Flores, A. Al Omar and J.M. Prado in "Fundamentals and Aplications of Microalloying Forging Steels", TMS, 1996, pp. 173-186.
[14] E. Flores, Doctoral Thesis, Universitat Politécnica de Catalunya, 1998.
[15] P. Manohar, D. Dunne, T. Chandra and C. Killmore, *ISIJ Int.*, 36, 1996, pp. 194-200.
[16] A.K. Koul and F.B. Pickering. *Acta Metall.*, 30, 1982, pp. 1303-1308

[17] C. J. Tweed, N. Hansen and B. Ralph, *Met. Trans.*, 14A, 1983, p. 2235.

[18] R. Elst, J. Van Humbeeck and L. Delay, *Acta Metall.*, 7, 1988, pp. 1723-1729.

[19] Ö. Dogan, G. Michal and H. Kwon, *Met. Trans.*, 23A, 1992, pp. 2121-2129.

[20] W.J. Liu and J.J. Jonas, in "Proc. International Conference on Physical Metallurgy of Thermomechanical Processing of Steels and other metals (THERMEC-88)", Tokyo, 1988.

[21] M. Guo and H. Suito, *ISIJ Int.*, 39, 1999, pp. 1289-1296.

[22] D.W. Tian, L.P. Karjalainen, B. Qian and X. Chen, *Met. Trans.*, 27A, 1996, pp. 4031-4038.

[23] P. Hellman and M. Hillert, *Scan. J. Met.*, 4, 1975, pp. 211-219.

[24] L. Anand and J. Gurland, *Met. Trans.,* 6A, 1975, pp. 928-931.

[25] S. Qing, J.M. Detraux, G. Touzot and D. Francois, *Mém. Sc. Rev. Métallurgie*, 1983, p. 183.

[26] D. Rousseau, D. Francois, M. Hecht and A. Gauthier, *Mém. Sc. Rev. Métallurgie*, 1978, p. 281.

[27] D. Casellas, S. Martínez, L. Llanes and M. Anglada, in "I Trobada Transfronterera sobre la Mecànica i la Física dels Materials Bifàsics", eds. A. Mateo, L. Llanes and M. Anglada, Barcelona, 1997, pp. 131-140.

Modelling of the Interaction of Precipitation and Grain Growth

M. Militzer

The Centre for Metallurgical Process Engineering, University of British Columbia, Vancouver, BC, Canada V6T 1Z4

Keywords: pinning, dissolution, coarsening, modelling.

Abstract. The presence of precipitates provides pinning thereby inhibiting grain growth. The pinning force decreases with annealing time when dissolution or coarsening of precipitates occurs, leading to potential for abnormal grain growth. When particle spacing is in the order of the grain size, pinning forces become a function of grain size. These time dependencies of pinning are incorporated into the statistical grain growth model of Abbruzzese and Lücke. The model is applied to describe observations of austenite grain growth in low carbon steels.

1 Introduction

Over the past few decades, modelling the microstructure evolution during hot rolling of low carbon steels has gained significant attention [1,2]. Austenite grain growth is the dominant microstructural feature during reheating and after completion of recrystallization. Frequently, empirical power law equations are employed to describe grain growth kinetics. It has long been recognized that precipitates limit grain growth by pinning [3,4]. The statistical grain growth model by Abbruzzese and Lücke [5] provides a more fundamental approach to explicitly incorporate pinning effects, and has been successfully applied to modelling of austenite grain growth in Al-killed low carbon and microalloyed high strength low alloy (HSLA) steels where a temperature-dependent pinning parameter was used [6, 7]. A potential time dependence of pinning due to dissolution or coarsening of precipitates was approximated by extremely simplified assumptions. More recently, the dissolution kinetics of precipitates, which occurs at the so-called grain coarsening temperature, was incorporated into the grain growth model by deriving the decrease in pinning from a dissolution model [8]. This approach was validated for a Ti-Nb HSLA steel where in addition to grain growth kinetics the particle size evolution was measured. However, the effects of precipitates on grain growth show more complexity. For example, time-dependent pinning forces are also associated with particle coarsening. Further, pinning can be a function of grain size and thus may be implicitly time dependent [9]. For example, Rabkin [10] described grain size dependent particle drag and its relevance for nanocrystalline materials.

In the present paper, various time dependencies of pinning forces are discussed as input for the Abbruzzese and Lücke grain growth model. In addition to dissolution, particle coarsening and grain size dependent pinning are incorporated. Three limiting cases; i.e., grain growth controlled by (i) dissolution, (ii) coarsening, and (iii) grain size dependent pinning, are described and applied to austenite grain growth kinetics in low carbon steels.

2 Pinning Force Models

The Abbruzzese-Lücke grain growth model is based on replacing the actual grain structure by spheres of the same volume. The grain size distribution is suitably subdivided into N size classes. Each grain is surrounded in a statistical way by grains of all other size classes. The growth (or shrinkage) rate of grains in size class i is given by

$$\frac{dR_i}{dt} = M_{gb} \sum_{j=1}^{N} w_{ij} F_{ij} \tag{1}$$

where R_i is the grain radius, M_{gb} the grain boundary mobility and w_{ij} the probability that grains of size classes i and j are neighbours. Provided a grain of size class i is sufficiently larger than a grain of size class j, the driving force F_{ij} for grain growth is given by

$$F_{ij} = \gamma_{gb} \left(\frac{1}{R_j} - \frac{1}{R_i} - P \right) \tag{2}$$

where γ_{gb} is the grain boundary energy and $F_{ij} = -F_{ji}$. When the interparticle spacing is small compared to the grain size the pinning parameter P is related to the particle size distribution by a Zener relationship; i.e.,

$$P = \frac{f}{Kr} \tag{3}$$

where K is a constant, f the volume fraction of the precipitates and r the mean particle radius. The analysis of Manohar et al. [4] suggests $K = 0.17$.

Based on the dissolution and coarsening model of Cheng et al. [11] and employing Eq. 3, the evolution of P during dissolution of a precipitate population can be expressed by [8]

$$P = (P_0 - P_r)\exp(-t/a) + P_r \tag{4}$$

where P_0 is the initial pinning force, P_r a residual pinning force, which, for example, can be associated with a second type of precipitates, and a represents a characteristic time for the dissolution kinetics.
Assuming a cubic coarsening law according to the Lifshitz-Slyozov-Wagner theory [12, 13], the time evolution of P during particle coarsening is given by

$$P = \frac{P_0}{(1 + t/b)^{1/3}} + P_r \tag{5}$$

where b is a characteristic time for the coarsening rate.
For the case where the interparticle spacing is not small compared to the grain size, Zener pinning; i.e., Eq. 3, has to be replaced by a pinning force which is a function of the average grain size, D, and, therefore, becomes an implicit function of time. Pinning is at its maximum, P_{max}, when the interparticle spacing, L, is equal to the grain size. For a stable particle population, this effect can be described in a first approximation by

$$P = \frac{P_{max}}{L} D + P_r \qquad \text{for } D < L \tag{6a}$$

and

$$P = (P_{max} - P_z)\exp(-(D - L)/l) + P_z + P_r \qquad \text{for } D > L \tag{6b}$$

where l is a characteristic length, P_z is the Zener limit for pinning and P_r a residual pinning force which may be present due to a second precipitate type.

3 Austenite Grain Growth in Low Carbon Steels

The above delineated limiting cases of time-dependent pinning forces are relevant for austenite grain growth in low carbon steels. The evolution of the average volumetric grain size is calculated assuming an initial log-normal grain size distribution with a width of 0.3, and values for grain boundary energy and mobility as described elsewhere [6]. Substantial grain growth is observed above the grain coarsening temperature when fine precipitates, e.g. AlN in Al-killed plain carbon steels or (Nb,Ti)C in Ti-Nb microalloyed steels, dissolve. The model for dissolution controlled grain growth had previously been verified for a Ti-Nb HSLA steel containing 0.06 wt% C, 1.37 wt% Mn, 0.068 wt% Nb, 0.038 wt% Ti and 0.0073 wt% N where dissolution of (Nb,Ti)C was confirmed with transmission electron microscopy (TEM) at 1150 and 1200 °C [8, 14]. Fig. 1 compares the predictions with the experimental grain growth data. Grain growth at 1150 and 1200 °C is described using the dissolution regime for P; i.e., Eq. 4, where $P_r = 11\ mm^{-1}$, P_0 decreases from 28 to 19 mm^{-1} and the rate constant a from 500 to 200 s as temperature increases from 1150 to 1200 °C. The decrease in P_0 can be attributed to the progress of dissolution during the additional heating period before the higher temperature is reached.

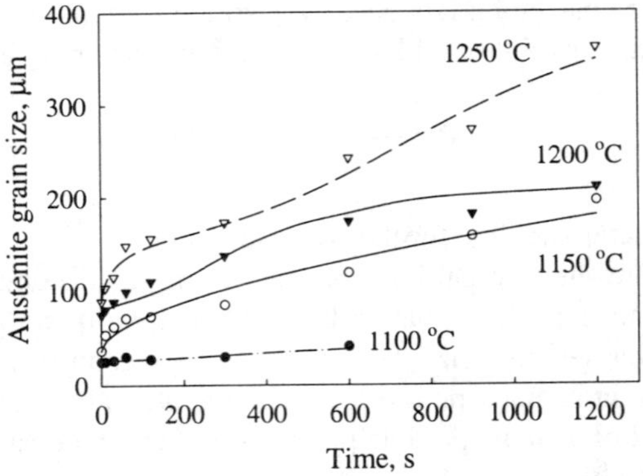

Fig. 1. Comparison of predictions for austenite grain growth with experimental data (symbols) of Lechuk et al. [8] for a Ti-Nb HSLA steel.

The grain growth behaviour at 1250 °C is also associated with a pinning parameter which decreases with time and was originally discussed in the framework of the dissolution relationship of Eq. 4. However, TEM studies did not reveal any changes in the particle size distribution at this temperature. The distribution is characterized by a stable population of TiN consisting of a significant number of smaller particles with an average diameter of approximately 180 nm and a few larger particles with diameters in the range of 1-2 μm [14]. The interparticle spacing of these larger particles is of the order of 100 μm; i.e., comparable to the observed grain sizes. Thus, pinning at 1250 °C consists of two components; i.e., Zener pinning due to the smaller particles and a grain size dependent pinning due to the larger particles. Consequently, grain growth at this temperature shall be described employing a grain size dependent pinning force according to Eqs. 6a and 6b. For simplicity, the Zener limit of the large particles can be neglected ($P_z = 0$) compared to Zener pinning of the small particles which is represented by the residual pinning, P_r. Then, employing $P_{max} = 6$ mm^{-1}, $P_r = 5$ mm^{-1}, L = 150 μm and l = 90μm, the grain growth behaviour can adequately be described as illustrated by the dashed line in Fig. 1. The maximum overall pinning force, $P_{max} + P_r$, due to TiN is with 11 mm^{-1} equal to the residual pinning force in the dissolution cases where grain sizes of approximately 200 μm are observed for the longest holding time of 20 minutes.

Both pinning force cases; i.e., dissolution and grain size dependence, lead to periods of significantly increased grain growth rates during the decrease of P. These increased growth rates are typical for abnormal growth which the Abbruzzese-Lücke model captures in an effective way, albeit it does not consider the heterogeneity of abnormal growth processes.

Grain growth below the grain coarsening temperature had previously been described with a time-independent pinning parameter which decreases as temperature increases [6, 8]. However, a more detailed inspection of the 1100 °C data for the Ti-Nb HSLA steel suggests a decrease in P with holding time due to coarsening, as illustrated with the dash-dotted line in Fig. 1 which was obtained by assuming a particle coarsening rate constant b of 250 s, $P_0 = 70\ mm^{-1}$ and $P_r = 10\ mm^{-1}$. The contribution of P_r is small compared to the overall pinning and was assumed to effectively reflect the pinning due to TiN in a consistent way with the high temperature data. Coarsening controlled pinned grain growth can be discussed in more detail using data by Jiao et al. [15] for austenite grain growth in an Al-killed Mn-Si TRIP steel containing 0.112 wt% C. These data include grain growth kinetics for up to one hour in the temperature range 960 – 1150 °C where limited grain growth due to AlN pinning is observed. Fig. 2 compares experimental data for three temperatures in this range with predictions from the Abbruzzese-Lücke model. Assuming a decreasing pinning parameter due to coarsening, grain growth kinetics can be well represented by taking $P_0 = 33\ mm^{-1}$, $P_r = 30\ mm^{-1}$ and b which decreases from 1000 s at 960 °C to 600 s at 1000 °C and 300 s at 1100 °C. This temperature dependence of b reflects the increase of coarsening rates as temperature increases.

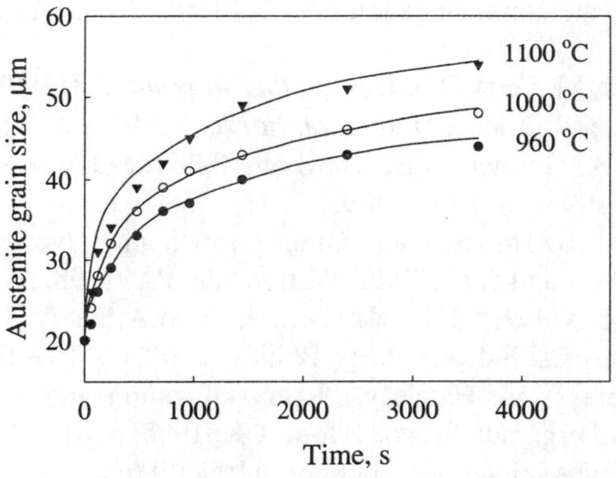

Fig. 2. Comparison of predictions with austenite grain growth data (symbols) by Jiao et al. [15] for a TRIP steel.

4 Conclusions

Simple time dependencies of pinning for dissolution and coarsening of precipitates, respectively, and the consideration of grain size dependent pinning permit the accurate description of austenite grain growth in low carbon steels. To track simultaneous dissolution and coarsening where simple analytical relationships for P are not available would require combining computer models for dissolution and coarsening with the grain growth model, leading to added complexity in terms of input parameters. Further, the current model does not explicitly account for solute drag which could be significant after dissolution. Residual pinning may effectively reflect this contribution, but the direct incorporation of a solute drag effect into the Abbruzzese-Lücke model would be crucial for developing a more versatile model which still can easily be applied to practical grain growth cases.

Acknowledgement

Financial support from the Natural Sciences and Engineering Research Council of Canada (NSERC) is acknowledged. The author would like to thank W.J. Poole for stimulating discussions.

References

[1] P. D. Hodgson, R. K. Gibbs, *ISIJ Internat.*, 32, 1992, pp. 1329-1338.
[2] M. Militzer, E.B. Hawbolt, T.R. Meadowcroft, *Metall. Mater. Trans.*, 31A, 2000, pp. 1247-1259.
[3] C. Zener, private communication to C.S. Smith, *Trans. AIME*, 175, 1948, p. 15.
[4] P.A. Manohar, M. Ferry, T. Chandra, *ISIJ Internat.*, 38, 1998, pp. 913-924.
[5] G. Abbruzzese, K. Lücke, *Mater. Sci. Forum*, 92-94, 1992, pp. 597-604.
[6] M. Militzer, A. Giumelli, E.B. Hawbolt, T.R. Meadowcroft, *Metall. Mater. Trans.*, 27A, 1996, pp. 3399-3409.
[7] M. Militzer, E.B. Hawbolt, in "Grain Growth in Polycrystalline Materials III", eds. H. Weiland et al., TMS, Warrendale, PA, 1998, pp. 639-644.
[8] S. Lechuk, M. Militzer, E.B. Hawbolt, in "HSLA Steels'2000", eds. G. Liu et al., Metallurgical Industry Press, Beijing, 2000, pp. 181-186.
[9] F. J. Humphreys, M. Hatherly, "Recrystallization and Related Annealing Phenomena", Pergamon Press, Oxford, UK, 1995, p. 81.
[10] E. Rabkin, *Scripta mater.* 42, 2000, pp. 1199-1206.
[11] L.M. Cheng, E.B. Hawbolt, T.R. Meadowcroft, *Metall. Mater. Trans.*, 31A, 2000, pp. 1907-1916.
[12] I.M. Lifshitz, V.V. Slyozov, *J. Phys. Chem. Solids*, 19, 1961, pp. 35-50.
[13] Z. Wagner, *Z. Elektrochem.*, 65, 1961, pp. 581-591.
[14] S. Lechuk, M.A.Sc. Thesis, University of British Columbia, Vancouver, BC, 2000.
[15] S. Jiao, J. Penning, F. Leysen, Y. Houbaert, E. Aernoudt, *Steel Res.*, 71, 2000, pp. 345-350.

Effect of Solute Copper on Grain Growth of Austenite in High Chromium Steels

Y. Futamura[1], T. Tsuchiyama[2], and S. Takaki[2]

[1] Graduate student of Kyushu University, 6-10-1, Hakozaki, Higashi-ku, Fukuoka 812-8581, Japan
[2] Department of Materials Science and Engineering, Kyushu University, 6-10-1, Hakozaki, Higashi-ku, Fukuoka 812-8581, Japan

Keywords: copper bearing steels, grain refining, austenite, grain growth, grain boundary segregation.

Abstract. The effect of Cu addition on the grain growth behavior of austenite was investigated in 9Cr (mass%) martensitic steels with various Cu contents. Austenite grain size of specimens linearly increased to the square root of annealing time at any solution-treatment temperatures. However, the growth rate was markedly lowered with increasing Cu content. For instance, 9Cr-4Cu steel had about four times lower growth rate than 9Cr-1Cu steel. Analysis by Auger Electron Spectrometry in 9Cr-4Cu steel revealed that Cu atoms segregated at the grain boundaries on the solution treatment at a high temperature, and this leads to the retardation of the grain growth of austenite in the high Cu steel resulting from the decrease in the mobility of grain boundary.

1 Introduction

Upon the application of high chromium martensitic steels to power generating plants, higher creep strength is required for energy saving and decrease of CO_2 emission. In the previous paper, we have tried to improve the creep strength of the 9Cr (mass%) steels by the addition of Cu which precipitates finely within the matrix of lath-martensite [1], and pointed out that the creep rupture time is prolonged with increasing Cu content due to the dislocation pinning effect by the Cu particles [2]. Moreover, another effect of Cu addition was also found in the study; 'grain refining effect'. When the 9Cr-6Cu steel was solution-treated at 1273K and then quenched to room temperature, a very fine martensitic structure could be obtained corresponding to the refining of prior austenite grains [3]. This result suggests that Cu addition to heat resistant martensitic steels is effective for not only strengthening but also toughening in these kinds of steels.
In this study, the grain growth behavior at elevated temperature was investigated in 9Cr-Cu steels, and then the mechanism of grain refining by Cu addition was

discussed in connection with the effect of grain boundary segregation of Cu atoms which occurs during the solution treatment.

2 Experimental Procedure

Fe-9Cr-(1~6)Cu alloys were used in this study. Chemical compositions of the steels are listed in Table 1. Ingots of 1.5kg, which were produced by induction melting in Ar gas atmosphere, were hot-rolled at 1273K, and then homogenized at 1423K for 18ks. The specimens were subjected to solution treatment under various conditions of (1223~1473)K-(0.6~14.4)ks in Ar gas atmosphere, followed by water-quenching to obtain martensite single structure. Microstructures were observed with optical and transmission electron microscopes. The prior austenite grain size of martensitic structure was determined by comparing optical micrographs with the ASTM grain size standards [4]. Grain boundary segregation of Cu atoms was detected by Auger Electron Spectrometry.

Table 1. Chemical compositions of steels used (mass%)

Steel	Cr	Cu	C	N	Mn	Si	P	S	Fe
9Cr-1Cu	8.93	1.03	0.006	0.0090	0.17	0.02	0.006	0.014	bal.
9Cr-2Cu	8.94	2.08	0.006	0.0102	0.16	0.02	0.007	0.013	bal.
9Cr-4Cu	8.99	4.19	0.006	0.0118	0.16	0.02	0.008	0.013	bal.
9Cr-6Cu	9.08	6.34	0.006	0.0143	0.16	0.02	0.008	0.013	bal.

3 Results and Discussion

3.1 Microstructure of as-quenched specimens

Fig. 1 represents optical and transmission electron micrographs (OM and TEM) of 9Cr-1Cu and -4Cu steels which were water-quenched after the solution treatment of 1273K-3.6ks. Both steels have typical lath-martensitic structure characterized by martensite-block and -packet structures (OM), high density of dislocation within laths (TEM). Cu addition gives no influence on the morphology of martensite-laths (TEM) but tends to make the block and packet structures fine (OM). Precipitates of Cu cannot be observed within the laths and even on the prior austenite grain boundaries (TEM). This indicates that all amount of Cu is in solution. Fig. 2 shows relation between Cu content and prior austenite grain size measured from the OM in Fig. 1. Prior austenite grains are markedly refined with increasing Cu content. In particular, it should be noted here that 9Cr-6Cu steel has fine grains of around 10 μm in diameter. It was also confirmed that the tendency of grain refining by Cu addition is found at any other solution treatment temperatures.

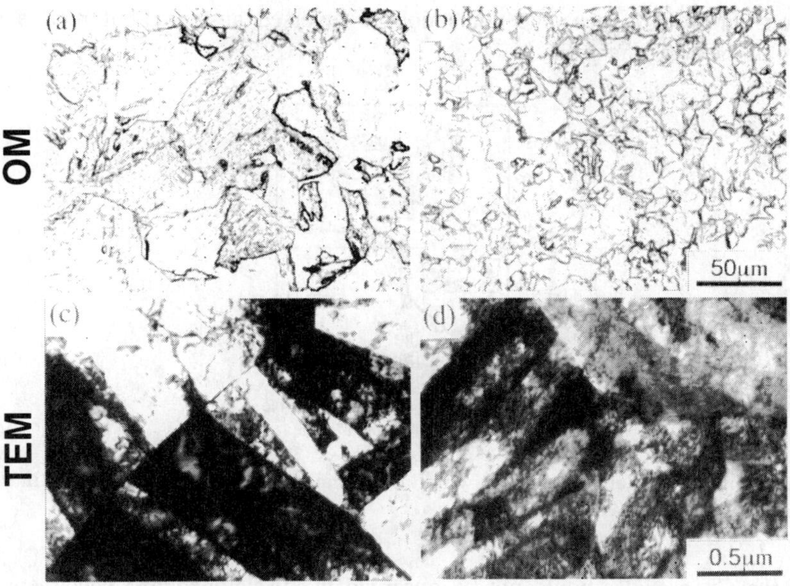

Fig. 1. Optical and transmission electron micrographs of 9Cr-1Cu (a), (c), and –4Cu steels (b), (d). Water-quenched after solution treatment at 1273K for 3.6ks.

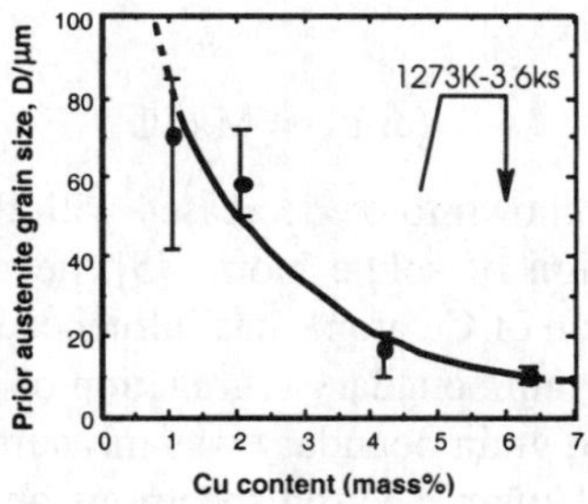

Fig. 2. Relation between prior austenite grain size and Cu content in 9Cr-Cu steels.

3.2 Grain growth behavior of austenite at elevated temperatures

Fig. 3 shows changes in austenite grain size with time in annealing at 1273K and 1423K in 9Cr-1Cu and -4Cu steels. Austenite grain size of each specimen linearly increases to the square root of annealing time at both temperatures. This means that the austenite grains grow based on the normal grain growth mechanism regardless of Cu content. However, the grain growth rate (slope of the line) is significantly different between these steels. The rate of 9Cr-4Cu steel is about four times smaller than that of 9Cr-1Cu steel. On the other hand, the initial grain sizes at time = 0, are almost same in the both specimens. These results suggest that Cu addition hardly influences the nucleation of reversed austenite but retards the growth of the austenite grains.

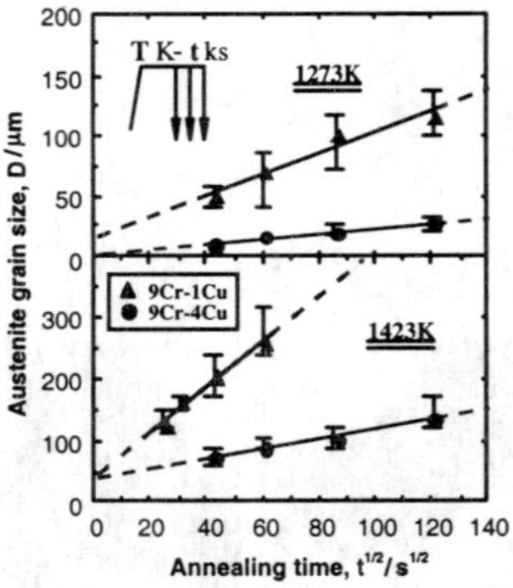

Fig. 3. Changes in austenite grain size with time of annealing at 1273K and 1423K in 9Cr-1Cu and –4Cu steels.

Since the Cu particles could not be observed in the as-quenched specimens as shown in Fig. 1, the retardation of grain growth of austenite should be only attributed to solute Cu. In general, the rate of grain growth; (dr/dt) is given by Eq. 1 as a function of mobility of grain boundary; M and driving force for the grain boundary migration; ΔG.

$$(dr/dt) = M \times \Delta G \tag{1}$$

The grain growth rate is known to be decreased with the reduction of mobility by grain boundary segregation of solute atoms [5], hence it is reasonable that the grain boundary segregation of Cu atoms has influenced the grain growth behavior. In order to examine the grain boundary segregation of Cu atoms in the specimens, the concentration of Cu at grain boundary was measured by Auger Electron Spectrometry. Fig. 4 displays Auger electron spectrums obtained from the transgranular (A) and intergranular fracture surface (B) in the 9Cr-4Cu steel which was fractured at 173K. This result demonstrates that Cu atoms segregate at the prior austenite grain boundaries. The concentration of the segregated Cu reaches 50at% or more after the solution treatment of 1273K-50.4ks. It is easy to image that such a large amount of segregated Cu retards the grain growth of austenite in the high Cu steels.

3.3 Mechanism for retardation of austenite grain growth caused by grain boundary segregation of Cu atoms

Integration of Eq. 1 gives the relation between grain size; D (=2r) and annealing time; t in a single phase structure as:

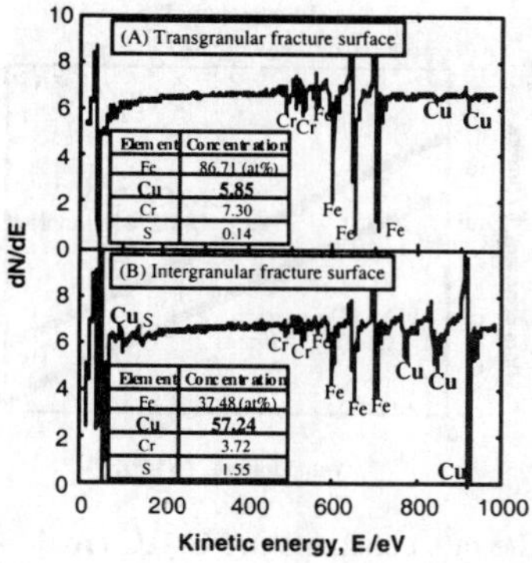

Fig. 4. Auger electron spectrums obtained for the transgranular (A) and intergranular fracture surface (B) in 9Cr-4Cu steel which was annealed at 1273K for 50.4ks after the pre-annealing at 1523K for 7.2ks (austenite grains were intentionally enlarged by this heat treatment to make the grain boundary fracture easy), and then water-quenched to room temperature. The specimen was fractured at 173K to obtain brittle fracture surface.

$$D^2 - Do^2 = K\,t \qquad (2)$$

where Do is the initial grain size, and K is the rate constant which depends on the mobility; M and grain boundary energy; σ ($K \propto M\sigma$). The mobility; M should be proportional to the diffusion coefficient of atoms which are required to diffuse for the grain boundary migration, thus the rate constant; K is expressed by the following equation,

$$K = A\exp(-Q/RT) \qquad (3)$$

where Q is the activation energy for grain boundary migration, T is the absolute temperature, R is the gas constant and A is a constant. Fig. 5 shows the arrhenius plot for the rate constants in 9Cr-1Cu and -4Cu steels. The rate constant of 9Cr-4Cu steel is 10~15 times smaller than that of 9Cr-1Cu steel at any temperatures. The activation energies (the slope of straight lines in this figure) are 274 and 277kJ/mol for the 9Cr-1Cu and -4Cu steels, respectively. These values approximately coincide with the activation energy of lattice diffusion of Cu atom in austenite (280kJ/mol [6]). This indicates that the lattice diffusion of segregated Cu atoms controls the rate of grain boundary migration. The exact mechanism of the grain boundary migration in Cu bearing steels has not been clear, but if it is assumed that the segregated Cu atoms have to keep up with the migrating grain boundary so as not to break down the equilibrium segregation state, the rate-controlling process of the grain growth should be the lattice diffusion of Cu atoms in austenite as indicated in the result of the arrhenius plot.

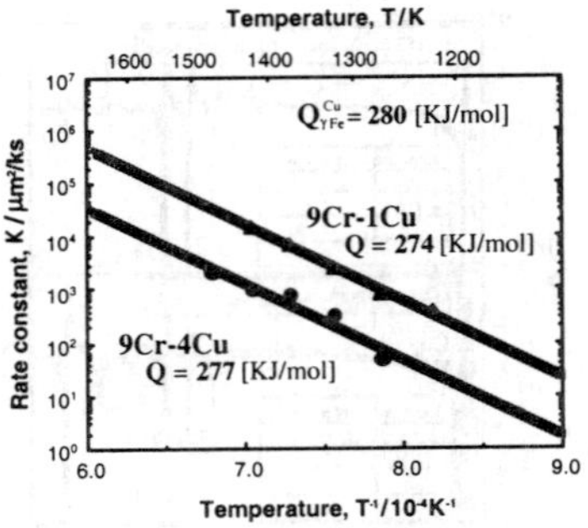

Fig. 5. Arrhenius plots for the rate constants in 9Cr-1Cu and –4Cu steels.

4 Conclusions

The effects of Cu addition on grain growth behavior of austenite and the grain refining mechanism were investigated in 9Cr-(1~ 6)Cu (mass%) martensitic steels, and the following findings were obtained:

(1) Prior austenite grains are markedly refined with increasing Cu content. For instance, the as-quenched 9Cr-6Cu steel had fine grains of around 10 μm in diameter.

(2) Cu atoms segregate at the austenite grain boundaries during the solution treatment at elevated temperatures in high Cu steels. Such a segregation of Cu effectively retards the grain growth of austenite due to the decrease in mobility of grain boundary.

(3) It was suggested that the segregated Cu atoms have to keep up with the migration of grain boundary with lattice diffusion and this process controls the rate of grain growth of austenite.

References

[1] N. Maruyama, M. Sugiyama, T. Hara and H. Tamehiro, *Mater. Trans. JIM*, 40, 1999, pp. 268-277.

[2] Y. Futamura, T. Tsuchiyama and S. Takaki, *ISIJ Int.*, 41, 2001, Supplement, pp. S106-S110.

[3] Y. Futamura, T. Tsuchiyama and S. Takaki, *Tetsu-to-Hagané*, 85, 1999, pp. 697-702.

[4] "Metals Handbook", 7, 8th ed. American Society for Metals, Materials Park, 1972.

[5] K. Lücke and K. Detert, *Acta Metall.*, 5, 1957, pp. 628-637.

[6] "Diffusion Date", ed. Diffusion Information Center, Solothurn, Switzerland and Cleaveland, Ohio, U.S.A.

Recrystallization and Grain Growth
Proceedings of the First Joint International Conference
Eds. G. Gottstein and D.A. Molodov, Springer-Verlag, 2001

Al-Ti Combination Effect on Austenite Grain Growing Control at Reheating Temperatures

S.F. Medina[1], M. Chapa[1], V. López[1] and B. Fernández[1]

[1] National Center for Metallurgical Research, CENIM-CSIC, D-28040 Madrid, Spain

Keywords: reheating temperature, austenite grain size, precipitate distribution.

Abstract. On the basis of steel 38MnSiVS5 a high number of castings were manufactured varying only the Ti and Al contents and obtaining ingots in which the Ti/Al ratio was different in each case. The specimens will be heated in air furnace to different austenitization temperatures, between 900ºC and 1200ºC, and subsequently quenched in water. The most important achievement would be that the Al plays a harmful role as austenite grain controller at high temperatures (>1050ºC), while at the same time obstructing the controlling action of Ti. The best austenite grain control was shown by steel with contents of Ti=0.044, Al=0.009, N=0.0131 (mass %) and a Ti/N ratio of 3.36. But the interpolation of the results allows it to be concluded that the best Ti/N ratio would be close to 2.5.

1 Introduction

Much work [1-4] has been carried out on austenite grain control by TiN particles, and though not all the results are in agreement, it can be said that in most steels which contain Ti the percentage of this element is relatively low, of the order of 0.020 (mass%), with N content which make the Ti/N ratio lower than the stoichiometric ratio of nitrides (3.42).

The low solubility of TiN in austenite makes Ti and N the ideal elements for controlling the austenite grain in processes where the austenite undergoes reheating at high temperatures [5]. Nevertheless, the interaction Ti-Al, both elements controlling the austenite grain size, has been rarely studied and the results found in this work are something different to others reported [6].

The requirements for a good grain refinement additive can be deduced of Gladman [7] and Wagner [8] expressions, respectively, and are: the particles should show low solubility and the equilibrium solute content in the austenite should be low in order to limit the diffusion flow of solute to growing particles.

The main aim of this work has been to study the influence of Ti and Al on austenite grain control at high temperatures, especially at the temperatures of thermomechanical treatments are carried out and finally to find an optimum Ti/N/Al ratio which improves the Ti/N above mentioned.

2 Experimental Procedure

The steel selected to study the control of the austenite grain was a steel 38MnSiVS5. This steel is known for its use in the manufacturing of automobile components, especially crankshafts. With this composition, manufacturers and users consider that the average austenite grain size may be controlled at least to an acceptable degree. However, abnormal growth of the austenite grain size is frequently seen, even when the austenitization temperature is relatively low.
In order to study in greater depth the control of the average grain size and especially abnormal grain growth, a series of chemical compositions were designed with regard to the elements known as grain controllers: Al, Ti. Table 1 shows the chemical composition of the steels manufactured, it being seen that Al and Ti vary in each one. The steels have been manufactured by Electroslag Remelting (ESR).

Table 1. Chemical composition (mass%) of the steels manufactured.

Steel	C	Si	Mn	S	P	V	Ti	Al	N
M1	0.33	0.82	1.53	0.050	0.015	0.115	0.07	0.011	0.0144
M2	0.32	0.79	1.40	0.056	0.016	0.097	0.03	0.013	0.0143
M5	0.34	0.6	1.35	0.060	0.018	0.092	0.044	0.009	0.0131
M6	0.34	0.58	1.41	0.016	0.014	0.091	0.005	0.008	0.0134
M7	0.32	0.83	1.38	0.019	0.017	0.100	0.01	0.037	0.0133
M8	0.35	0.79	1.40	0.020	0.017	0.100	0.013	0.031	0.0135
M9	0.33	0.78	1.43	0.042	0.016	0.100	0.033	0.036	0.0143
M10	0.33	0.81	1.40	0.042	0.017	0.100	0.044	0.029	0.0145
M11	0.34	0.75	1.38	0.042	0.014	0.098	0.024	0.028	0.0140
M12	0.34	0.78	1.33	0.042	0.020	0.098	0.055	0.02	0.0146

The specimens will be heated in air furnace to different austenitization temperatures, between 900ºC and 1200ºC, with holding time of 75 min, and subsequently quenched in water. Analysis of the microstructure will make it possible to study the evolution of the austenite grain (average size and abnormal growth) as a function of the temperature. The specimens were prepared for metallographic observation. The study of precipitates (TiN) was carried out using SEM and TEM.

3 Results and Discussion

3.1 Evolution of the austenite grain

The austenite grain size at different temperatures was determined for all the steels in Table 1, except for steel M6 due to its low Ti and Al contents. All the steels showed a duplex microstructure (Fig. 1) after a certain temperature, except for steel M1, which for the majority of the steels was 1000°C. The presence of the duplex microstructure makes it necessary to simultaneously estimate the finer grain and coarser grain fractions, respectively. This is the reason why the austenite grain size has been evaluated taking into account the different grain fractions. However, in order to facilitate the comparison between the different steels, the weighted mean size has been determined. As an example, Fig. 2 shows the evolution of the austenite grain size versus temperature for the steel M12. The final comparison between the weighted mean values is shown in Fig. 3, it being seen that steel M5 presents the best behaviour, followed by steel M2.

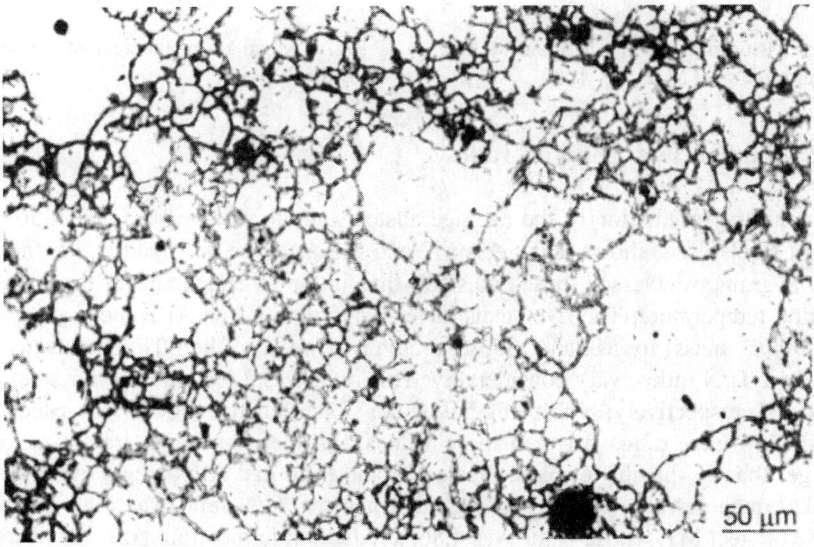

Fig. 1. Optical micrograph of steel M7. Austenitizing temperature: 1000°Cx75 min.

3.2 Austenite grain size versus time

With the intention of checking the "solidity" (or consistency) of the austenite grain, the grain size versus time was determined for steels M5, M8 and M10 at the temperatures of 900°C and 1100°C, respectively. These three steels were selected as the best representatives of three compositional variants: M5 as a high Ti content and low Al content steel; M8 as a low Ti content and high Al content steel; M10 as a high Ti content and high Al content steel.

In general terms, steel M5 is once again seen to be the steel with the finest structure, though steel M8 presented a somewhat better control when the temperature was 1100ºC, possibly because the effects of the high Al content and the presence of AlN still seem to have an influence at this temperature.

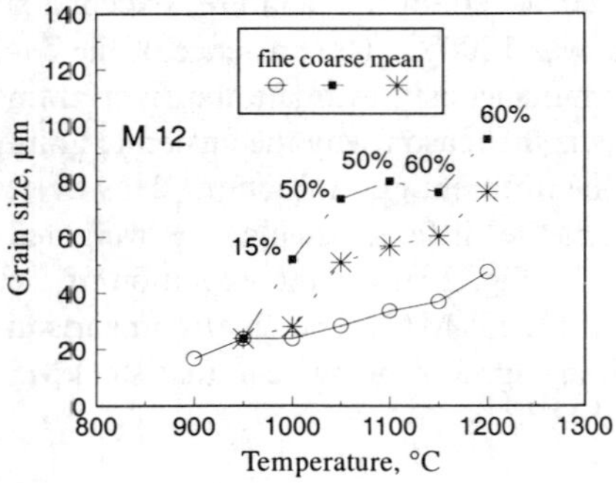

Fig. 2. Austenite grain size against temperature, for steel M12.

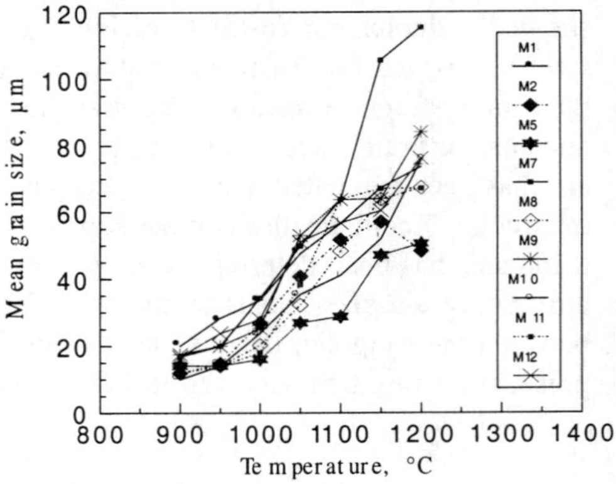

Fig. 3. Mean grain size against temperature in steels used.

3.3 Optimun Ti/Al and Ti/N ratios

Graphic representation of the average austenite grain size versus Ti/N ratio (Fig. 4) of steels used, shows that steels M2 and M5 gave the best results, i.e. the austenite grain grew less in these two steels than in the others, whatever the austenitization temperature. The two steels have in common a low Al content, 0.013 and 0.009 (%, mass) for M2 and M5, respectively. However, their Ti contents, as well as their Ti/N ratios, vary considerably. This ratio was 2.09 and 3.36 for steels M2 and M5, respectively; both being below the stoichiometric ratio (3.42). Steels M9 and M10 have ratios between these values, but their austenite grain sizes were larger than in the above cases. Steels M9 and M10 had Al contents of 0.036 and 0.029, respectively. On the other hand, when the Ti/N ratio was high, as in the case of steel M1, whose value was 4.86, a reduction in the grain size was observed compared with its immediate predecessor in the order of Ti/N ratios, steel M12 (Ti/N=3.77). While steel M12 had an Al content of 0.020 (%, mass), the value corresponding to steel M1 was 0.011 (%, mass).

In other words, at low temperatures (≤950ºC) the grain size can be considered to be independent of the Ti/N ratio and also of the Al content. When the temperature of 1000ºC is reached the influence of the Ti/N ratio starts to be noted, as does the influence of Al, it being observed that steels M7 and M1, which have the lowest and highest Ti/N ratios, respectively, below and above the stoichiometric ratio, give the largest grain sizes. Between these two steels, and observing the line corresponding to 1000ºC, it is also possible to start seeing the formation of peaks, or relative maximums, that correspond to the steels with the highest Al contents (M9, M10, M12). As the temperature rises, the formation of peaks becomes accentuated, corresponding to the same steels. Between steels M7 and M1 the formation of relative minimums is seen, corresponding to steels M2 and M5,

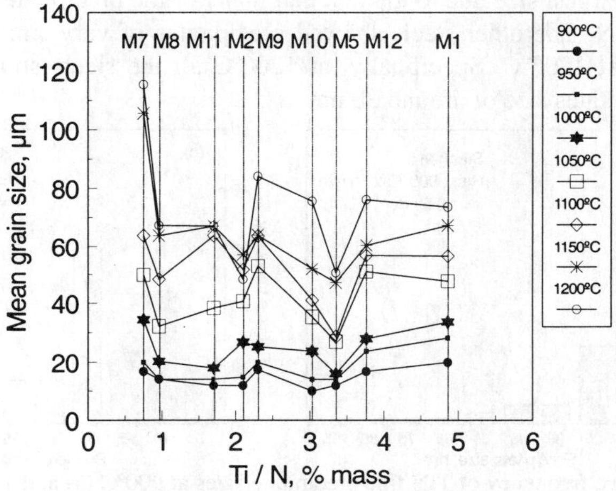

which are precisely those with the lowest Al contents.

Fig. 4. Mean grain size against Ti/N ratio for steels used.

These results indicate that Al has been a harmful element in austenite grain control at high temperatures.

3.4 Precipitates

Regarding to **coarse precipitates**, large TiN precipitates were first seen by light microscopy. It was observed that the relative abundance of precipitates and their size at a particular temperature is shown to be a function of the Ti content.

A second study of precipitates was carried out by SEM, determining compositional and stoichiometric spectra for the precipitates. The precipitates observed were always of TiN with a stoichiometric ratio close to 1/l. In high Al content steels, such as steel M10, some precipitates showed a core containing Al and occasionally also Ca, which leads us to suspect that these precipitates nucleated on particles of alumina or calcium-aluminates.

By means TEM the **fine precipitates** of steels M5, M8, M10 and M11 were analysed. Fine precipitates were considered to be those with a precipitates size of less than 100 nm. Steels M8, M10 and M11 showed AlN precipitates at 900ºC, espe-

cially in the cases of steels M8 and M11. AlN precipitates were normally larger than TiN precipitates.
In steels M8, M10 y M11 the count of precipitates and the measurement of their sizes was carried out on the TiN precipitates present. At 900°C, the results were favourable for steel M5, though at the temperature of 1200°C the sizes were similar. As an example, Fig. 5 shows the distribution of fine precipitates at 900°C (a) and 1200°C (b) for the steel M5, respectively. At 900°C steel M8 showed the precipitates of largest size due to the fact that many of the precipitates measured were in fact TiAlN. The other steels showed precipitates of very similar sizes at both 900°C and at 1200°C. Specifically, at 1200°C all the steels showed a weighted mean precipitates size of around 30 nm.

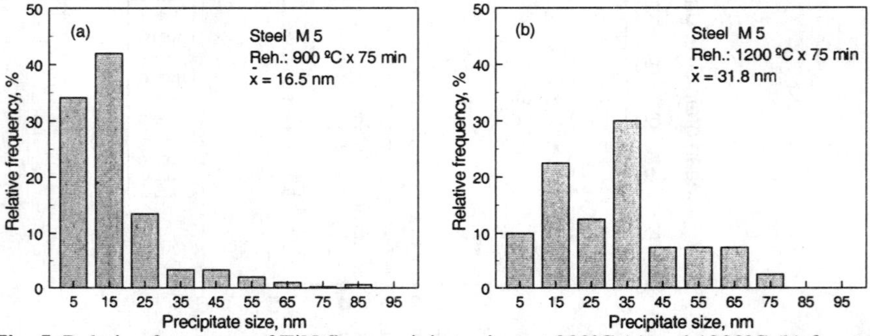

Fig. 5. Relative frequency of TiN fine precipitate sizes at 900°C (a) and 1200°C (b) for steel M5.

The above demonstrates that there is not an optimum Ti/N ratio but an interval, possibly between 2 and 3.4, which would give a very similar average precipitate size, at least up to 1200°C. The study and determination of the distribution of precipitate sizes leads to the conclusion that AlN particles are not harmful only due to their greater size than TiN particles, but especially due to fact that their quick dissolution at temperatures below 1100°C causes a notable drop in pinning forces, leading to a considerable expansion of the grain boundaries where these precipitates are found.

References

[1] S.F. Medina and J.E. Mancilla, *Scripta Metall. Mater.*, 30, 1994, pp. 73-78.
[2] H. Baumgart and C. Strassburger, *Steel Res.*, 56, 1985, pp. 645-650.
[3] R.I. Bodnar, *Iron Steelmaker*, 21, 1994, pp. 19-24.
[4] J.R. Paules, *Iron Steelmaker*, 18, 1991, pp. 41-44.
[5] J. Kunze, B. Beyer, S. Oswald, W. Gruner, *Steel Res.*, 66, 1995, pp. 161-166.
[6] P. Reynolds, *Ironmaking Steelmaking*, 18, 1991, pp. 52-58.
[7] T. Gladman, *Ironmaking Steelmaking*, 16, 1989, pp. 241-245.
[8] C. Wagner, *Z. Elektrochem.*, 65, 1961, pp. 581-591.

Comparison between Pinned and Non-Pinned Initial Grain Growth in a Two-Dimensional Soap Froth

C. H. Wörner[1], A. Olguín[1], M. Ortíz[1*] and O. Herrera[2]

[1] Instituto de Física, Universidad Católica de Valparaíso, Casilla 4059, Valparaíso 02, Chile
[2] Departamento de Física y Meteorología, Universidad de Valparaíso, Casilla 5030, Valparaíso 01, Chile
[*] Present address: Departamento de Física, Universidad de Playa Ancha, Casilla 34V, Valparaíso 01, Chile

Keywords: Zener pinning, grain growth, foams.

Abstract. Kinetic data concerning two-dimensional soap froth growing in a free environment appears in the literature. Recent studies on Zener pinned grain growth shows different kinetics. It is the main purpose of this note to experimentally study their similitudes and differences in the initial states of development. Our outcomes found that - analogous to free growth – the pinned foam present a initial ordered state and later on a stationary (disordered) growing period. Finally, the system reaches the stagnated state. We discuss our experimental results in terms of the current theories for grain growth and pinning.

1 Introduction

One of the well-known mechanisms to control grain size in polycrystalline (mainly metals and ceramics) materials is the introduction of some amount of dispersed second phase precipitates. These particles inhibit grain boundary growth and this action is commonly known as Zener pinning or Zener drag [1]. For a review of its principles and applications the paper of Wörner and Hazzledine [2] can be consulted. The analysis of this phenomenon is a subject of contemporary interest as shown by recent papers on this matter [3-5]. It is possible to analyze the phenomenology of this subject by studying theoretical models, simulation approaches and direct experimental data, or using the analogy between soap froth coarsening behavior and grain growth. The present work will use this last experimental approach. Several papers validate this analogy (see i.e., ref. 6 and 7), and some others particularly treat the case of pinned growth [8-11]. The data we will inform in this note correspond to the growing of two-dimensional foams within anchored rounded pins.

As universally accepted, free grain growth follows -in the scaled state- a so-called "parabolic law" [12]:

$$\frac{dR}{dt} = \frac{A}{R} \tag{1}$$

In an attempt to improve this kinetic equation for inhibited grain growth, Burke [13] introduced in this equation a braking term proportional to $1/R_m$. In these formulae, R(t) stands for the mean grain radius, R_m is the final mean radius and A is a constant involving geometrical factors and physical parameters (mobility and grain boundary energy). On the other side, by modify the equation for free grain growth considering two types of grain behavior, the following equation has been proposed [14]:

$$\frac{dR}{dt} = A\left(\frac{1}{R} - \frac{R}{R_m}\right). \tag{2}$$

This last equation was first proposed by Michels et al. [15], to rule grain boundary inhibited growth by solute drag.
It is the purpose of this note to examine the validity of these models for the initial stages of grain growth, both for the inhibited and free grain growth.

2 Experimental

Two-dimensional soap froths were allowed to grow between two glass plates 20 x 20 cm^2, as described elsewhere [9]. By taking photographic snapshots, their kinetic behavior was followed. After that, the images were processed so obtaining their relevant kinetic parameters. We report here two experimental runs. One of the samples grows freely and the other one in a pinned environment. Both samples started simultaneously in order to avoid artifacts due to different ambient conditions. We focus in the initial stages of the foam growth. Typically, the system reaches the stagnated state in a couple of months say, 2500 hours [9], but we report here results only until 110 h, far away from the stagnated state. Our rounded pins have 1.5 mm in diameter and cover a surface fraction f=0.022. The obtained results appear in Fig.1, which shows the temporal behavior of the mean area. Our estimated experimental errors are included in the symbols used for the data points.

3 Discussion and Conclusions

As noted by several authors (see i.e. Glazier et al. [16]) the kinetic equation for free grain growth follows the parabolic law (Eq. (1)) only in the so-called scaled state. Before this state it undergoes a transition between the (quasi) ordered state and a disordered one.

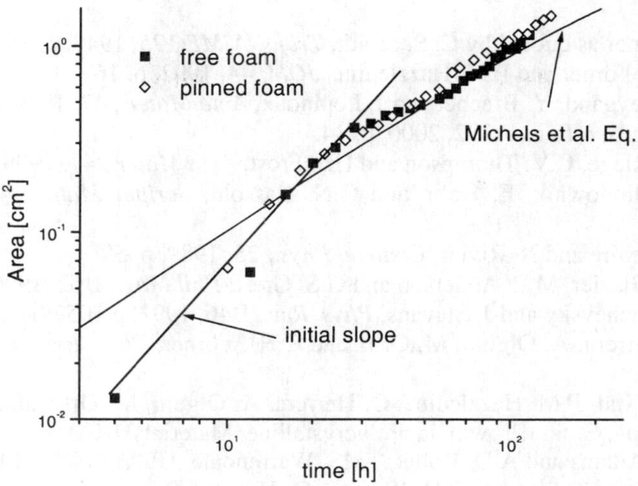

Fig. 1. Kinetic data for pinned and non-pinned growth.

After evolving from the ordered to disordered state, it reaches the scaled state. Pinned growth presents a third regime, the stagnated state. As noted in section 2, we do not present data on this last state.
This note points to answer the question about the existence of these two states in the pinned case. According to the data displayed in Fig. 1, it appears in both cases the two regimes. In fact, in the considered time scale, the are not differences (between the experimental errors) in the behavior of both samples. This fact is consistent with the hypothesis that in the initial stages of the pinned growth, pins do not impede grain growth, that is, the foam grows freely.
On the other side, in the same figure it is displayed the prediction of Eq. (2). As expected, this equation does not fit the data for the whole experimental range of measurements, but fits surprisingly well within the scaled state. The time for transition between these two regimes can be estimated in about 10.5 hours. Also shown in the figure is the initial slope. This datum must be included in the theoretical models in order to fully explain the experimental outcomes.

Acknowledgement

This work was partially financed by CONICYT (Chile) Grant 1990800.

References

[1] C. Zener as quoted by C. S. Smith, *Trans. AIME* 175, 1948, p. 15.
[2] C.H. Wörner and P.M. Hazzledine, *JOM,* 44, 1992, p. 16.
[3] D. Weygand, Y. Brechet and J. Lepinoux, *Acta Mater.*, 47, 1999, p. 961; *Mater. Sci. Eng.*, A292, 2000, p. 34.
[4] S.P. Riege, C.V. Thompson and H.J. Frost, *Acta Mater.*, 47, 1999, p. 1879.
[5] M. Miodownik, E. Holm and G.N. Hassold, *Scripta Mater.*, 42, 2000, p. 1173.
[6] D. Weaire and N. Rivier, *Contem. Phys.,* 25, 1989, p. 59.
[7] J.A. Glazier, M.P. Anderson and G.S. Grest, *Phil.Mag.,* B62, 1990, p. 615.
[8] O. Krichevsky and J. Stavans, *Phys. Rev.,* B46, 1992, p.10579.
[9] O. Herrera, A. Olguín, M. Ortíz and C.H. Wörner, *Phys. Rev.,* E55, 1997, p. 4475.
[10] B.K. Kad, P.M. Hazzledine, O. Herrera, A. Olguín, M. Ortíz and C.H. Wörner, in „Grain Growth in Polycrystalline Materials III“, eds. A. Weiland, B.L. Adams and A.D. Rollet, TMS, Warrendale, USA, 1998, p.147.
[11] A. Olguín, M. Ortíz, C.H. Wörner, O. Herrera, B.K. Kad, P.M. Hazzledine, *Phil. Mag.*, to be published.
[12] H.V. Atkinson, *Acta Mater.,* 36, 1988, p. 469.
[13] J.E. Burke, *Trans. AIME,* 175, 1949, p. 73.
[14] C.H. Wörner, A. Olguín, M. Ortíz and O. Herrera, to be published.
[15] A. Michels, C.E. Krill, H. Erhardt, R. Birringer and D.T. Wu, *Acta Mater.,* 47, 1999, p. 2143.
[16] J.A. Glazier, S.P. Gross and J. Stavans, *Phys. Rev.,* 36, 1987, p. 306.

Simultaneous Nitriding and Decarburization as a Way of Acquiring Grain Growth Inhibition in 3%Si-Fe

C.H. Han, J.S. Woo, K.S. Han, and S.Y. Cha

Technical Research Lab., POSCO, P.O.Box 36, Pohang, Kyoungbuk, Korea

Keywords: nitriding, decarburization, Goss texture, secondary recrystallization.

Abstract. A new nitriding process i.e., simultaneous nitriding and decarburization(SDN) in the same furnace, was experimented. 0.3mm thick specimens containing 3%Si, 0.2%Mn, 0.015%Al, 0.5%Cu were prepared by vacuum induction melting, 1,200°C slab reheating, hot rolling and one stage cold rolling. These specimens were nitrided simultaneously with decarburization in a mixed wet atmosphere of H_2, N_2 and NH_3 and then batch-annealed for secondary recrystallization. After the nitriding treatment, it was found that the specimens contained several hundreds ppm of nitrogen and the nitrining process did not interfere with the decarburization. In addition, the specimens developed stable secondary recrystallization with sharp Goss texture after batch annealing.

1 Introduction

There have been several reports concerning on the formation of grain growth inhibitors in the later stage of the manufacturing process of grain oriented silicon steel (GO). In the early 1980s, Horky and Pacl proposed that by nitriding the decarburized 3%Si-Fe with ammonia either at 450°C or 550°C and further annealing it at 970°C, they were able to form Si_3N_4 precipitates effective in inhibiting normal grain growth and, consequently, obtain stable secondary recrystallization with sharp Goss orientation [1,2]. Chernavtseva et. al. experimented on a possibility of the formation of excess AlN phase in 3%Si-Fe by using an atmosphere of fully dissociated ammonia and demonstrated the formation of AlN by nitriding [3]. Quite recently, Takahashi and Harase announced that they had developed a new method to reduce the slab reheating temperature of high permeability GO [4]. They reported that when the decarburized 3% Si-Fe was nitrided with ammonia, (Si,Mn)N and Si_3N_4 were formed after nitriding and these were dissolved and reprecipitated into fine (Al,Si)N particles in subsequent batch annealing.
In this study, the authers experimented a new nitriding process where 3%Si-Fe was simultaneously decarburized in the same furnace. 0.3mm thick cold-rolled specimens of 3%Si-Fe were prepared by vacuum induction melting, slab reheating, hot rolling and cold rolling. These specimens were nitrided simultaneously

with decarburization in a mixed wet atmosphere of H_2, N_2 and NH_3. In the present work, we will discuss how the nitriding process affect the primary structure and its nitrogen content as well as the precipitate distribution and secondary recrystallization behavior.

2 Experimental Procedure

An ingot containing 3%Si, 0.2%Mn, 0.015%Al and 0.5%Cu was prepared by vacuum induction melting. The ingot was reheated at 1,200°C and hot-rolled to 2.3 mm in thickness. The hot-rolled plate was annealed at 1,100°C followed by cold rolling down to 0.3 mm. The cold-rolled sheets were annealed at 850°C in a mixed wet atmosphere of H_2, N_2 and NH_3 to be nitrided and decarburized simultaneously in the same furnace. For comparisons, the cold-rolled sheets were only decarburized at 850°C in a mixed wet atmosphere of H_2 and N_2.
For secondary recrystallization and removing impurities, the batch annealing of nitrided specimens was carried out as follows; the nitrided specimens were heated up to 1,200 °C in a mixed H_2 and N_2 atmosphere at the rate of 15°C /hr, and held for 10 hours at this temperature in 100% H_2 atmosphere.
To investigate the changes in the primary microstructure and the C and N_2 contents during the SDN treatment, the samples were taken out from the furnace at 650°C, 750°C and 850°C. The specimens annealed at 850°C for 1 minutes were also used for the above mentioned purpose and the observation of precipitates formed during nitriding.
The extraction replica method was used for the observation of precipitates. Extraction replicas were taken at the near surface (t/8) and the center (t/2) layer of each sample. The magnetic induction of the batch annealed specimens was measured with 6x30 cm samples by a single strip tester.

3 Results

The differences between the SDN process and decarburization-only process are investigated by comparing the behavior of primary recrystallization and decarburization. Fig. 1 shows the microstructural changes in the only-decarburized sheet and the SDN treated sheet. When annealed to 750°C, both specimens show primary recrystallized microstructure. As the temperature goes higher and the annealing time gets longer, both microstructures become more homogeneous. After annealing for 1 minutes at 850°C, the microstructures formed by the two processes become almost similar in terms of size and morphology except that the grain growth in the simultaneously decarburized and nitrided specimen is retarded slightly, especially, in the center region.
Fig. 2 is the changes in average grain size with the annealing temperature and time. The grain size is nearly the same at 750°C in both samples. But, when the

temperature is raised to 850°C, the average grain size of the only-decarburized sheet is larger than that of the simultaneously decarburized and nitrided sheet. Even if the annealing time is prolonged to 1 minutes at this temperature, the difference in the average grain size is maintained the same.

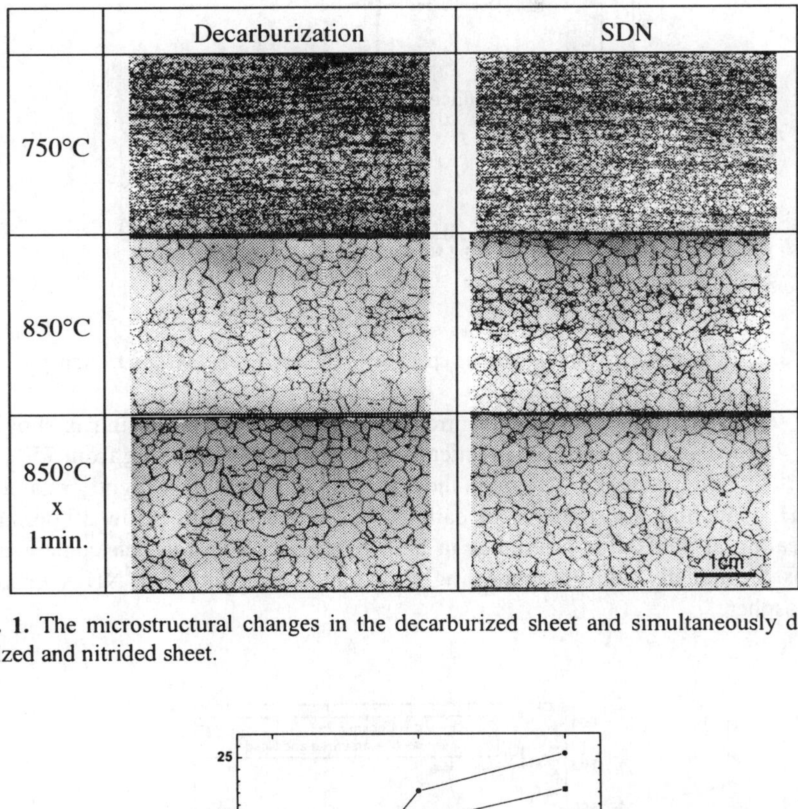

Fig. 1. The microstructural changes in the decarburized sheet and simultaneously decarburized and nitrided sheet.

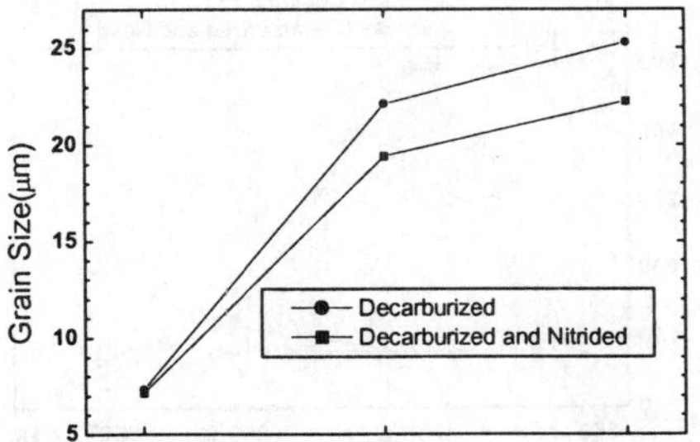

Fig. 2. The average grain size changes with the annealing temperature and time.

Fig. 3 shows the grain size distribution of the specimens annealed at 850°C for 1 minutes. There is no difference between the distributions at the surface region. On the other hand, in the center region, the quantity of relatively small grains is much more in the simultaneously decarburized and nitrided sheet than in the decarburized sheet.

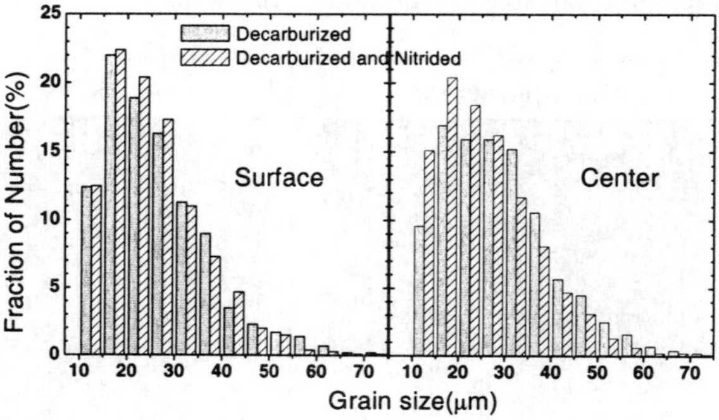

Fig. 3. The grain size distribution of the specimens annealed at 850°C for 1 minutes.

The variation of the carbon and nitrogen contents during annealing is shown in Fig. 4. The carbon is removed drastically at the temprature ranging from 750 °C to 850°C in both processes. Although the elimination of carbon is slightly retarded in a NH_3 containing atmosphere, the carbon content is remained below 30 ppm after annealing at 850°C for 1 minutes in both processes. The nitrogen content is increased gradually with the annealing temperature and time in a NH_3 containing atmosphere.

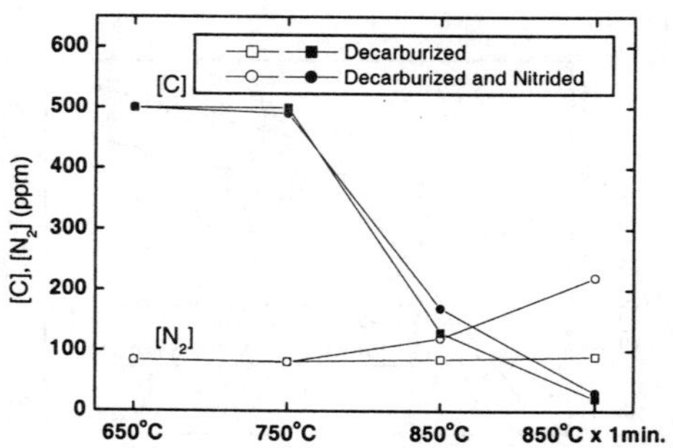

Fig. 4. The variation of the carbon and nitrogen contents during annealing.

Fig. 5 shows the precipitates formed in the simultaneously nitrided and decarburized specimen annealed at 850°C for 1 minutes. The precipitates in the surface region are fine and dispersed uniformly. They are identified as (Al,Si)N and

(Al,Si,Mn)N. Differing from the previous researches, the precipitates without Al such as Si_3N_4 and (Si,Mn)N are not found when the specimen is nitrided at 850°C. Even though the number of precipitates are less in quantity in the center of the specimen, considerably many precipitates identified as (Al,Si)N are observed.
Fig. 6 shows the precipitates in the specimen extracted at 1,000°C during batch annealing.

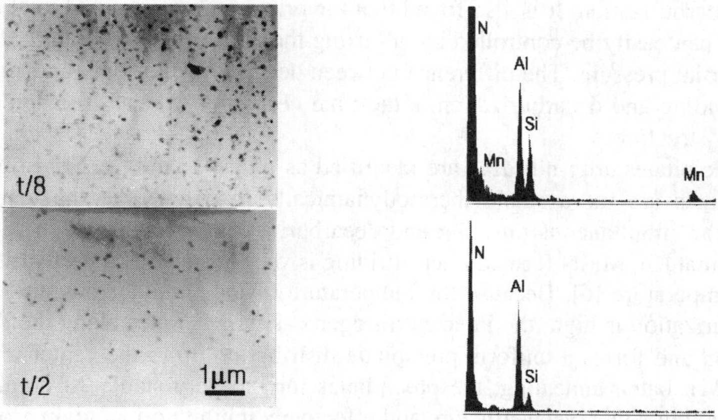

Fig. 5. The precipitates formed in the simultaneously nitrided and decarburized specimen annealed at 850°C for 1 minute.

The precipitates existing around this temperature play an important role in inhibiting normal grain growth and developing secondary recrystallization. The precipitates are identified as (Al,Si)N and (Al,Si,Mn)N as shown in Fig. 6 and their crystal structure are same as AlN crystal. The magnetic induction of the simultaneously decarburized and nitrided samples showed 1.92 Tesla with the magnetizing force of 1,000A/m at 50Hz.

Fig. 6. The precipitates in the specimen extracted at 1,000°C during batch annealing.

4 Discussions

The formation of grain growth inhibition in the later stage of manufacturing process make it possible to lower the slab reheating temperature in producing GO.

Therefore, there have been many efforts to form grain growth inhibitors by nitriding the steel after cold rolling. The authors carried out a series of experiments to develop a new process where the injection of nitrogen and removal of carbon can be possible simultaneously in the same furnace. As a result, it is found that the cold-rolled specimen annealed in a mixed wet atmosphere of H_2, N_2 and NH_3 contains more than 200 ppm of nitrogen and the nitriding process did not interfere with decarburization. It is also found that the primary grain size and total nitrogen content can easily be controlled by adjusting the annealing temperature, time and NH_3 partial pressure. The difference between decarburization-only and simultaneous nitriding and decarburization is the time necessary to reach the homogenous primary structure.

The precipitates after nitriding are identified as (Al,Si)N and (Al,Si,Mn)N. These precipitates are more stable thermodynamically than Si_3N_4 at the temperature where the simultaneous nitriding and decarburization is conducted [5]. It is also known that (Si, Mn)N formed after nitriding is very unstable and easily dissolve at high temperature [6]. Because the temperature of the simultaneous nitriding and decarburization is high, the injected nitrogen easily penetrates along the thickness direction and forms a uniform precipitate distribution up to the center region. At 1000°C in batch annealing, the precipitates turn to very stable Al dominant nitrides having AlN crystal structure and effectively inhibit normal grain growth.

5 Conclusion

A new nitriding process where the injection of nitrogen and removal of carbon is done simultaneously in the same furnace has been developed. In the new process, nitrogen can be injected up to several hundreds ppm without interfering decarburization. The precipitates formed after the nitrogen injection are (Al,Si)N and (Al,Si,Mn)N which are more stable than (Si,Mn)N or Si_3N_4. During batch annealing the precipitates become Al dominant nitrides having AlN crystal structure and, consequently, stable enough to effectively inhibit normal grain growth before secondary recrystallization.

References

[1] P. Horky and P.Pacl, *J. of Mag. and Mag. Mat.*, 41, 1984, pp. 14-16.
[2] P. Horky and P.Pacl, *J. of Mag. and Mag. Mat.*, 46, 1984, pp. 87-90.
[3] G.A. Chernavtseva, L.P. Emelyanenko, B.M. Mogutnov and L.A. Shvartsman, *Russ. Metall.*, 4, 1983, pp. 116-118.
[4] N. Takahashi and J. Harase, *Mat. Sci. Forum*, 204-206, 1996, pp. 143-154.
[5] Private communication with Dr. B.J. Lee
[6] Y. Ushigami et. al, *Mat. Sci. Forum*, 204-206, 1996, pp. 593-598.

Grain Growth and Intergranular Carbide Precipitation in Nickel-Base Alloy 690

Y.S. Lim, S.S. Hwang, J.H. Suh and J.S. Kim

Steam Generator Materials Project, Korea Atomic Energy Research Institute,
P. O. Box 105, Yusong, Taejon 305-600, Korea

Keywords: alloy 690, grain growth, Cr carbides, grain boundary, TEM

Abstract. The grain growth and precipitation behavior of intergranular carbides in Nickel-base alloy 690 were studied through microscopic equipment. The experimental grain growth data showed a classical growth equation, and carbon atoms in solid solution were found to have a strong drag effect on grain boundary migration. The precipitation of intergranular carbides, identified as Cr-rich $M_{23}C_6$, was retarded (or suppressed) on the low angle grain boundaries and the special boundaries expressed by the coincidence site lattice model, compared to that on the random high angle grain boundaries. The results of intergranular $M_{23}C_6$ carbide precipitation are explained in terms of the influence of grain boundary energy.

1 Introduction

Alloy 690 (Ni-30wt%Cr-11wt%Fe) has been used as a new steam generator tubing material in nuclear power plants due to its superior intergranular stress corrosion cracking (IGSCC) resistance to alloy 600 (Ni-16wt%Cr-8wt%Fe) in various corrosive environments [1]. A great deal of research has led to the current understanding that many factors such as compositional, metallurgical and environmental variables affect steam generator tube performance. In particular, changes of grain size and intergranular carbide precipitation in a given alloy were found to cause considerably different results on IGSCC properties under some environmental factors. The material that has a large grain size with fully developed intergranular carbides showed better resistance to IGSCC than the others [2]. Hence, the steam generator tubing materials are specified to have an average grain size above 32 μm [3]. Although no clear explanation exists for the improved IGSCC resistance in caustic and de-aerated neutral solutions, it is generally accepted that precipitation of a high density of intergranular chromium carbides by thermal treatment improves the IGSCC resistance. The intergranular carbides are believed to act as barriers to the crack propagation during IGSCC [4]. Precipitation of intergranular chromium carbides, is known to selectively occur, and to be related to the grain boundary energy [5].

The present work is an attempt to describe the microstructural changes of alloy 690 for a high resistance to IGSCC, i.e., variation of average grain size and intergranular carbide precipitation by thermal treatment. The effects of annealing temperature/time and carbon content on grain growth were investigated, and the behavior of intergranular carbide precipitation was examined by transmission electron microscopy (TEM). The correlation between grain boundary character and precipitation of intergranular carbides were characterized according to the coincidence lattice site (CSL) theory [6].

2 Experimental Procedure

Alloy 690 tubes were obtained in a 74.2 % as-pilgered condition. (hereafter, alloy 690AP). The heat bulk compositions of the major elements are given in Table 1. For the two alloys under consideration (690HC and 690LC in Table 1), carbon which is the major non-metallic element of chromium carbides, is about 0.011 wt% higher in 690HC than in 690LC.

Table I. Chemical compositions of the alloys used (wt%)

Alloys	Ni	Cr	Fe	C	Si	Ti	Al	Co	S	P
600HC	Bal.	29.15	8.19	0.030	0.16	0.16	0.18	0.01	0.001	0.004
600LC	Bal.	29.20	8.68	0.019	0.15	0.23	0.28	0.01	0.001	0.004

TEM specimens were prepared by cutting longitudinal strips from the tubes, grinding to flat slabs approximately 60 μm thick, and finally electropolishing the 3 mm discs. A 7 % perchloric acid + 93 % methanol solution cooled to -40 °C was used, and a current of approximate 50 mA was applied for jet polishing. TEM observation was carried out with a JEOL 2000 FXII (operating voltage 200 kV). Specimens for optical and SEM examination were made by electrolytic etching with a solution of 10 % HNO_3 + 90 % methanol (nital solution), and chemical etching with a solution of 2 % bromine + 98 % methanol. JEOL 5200 (operating voltage 25 kV) was used for scanning electron microscopy (SEM).
The grain orientation analysis was performed by Kikuchi diffraction working in a convergent beam electron diffraction mode, which results in more accurate orientation than the conventional selective area diffraction (SAD) mode. An analytic approach by Young *et al.* [7] was adopted to get the misorientation angle/axis pair. Each misorientation was represented in the form of the smallest-angle description (or, 'disorientation' representation) of the angle/axis pair. In the present study, an analytic method [8] based on matrix algebra was used to experimentally determine the deviation angle ($\Delta\theta_d$) from the exact CSL orientation, Finally, as a guide to set

up a significative upper limit (i.e., maximum allowable deviation angle, $\Delta\theta_m$) for $\Delta\theta_d$, the Brandon formula [9] was used, following that $\Delta\theta_m = 15 \times \Sigma^{-1/2}$.

3 Results and Discussion

3.1 Grain Growth

The microstructures of alloy 690AP, as starting materials, are shown in Fig. 1. In the figures, it can be seen that the grains were elongated along the rolling direction, and chromium carbides were precipitated on grain boundaries during tube fabrication. Since the pre-existing carbides can have an additional effect on grain boundary motion, they were completely dissolved during heat treatment for grain growth.

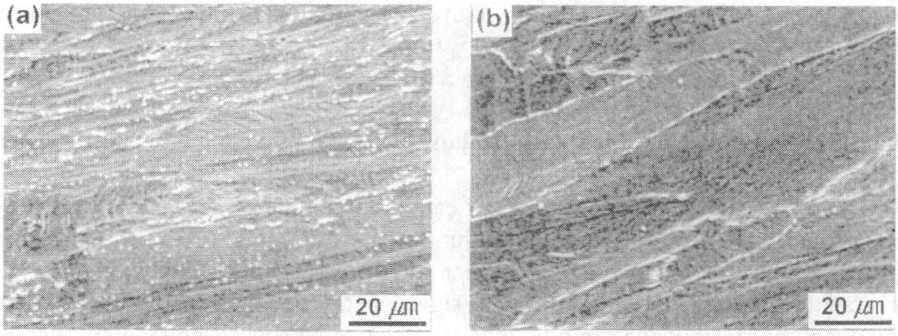

Fig. 1. SEM micrographs of as-pilgered (a) Alloy 690HC and (b) Alloy 690LC

The dissolution temperature of chromium carbides (or carbon solubility temperature) of alloy 690HC was determined as 1107°C, and that of alloy 690LC was 1056°C. It should, therefore, be noticed that even a small increase of carbon content significantly raises the carbon solubility temperature at high temperatures (i.e., temperature difference of 51°C by an increase of 0.011 wt% carbon content in the present study). The heat treatment for grain growth was done above carbon solubility temperature to completely dissolve the pre-existing chromium carbides, and the results are shown in Fig. 2. The average grain sizes in Fig. 2a and 2b were estimated by the mean linear intercept method [10]. Since grain boundary migration is a thermally activated process, its kinetics followed an Arrhenius type temperature dependence in both alloys. The activation enthalpies of grain boundary migration in alloy 690HC and alloy 690LC were calculated as 1.27×10^5 J/mol.K and 1.65×10^5 J/mol.K, respectively. As can be seen readily from the figures, a small increase of carbon content also considerably inhibited grain growth. For example, the average grain size of alloy 690LC was about 42 μm by heat treatment at 1120 °C for 10 min., however, that of Alloy 690HC was only 28 μm under the same heat treatment condition. And, to get an average grain size of 32 μm during annealing time of 10 min., annealing temperature of 1130°C is necessary for alloy 690HC, however, only 1080°C is sufficient for alloy 690LC.

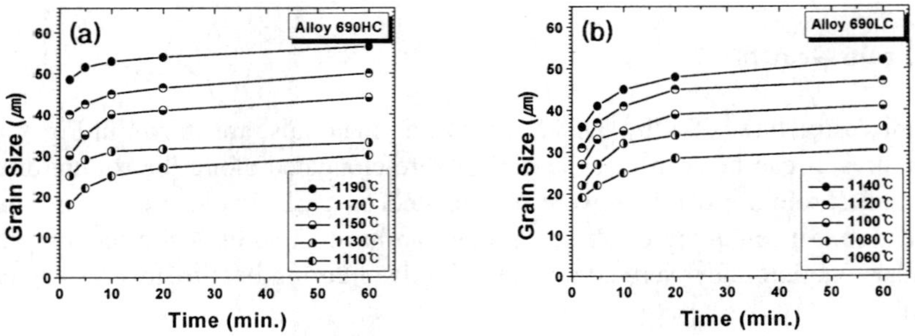

Fig. 2. Variations of average grain sizes of (a) alloy 690HC and (b) alloy 690LC as a function of annealing temperature and time.

3.2 Intergranular Carbide Precipitation

The thermal treatment for intergranular carbide precipitation was done with alloy 690LC at 720°C for 10 hours after solution annealing at 1110°C for 5 minutes, which brought into the average grain size of 35 μm, as shown in Fig. 2b. The intergranular chromium carbides were significantly developed, by the thermal treatment, on most of the grain boundary except some special boundaries such as coherent twin boundaries. Fig. 3a is a TEM bright field image, and Fig. 3b is the related SAD pattern from which the intergranular chromium carbides were identified as chromium-rich (Cr-rich) $M_{23}C_6$. It is well known that Cr-rich $M_{23}C_6$ has a cube-cube orientation relationship such as $\{100\}_\gamma//\{100\}_{M23C6}$, $<100>_\gamma//<100>_{M23C6}$ with one grain, as shown in Fig. 3b.

Fig. 4 shows some typical morphologies of intergranular carbides precipitated on different grain boundaries. Fine and faceted carbides were densely distributed on the low angle grain boundaries (Fig. 4a), from which it is believed that the carbide morphology corresponds to the 'early stage of growth' after nucleation. Coarse and irregular carbides were discretely distributed on the random high angle (or, general) grain boundaries (Fig. 4b). In the figure, some faceted carbides appeared to be combined with each other, to form a large irregular shape. Therefore, the time development of intergranular carbide precipitation on this boundary corresponds to the "stage of concurrent coalescence and growth". On the other hand, carbides were never found on the coherent twin boundaries in Fig. 4c (i.e., $\Sigma 3$ special boundaries). The grain boundary in Fig. 4d was analyzed as a $\Sigma 15$ special boundary, and tiny carbides were densely distributed, similar to the case of low angle grain boundary (Fig. 4a).

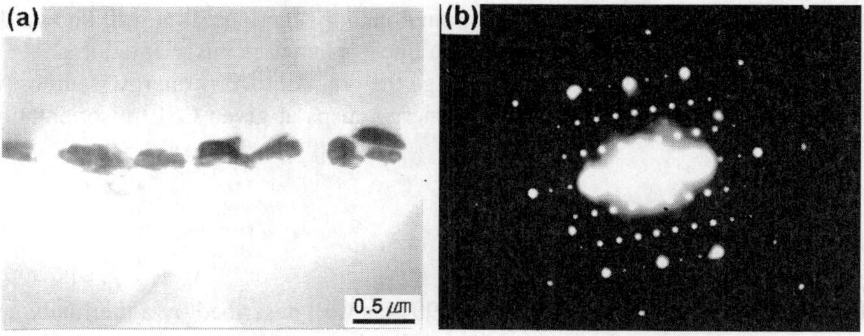

Fig. 3. (a) Grain boundary Cr-rich $M_{23}C_6$ carbides, and (b) the related SADP which shows the cube-cube orientation relationship with one grain.

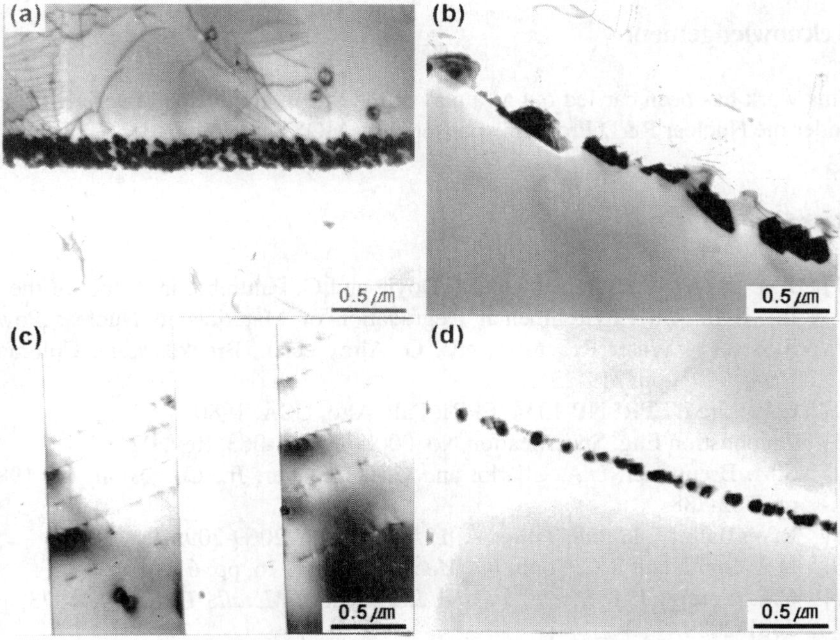

Fig. 4. TEM bright field images showing carbide morphologies on the (a) low angle boundary, (b) random high angle boundary, (c) coherent twin boundary, and (d) Σ15 special boundary.

The above results can be understood, in part, in terms of the influence of grain boundary energy. The extremely low energy of coherent twin boundary in a nickel-base alloy [11] could be a reason for the absence of intergranular carbides on the boundary. Similarly, their suppression on the low angle and CSL boundaries could also be attributed to their grain boundary energies. It is well known that the grain boundary energy increases in a linear manner as misorientation angle increases up to 10 – 15°, and after this stage the grain boundary energy is almost independent of misorientation, with the energy cusps at given CSL misorientations [5].

4 Conclusions

The kinetics of grain growth in Alloy 690 was well described by a thermally activated process, and carbon atoms in solid solution had a strong drag effect on boundary migration, and therefore, suppressed grain growth. Precipitation of intergranular chromium carbides is retarded on low angle grain boundaries and CSL boundaries, in comparison with that on random high angle grain boundaries.

Acknowledgement

This work has been carried out as a part of the Steam Generator Materials Project under the Nuclear R&D Program sponsored by MOST in Korea.

Reference

[1] J.M. Sarver, P.E. Doherty, D.M. Doyle and G. Palumbo, in "Proc. of the 7th Int. Sym. On Environmental Degradation of Materials in Nuclear Power Systems – Water Reactors", eds. G. Airey et al., Breckenridge, Colorado, USA, 1995, pp. 219-231.

[2] G.P. Airey, EPRI NP-1354, EPRI, Palo Alto, USA, 1980.

[3] Combustion Eng. Specification No. 00000-MCM-063, Rev. 02.

[4] S.M. Breummer, L.A. Charlot and C.H. Henager, Jr., *Corrosion*, 44, 1988, pp. 782-788.

[5] R.W. Balluffi, *Metall. Trans*. A, 13A, 1982, pp. 2069-2095.

[6] D.A. Smith and R.C. Pond, *Int. Met. Rev*., 21, 1976, pp. 61-74.

[7] C.T. Young, J.H. Steele, Jr. and J.L. Lytton, *Metall. Trans*., 4, 1973, pp. 2081-2089.

[8] G.L. Bleris, J.G. Antonopoulos, Th. Karakostas and P. Delavignette, *Phys. Stat. Sol. (a)*, 67, 1981, pp. 249-257.

[9] D.G. Brandon, *Acta Metall*., 14, 1966, pp. 1479-1484.

[10] ASTM E112-88, "Standard Method for Determining Average Grain Size",Vol. 03.01, ASTM, Philadelphia, USA, 1992, pp. 294-319.

[11] L.E. Murr, "Interfacial Phenomena in Metals and Alloys", Addison-Wesley, Mass., USA, 1975, pp. 138-142.

On the Effect of Particle Volume Fraction on Grain Growth Inhibition

V.Yu. Novikov

Treptower Str. 74d, D-22147 Hamburg, Germany

Keywords: grain growth, inhibition, theory, modelling.

Abstract. An analysis of the known experimental data on the reduced limiting grain size shows that it can be described by a f^{-1} dependence up to the particle volume fraction f = 0.12 if the particles are distributed randomly. To find a microstructural parameter which can be used for characterization of the grain growth stagnation, computer modelling of the growth process in polycrystals with randomly distributed particles is undertaken, the size and volume fraction of the particles being assumed independent of time. It has been shown that the most probable grain size is the parameter best conforming with Zener theory.

1 Theory vs. Experimental Data

In materials with randomly distributed disperse particles the mean grain size should reach eventually the limiting value, L, connected with the volume fraction, f, and the mean size of the particles, *d*, as follows:

$$L/d = af^n \tag{1}$$

where n = −1 and *a* is a constant [1]. No restrictions on the magnitudes of d and f are imposed in the theory. According to the recent analysis [2,3] of the experimental data this dependence persists only up to $f \cong 0.01$, and at $f > 0.01$ n= −1/3 which is contrary to [1].

A thorough inspection of the experimental data discussed in [2,3] shows that some of the data relate to materials with the majority of particles at grain boundaries. These data are incomparable with the theory [1] but only those which relate to materials with randomly distributed particles [4-6]. Since the limiting grain size was found in [5] only, we were compelled to estimate this size from [4,6] approximately. The results of our analysis presented in Fig. 1 (curve 1) can be described by Eq. 1 with n = −0.95 at $0.00016 \leq f \leq 0.12$ [7]. Taking into account a wide scatter of the experimental results and the fact that they are averaged and approximated, it may be concluded that it is in a satisfactory agreement with the f^{-1} dependence predicted in [1]. It may be supposed that n = −1/3 found in [2,3] relates to the case of particles at grain boundaries [8].

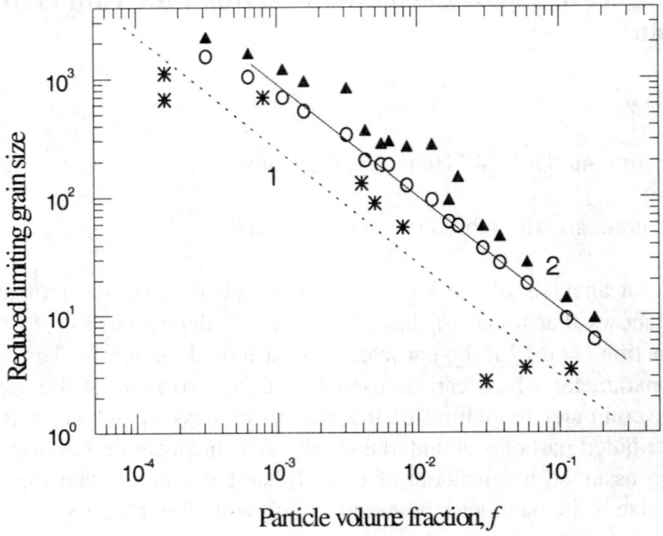

Fig. 1. (1) Reduced limiting grain size vs. f estimated from the experimental data [4-6] (stars), and (2) $\langle D \rangle_{lim}/d$ (triangles) and M_{lim}/d (circles) found by modelling.

A very small number of the experimental data and their wide scatter (see Fig. 1) make it necessary to investigate in more detail the effect of disperse particles on grain growth stagnation. The majority of the experimental data on grain size are obtained by the random intercept method which yields a volume-averaged quantity [9] dependent strongly on grain size non-homogeneity typical of grain growth stagnation. As has recently been shown this quantity cannot be used for the description of the grain growth kinetics even in a well homogeneous microstructure [10]. So a more reliable grain size characteristic should be found for analysing grain growth stagnation. This has been done by computer modelling.

2 Modelling of Grain Growth Stagnation

2.1 Model description

The growth suppression by disperse particles was studied by means of computer simulation using a statistical model of grain growth [11]. It is assumed in the model that all the grain boundaries are of equal and constant energy and mobility. The disperse particles are supposed to be of a spherical shape, an identical size, and a random spatial arrangement. The energy of the interfaces particle/grain is assumed equal to the grain boundary energy. All these assumptions regarding the particles correspond to [1]. The drag force exerted by the particles upon the moving grain boundary can be described as:

$$F_Z = c\gamma f/d \tag{2}$$

where c is a constant and γ the grain boundary energy. F_Z is assumed equal for all grain boundaries. No suppositions about the shape of the grain boundary in the vicinity of the adjacent particles are made.

The grain growth process was described as follows. According to [11], grains of different sizes are arranged randomly, and, hence, the number of the adjacent grains of a given size is proportional to their total number fraction in the grain ensemble at a given point in time. A crystallite of size D_i can consume the adjacent grains of size D_j provided the driving force for growth of the former is positive. The net driving force, Ψ_{ij}, for the growth of a crystallite of size D_i is:

$$\Psi_{ij} = c\gamma\,(1/D_j - 1/D_i - F_Z) \tag{3}$$

and, the displacement of the boundary between these grains is:

$$\Delta D_{ij} = m\Psi_{ij}\Delta t \tag{4}$$

where Δt is a unit time step. The volume increment, ΔV_{ij}, of a crystallite of size D_i due to the consumption of an adjacent grain of size D_j is:

$$\Delta V_{ij} = s_{ij}\,\Delta D_{ij}\,n_i\,n_j \tag{5}$$

(s_{ij} is the contact area of these grains, and n_i and n_j their number fractions in the grain ensemble at a given point in time), and the volume decrement, ΔV_{ji}, of the consumed grains of size D_j is equal to ΔV_{ij}. The number of grains, ΔN_i and ΔN_j, transferring into the size classes i+1 and j−1, respectively, is derived from the volume increment of the growing crystallites and from the volume decrements of the consumed grains of various sizes. Owing to this, the total volume of the grain ensemble is maintained constant with an accuracy better than 0.1%. Alterations in the total grain size distribution with time are modelled by applying this procedure subsequently to grains of different sizes, D_i, in the range from D_m to D_2 and $D_j \leq 1/(1/D_i + F_Z)$ where D_m is the maximum grain diameter.

The growth kinetics was modelled on 3-D polycrystals with an initial grain size distribution close to log-normal. Microstructure evolution during grain growth was traced by the variation of the grain size distribution, time being expressed in arbitrary units equal to 60 unit time steps.

2.2 Limiting grain size vs. particle volume fraction

In order to determine the limiting grain size, the grain growth kinetics was studied at different combinations of d (from 0.1 to 3 μm) and f (from 0.0003 to 0.175) in model polycrystals with the initial mean grain diameter $\langle D\rangle_0$ = 22.5 μm (it is proportional to the mean grain intercept), the most probable grain diameter M_0 = 13.9 μm, and D_m = 42.5 μm. It should be noted that M, in contrast to $\langle D\rangle$, is not affected by microstructural non-homogeneity. The magnitude of F_z was assumed independent of time.

It was observed that the grain growth process develops as normal grain growth at relatively low values of F_z and as abnormal grain growth (AG) at higher F_z (see Fig. 2). Since F_z was constant, AG did not develop completely [12] and the maximum volume fraction of abnormal grains did not exceed ~40% at intermediate F_z values. At very high F_z, the growth process was strongly suppressed, but it may be described as the AG incubation period because the grain size non-homogeneity, characterized by the ratio D_m/M, increased from ~3 in the initial microstructure to >10. In the case of normal grain growth magnitudes of $\langle D\rangle$ and M increase steadily, but at high growth duration their increase slows down. In the case of AG, magnitudes of $\langle D\rangle$ and M increase very slowly at the stage where AG stops to evolve (at intermediate F_z) or at the AG incubation period (at high F_z). The limiting grain size was defined basing upon the magnitude of the relative growth rate $v = (\Delta x/x)/t$ where x is either M or $\langle D\rangle$, and t is time in arb.units (au). It was assumed that the limiting grain size is attained if $v \leq 0.002$ 1/au in the case of normal grain growth or ≤ 0.001 1/au in the case of AG.

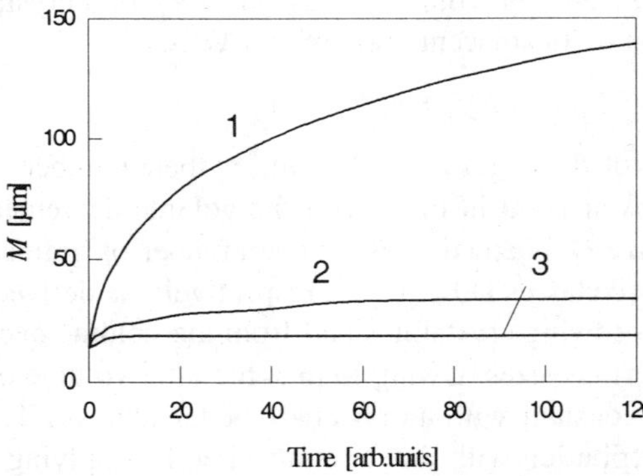

Fig. 2. Grain growth kinetics at (1) f = 0.00031, d = 0.1 μm (normal grain growth), (2) f = 0.0031, d = 0.1 μm (suppressed AG), (3) f = 0.0286, d = 0.5 μm (AG incubation period).

The results obtained in the simulations are presented in Fig. 1 (curve 2). It is seen that the data for M_{lim}/d are characterized by a very narrow scatter, whereas the points for $\langle D\rangle_{lim}/d$ are noticeably scattered, as might have been expected. So the usage of M_{lim} instead of $\langle D\rangle_{lim}$ seems more rewarding in this case. The regression analysis of the modelling results was undertaken on the data obtained in the case of the AG development only. In the range $0.003 \leq f \leq 0.175$ the dependence M_{lim}/d vs. f is described by Eq. 1 with n = −0.95, in an good agreement with [1].

Thus, the simulation data prove convincingly the applicability of the theory [1] in a wide range of the pinning forces exerted by randomly distributed particles. Besides, as shown by the simulations, the most probable grain size is the micro-structural parameter best suited to the description of grain growth stagnation.

References

[1] K. Zener, quoted in: C.S. Smith, *Trans.AIME*, 175, 1948, pp. 47-58.

[2] D.G. Olgaard and B. J. Evans, *Am.Ceram.Soc.*, 69, 1986, pp. C272-C277.

[3] P.M. Hazzledine and R.D.J. Oldershaw, *Phil.Mag.A*, 61, 1990, pp. 579-589.

[4] P. Hellman and M. Hillert, *Scand.J.Metall.*, 4, 1975, pp. 211-219.

[5] A.K. Koul and F.B. Pickering, *Acta Metall.*, 30, 1982, pp. 1303-1308.

[6] C.J. Tweed, N. Hansen, B. Ralph, *Metall.Trans.*, 14A, 1983, pp. 2235-2243.

[7] V.Yu. Novikov, *Scripta Mater.*, 42, 2000, pp. 439-443.

[8] V.Yu. Novikov, "Grain Growth and Control of Microstructure and Texture in Polycrystalline Materials", CRC, 1996.

[9] G.F. Van der Voort and J.J. Friel, *Mater.Char.*, 29, 1992, pp. 293-302.

[10] V.Yu. Novikov, *Scripta Mater.*, 39, 1998, pp. 945-949.

[11] V.Yu. Novikov, *Acta Metall.*, 26, 1978, pp. 1739-1744.

[12] V.Yu. Novikov, *Scripta Mater.*, 37, 1997, pp. 463-469.

Grain Boundary and Triple Junction Properties

Migration of Low Angle Grain Boundaries during Secondary Recrystallisation in Silicon Iron

H. Homma[1], B. Hutchinson[2]

[1] Steel Research Laboratories, Nippon Steel Corporation, 20-1 Shintomi, Futtsu, Chiba, 293-8511, Japan
[2] Swedish Institute for Metals Research, Drottning Kristinas väg 48, S-114 28 Stockholm, Sweden

Keywords : grain boundary migration, low angle grain boundary, secondary recristallisation.

Abstract. The sluggish migration of low angle grain boundaries was examined with regard to the progress of secondary recrystallisation in silicon iron processed by the two stage cold rolling method. The Goss secondary grains are frequently surrounded by low angle grain boundaries during their growth. In the present work, the retarded migration was demonstrated by measuring the frequency of grains at the perimeter of the secondary grains that could stunt growth through 'orientation pinning'. The poor migration of Goss secondary grains is concluded to be caused by their reduced mobility. The orientations of the stunting grains were observed to deviate less than 7° from those of the growing secondary recrystallisation grains. Such low angle grain boundaries are expected to be comprised of dislocation arrays and be much less mobile than general boundaries having misorientations of more than about 15°.

1 Introduction

The characteristics of grain boundaries is a subject of continuing interest in relation to their migration and the phenomenon of grain growth. On the assumption of a single character for the all boundaries, Hillert [1] and Srolovitz et al. [2] concluded that only normal grain growth is achieved. However, it is not clear how the variation in grain boundary properties can be related to the occurrence of abnormal grain growth.

The present work, therefore, is an attempt to investigate the effects of grain boundary character on their migration behaviour. The low angle grain boundary criterion was utilised for the evaluation of migration behaviour, as mobility in these cases is widely accepted to be explicitly sluggish. Movement of the boundaries was provided by secondary recrystallisation in silicon iron processed by two stage cold rolling [3]. In this technique, the matrix grains that are consumed by the abnormally growing {110}<001> oriented (Goss) secondaries possess a texture having an intensity accumulation of similar orientation.

2 Experimental

Secondary recrystallisation is an abnormal grain growth which is prompted by the existence of dispersed second phase particles and specific conditions of the texture. In the two stage cold rolling method for grain oriented silicon iron production, the reduction of the final cold rolling is 60% and MnS is employed as the inhibitor. Precipitates of MnS in the alloy generate a Zener pinning force resisting the grain boundary migration and is reported to cause the selectivity of the grain growth [4]. Goss oriented grains grow preferentially to visible size in the final annealing. The texture of the matrix grains also has intensity peaks at {110}<001>, as is expected after moderate cold rolling and subsequent recrystallisation annealing [5]. Figure 1 shows both the orientation of secondary recrystallisation grains and the texture of the matrix. During the process, the growing Goss grains have a high probability of encountering grains with small misorientations. Therefore, the progress of the secondary recrystallisation inevitably comprises some migration of low angle grain boundaries.

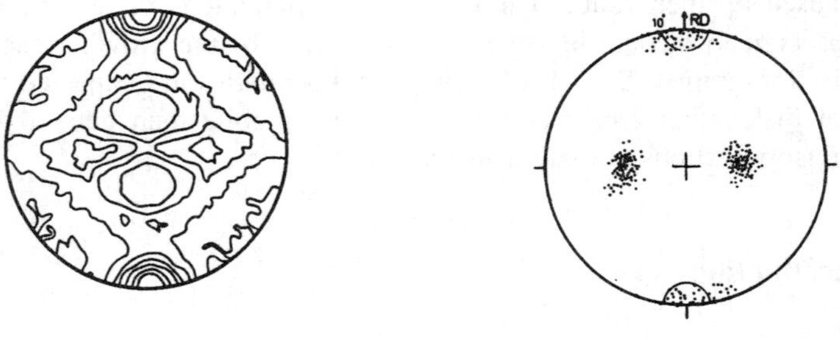

a b

Fig. 1. {100} pole figures for (a) matrix texture, (b) secondary grain orientations.

The presence of numerous low angle grain boundaries means that these should be considered both in relation to the Zener pinning force and to the driving force for migration. The latter is provided by the product of grain boundary tension and the curvature which results from the grain boundary geometry at the triple junctions.

The grains adjacent to the growing secondaries were examined with regard to their boundary migration. Mobile grain boundaries travel quickly across the grains and vanish, whereas immobile grain boundaries should remain more frequently in the intermittent state and so persistent grains will be more commonly observed.

3 Results

Firstly, the state of the neighbouring grains around the progressing secondaries was examined. Considering first the grain size, large grains are resistant to the advancing secondary grains in the matrix and have a tendency to persist at the perimeter. Figure 2 shows a micrograph of the front of a Goss secondary grain and the matrix at the stage where secondary recrystallisation was interrupted. The average grain diameters of the matrix and the secondary neighbourg grains were 13.2 μm and 17.3 μm respectively. The assumption that the grains having slow moving boundaries with the secondary grain are easily observed was confirmed here, so that a similar approach was applied for the orientation analysis.

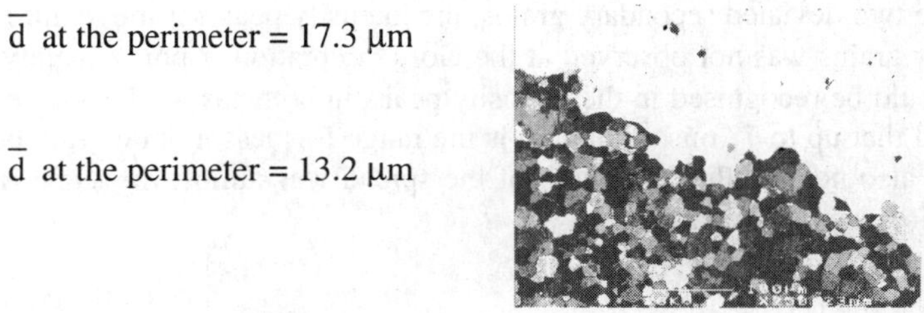

Fig. 2. The micrograph of the secondary recrystallisation front indicating the average diameters of the matrix and the perimeter grains.

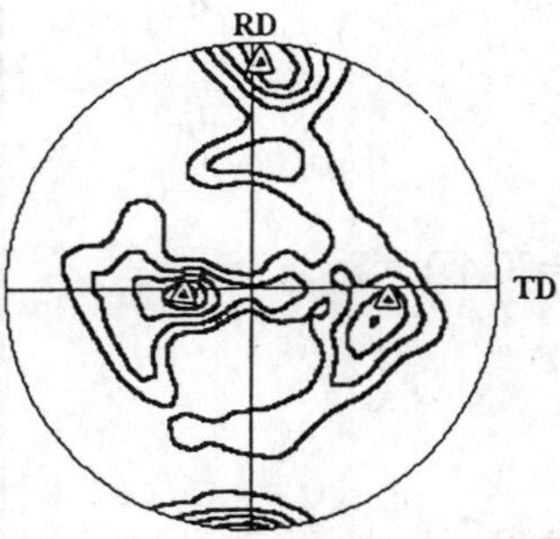

Fig. 3. The texture of the neighbouring grains to the deviated Goss secondary grain, whose orientation is indicated with the triangle.

Orientation measurements of the neighbouring grains to the secondaries was carried out. Their orientations were analysed with EBSP and manipulated as a texture representation, summing up the 10° gaussian spreads centred on the each measured orientation. Figure 3 shows a {100} pole figure of an orientation of a slightly deviated Goss secondary grain and the texture of the neighbouring grains. The orientation of the secondary grain is indicated by the triangle. The texture is clearly shifted from the matrix texture and the intensity is accumulated around the orientation of the secondary grain. Therefore the low angle grain boundaries are demonstrated from the analysis to have low mobility during the secondary recrystallisation in silicon iron.

Further quantitative analysis was carried out on three secondary grains chosen so that one was close to the ideal Goss orientation while the others were rotated around the RD axis and ND axes. The Euler angles were (φ_1, Φ, φ_2)=(2.1°, 44.3°, 5.4°), (186.2°, 32.9°, 80.4°) and (192.4°, 44.5°, 86.7°) respectively. The manipulated textures from the EBSP measurement are represented on the reduced ODF map as the section at φ_2=0° with the range of φ_1=0 – 90° and Φ=0° – 45°. The ideal Goss orientation is at φ_1=0° and Φ=45° on the map. The orientations of the secondary grains are indicated by the open triangles. In each case, the same result is seen that the textures of the neighbouring grains have maxima around the orientation of the adjacent secondary grains.

For the two deviated secondary grains, the intensity peak for the texture of the perimeter grains was not observed at the Goss orientation. Approximately 7° deviation could be recognised in the intensity peaks in both cases. Therefore, it is considered that up to 7° misorientation is the range for less mobile grain boundaries. It was also notable that the width of the spread was almost the same for the two deviation axes, RD and ND.

(a) ideal Goss
(φ1=-2.1°Φ=44.3°φ2=5.4°)

(b) rotation around RD
(φ1=186.2°Φ=32.9°φ2=80.4°)

(c) rotation around ND
(φ1=192.4°Φ=44.5°φ2=86.7°)

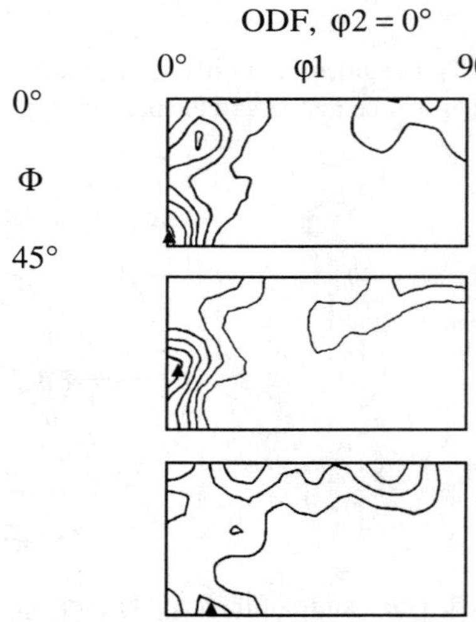

Fig. 4. The textures of the neighbouring grains to the Goss secondary grains (a) almost ideal, (b) rotated around RD axis and (c) rotated around ND axis. The triangles indicate the orientation of the secondaries.

4 Discussion

The migration behaviour of the low angle grain boundaries observed in the secondary recrystallisation of silicon iron indicated quantitatively the role of grain boundary character. The atomic structure of low angle grain boundaries is usually modelled as dislocation arrays. Each dislocation interacts through elastic energy to form the array, and so the grain boundary energy etc. could be reasonably calculated [6]. However, such models usually conclude that the maximum misorientation angle is of the order of 15°, which is larger than the result obtained from the present migration analysis; the slow migration of the low angle grain boundaries is realised within half of the misorientation range expected from the dislocation structure model.

Different choices of the rotation axis of the deviation provided the same deviation range for the immobile boundaries suggesting that the influence of the dislocation arrangements in the boundary on the migration property is not great. Furthermore, the micrograph of the secondary front in Fig. 2 shows equiaxed or round shapes for the grain boundaries which do not represent the obvious facetted structure of a dislocation array. Therefore the migration properties are dominated not only by the atomic structure but also by more macroscopic interactions of grain boundaries.

The grain boundary migration here was provided as grain growth where grain boundary tension together with curvature is generally regarded as the origin of the driving force. The curved shape of the boundary is generated from the geometry at the triple junction, which is determined by the tension of the grain boundaries gathering at the point (7). The assumption above was formulated by May and Turnbull for the case of secondary recrystallisation (4) base on two dimensional model. At a sufficiently late stage, the secondary grain diameter can be regarded as extremely large compared to the matrix grains, and they obtained the following equation,

$$F=2V(\sigma_m/c-\sigma_s I) \tag{1}$$

Here, F is the driving force for the migration, V is the volume unit, c is the size of the side of a secondary grain towards each matrix grain, I is a particle shape factor, σ_m is the grain boundary tension between the matrix grains and σ_s is the tension between the secondary grain and the matrix. This equation suggests that the driving force due to the grain boundary geometry is independent of grain boundary character and also that low energy boundaries experience less pinning force than general boundaries. Therefore, from the driving force analysis, it does not appear that the low angle grain boundaries feel a small driving force, but rather, the opposite.

Table 1. The grain diameters of the matrixes and the neighbours to the secondary grains.

Secondary recrystallisation	Neighbours (A)	Matrix (B)	A/B
2 stage cold rolling	17.3 μm	13.2 μm	1.31
1 stage cold rolling	13.1 μm	10.8 μm	1.21

An analysis of grain diameters confirms the effect of driving force at the secondary grain front. The ratio of the average grain diameters at the perimeter and

within the matrix was compared to HI-B material which is produced by a single stage cold rolling method [8] and possesses a totally different matrix texture. In HI-B material, the secondary grains are surrounded by general boundaries during the growth [9]. The ratios shown in Table 1 are not significantly different, therefore almost the same pressures were applied on the secondary grains by the low angle grain boundaries as the general boundaries.
The migration rate is expressed as the velocity for a given driving force,

$$V=MF \tag{2}$$

where, M is defined as the mobility. According to the discussion above, the small velocity of low angle grain boundary is due to the small value of their mobility. Goss orientation suffers 'orientation pinning' from much of the matrix texture which reduces the velocity of an abnormally growing grain but maintains it at positive value. The mobility must always be positive although the driving force can be zero. The mobility is explained as the frequency of atom transfers across the boundary, and it is natural to think that the structure of low angle grain boundaries does not promote this mechanism. The present experimental results also show that the effects are not dependent on the direction of the dislocation array, i.e. the diffusion of atom across the boundary was scarcely affected by the boundary orientation.

5 Conclusions

The migration behaviour of low angle grain boundaries was investigated in the case of secondary recrystallisation in silicon iron. Slow migration of Goss secondary grains is concluded to be caused by their reduced mobility due to orientation pinning. The orientations of the stunting grains were observed to deviate less than 7° from those of the growing secondary recrystallisation grains. Such low angle grain boundaries are expected to be comprised of dislocation arrays and be much less mobile than general boundaries for which misorientations are more than about 15°.

References

[1] M. Hillert, Acta Metall. 13, 1965, p. 227.
[2] D.J. Srolovitz, G.S. Grest and M.P. Anderson, *Acta Metall.*, 33, 1985, p. 2233.
[3] N.P. Goss, U.S. Pat. 1965559, 1934.
[4] J.E. May and D. Turnbull, *Trans. Met. Soc. AIME*, 212, 1958, p.769.
[5] W. B. Hutchinson, *Int. Met. Rev.*, 29, 1984, p. 25.
[6] W.T. Read and W. Shockley, *Phys. Rev.*, 78, 1950, p. 275.
[7] C. Herring, "Structure and Properties of Solid Surface", Univ. Chicago Press, 1953, p. 31.
[8] S. Taguchi, A. Sakakura and H. Takashima, U. S. Pat. 3287183, 1966.
[9] W.B. Hutchinson and H. Homma, G.G. Plyc, in"Proc. ICGG-3", 1998, p. 387.

The Effect of Solute Elements on Grain Boundary Mobility during Recrystallization of Single-Phase Aluminium Alloys

Y. Huang and F. J. Humphreys

Manchester Materials Science Centre, Grosvenor Street, Manchester M1 7JL UK

Keywords : aluminium, solute effect, boundary mobility, recrystallization, EBSD.

Abstract. A combination of in-situ annealing and electron backscattered diffraction in the SEM has been used to investigate the effect of solute elements on the grain boundary mobility during the annealing of deformed single crystals of high purity aluminium and of a series of single-phase aluminium alloys. The relationship between boundary velocity and driving pressure was close to linearity in all materials and the activation energies for boundary migration were consistent with control by lattice diffusion of the solutes. The addition of solutes strongly reduced the boundary mobilities which maximized at 40°<111> to the matrix with a broad peak. The experimental results and the solute effects on the grain boundary mobility are discussed in terms of theories of boundary migration.

1 Introduction

It has long been recognised that solute elements segregate to grain boundaries (GB) in order to lower the energy. During boundary migration, the solute atmosphere diffuses with, but lags behind the boundary, resulting in a dragging pressure on the boundary. Small additions of solute elements to high purity metals can drastically reduce the mobility (M) of boundaries. In order to understand and control the processes of recovery, recrystallization and grain growth during thermomechanical processing [1-3], it is important that the effects of solute drag are quantified. Although the general effects of solute on boundary migration have long been known, quantitative measurements have been made only on a limited number of systems. Careful research carried out on curvature-driven boundary migration in bicrystals of known orientation, has clearly demonstrated the effects of boundary character, orientation and purity on boundary mobility, particularly at very high temperature [4-8]. However, recent work [9] has suggested that there may be differences in boundary behaviour during recrystallization and curvature-driven growth, which could be attributed to differences in the type and magnitude of the driving pressure, the lower temperatures involved and the fact that a recrystallization front generally comprises small areas of boundary of different character, rather than a single boundary type.

The present work involved a study of the effect of type and amount of solute elements on the GB mobility during recrystallization of deformed single crystals of some single-phase aluminium alloys.

2 Experiments

In-situ annealing experiments of channel-die plane strain compressed Goss-orientated {110}<100> single crystals of high purity aluminium (5N) (HP-Al) and a serious of the HP-Al based single-phase aluminium alloys, including Al-0.05Si, Al-0.05Mn, Al-0.1Mg and Al-1Mg, were carried out in a JEOL JSM6300 SEM using the procedures detailed by Huang and Humphreys [9]. Disk specimens (~ϕ3×1mm) were annealed isothermally at temperatures up to 420°C for 30 s-4 hrs and recrystallization was initiated from a fine scratch on the sample surface. The GB velocities were measured from video recordings in the direction of fastest growth, and the annealing was halted when grains had grown a distance of ~500 μm. The partly recrystallized samples were analyzed by electron backscatter diffraction (EBSD) and the orientations of the recrystallized grains and their relationship to the deformed matrix were determined. The subgrain sizes and misorientations were measured at the end of the experiment and used to calculate the stored energy of the recovered matrix and hence the driving pressure [9].

3 Results

The Goss orientation is stable during channel die plane strain compression [10] and the resultant subgrain structure with no long-range orientation gradients provided an deal matrix into which recrystallizing grains could grow. Fig. 1 shows a typical microstructure of a partly recrystallized specimen after in-situ annealing. Grains originating at the scratch have grown into the deformed matrix, and it may

Fig. 1. Electron backscattered micrograph of an in-situ annealed Al-0.05Mn sample, showing the basic microstructural features of the recrystallization.

be seen that many grains are elongated, and only a few have grown extensively, showing that the GB velocities are anisotropic. In agreement with previous work [11] it was found that the fast moving GBs were of tilt character. The overall orientation distributions of the recrystallized grains was found to be almost random and the fast grains were found to be misoriented from the matrix by 35-45° about a <111> axis. GB velocity is shown as a function of misorientation angle in Fig. 2a and the deviation from a <111> axis in Fig. 2b. Broad peaks are seen in both cases and the half width of the peaks for the alloys (9.5-10°) in Fig. 2a are smaller than that that (11.2°) of the HP-Al. However, for the axis deviation shown in Fig. 2b, the HP-Al shows a narrower peak (7.3°) than the alloys (8-13.4°). Detailed analysis of the data in Fig. 2 is described elsewhere [12].

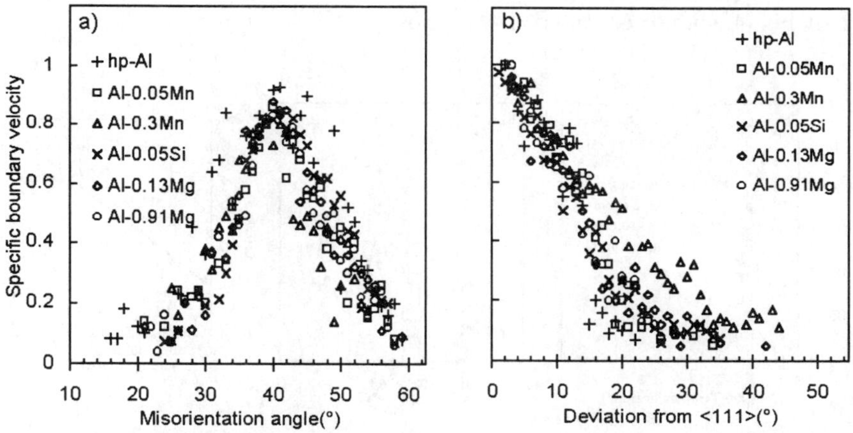

Fig. 2. The specific GB velocity (defined as the growth distance of a grain relative to that of the largest grain in the specimen) as a function of (a) misorientation angle (for grains with misorientation axes within 10° of <111>) and (b) deviation from a <111> axis (for grains with misorientation angles between 30° and 50°).

An exception occurred for the HP-Al where it was found that at temperatures above 200°C grains with a 30~50°/<111> relationship were dominant as usual, but below 150°C grains misoriented from the matrix by 28-38° about <100> axes grew preferentially. The results of 28-38°/<100> grains are not included in Fig. 2 but are described in ref. [13].

The GB velocity was found to be proportional to the driving pressure. GB mobilities were determined from the slopes of pressure-velocity plots and Fig. 3a shows the effect of temperature and solute elements on GB mobility including the results of HP-Al <100> boundaries. It is seen that whilst all the solute elements have strongly reduced the GB mobility in the alloys, the mobility values for the alloys are relative close to each other compared to the big difference from the mobilities in HP-Al. In Fig. 3b the mobility is correlated with the terminal solubility of the solute (X_0/k), X_0 is the solute concentration in the matrix and k the partition coefficient of the alloy, and a logarithm linear relationship was obtained if the single data of the HP-Al (impurity level 0.001%) is neglected. The mobility is then described as a function of X_0: $M=A\ (X_0/k)^{-B}$, where $A=1\sim200\times10^2\ m^4/J/s$ and $B=0.28\sim0.36$ are temperature dependent constants. The temperature dependence of the mobility in all the materials is consistent with an equation of form $M=M_0\exp(-Q/RT)$ and the derived activation energies (Q) and pre-exponentials (M_0) for GB migration are shown in Fig. 4. In contrast to the predictions of earlier solute drag theories [14-16], but in agreement with the experiments of Gordon and Vandermeer [18], the decrease in mobility with increasing solute content is not due to a reduction of M_0, but to a considerable rise of both M_0 and Q. As a result, the effect of solutes on mobility is less at higher temperatures. A relation-ship, $Q=\alpha\ln(M_0)+\beta$, which is in agreement with previous work [7,17], was found for the data in Fig. 4, with $\alpha=6.9$ and $\beta=109$ kJ/mol.

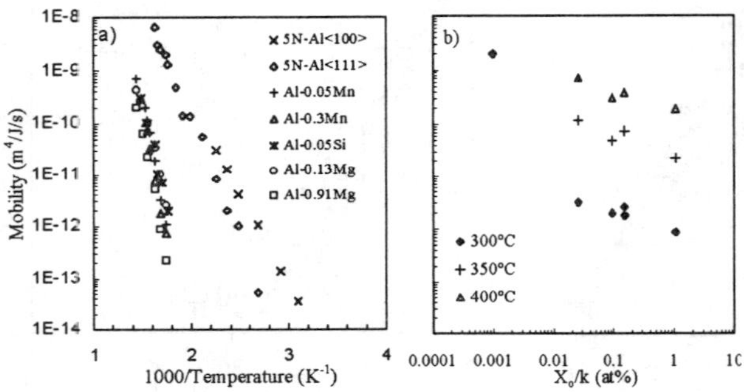

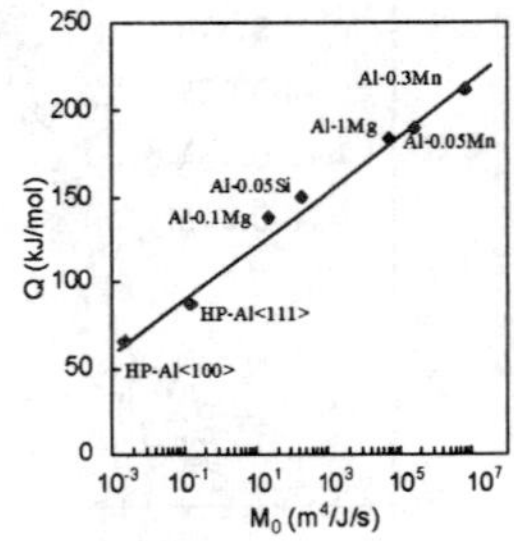

Fig. 3. The dependence of GB mobility on temperature and solute elements: (a) the effect of temperature and solute species and (b) as a function of X_0/k.

Fig. 4. The correlation between the activation energy (Q) and pre-exponential factor (M_0).

4 Discussion

The results confirm earlier work [19] on the orientation dependence of GB mobility in Al alloys during recrystallization. The extent of the data in the present investigation clearly shows that under these conditions there is no fast growing orientation relationship other than that approximating to 40°<111> for the alloys, and also shows that the broad mobility peak of Fig. 2 is a genuine effect for boundary migration during recrystallization. This differs from curvature driven boundary migration in which distinct, very sharp mobility peaks occur in association with special GBs such as the Σ7 (38.2°<111>) boundary [7].

The decrease in boundary mobility with increasing deviation from a 40° misorientation is seen in Fig. 2a to be sharper for the alloys than for the HP-Al, suggesting that the mobility decrease of random boundaries due to solutes is more than that of boundaries near 40°<111>. This is similar to the result obtained in tin by Aust and Rutter [4] and may be explained by the selective segregation effect [3].
However, the effect of deviation from a <111> axis on mobility shown in fig.2b is different. The mobility peak of the alloys is broader than that of the HP-Al and the broadest peak occurs in Al-0.3Mn, which is contrary to the predicted effect of selective segregation. The broad mobility peaks found in the present work cannot readily be correlated with the crystallography and structure of Σ7 boundaries because the boundary misorientation is not constant either along its length or during its growth and because of the large deviation of the high mobility regime from Σ7.
The interaction energy between a solute and boundary increases approximately as the size misfit between the solute and matrix increases and as the solute-solute bond strength decreases [1]. However, no direct correlation between the mobility variation and these two parameters was found in this investigation. For a given solvent, different solutes are thought to produce different effects on mobility and Aust and Rutter [5] found, in Pb alloys, that the faster moving solutes had the greater effect on mobility. However, in the present the mobility was found to be related to the terminal solubility of solutes as shown in Fig. 4, irrespective of the species. This might be an indication that the individual interaction between a solute element and GB is less important than the group behaviour of any solute species, which may be associated with free volume associated with the boundary.
The boundary velocities were found proportional to the driving pressures for all the conditions investigated and the mobilities appear to be controlled by the same thermally activated process, although it is arguable that the small decrease in slope for the alloys at the highest temperatures in Fig. 3a is indicative of some change in mechanism.
There is no evidence in the materials and conditions of the present investigation of a transition to high boundary velocity associated with the predicted breakaway of a boundary from its solute atmosphere [14-17], although such a transition has been reported in aluminium alloys containing trace amounts of solute [see e.g. 2]. However, the driving pressures and annealing temperatures in the present work are typical of those encountered during recrystallization of aluminium, and the type and concentration of solutes are similar to some industrial alloys. This suggests that during the recrystallization of most aluminium alloys, boundary breakaway is confined to very dilute alloys, and unlikely to play a significant role in the recrystallization of commercial alloys.
The values of Q obtained show that the GB migration in the alloys is a process controlled by the lattice diffusion of solutes [8,20,21]. The fact that the mobility reduction due to solute elements occurs by the increase of both M_0 and Q and that the value of Q increases with increasing solute contents (Fig. 4) may suggest that there is a continuous change of the structure and/or thermodynamic properties of GB and the nearby diffusion region as the species and content of solutes change. In view of the increasing interest in modelling recrystallization in industrial alloys, there is an urgent requirement for further experimental investigation of the structu-

re of moving GBs and of the dynamic interaction between solutes and boundaries in alloys containing significant amounts of solute.

Acknowledgements

The authors wish to acknowledge EPSRC for financial support, and Alcan International, Banbury, for the provision of materials and helpful discussions.

References

[1] D.A. Smith et al. in "Grain-boundary structures and kinetics", ed. R. W. Balluffi, ASM, Metals Park, Ohio, 1979, pp. 337-371.

[2] F. J. Humphreys, M. Hatherly, "Recrystallization and related annealing phenomena", Pergamon press, Oxford, 1995.

[3] G. Gottstein, L. S. Shvindlerman, "Grain Boundary Migration in Metals: Thermodynamics, Kinetics, Applications", CRC Press, 1999.

[4] K. T. Aust, J. W. Rutter, *Trans. of AIME,* 215, 1959, pp. 119-127.

[5] K. T. Aust, J. W. Rutter, in "Ultra-High purity Metals", AMS, pp. 115-148.

[6] B. B. Rath, H. Hu, *Trans. of AIME,* 245, 1969, pp. 1577-1585.

[7] D. A. Molodov, U. Czubayko, G. Gottstein and L. S. Shvindlerman., *Acta Mater.*, 46, 1998, pp. 553-564.

[8] M. Winning, G.Gottstein, L.S. Shvindlerman, *Acta Mater.*, 49, 2001, pp. 211-219.

[9] Y. Huang and F. J. Humphreys, *Acta Mater.*, 47, 1999, pp. 2259-2271.

[10] M. Ferry and F. J. Humphreys, *Acta Mater.*, 44, 1996, pp. 1293-1308.

[11] H. Gleiter abd B. Chalmers, *Prog. Mater. Sci.*, 16, 1972, pp. 1-282.

[12] Y. Huang and F. J. Humphreys. (to be published).

[13] Y. Huang and F. J. Humphreys, in "Recrystallization and related annealing phenomena", eds. T. Sakai and H. G. Suzuki, JIM, 1999, pp. 445-450.

[14] K. Lücke and K. Detert, *Acta Metall.,* 5, 1957, pp. 628-637.

[15] K. Lücke and H. Stüwe, *Acta Metall.,* 19, 1971, pp.1087-1097.

[16] J. W. Cahn, *Acta Metall.*, 10, 1962, pp. 789-798.

[17] D. A. Molodov, B. B. Straumal, L. S. Shvindlerman, *Sov. Phys. Solid State*, 26, 1984, pp. 629-633.

[18] P. Gordon, R. A. Vandermeer, *Trans. of AIME*, 224, 1962, pp. 917-928.

[19] B. Liebmann, K. Lücke and G. Masing, *Z. Metallk.*, 47, 1956, pp. 57-63.

[20] L. F. Mondolfo, "Aluminium Alloys", Butterworth, London, 1976.

[21] C. J. Smithells, "Smithells Metals Reference Book", Butter-Heinemann, Oxford, 1992.

Geometrical Theory of Elements of Polycrystals Consisting of CSL Boundaries

V.Y. Gertsman

Pacific Northwest National Laboratory, P.O.Box 999, Richland, WA 99352, USA

Keywords: grain boundaries, triple junctions, quadruple points, polycrystal.

Abstract. The coincidence site lattice theory of grain boundaries is extended to multicrystal ensembles, such as triple junctions, quadruple points and multiple junctions of grains. It provides a basis for analyzing the junction structures as well as polycrystalline microstructure. Some hypothetical examples and a case study of an experimental microstructure of recrystallized pure nickel sample are presented.

1 Introduction

Polycrystals are not just agglomerates of grains since the grain joints have their own structure and influence the polycrystal properties. This was realized long ago with respect to grain boundaries and, recently, triple junctions have also attracted a considerable amount of attention [1]. Most real polycrystalline materials also contain other types of grain boundary joints. Thus, if there is more than one layer of grains in the polycrystal, then it must contain grain vertices. The simplest of such objects is the quadruple point, which is the point of contact between four grains and six grain boundaries or, equivalently, the junction of four triple junctions. Also, quite often multiple junctions of grain boundaries exist, where more than three grains (and grain boundaries) meet along a common line [2,3].

Progress in studying complex grain joints is hindered by the lack of an accredited crystallographic theory that would help analyze experimental data and could serve as a framework for phenomenological models. In the case of grain boundaries, the coincidence site lattice (CSL) theory has provided such a crystallogeometrical framework [4]. The present paper offers similar tools for triple junctions and other ensembles built of a limited number of grains having a cubic crystal lattice. This may be a step toward a comprehensive concept of the structure of the polycrystal.

2 Theory Formulation

This section briefly describes the relationships among the parameters of CSLs of multicrystalline assemblages and constituent grain boundaries. Only the joints of unstressed crystallites are considered here. Due to limited space allocated to the paper the formulae are given without proof. Rigorous derivations of them are presented in the two-part work [5].

CSL misorientation between two crystals of the cubic system can be described as $R = \{a\}/\Sigma$, where $\{a\}$ is an integral 3×3 matrix and Σ is the CSL multiplicity factor, the so-called reciprocal density of coincident sites [4].

There is a widely spread misconception concerning the Σ combination rule in the triple junction: it is often assumed that one of the Σ's (e.g. Σ_3) is always greater than the other two and the formula is

$$\Sigma_3=\Sigma_1\Sigma_2. \tag{1}$$

Actually, the general formula for the Σ combination rule at a CSL triple junction in the cubic system can be written as:

$$\Sigma_3=\Sigma_1\Sigma_2/\alpha, \tag{2}$$

where α is the greatest common divisor of the matrix $\{b\}=\{a\}_1\{a\}_2$.

The following relationship links α with the generating Σ values:

$$\alpha = \beta^2, \tag{3}$$

where β is a common divisor of Σ_1 and Σ_2.

The corresponding formula for the multiple junction, i.e. for the n-fold junction of n CSL boundaries:

$$\Sigma_n=\prod_{i=1}^{n-1}\Sigma_i/\alpha_n, \tag{4}$$

where α_n is the greatest common divisor of the matrix $\{b\}=\prod_{i=1}^{n-1}\{a\}_i$.

Instead of Eq. 3 a less restrictive relationship holds for multiple junctions:

α_n is the square of a common multiple of $\prod_{i=1}^{n-1}\Sigma_i$.

Superposition of three or more cubic lattices, mutually rotated in such a way that each pair of the lattices creates a CSL, generates a common CSL (multi-CSL) with the multiplicity factor calculated as:

$$\Sigma^{MJ} = \text{greatest common divisor } \left(\prod_{i=1}^{n}\Sigma_i\right)_k^{1/2}, \tag{5}$$

where $\left(\prod_{i=1}^{n}\Sigma_i\right)_k$ represents all k possible multiple junction arrangements of the given crystal orientations.

For the triple junction this formula becomes:

$$\Sigma^{TJ} = (\Sigma_1\Sigma_2\Sigma_3)^{1/2}. \tag{6}$$

It is not essential that all crystallites meet along a common line, thus forming a multiple junction. The above formulae depend only on the lattice orientation of crystallites, but not on how they are spatially arranged and connected to each other. If we have a multicrystal and not a multiple junction, then the system is still described by Eqs. 4 and 5. However, in this case not all grain boundaries present in the system are to be counted among the n boundaries. The minimum necessary number of boundary CSLs required to calculate Σ^{MJ} is equal to the number of different crystal orientations. That is, for calculating Σ^{MJ}, every system of grains can be reduced to an equivalent multiple junction.

3 CSL Approach to Multigrain Ensembles

One of the difficulties in triple junction studies has been the lack of convenient and simple categorization of such objects. Existence of the triple junction CSL provides a possible means for such a classification. An example of analysis of a real microstructure is presented below in Section 4. Certainly, as with grain boundaries, the Σ-description of triple junctions may simply be a classification of convenience and we are not hinting at any link of physical properties of the junction with the magnitude of Σ^{TJ}. Nevertheless, it is remarkable that such a complex object can be characterized by a single scalar parameter. The relationships given in the previous section allow calculation of any possible combination of CSL misorientations at triple (and other) junctions of cubic crystallites. However, these relationships provide only the necessary condition and their fulfillment is not enough for the junction to be in fact possible. To find whether a certain combination can exist, it is necessary to directly check the misorientation multiplication. Possible combinations have been listed in [5].

The proposed geometric theory allows singling out at least one class of triple junction, namely, $\alpha \neq 1$ junctions. By this term we mean the triple junctions for which Eq. 1 is not fulfilled. Of course, in every junction $\alpha \neq 1$ for at least one of the boundaries. So, it is emphasized that the term "$\alpha \neq 1$ junction" refers to cases when none of the α's are equal to 1. That such triple junctions do exist was first shown in [6]. At present, we do not know how frequent such junctions are in real polycrystals. In the sample described below in Section 4 $\alpha \neq 1$ junctions are not very frequent. However, "rare" does not necessarily mean "unimportant". Recently, it has been suggested [7] that $\alpha \neq 1$ triple junctions may have special properties, different from those of $\alpha=1$ junctions, in particular, with regard to dislocation transmission between grain boundaries. This hypothesis is based on the consideration of the Displacement Shift Complete (DSC) lattice, which defines the Burgers vectors of grain boundary dislocations. In $\alpha=1$ junctions the DSC lattice of the boundary with the largest Σ contains DSC lattices of the other two boundaries as sublattices. When $\alpha \neq 1$, the triple junction CSL is a sublattice of all three grain boundary CSLs. Fig. 1 gives a graphic illustration of this statement. It shows three superimposed lattices with the $\Sigma=9$ misorientation between each pair. In the plane of the drawing (Fig. 1a) every third lattice site is common for each of the three crystal lattices, but such a situation takes place only in every ninth (511) plane. Each pair of the crystal lattices also has coincident sites in every third (511) plane, but these are not common for all the three lattices (see Figs. 1b and c). Hence, in this example all three grain boundary CSLs have Σ=9, but Σ^{TJ}=27 (see Eq. 6). In such cases, the DSC lattice of the triple junction contains DSC lattices of all three boundaries as sublattices.

For comparison, Fig. 1d displays two lattices having the 60°[511] misorientation, which is one of the equivalent descriptions of the Σ27b misorientation between two cubic crystals.

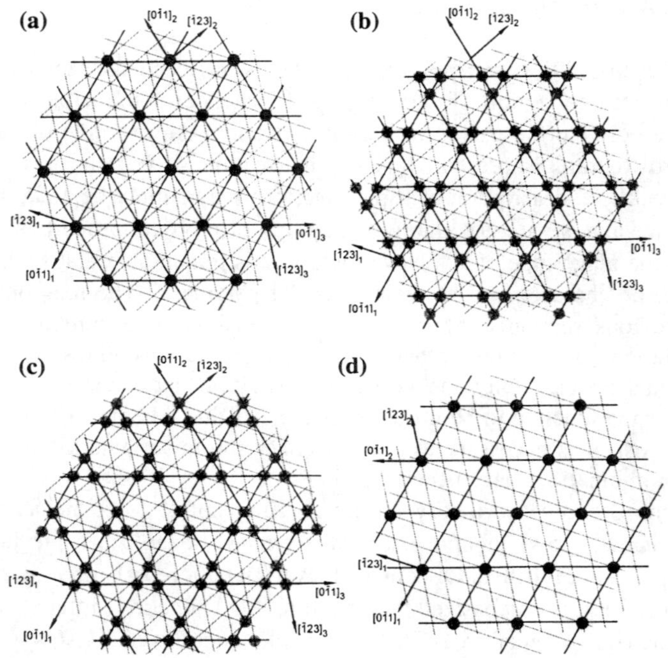

Fig. 1. (a-c) Superposition of three simple cubic lattices forming triple junction CSL Σ^{TJ}=27. (d) Superposition of two simple cubic lattices forming grain boundary CSL Σ27b.

The projection direction is [511]. Three layers of lattice sites are shown: (a) and (d) at zero height, (b) at height $1/\sqrt{3}$, and (c) at height $2/\sqrt{3}$. Black circles in fig.(a) indicate the lattice sites common for all three crystal lattices, while gray circles in figs.(b) and (c) denote the lattice sites common only to pairs of the crystal lattices. It is possible to formally introduce the CSL of the polycrystal with the multiplicity factor calculated according to Eq. 5. Of course, in most practical cases Σ would be an astronomical number. However, this concept might be useful in some special cases, such as certain grain clusters or specially prepared thin film structures. For example, hypothetically one can fabricate a polycrystalline film with only three different orientations of crystallites if the film consists only of triple junctions. If these orientations are such that two of the mutual misorientations are Σ=3, then each triple junction will contain two Σ3 and one Σ9 boundary, and with the appropriate topological arrangement of the triple junctions the CSL of the whole system will have Σ=9. The multi-CSL model can also be applied to more complex hypothetical microstructures considered in [8].

When considering a common CSL, the crystallite assemblage does not have to be two-dimensional. Thus, the same formula (5) is applicable to grain vertices, such as quadruple points. Six grain boundaries meet at such a point, but the misorientations of only three of them are independent. The CSL multiplicity factor of the quadruple point can be calculated e.g. as:

$$\Sigma^{QP}==\Sigma_1\Sigma_2\Sigma_3/\text{g.c.d.}(\beta_{12}\beta_{13},\ \beta_{12}\beta_{23},\ \beta_{13}\beta_{23}) \qquad (7)$$

where Σ_1, Σ_2 snd Σ_3 are the Σ values of any three "pyramidal" boundaries, i.e. boundaries of the same grain converging at the quadruple point. The quadruple point is analyzed in more detail in [5].

4 Statistical Examination of Experimental Microstructure

To demonstrate an application of the theory to statistical analysis of a microstructure it is easier to choose a system with an abundance of CSL boundaries. Recrystallized f.c.c. materials with low stacking fault energy satisfy this condition since multiple annealing twinning produces large amounts of $\Sigma3^n$ misorientations. A pure nickel sample with an average grain size of 14 μm (determined as an equivalent sphere diameter including twins) was examined in a scanning electron microscope. Misorientations at the grain boundaries were measured by the electron backscatter diffraction technique. Boundaries were classified into $\Sigma3^n$ categories using the "flat" nearness criterion of 5°. The upper limit for Σ was set at n=6, i.e. all grain boundaries with misorientations deviated from any $\Sigma3^n$ (n<7) by more than 5° were classified as "random".

The examined area contains 96 grains, 227 grain boundaries, 137 triple junctions and 8 quadruple junctions. A summary of the grain boundary statistics is presented in the following table:

Table 1. Statistics of grain boundaries

GB Type	Σ3	Σ9 (3^2)	Σ27 (3^3)	Σ81 (3^4)	Σ243 (3^5)	Σ729 (3^6)	Other
Number	87	33	13	24	21	18	31
%	38.3	14.5	5.7	10.6	9.3	7.9	13.7

The above table gives the distribution of grain boundaries by their frequency of occurrence and not by their length as in some publications. As expected (see e.g. [3]), the relative length exceeds the number fraction only for Σ3 boundaries: the relative lengths are 69.2, 6.2 and 1.9% for the Σ3, Σ9 and Σ27 boundaries, respectively.

Out of the 137 triple junctions, 99 are such that all constituent boundaries can be described as $\Sigma 3^n$ with n<7. Among 38 other triple junctions: 20 have one and 18 have two random (n>6) boundaries. Summary statistics of the triple junctions are given in the following table:

Table 2. Statistics of triple junctions

TJ Type	$\Sigma 3^2$	$\Sigma 3^3$	$\Sigma 3^4$	$\Sigma 3^5$	$\Sigma 3^6$	Other
Number	53	8	12	13	13	38
%	38.7	5.8	8.8	9.5	9.5	27.7

Five of the triple junctions are $\alpha \neq 1$ junctions, namely, Σ9-Σ243b-Σ243i ($\Sigma^{TJ}=729=3^6$), Σ81a-Σ81d-Σ81d ($\Sigma^{TJ}=729=3^6$), Σ81b-Σ81c-Σ729r ($\Sigma^{TJ}=2187=3^7$), Σ81b-Σ81b-Σ729w ($\Sigma^{TJ}=2187=3^7$) and Σ81a-Σ729r-Σ729t ($\Sigma^{TJ}=6561=3^8$).
Only one of the eight quadruple junctions does not contain at least one random (n>6) boundary: Σ3-Σ81b-Σ3-Σ81c ($\Sigma^{QJ}=243=3^5$). One can easily check that this is an $\alpha \neq 1$ junction.

Acknowledgments

This work was supported by the Office of Basic Energy Sciences, U.S. Department of Energy under contract DE-AC06-76RLO 1830. Partial financial support for participation of the author in the conference was provided by the Alexander von Humboldt Foundation.

References

[1] G. Gottstein, A. King and L.S. Shvindlerman (eds.), Special Issue on Triple Junctions of *Interface Science*, 7 (3/4), 1999.
[2] C.V. Kopezky, A.V. Andreeva and G.D. Sukhomlin, *Acta Metall. Mater.*, 39, 1991, pp. 1603-1615.
[3] V.Y. Gertsman and J.A. Szpunar, *Scripta Mater.*, 38, 1998, pp. 1399-1404.
[4] H. Grimmer, W. Bollmann and D.H. Warrington, *Acta Cryst.*, A30, 1974, pp. 197-207.
[5] V.Y. Gertsman, *Acta Cryst.*, A, 2001, in press.
[6] V.Y. Gertsman and K. Tangri, *Scripta Metall. Mater.*, 32, 1995, pp. 1649-1652.
[7] K. Owusu-Boahen and A.H. King, *Scripta Mater.*, 43, 2000, pp. 175-179.
[8] K. Miyazawa, Y. Iwasaki, K. Ito and Y. Ishida, *Acta Cryst.*, A52, 1996, pp. 787-785.

In-Situ Measurement Technique of Grain Boundary and Triple Junction Mobility

D. Mattissen[1], D.A. Molodov[1], G. Gottstein[1] and L.S. Shvindlerman[1,2]

[1] Institut für Metallkunde und Metallphysik, RWTH Aachen, D-52056 Aachen, Germany

[2] Institute of Solid State Physics, Russian Academy of Science, Chernogolovka, Moscow, distr., 142432 Russia

Keywords: in-situ experiments, hot stage, triple junction, grain boundary, mobility, activation enthalpy.

Abstract. The paper is dedicated to a special technique for in-situ observation and recording of moving grain boundary systems. The orientation contrast in an SEM was utilized to reveal grains with an Electron Back Scatter Detector (BSD). A special hot stage for the SEM was designed and manufactured. An experiment of a grain boundary system containing a triple junction was carried out. The results are discussed with regard to the influence of the triple junction on the movement of the grain boundary system.

1 Introduction

Grain boundaries and triple junctions comprise the granular structure of a polycrystalline material. Even so the number of triple junctions in polycrystals is comparable in magnitude with the number of boundaries, all attributes in the behavior of polycrystals during grain growth have been solely referred to the motion of grain boundaries so far. First Fradkov and Shvindlerman [1] introduced that a triple junction may have a finite mobility. Experimental work on tricrystals confirms this assumption [2]. Nowadays advanced computer simulation approches like vertex models have incorporated triple junction mobility as an additional kinetic parameter and thus were successful in reproducing experimental results of triple junction and grain boundary motion in two-dimensional systems [3]. The main reasons for the lack of experimental results on triple junction motion are the difficulties in performing in-situ experiments, as not only the velocity of the boundary system motion must be measured but also the shape of the intersecting grain boundaries during the motion. Moreover, experiments have to be conducted in the course of which the grain boundaries and triple junctions move steadily.
In recent research it was shown that triple junctions influence the motion of the adjoining grain boundaries. So their role is not reduced to control the thermody-

namic equilibrium angles at the junction during boundary motion. The steady-state motion of a grain boundary system with a triple junction is possible only in a very limited set of geometrical configurations. One of them is shown in Fig. 1.

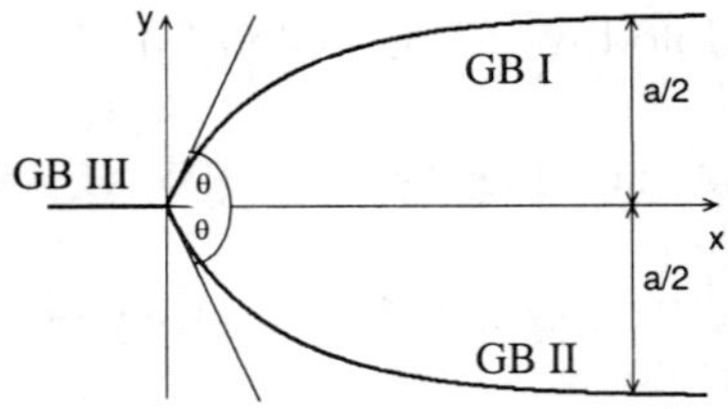

Fig. 1. Geometry of the grain boundary system with triple junction in the course of steady state motion.

The boundaries of this system are perpendicular to the plane of the diagram, and far from the triple junction they run parallel to one another and to the x-axis. This makes the problem quasi two-dimensional. The three boundaries of the system are considered identical, in particular their surface tension σ and their mobility m^{GB}. So it is a uniform system. Furthermore, it is assumed that m^{GB} and σ are independent of the inclination of the grain boundaries. These assumptions define the problem to be symmetric with regard to the x-axis. A steady state motion of the whole system is possible indeed. The dimensionless criterion (m^{TJ} triple junction mobility)

$$\Lambda = \frac{m^{TJ} a}{m^{GB}} = \frac{2\theta}{2\cos\theta - 1} \tag{1}$$

describes the drag influence of the triple junction on the motion of the entire boundary system. For $\Lambda > 1$ the junction does not drag the motion of the boundary system and the angle θ tends to the equilibrium value $\pi/3$. In such a case the velocity V of the motion of the entire boundary system is independent of the mobility of the triple junction and is determined by the grain boundary mobility and the acting driving force:

$$V = \frac{2\theta m^{GB} \sigma}{a}. \tag{2}$$

In contrast, for $\Lambda < 1$ the motion of the system is controlled by the motion of the triple junction and the angle θ tends to zero. The velocity depends only on the triple junction mobility m^{TJ} and the grain boundary surface tension σ

$$V = \sigma m^{TJ}. \tag{3}$$

An experimental proof of this behavior is a challenge.
A technique for in-situ measurements in an SEM will be introduced and an example will be presented.

3 Technique

The measurement of the movement of grain boundary systems is not a trivial problem, since the velocity should be recorded without interference by an etched surface or by grain boundary grooves. Therefore, boundary velocity measurements by stepwise annealing can give erroneous results because it is influenced by grain boundary grooving. Moreover not only the velocity of the system has to be measured but also the geometry of the triple junction. The XICTD method (X-Ray Interface Continuous Tracking Device) [4] can be utilized to measure the movement of the triple junction. However, the shape cannot be measured concurrently using this technique. Also the method of optical orientation contrast [2] cannot be applied for optically isotropy materials like Al. For optically isotropic materials the orientation contrast of adjacent grains by back scattered electrons can be utilized to image their common grain boundary [5]. In order to observe boundary motion at elevated temperature a special hot stage for the SEM (LEO 1530 and JEOL 820) was designed and manufactured. Several important aspects had to be realized: the electron beam must not be influenced by the hot stage, the detectors must not get to hot. As the furnace is shielded, the hot stage has a very fast heating rate. It heats up the specimen to 200°C in 4 min. From this temperature a specimen can reach the desired temperature in 2 min. It easily reaches 600°C without using full power. The specimens have a size from 10 x 15 x 2 mm^3. Images were taken with a frame grabber and stored onto a disk.
Experiments were carried out on tricrystals of high purity aluminium (99,999%) with a grain boundary geometry as shown in Fig. 1. The tricrystals were grown from specially oriented seeds in a horizontal Bridgman furnace. After crystal growth and the triple junction was formed with a special technique.

4 Measurements

4.1 Orientation contrast

Humphreys et al. used a technique where the orientation contrast of polycrystalline material was revealed by an Electron Back Scatter Detector (BSD) [5]. Because the boundaries of the analyzed material were high angle grain boundaries the orientation contrast was sufficiently pronounced. As in this work the experiment were carried out on specially oriented crystals, e.g. crystals with tilt grain boundaries, the resolution for the misorientation contrast of the BSD had to ascertained. Therefore measurements were carried out on various grain boundaries. Two sets of small angle grain boundaries were compared. In both samples each grain had a <111> plane parallel to the surface of the sample. As the orientations of the grains are hardly perfect there were always small differences between the surface orientations. For example, the <111> planes of a 9.7° <111> grain boundary had a deviation of 1.5° and 2.1° in two directions from the perfect orientation.

Therefore the grain boundary could be detected very well with the BSD. The surface planes of a 12.2° <111> grain boundary had a deviation of 0.5°. Correspondingly orientation contrast was less distinct than in the previous sample. In another SEM (LEO 1530) with another BSD the boundary could not be detected at all. Therefore, we conclude that the smallest deviation that can be detected with our equipment is about 0.5° (JEOL 820). These experiments were conducted at room temperature. The orientation contrast obtained at higher temperatures is shown in Fig. 2. Apparently, with increasing temperature the orientation contrast deteriorates.

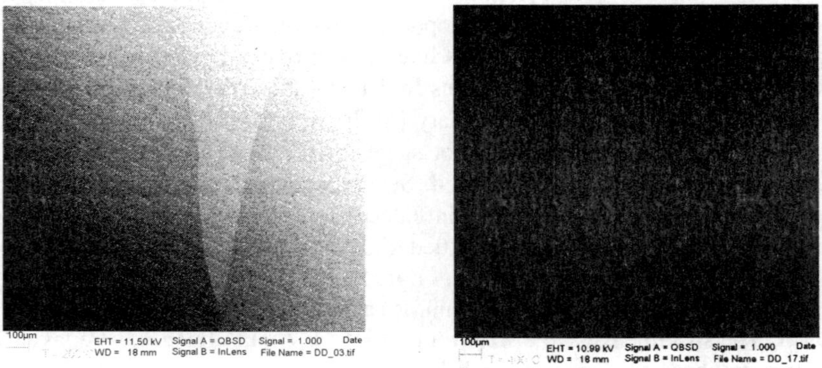

a) b)

Fig. 2. Orientation contrast of high angle grain boundaries, (a) at 200°; (b) at 400°C.

4.2 In-situ heating experiment on an Al tricrystal

The motion of a grain boundary system with a triple junction was investigated in the temperature range between 420° and 500°C. The investigated triple junction consisted of two asymmetrical 40° <111> tilt grain boundaries (GB I and II) and one symmetrical 80° <111> tilt boundary, which is crystallographically equivalent to the symmetrical 40° <111> tilt boundary. The mobility and energy of the ± 40° <111> boundaries are the same, the system can be regarded as a uniform system [2]. For this specimen the system velocity was found to remain constant during the experiment at a given temperature and increased with rising temperature (Fig. 3). The angle θ was seen to increase with rising temperature. Using the measured values of θ, the dimensionless parameter Λ was calculated for each annealing temperature (Eq. 1), and the temperature dependency of Λ was determined, which is shown in Fig. 4. Due to the temperature dependence of θ, the criterion Λ was found to increase with increasing temperature as well. At relatively low temperatures up to 440°C the value of Λ is of the order of unity.

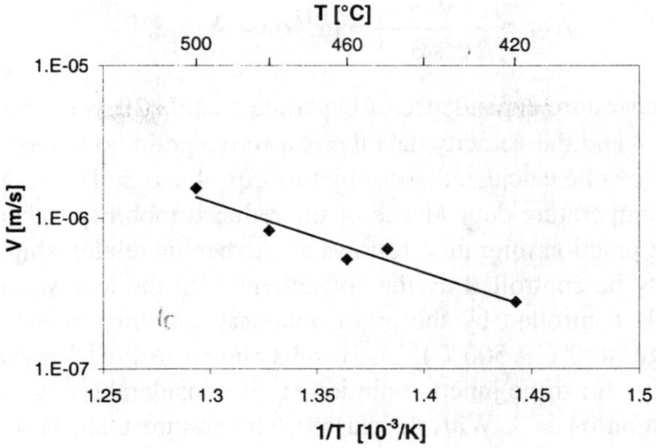

Fig. 3. Temperature dependence of the velocity V of the uniform 40° <111> grain boundary system.

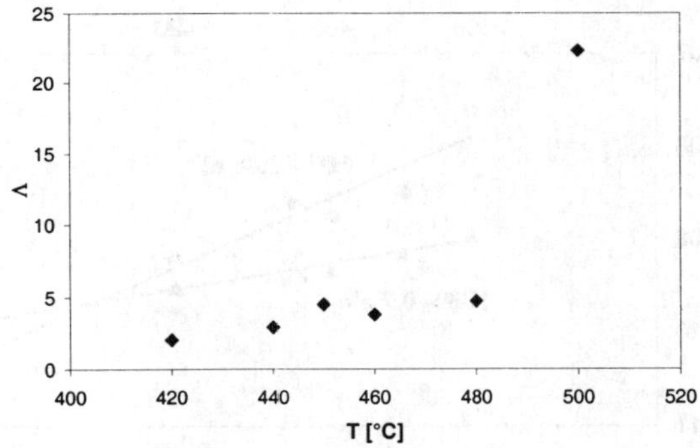

Fig. 4. Temperature dependence of the criterion Λ for the uniform 40° <111> grain boundary system.

Low values of Λ means that the triple junction mobility is comparable to the grain boundary mobility since the criterion Λ specifies the ratio of both mobilities (Eq. 1). Therefore the motion of the boundary system in the low temperature regime will be affected by the triple junction. The rise of Λ with increasing temperature indicates that the motion of the system becomes increasingly less affected by the triple junction and the boundary kinetics becomes predominant.
To estimate the influence of the triple junction on the grain boundary system we determined the temperature dependence of the reduced mobility A of the involved triple junction and the grain boundaries. Using Eqs. 2 and 3 we obtain

$$A_{GB} \equiv \frac{V \cdot a}{2\theta} = m^{GB}\sigma = A_{0B}e^{-H^{GB}/kT} \qquad (4)$$

$$A_{TJ} \equiv \frac{V \cdot a}{2\cos\theta - 1} = m^{TJ}\sigma a = A_{0TJ} e^{-H^{TJ}/kT} . \quad (5)$$

From the temperature dependence of the contact angle 2θ as expressed in terms of Λ (T) in Fig. 4 and the velocity data the respective grain boundary and triple junction mobility can be calculated according to Eqs. 4 and 5. The results are given in Fig. 5. The temperature dependence of the reduced mobility for grain boundaries and for triple junction migration follows an Arrhenius relationship. As the system will obviously be controlled by the constituent with the lowest mobility, this actual system is controlled by the grain boundary mobility in the measured temperature range (420°C - 500°C). The results shown in Fig. 5 reveal that the activation enthalpy for triple junction motion H^{TJ} is considerably higher than for grain boundary migration H^{GB}. With decreasing temperature there is a transition from grain boundary kinetics to triple junction kinetics for the measured system. The transition temperature is estimated to be about 400°C. Below 400°C the system is obviously controlled by the reduced triple junction mobility.

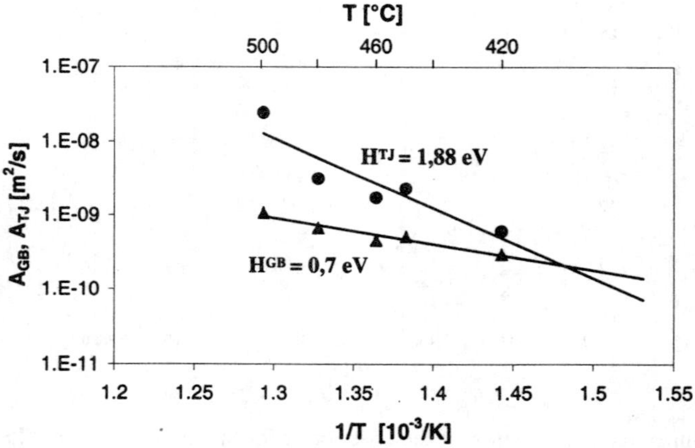

Fig. 5. Temperature dependence of triple junction and grain boundary mobility, triangles: (reduced) grain boundary mobility, circles: triple junction mobility.

6 Conclusion

A method was presented, which allows in-situ investigations of the motion and geometry of a grain boundary system containing a triple junction and the measurement of its mobility. Thus, boundary mobility, triple junction mobility and boundary shape could be measured concurrently. In an example of an Al tricrystal it was demonstrated that the motion of a grain boundary system at high temperatures was less affected by the triple junction, and therefore effectively controlled by the grain boundary mobility. But in a low temperature regime the motion of the grain boundary system will be controlled by the slowly moving triple junction. As expected the influence of triple junctions depends on temperature.

Acknowledgdement

The authors are grateful to the Deutsche Forschungsgemeinschaft for financial support of the program through contracts MO 848/3-1.

References

[1] V.E. Fradkov, L.S. Shvindlerman, *Phys. Chem. Mech. Surfaces*, 1, 1982, pp. 180-191.

[2] U. Czubayko, V.G. Sursaeva, G. Gottstein, L.S. Shvindlerman, *Acta mater.*, 46, 1998, pp. 5863-5871.

[3] D. Weygand, Y. Brechet, J. Lepinoux, *Acta mater.*, 46, 1998, p. 6559.

[4] U. Czubayko, D.A. Molodov, B.-C. Petersen, G. Gottstein, L.S. Shvindlerman, *Meas. Sci. Technol.*, 6, 1995, pp. 947-952.

[5] F.J. Humphreys, M. Ferry, I. Brought, C.P. Johnson, *Textures Microstruct.*, 26-27, 1996, pp. 281-301.

Capillarity Vector Reconstruction of the Relative Energies of [0001] Tilt Boundaries in Alumina

D.M. Saylor, K. W. Cherry, and G.S. Rohrer

Department of Materials Science and Engineering, Carnegie Mellon University, Pittsburgh, PA, 15213-3890, USA

Keywords: grain boundaries, capillarity vector, Al_2O_3, grain boundary energy.

Abstract. We have measured the geometry and crystallography of the grain boundaries that form 124 triple junctions in a specimen of polycrystalline alumina. Each of these triple junctions is comprised of [0001] tilt boundaries which were assumed to be in local thermal equilibrium. By using the capillarity vector reconstruction method, we determined the energies of these boundaries as a function of the tilt angle and the grain boundary plane. The results show that for misorientations less than 25°, the relative energy increases with the misorientation angle, reaches a maximum, and then decreases as the Σ3 misorientation is approached at 60°. Furthermore, the energy anisotropy caused by changes in the grain boundary plane is nearly as large as that caused by changes in misorientation.

1 Introduction

The physical and chemical properties of grain boundaries are known to be anisotropic. It has been pointed out that for a microstructure whose interfacial junctions are in local equilibrium, the mesoscale structure of the grain boundary network contains an imprint of the grain boundary properties [1]. By measuring the geometry and crystallography of the three interfaces that meet at tri-grain junctions, and assuming that Herring's [2] condition for local equilibrium is satisfied, it is possible to determine the anisotropy of the grain boundary free energy. In this paper, we reconstruct the relative grain boundary energies from microstructural observations using the capillarity vector reconstruction method, as originally described by Morawiec [3]. This method has previously been used to reconstruct the anisotropy of the surface energy of magnesia from thermal groove data [4]. In the present paper, we apply the same technique to extract relative grain boundary energies from observations of tri-grain junctions, referred to as triple junctions.
We consider a grain boundary's character to be determined by its crystallographic lattice misorientation and the boundary plane. The number of distinct grain boundary characters expected in a sample with random texture is very large, and this has made it challenging to characterize their diversity and establish character-

property correlations [5]. In the present case, we have avoided the problem by considering a sample with strong crystallographic and morphological texture so that there are only two degrees of freedom and the space of populated characters is significantly reduced. The sample has platy grains, with the basal plane aligned perpendicular to the sample normal so that there is a [0001] axial texture that is approximately 100 times random at maximum and most of the boundary planes are parallel to the sample normal [6]. Because of this, there are a significant number of boundaries that can be represented by a single <0001> misorientation axis and a boundary plane whose normal is perpendicular to this axis. Therefore, the boundaries we consider here are all of tilt character so that the misorientation and the boundary plane are each described by a single angular parameter. Because of the triad parallel to <0001>, all distinct misorientations (tilts) are between 0 and $\pi/3$ and all boundary plane normals (inclinations) are between 0 and $2\pi/3$.

2 Experimental Details

The sample used in this study was produced by Brandon's group as described previously [6]. Secondary electron (SE) images of grain boundary triple junctions were obtained in a Philips XL40 FEG scanning electron microscope (SEM) with the stage at 0° tilt. The images were captured at 20,000x magnification such that only one triple junction appeared in each image (an example is shown in Fig. 1). After collecting the SE images, the sample stage was tilted to an angle of 60° and electron backscattered diffraction patterns (EBSPs) were obtained for the grains bounding each characterized triple junction. The EBSPs were automatically indexed using TSL analysis software to obtain the grain orientations, and the indexing solution for each pattern was checked manually. Grain orientations from OIM tend to have systematic errors associated with them for a variety of experimental reasons, such as misalignment in the SEM or incorrect pattern center. The systematic errors were corrected by rotating the data such that the peak of the <0001> pole figure was centered about the sample normal. Each image was corrected such that the pixel to micron ratio was identical along the x and y axes. The corrected images were then used to manually measure the in-plane dihedral angles, as illustrated in Fig. 1.

Grain orientations were parameterized in terms of Eulerian angles, (ϕ_1,Φ,ϕ_2), from which the misorientation for any pair of grains can be determined. Because alumina is trigonal, there are 6 x 6 x 2 = 72 symmetrically equivalent grain misorientations. Thus, for grains with orientations g_1 and g_2, we consider misorientations $\Delta g=(C_i g_1)(C_j g_2)^{-1}$ and $\Delta g=(C_j g_2)(C_i g_1)^{-1}$, where the C_i are the symmetry operators for this crystal system. Therefore, the misorientation, θ, is $\cos^{-1}[(tr(\Delta g)-1)/2]$ and we define the fundamental zone of misorientation space as that for which θ is as small as possible and the misorientation axis, $n_i=\varepsilon_{ijk}\Delta g_{jk}/2\sin(\theta)$, is such that $n_3 \geq 0$, $n_1 \geq 0$, and $(n_2/ n_1)^2 \leq 1/3$. To characterize the boundary plane, we define the tilt angle, ρ, to be the angle between the boundary plane normal and <1000>. When deter-

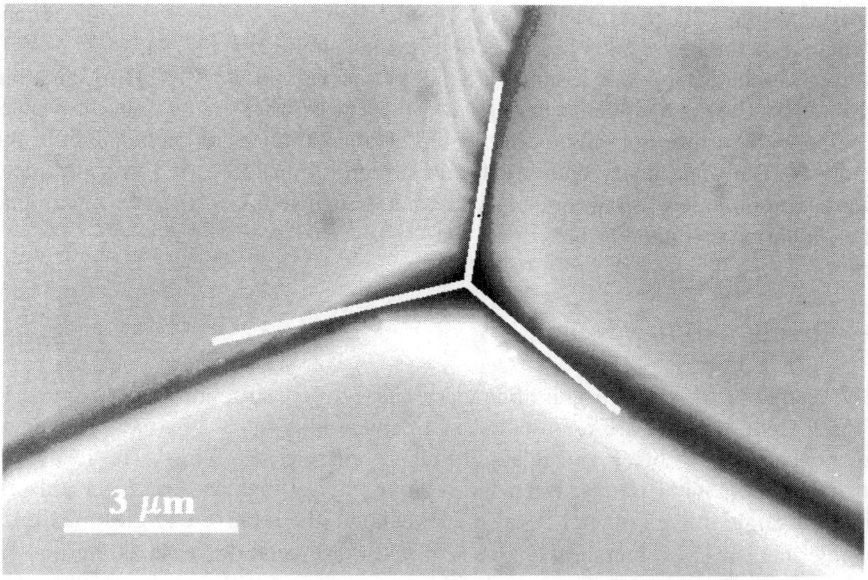

Fig. 1. Secondary electron image typical of those used in the analysis. Lines drawn on the image are used to determine the dihedral angles and the in plane angles.

determining a tilt angle for each boundary, we are obliged to select one of the two adjoining crystal orientations as the reference frame. Since the calculated values of Δg that fall in the fundamental zone of misorientation space always arise from the same combinations of non-inverted and inverted orientations, we select the non-inverted orientation as the reference frame.

To reconstruct the grain boundary energy, we assign to each interface a capillarity vector, $\vec{\xi}$, as defined by Hoffman and Cahn [7,8]. The component of the capillarity vector normal to the interface is equal in magnitude to the boundary energy, γ, such that $\vec{\xi} \times \hat{l} = \gamma\,\hat{t}$, where $\hat{t}$ is the unit vector that lies in the interface plane and points in a direction perpendicular to and away from the line of intersection of the three interfaces ($\hat{l}$). The component of the capillarity vector in the plane tangent to the interface that is related to the change in γ with orientation is given by $(\partial\gamma/\partial\beta)_{max}\,\hat{t}_0$, where $\hat{t}_0$ points in the direction of maximum increase of γ and, β is the right handed angle of rotation about $\hat{l}$, which is measured from a reference direction. This is a measure of the so-called torque force that is normal to the interface and urges it to rotate about the line of intersection. In other words, $\vec{\xi} \times \hat{l} = (\partial\gamma/\partial\beta)\hat{n}$. Using our experimental observations of $\hat{n}$, $\hat{l}$, and the dihedral angles, we can assume that the Herring equilibrium equation applies $(\vec{\xi}^1 + \vec{\xi}^2 + \vec{\xi}^3) \times \hat{l} = 0$, where the superscripts label the three interfaces at the triple junction) and determine a set of capillarity vectors that most nearly satisfies

this condition at each triple junction. The details of the capillarity vector reconstruction method have been described in previous publications [3, 4].
In treating the data, it was assumed that all grains had a perfect [0001] orientation with respect to the sample normal. In fact, there was some scatter about this pole. Grains with surface normals more then 15° from [0001] were excluded from the analysis. The preliminary data set discussed here is made up of 124 triple junctions. The domain of grain boundary characters is discretized in such a way that the resolution is limited to 12°.

3 Results and Discussion

The results of the relative grain boundary energy reconstruction are illustrated in Fig. 2. Here, the data are represented in a density plot where relatively low energies correspond to black and relatively higher energies are white. The results illustrate two general trends. First, the relative energy increases with the misorientation angle in the range of $0 < \theta < 25°$. Second, the relative energies maximize at intermediate misorientation angles ($25° < \theta < 40°$) and decrease as the misorientation approaches 60°. The two trends are consistent with conventional theory regarding low angle boundaries and boundaries near coincident site lattice (CSL) orientations [9,10]. At low misorientations, the tilts can be accommodated by edge dislocations parallel to [0001] and, the relatively lower energy of boundaries with a near 60° misorientation are expected because there is a $\Sigma 3$ CSL boundary at this point.
Perhaps the most interesting feature of the results is that the variation of the energy along the boundary plane axis is almost as large as those along the misorientation axis. For boundaries with misorientations less than 25°, as the boundary plane rotates away from <1000>, the energy increases. Previously, relative grain boundary energies for alumina were determined using a thermal groove technique and the dispersion of the energies for boundaries with similar misorientations was nearly as large as the total variation in energy [11,12]. The current observations confirm the idea that the energy of a boundary varies with both its misorientation (θ) and its inclination (ρ). At the $\Sigma 3$ misorientation ($\theta = 60°$), we expect minima in the relative energies at $\rho = 30°$ and $\rho = 90°$, where lattice points that actually fall on the boundary plane maximize coincidence. There are minima near both of these orientations that, considering the limited resolution of this reconstruction (12°), might be ascribed to this special boundary. By accumulating more data, we plan to increase the resolution of the reconstructed grain boundary energy and establish a more detailed relationship between grain boundary character and energy.

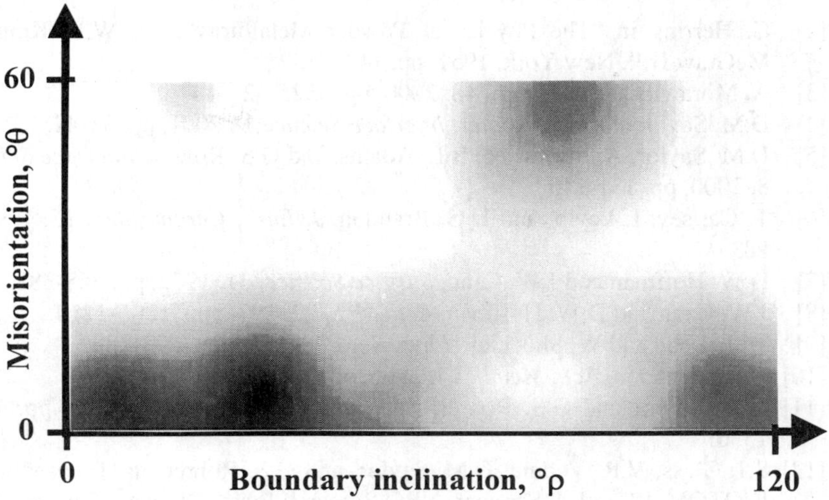

Fig. 2. The reconstructed relative grain boundary energy, as a function of misorientation and inclination. The darkest points represent the lowest relative energy (0.9) and the lightest represent the highest (1.02).

4 Summary

Geometric and crystallographic data obtained from triple junctions in a textured alumina polycrystal have been used to determine the relative energies of [0001] tilt boundaries using the capillarity vector reconstruction method. The data show anticipated variations both at low misorientations and near the Σ3 misorientation. Furthermore, the energy anisotropy derived from changes in the boundary inclination are almost as large as that derived from the lattice misorientation.

Acknowledgement

This work was supported primarily by the MRSEC program of the National Science Foundation under award number DMR-0079996.

References

[1] B.L. Adams, D. Kinderlehrer, W.W. Mullins, A.D. Rollett, and S. Ta'asan, *Scripta Mater.*, 38, 1998, pp. 531-536.

[2] C. Herring in "The Physics of Powder Metallurgy", ed. W.E. Kingston, McGraw Hill, New York, 1951, pp. 143-79.
[3] A. Morawiec, *Acta Mater.*, 48, 2000, pp. 3525-32.
[4] D.M. Saylor and G.S. Rohrer, *Interface Science*, 9, 2001, pp. 33-44.
[5] D.M. Saylor, A. Morawiec, B.L. Adams, and G.S. Rohrer, *Interface Science*, 8, 2000, pp. 131-140.
[6] T. Carisey, I. Levin, and D.G. Brandon, *J. Euro. Ceram. Soc.*, 15, 1995, p. 283.
[7] D.W. Hoffman and J.W. Cahn, *Surface Science*, 31, 1972, pp. 368-388.
[8] J.W. Cahn and D.W. Hoffman, *Acta Met.*, 22, 1974, pp. 1205-1214.
[9] W.T. Read and W. Shockley, *Phys. Rev.*, 78, 1950, pp. 275-289.
[10] D. G. Brandon, *Acta Metall.* 14, 1966, pp. 1479-84.
[11] D.M. Saylor and G.S. Rohrer, *J. Amer. Ceram. Soc.* 82, 1999, pp. 1529-1536.
[12] S.J. Glass, V.R. Vedula, D.M. Saylor, and G.S. Rohrer, in "Proceedings of ICOTOM-12", ed. J. Szpunar, NRC Research Press, Ottawa, 1999, pp. 1489-1494.

Recrystallization and Grain Growth
Proceedings of the First Joint International Conference
Eds. G. Gottstein and D.A. Molodov, Springer-Verlag, 2001

Grain Boundary Properties: Qualitative Analysis of the Relationship "Misorientation-Structure-Energy"

V.P. Yashnikov

Institute of Solid State Physics RAS, Chernogolovka, Moscow distr., 142432, Russia

Keywords: GB energy, relaxation, misorientation, critical points, structural multiplicity.

Abstract. The structure of GB energy functional in the space of atomic equilibrium configurations in a neighborhood of an ideal (non-relaxed) bicrystalline configuration is discussed in the connection with the problem of stuctural multiplicity of the relaxed grain boundary. The uniformization procedure for the multi-valued mapping "misorientation of ideal bicrystalline structure-relaxed grain boundary atomic configuration" over the misorientation space is considered. The existence of topological singularities such as maxima, minima and saddle points of the GB energy on the distal surface of the grain orientation space are derived from the Morse inequalities.

1 Introduction

Many measurements and computer simulations what are made to the present time for bicrystals of metals, dielectrics and semiconductors show that the dependence of grain boundary energy on the grain misorientation angle and axis may have complicated, non-monotonic and even oscillating character [1-9]. The latter is connected with the non-monotonicity of the degree of structural coherence by fitting two misoriented crystalline half-spaces generating the bicrystalline system.
Although the problem is of great fundamental and applied importance, there is no theoretical treatment of it which is developed so far as to provide us with an explicit analytic expression of the GB energy as the function of GB misorientation parameters in the whole space of geometricaly possible orientations. Therefore the qualitative methods using topological argumentation and crystallographic symmetry reasons are especially of interest in the setting.
Before to start the analysis of topological singularities of the GB energy as the function of grain misorientation we consider the more complicated primery geometric construction such as the relaxed GB atomic equilibrium positions configuration as the function of misorientation. Thereby the principle of the spontaneous symmetry breakdown plays the key role. In according to that any stable GB atomic configuration may be obtained as the result of elastic relaxation process

from am appropriate ideal (i.e. elastically not distorted) bicrystalline configuration.

2 Imbedding of a Bicrystalline Configuration in the Complete Family of the Generic Ones

We start from the more simple problem of the imbedding of an ideal bicrystalline pattern (configuration) with arbitrarily chosen and fixed misorientation. The interface in the initial state of the system will be assumed planar one for the further simplification of the analysis. The use of a curved surface is possible under additional restriction that the radius of curvature at every point is essentially greater than the lattice parameter of the crystal pattern, This "marked" or "seed" configuration is denoted by {K(o); S(o); g(o)K(o)}. Thereby one of the two half-spaces, in which the plane S(o) divides the three-dimensional space, is filled by points of the pattern K(o) and the other complimentary half-space is occupied by points of the pattern g(o)K(o).
If the pattern K(o) and the plane So leaves to be unchanged and we apply to K(o) all possible rotations g having the same center as g(o) then we obtain the complete three-parametric family of the bicrystalline patterns which are, by definition, generic to the "seed" configuration. If the pattern K(o) belongs to the triclinic class the family is topologically equivalent to the group space of three-dimensional rotations. In the alternative case, when K(o) has proper rotational symmetries, the family of generic bicrystalline patterns has the topological structure of the two-sided homogeneous space P\SO(3)\P with respect to the proper point group of the pattern K(o).
It is useful to represent the family G{K(o); S(o); g K(o)) of generic bicrystalline patterns by points of the infinite-dimensional configuration space Con (E) over the conventional three-dimensional space E. The complete family of generic patterns is mapped then by the three- dimensional closed surface

$$g \Longrightarrow \mathrm{Con}\ (B\{K(o);\ S(o);\ g\ K(o)\} \qquad (1)$$

in the space of configurations. It should be noted on the step of analysis that the correspondence between misorientations [P g P and points {K(o); S(o); g K(o)} of the configuration space Con(B) is one-to one.

3 Spontaneous Symmetry Breakdown, Elastic Relaxation and Structural Multiplicity of a Grain Boundary

Let us assume that identical particles interacting by a pair potential V(x) are placed at every point of a bicrystalline pattern. It is assumed that V(x) is spherically symmetric one with the range of repulsion by $r < ro$ and with the range of attraction by $r > r(o)$. Details concerning various modifications of inter-atomic potentials in computer simulations of crystal structures and interfaces can be found for example in [10-12]. One consider as a rule (in order to avoid the difficulties connected with the correct description of the behavior of a bicrystalline system "at the infinity") the portion of the infinite system contained in a finite domain D(s) where the parameter s characterizes the size of the macroscopic cutoff domain D(s).

The periodic boundary conditions can be used if the cutoff domains D(s) are chosen to be identical to the polyhedra which are homotetic to the Wigner-Seitz cell of the one of two crystals in the bicrystaline pattern. The parameter s is equal in the case to the diameter of the macroscopic cutoff polyhedron. The D(s)-portion of the bicrystalline pattern {K(o); S(o); g K(o)} must be extended on the complete space E by the appropriate translations.

The system of identical particles placed at the points of the "finite" ideal bicrystalilne configuration {K(o); S(o); g K(o); D(s)} is elastically stressed independently on the choice of the conditions on the geometric boundary of D(s). Therefore, the relaxation process starts and minimizes the energy of the system via atomic displacements . There is a series problems which are of the importance for our understanding the nature and structure of interfaces:

(1) Existance of the limite interfacial structure (atomic equilibrium positions configuration) in the relaxation process in the case where macroscopic cutoff domain D(s) is fixed.
(2) Uniqueness conditions of the relaxed configuration.
(3) Effect of the shape and size of the macroscopic cutoff domain D(s) on the relaxation process and the limit GB configuration.
(4) Effect of the conditions which are prescribed on the geometric boundary of the cutoff domain D(s).
(5) Effect of the choice of inter-atomic potential and effect of the shape and size of the microscopic cutoff domain for the chosen inter-atomic potential V(x).

Rigorous analytical results concerning the questions are unknown in the literature. However, computer simulations show that the fact of existence of the relaxed interface is sensitive to the choice of inter-atomic potential [8]. It should be noted that answering the questions (1) - (5) is conditioned by the structure of the energy functional in a neighborhood of the surface (1) in the configuration space. The a priory analysis of possible cases is given below.

4 Local and Global Analysis of the Mapping "Seed Pattern Misorientation-Relaxed GB Configuration-Energy"

Qualitative analysis of the relationship "GB atomic structure-energy" is most simple when energy functional has one and only one local minimum in a neighborhood of the "seed" configuration {K(o); S(o); g K(o); D(s)} for all sufficiently large values of the parameter s. In the frame of the global analysis the variety of the probe (or test) configurations using in the minimization procedure of the energy functional should be described rigorously and in details because the qualitative approach is purposed on an analysis of the problems 1)-5). using topologic and geometric structure of the variety of all bicrystalline configurations which are admissible in the extremal problem with the prescribed data {K(o); S(o); g K(o); D(s)}.

Besides that an appropriate analog of the notion of the variational derivation of energy with respect to localized changes of any virtual interface in the bicrystalline system of N particles should be introduced. It is needed to show that variational derivation of the energy functional is equal to zero at any point of the stationary interface.

We omit here many mathematical details and come back to the case when the problem of minimum of the energy functional on the variety of admissible configurations has one and only one solution for all sufficiently large values of the macroscopic cutoff size s. It will be assumed additionally that the stabilization condition is fulfilled, i.e. all changes in relaxed configurations have the character of small corrections with growing s. Then GB energy is a single-valued function of misorientation and possible dependence of it on the parameter s may be neglected for large values of the one.

The basic question weather or not the correspondence “misorientation-relaxed GB configuration” may be considered .as a smooth mapping of the misorientation space into the configuration space Con(E) is complicated and although it is of the basic importance there is no clear answer on it at the present time. The possibility of the existence on GBs of “rough” substructures of mesoscopic scale shows that the property of continuity and smoothness of the above mentioned correspondence may be fulfilled only after appropriate averaging procedure.

Nevertheless, there are reasons to consider the GB energy as a more smooth function of misorientation than the GB atomic configuration. This may be shown via introducing the ordering parameter such as the GB electron density in the relaxed state. So, we assume here that GB energy after relaxation is a smooth function correctly defined on the misorientation space P\SO(3)\P. Therefore, the theory of M. Morse [13,14] may be used for the analysis of the non-degenerate singularities of the GB energy as the function of misorientation.

If we use the Morse inequalities we obtain, in particular, the lower estimation of the total number of critical points of GB energy and, separately, the number of non-degenerate maxima, one of minima and one of the saddle points. It should be noted that the Morse theory is one of general mathematical character. The theory demonstrates that the minimal possible number of critical points of a smooth function defined on a smooth variety depends on the “topological complexity” of the variety.

Let us analyze the case where GB energy has finite number of different minima. on the variety of bicrystalline configurations which have the same board as the ideal “seed“ configuration {K(o); S(o); g K(o); D(s)}. If this is valid for a misorientation P g P, then it is valid in a topologically open neighborhood U(g) of the point P g P in the homogeneous space P\SO(3)\P. In the other words the property of the mapping "misorientation- relaxed configuration " to be multi- valued (i.e. n-valued by n>1) is stable relative to sufficiently small perturbations of the misorientation P g P. Therefore, the set of the multiple points is topologically open in the misorientation space. The set of all possible multiple misorientations is denoted by the symbol M=M{K(o); S(o); D(s)}. The following three cases are possible:

I M is a connected domain of the misorientation space.

II M is the union of a finite family of mutually disjoint and connected domains.

III M is the union of infinite (countable) family disjoint and connected domains.

Let us note that the angular diameter of the misorientation space is finite, thus in the case III, there is for any real number d>0 only finite family of the domains whose angular diameters are greater than d. The set U=U{K(o); S(o); g K(o); D(s)} of the structurally simple (i.e. non-multiple) misorientations is the complement (in the set-theoretical sense) to the open set of multiple misorientations. Thus the set U is a closed subset of the misorientation space. Let us mention that, in general, a closed set in P\SO(3)\P may be of a very complicated structure. It is discussed below only one case which is most simple from topological viewpoint. Let us assume that there is no interior points in the set U, i.e. the case I is realized. It is well known from the general topology that a closed set in the space such as P\SO(3)\P may be extremely complicated in the structural sense. We will have confined in the present paper by analysis of the simplest topologic model of the multi-valued relationship "misorientation-relaxed GB configuration". Therefore it will be assumed, in addition to the above, that U is the intrinsically-branched two-dimensional surface that subdivides the misorientation space into a finite family of simply- connected domains. The domains are filled by multiple misorientations. So, the existence of several relaxed modification for GBs is the "typical event in this model. At the same time the existence of simple (non-multiple) GB structures is the "exceptional event" in the space of misorientations.
Let us note that theoretically the GB energy in the case I can be single-valued function of misorientation. This means that all structural modifications of the relaxed GB have one and the same value of the energy. Then the Morse inequalities may be directly applied to the dependence "misorientation- GB energy". However. In general case, we must assume that the energies of different modifications are not equal. Then GB energy is a multi-valued function of misorientattion and the Morse inequalities must be used after a correct procedure of uniformization in the whole misorientation space.

5 Conclusions

It is given and analyzed a general formulation of the problem of the dependence of GB relaxed atomic configurations and GB energy in bicrystals on the so important exterior, properties controlling parameter as the grain misorientation.

It is shown that the correct imbedding of the relaxed interface between two misoriented grains of a bicrystal in the complete three-parametric family of the generic bicrystalline structures may lead to the topologically-branched covering of the related misorientation space by the variety of elastically relaxed GB configurations.

The energy of relaxed GB atomic configurations versus grain misorientation is the crystallographically symmetric) function on the group space SO(3) of the three-dimensional rotations. Symmetry properties of the GB energy are described by the discrete group of the motions generated by left and right shifts of the group space on the elements belonging to the proper point group of the grain lattice. Topological singularities (critical points) of the GB energy located on the distal surface of the primitive orientation polyhedron. The total number of topological maximum, minimum and saddle points of GB energy are estimated by the Morse inequalities.

References

[1] "Grain Boundary Structure and Properties", eds. G. A. Chadwick, D. A. Smith, Academic Press, 1976.

[2] E. V . Fridman , Ch. V. Kopezky and L. S. Shvindlerman, *Z. Metallkunde*, 66, 1975, pp. 535.

[3] "Interfaces in Crystalline Materials", eds. A.P. Sutton and R. W. Baluffi, Clarendon Press, 1995.

[4] R. W. Baluffi and A. P. Sutton, *Mater. Sci. Forum*, 207-209.1996, pp.1-12.

[5] G. Gottstein and L. S. Shvindlerman, "Grain Boundary Migration in Metals: Thermodynamics, Kinetics, Applications", CRC Press, 1999.

[6] D. Wolf and K. L. Merkle, in "Materials Interfaces", eds. D. Wolf and S. Yip, Chapman and Hall, London,1992, pp.87-150.

[7] A. Otsuki, *Materials Science Forum*, 207-209, 1996, pp. 413-416.

[8] V. Pontikis, *J. Physique*, C5. Suppl. no. 10, 1988, pp. 327-335.

[9] J. F. Lutsko, D. Wolf and S. Yip, *Ibid.* pp. 375-379.

[10] M. I. Baskes and C.F. Melius, *Phys. Rev.* B, 20, 1978, pp. 3197-3204.

[11] M. Parinello and A, Rahman, *Phys. Rev. Letters*, 45, 1980, pp. 1196-1199.

[12] D.W. Heerman, "Computer Simulation Methods in Theoretical Physics", Springer-Verlag, 1988.

[13] M. Morse, *Trans. Amer. Math. Soc.*, 27, 1925, pp. 345-396.

[14] M. Morse, *Trans. Amer. Math. Soc.*, 33, 1931, pp. 72-91.

Motion of Grain Boundaries with Triple Junctions in Aluminium

D. A. Molodov[1], G. Gottstein[1], S. G. Protasova[2], V. G. Sursaeva[2], and L. S. Shvindlerman[1,2]

[1] Institut für Metallkunde und Metallphysik, RWTH Aachen, D-52056 Aachen, Germany

[2] Institute of Solid State Physics, Russian Academy of Science, Chernogolovka, Moscow, distr., 142432 Russia

Keywords: grain boundary, tricrystals, mobility, activation enthalpy.

Abstract. The results of an investigation of the steady state motion of grain boundary systems with triple junctions in high purity Al are presented. In particular, the migration of systems with <111> and <110> tilt boundaries was studied. The experimental results demonstrate that the motion of grain boundary systems with triple junctions in Al can be controlled by slowly moving triple junctions. The influence of triple junctions depends on temperature. It is particularly strong at low temperatures. In the high temperature regime the motion of a connected grain boundary system is less affected by the triple junction, and, therefore, effectively controlled by the grain boundary mobility. The experiments reveal a drastic difference between activation enthalpy of grain boundary and triple junction motion. Therefore, there is a temperature, below which triple junctions govern the motion of the connected boundary system. This temperature was found to depend on the particular grain boundary and triple junction geometry.

1 Introduction

To determine the effect of triple junctions on the kinetics of grain growth quantitatively, the mobility of a triple junction has to be measured. However, a steady-state motion of a grain boundary system with a triple junction, where the curved boundaries maintain their shape during motion so that the entire system moves with the same and constant velocity v, is only possible in a very small set of geometrical boundary configurations. One of them is shown in Fig. 1. This model grain boundary system can move steadily. For 2D-systems, assuming that all grain boundaries possess equal properties, irrespective of grain misorientation and crystallographic orientation of the boundaries (so called uniform grain boundary model), the shape and velocity of moving grain boundaries with a triple junction can be calculated analytically. The complete analysis of the motion of such grain boundary system is given in Ref. [1,2].

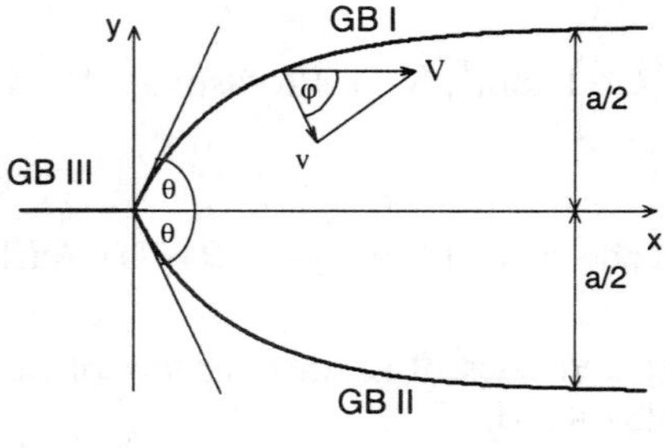

Fig. 1. Geometry of the grain boundary system with triple junction during steady-state motion.

The shape of the GB system in the steady-state motion was predicted for both uniform grain boundary model and the case when the system is symmetric relative to the x-axis, i.e. the grain boundaries I and II are the same, but different from boundary III. The respective surface tensions and mobilities of the boundaries are : $\sigma_1 = \sigma_2 \equiv \sigma \neq \sigma_3$, $m_{b1} = m_{b2} = m_b \neq m_{b3}$. For the situation given in Fig. 1 the velocity of the triple junction v_{tj} can be expressed as [1,2]

$$v_{tj} = m_{tj}(2\sigma\cos\theta - \sigma_3) \qquad (1)$$

the steady-state value of the angle θ is

$$\frac{2\theta}{2\cos\theta - \sigma_3/\sigma} = \frac{m_{tj}a}{m_b} = \Lambda \qquad (2)$$

The criterion Λ reflects the drag effect of the triple junction on the motion of the system. For a low mobility of the triple junction ($\Lambda \to 0$) the motion of the system is controlled by the mobility of the triple junction. For the opposite limiting situation ($\Lambda \to \infty$) the motion of the system is governed by the grain boundary mobility. The velocity of motion of the boundary system is then given by

$$V = \frac{2\theta_{eq} m_b \sigma}{a} \qquad (3)$$

where the equilibrium triple junction angle θ_{eq} is equal to

$$\theta_{eq} = \arccos(\sigma_3/2\sigma) \qquad (4)$$

The state of motion of the entire grain boundary system can be distinguished experimentally for a given ratio σ_3/σ by measuring the contact angle θ.

The motion of a grain boundary system with a triple junction with a geometry as shown in Fig. 1 was measured in zinc-tricrystals [3]. From a considerable number of experiments on samples with different grain boundary geometries it was evidenced that triple junctions can retard effectively grain boundary motion. This result is of principal importance for our fundamental understanding of microstructure evolution.

The current study for the first time addresses the effect of triple junctions on grain boundary motion in Al by measurement of the motion of specific grain boundary systems with triple junctions in tricrystals. Since the mobility of grain boundaries in Al has been well investigated in bicrystal experiments, such measurements permit a direct comparison of grain boundary and triple junction properties.

2 Experimental

The experiments were carried out on tricrystals of high purity (99,999%) aluminium with a grain boundary geometry as shown in Fig. 1. The crystallographic characteristics of the studied tricrystals are given in Table 1. The orientations of the three contiguous grains of each sample were determined by the Laue technique and electron back scatter diffraction (EBSD).

Table 1. Misorientation of the investigated tricrystals.

Sample	GBI	GBII	GBIII
SI	21°<111>	18°<111>	3°<111>
SII	27°<110>	22°<110>	5°<110>
SIII	44°<110>	29°<110>	15°<110>

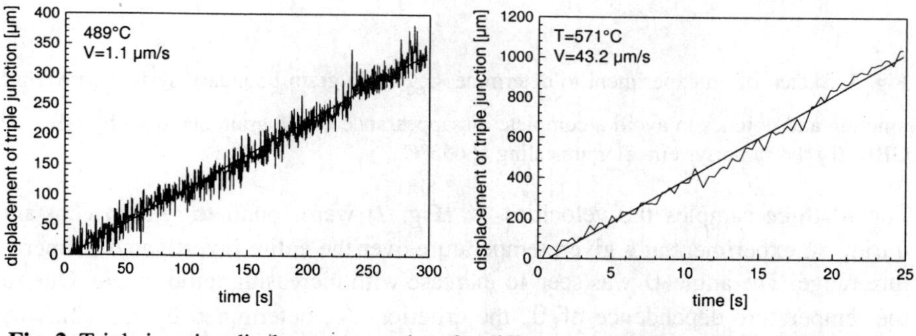

Fig. 2. Triple junction displacement vs. time for different temperatures for sample SII.

The tricrystals were grown by directional crystallization. Prior to the measurements the samples were electrolytically polished to improve the surface quality and to remove any grain boundary grooves. For measuring the migration rate of the grain boundary system an X-ray diffraction technique was utilized [4]. Fig. 2 shows the time–displacement diagrams for the motion of one of the investigated grain boundary systems at different temperatures. For each temperature the velocity v, the angle θ, and the width of the shrinking grain a were determined. For this the sample was rapidly cooled down after measurements of the migration rate at each temperature. The position of the grain boundary system and the angle θ were recorded from the grain boundary grooves. Before subsequent heating to measure the migration rate the specimen was polished to remove the grooves.

3 Results

The motion of three grain boundaries systems with triple junctions (Table 1) was investigated in the temperature range between 400 and 590°C. The investigated triple junctions consisted of two high angle tilt grain boundaries (GBI and GBII) and a low angle tilt boundary (GB III). Due to different properties of low angle and high angle grain boundaries and under the assumption that the properties of high angle grain boundaries vary only little with changing misorientation, the investigated grain boundary systems can be regarded as symmetrical junctions.

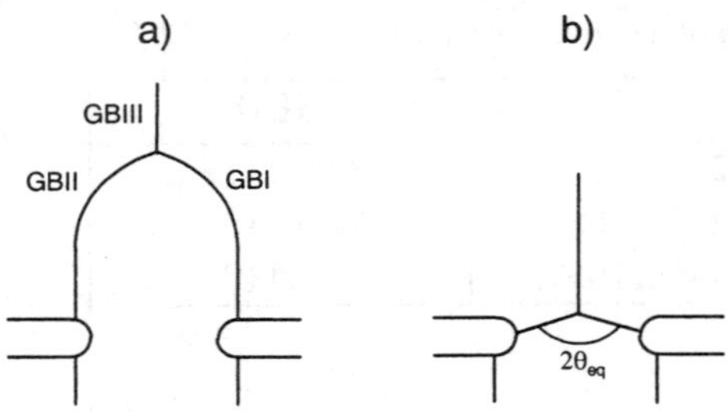

Fig. 3. Sketch of an experiment to determine θ_{eq}. (a) grain boundary system with triple junction and notches to avoid a complete dissappearance of the grain enclosed by GBI and GBII; (b) the same system after annealing at 655°C.

For all three samples the velocities v (Fig. 2) were found to remain constant during an experiment at a given temperature over the entire investigated temperature range. The angle θ was seen to increase with increasing temperature. Due to the temperature dependence of θ, the criterion Λ, determined by Eq. (2), was found to increase with increasing temperature as well. For a calculation of Λ for the motion of grain boundaries in the sample SI the ratio σ_3/σ was determined under the assumption that for temperatures near the melting point the value of θ

reaches the thermodynamic equilibrium value (see Eq. 4). It is obvious that $\theta \rightarrow \theta_{eq}$ when the kinetics of the system become grain boundary controlled. In other words, $\theta_{measured} \approx \theta_{eq}$ can be expected near the melting point. The measurement of θ near the melting point was obtained by the following experiment. A sample with grain boundaries in configuration SI was annealed at the temperature T=655°C for 5 minutes. To avoid the complete vanishing of the shrinking grain between GBI and GBII the sample was notched as shown in Fig. 3a. After annealing the sample was rapidly cooled down, and the position of the grain boundaries and the angle $2\theta \approx 2\theta_{eq} = 165°$ (Fig. 3b) was measured by optical microscopy. To determine Λ for the samples SII and SIII the data of the calculation of grain boundary energy calculated by Hasson and Goux were used [5].

4 Discussion

The measured quantities in the current experiments were the migration rate v of the grain boundary system with triple junction and the triple junction angle θ. Using the measured values of θ, the dimensionless parameter Λ was calculated for each annealing temperature (Eq. 2), and the temperature dependency of Λ was determined. Fig. 4 shows this dependency for the samples SI and SII. As follows from the approach given above, at low temperatures, where Λ is of the order of unity, the motion of the grain boundary system is controlled by the mobility of the triple junction. A rise of Λ with increasing temperature reflects the transition from triple junction- to grain boundary controlled motion of the grain boundary system.

The used technique for measuring the angle θ, raises the important question whether the angle measured after the sample was cooled down corresponded to the real triple junction angle at annealing temperature. There are two main factors, which may affect the angle θ: a thermodynamical influence, which reflects the temperature dependence of the grain boundary surface tension, and a kinetic influence, defined by the relation between grain boundary and triple junction mobilities and the width of the vanishing grain a. The thermodynamical effect is negligible, since the boundary surface tension depends only slightly on temperature, although how slightly is not exactly known. In any event, if the grain boundary surface tensions and their temperature coefficients are comparable, then the angle θ should be essentially temperature independent. The kinetic factor can really affect the magnitude of the angle θ. Taking into account that the triple junction kinetics predominates at relatively low temperatures, as observed in [3], the observed value of θ may be reduced compared to the real angle at annealing temperature. However, due to the strong temperature dependence of grain boundary mobility and the high cooling rate of the sample this change of the real value of θ is likely to be sufficiently small.

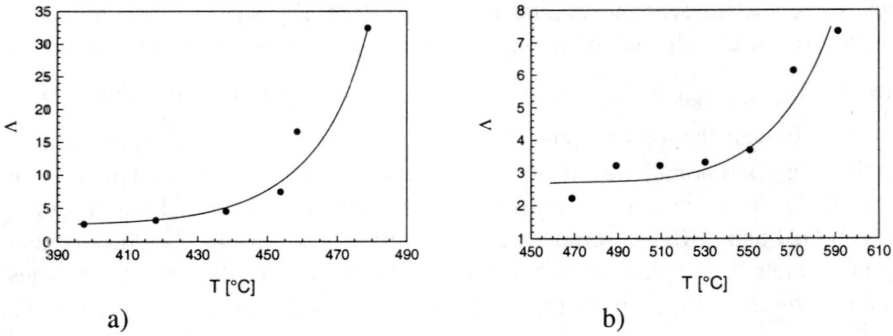

Fig. 4. Temperature dependence of the criterion Λ for triple junctions in samples SI (a) and SII (b).

As it can be seen in Fig. 4, the measurement of the temperature dependency of Λ can be utilized as a method to analyze the experimental data and to separate different regimes of motion of the system, i.e. triple junction and boundary controlled kinetics. At relatively low temperatures, up to 430°C for sample SI and up to 510 °C for sample SII, the value of Λ is of the order of unity. Since the criterion Λ specifies the ratio of triple junction mobility to grain boundary mobility (Eq. 2), low values of Λ mean that the triple junction mobility is comparable to the grain boundary mobility. Therefore, the motion of the system in the low temperature regime can be interpreted as a triple junction controlled motion. The rise of Λ with increasing temperature indicates that the system motion becomes increasingly less affected by the triple junction, and boundary kinetics become predominant.

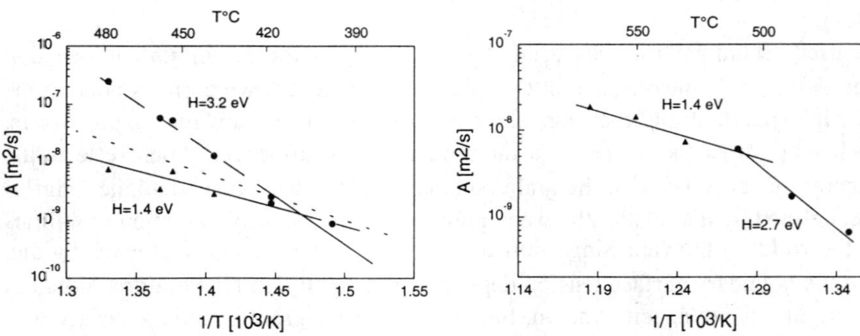

Fig. 5. Temperature dependence of triple junction (●) and grain boundary mobility (▲) in samples SI (a) and SII (b). Dotted line represents the mobility of a 21°<111> tilt boundary in Al as reconstructed from literature data [2] ($H = 1.6$ eV; $A_0 = 10^3$ m²/s).

For a comparison of the current results with measured mobilities on single grain boundaries we determined the temperature dependence of the reduced mobility A

of the involved triple junctions and grain boundaries (Fig. 5). Using Eqs. 1 and 3 we obtain

$$A_b \equiv m_b\sigma = \frac{v \cdot a}{2\theta} = A_{0b}e^{-H_b/kT} \text{ and} \tag{5}$$

$$A_{tj} \equiv m_{tj}\sigma a = \frac{v \cdot a}{2\cos\theta - \sigma_3/\sigma} = A_{0tj}e^{-H_{tj}/kT} \tag{6}$$

From velocity data and the temperature dependence of the contact angle 2θ as expressed in terms of Λ (T) in Fig. 4 the respective grain boundary and triple junction mobility can be calculated according to Eqs. 5 and 6. The results are given in Fig. 5 where the triangles represent the (reduced) grain boundary mobility as determined by Eq. 5, while the circles denote the respective triple junction mobility determined by Eq. 6. The system will obviously be controlled by the slowest moving constituent, i.e. the triple junction at low temperatures and the grain boundary at high temperatures. According to their respective linear behavior in the Arrhenius plot the activation enthalpy can be determined as indicated in the figure. Obviously, there is a distinct transition from triple junction kinetics at low temperatures to grain boundary kinetics at elevated temperatures and the activation enthalpy for triple junction motion H_{tj} is considerably higher than for grain boundary migration (H_b). The behavior of the grain boundary controlled branch in Fig. 5 compares well with measurements of the reduced grain boundary mobility obtained from literature data [2] of independent bicrystal experiments. The corresponding evaluation for system SI is given in Fig. 5a and yields comparable behavior.

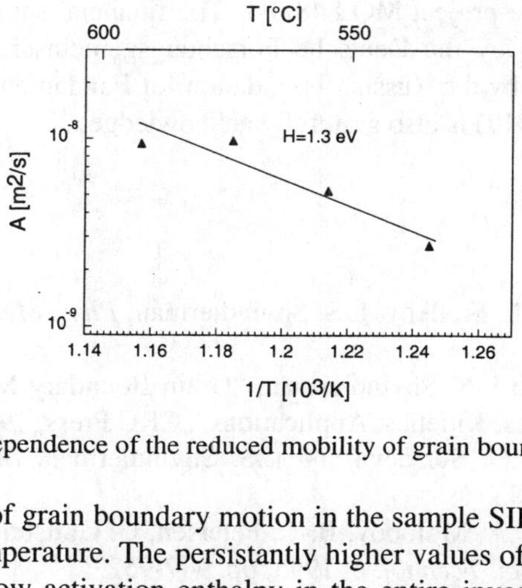

Fig. 6. Temperature dependence of the reduced mobility of grain boundaries in sample SIII.

The measurements of grain boundary motion in the sample SIII were performed at a relatively high temperature. The persistantly higher values of Λ and the constant but comparatively low activation enthalpy in the entire investigated temperature range (Fig. 6) indicate that the system always moves in the boundary controlled regime. It is worthy to note that the motion of system SII, also consisting of <110> tilt boundaries, in this temperature range is controlled by the boundary motion as well.

5 Conclusions

The experimental results demonstrate that the motion of grain boundary systems with triple junctions in Al can be controlled by slowly moving triple junctions. The influence of triple junctions depends on temperature. It is particularly strong at low temperatures and small grain size.

A drastic difference between activation enthalpy of grain boundary and triple junction motion was revealed by the current experiments. Therefore, there is a temperature, below which triple junctions govern the motion of the connected boundary system and grain boundaries simply follow the triple junction.

It has been shown, that the transition range between grain boundary and triple junction control of migration kinetics may depend on grain boundary and correspondingly triple junction geometry (<111> and <110> tilt boundaries in the current study). Hence, particular grain boundary systems consisting of structurally different grain boundaries need to be adressed to probe the influence of triple junctions and their geometry on texture development during grain growth.

Acknowledgements

The authors express their gratitude to the Deutsche Forschungsgemeinschaft for financial support of the project MO 848-3-1. The financial support of the authors' cooperation program by the Deutsche Forschungsgemeinschaft (Grant No. 436 RUS 113/539/0) and by the Russian Foundation for Fundamental Research (Grant RFFI-DFG 99 02 04017) is also gratefully acknowledged.

References

[1] A.V. Galina, V.E. Fradkov, L.S. Shvindlerman, *Phys. Met. Metall.*, 63, 1987, pp. 165-168.

[2] G. Gottstein and L.S. Shvindlerman, "Grain Boundary Migration in Metals: Thermodynamics, Kinetics, Applications", CRC Press, 1999.

[3] G. Gottstein, V.G. Sursaeva and L.S. Shvindlerman, *Interface Science*, 7, 1999, pp. 273-283.

[4] U. Czubayko, D.A. Molodov, B.-C. Petersen, G. Gottstein, and L.S. Shvindlerman, *Meas. Sci. Technol.* 6, 1995, pp. 947-952.

[5] G. Hasson and C. Goux, *Scripta Metall.*, 5, 1971, pp. 889-894.

Crystallographic Distribution of Grain Boundaries in Magnesium Oxide

D.M. Saylor[1], A. Morawiec[2], K.W. Cherry[1], F.H. Rogan[1], G.S. Rohrer[1], S. Mahadevan[3], and D. Casasent[3]

[1] Materials Science and Engineering Dept., Carnegie Mellon University, Pittsburgh, PA 15213, USA

[2] Instytut Metalurgii i Inzynierii, Materialowej PAN, Reymonta 25, 30-059 Krakow, Poland

[3] Electrical and Computer Engineering Dept., Carnegie Mellon University, Pittsburgh, PA 15213, USA

Keywords: grain boundaries, misorientation, microstructure, grain boundary distribution, magnesium oxide.

Abstract. We have developed a technique that allows the geometry and crystallography of large quantities of contiguous crystallites to be accurately characterized. Using this technique, in concert with high precision serial sectioning, we have extracted the geometric and crystallographic configuration of approximately 2.5 mm^2 of grain boundary area in a 0.07 mm^3 volume of polycrystalline magnesium oxide that contained 2399 unique grains. Using these data, we have specified the distribution of grain boundaries within the five-dimensional space of misorientation and inclination types. The data show that in addition to the texture in the misorientation space induced by processing, there is texture in the space of grain boundary planes. For example, for 5° misorientations about <111>, pure twist and tilt boundaries occur more frequently than general boundary plane configurations.

1 Introduction

There are five mesoscopically observable characteristics of a grain boundary: three that describe the lattice misorientation and two that describe the boundary inclination. While the distribution of lattice misorientations is frequently derived from the analysis of planar sections [1,2], grain boundary inclinations are rarely reported for more than a handful of boundaries. In a recent paper, we characterized the five mesoscopically observable parameters for 4665 grain boundaries in a magnesia polycrystal [3]. However, the grain boundaries in this data set were restricted to those meeting at triple junctions, and did not represent the true distribution of grain boundaries. The purpose of the present paper is to describe an automated technique that allows the geometry and crystallography of large quantities of contiguous crystallites to be accurately characterized. Using this technique, along with high precision serial sectioning, we have extracted the crystallographic and geometric configuration of the three-dimensional grain boundary network in a magnesia polycrystal and specified the distribution of grain boundaries in the space of the five mesoscopically observable parameters.

2 Experimental Procedure

Acquisition of the geometric and crystallographic data required to accurately characterize the mesoscopic grain boundary parameters requires orientation measurements with high spatial resolution from multiple sections through the sample. This was accomplished using an automated scanning electron microscope (SEM) mapping system that controls both the stage and beam position as well as the acquisition of images and electron backscattered diffraction patterns (EBSPs). When the mapping is conducted, the sample surface is initially divided into sectors. At each sector, a secondary electron image is recorded, and EBSD measurements of crystallite orientation are made at regular intervals within the sector. When the sector characterization is complete, the microscope stage is automatically moved to the next sector, which has a small overlap with the previous one. The entire procedure is carried out under computer control. Because we use SEM images to determine grain boundary positions, the orientation measurements can be conducted at relatively coarse intervals. Thus, compared to the conventional OIM methods, we are able to resolve the boundary positions accurately without accumulating overly redundant orientation data [4]. After an area is mapped, high precision serial sectioning is used to remove a thin layer and the process is repeated so that the three-dimensional characteristics of the grain boundary network can be determined.

The experiment described here was carried out on polycrystalline magnesia with a 109 μm grain size that was described in detail elsewhere [3]. For the present paper, it is useful to note that hot pressing during the processing of this material induced a <111> axial texture that was 11 times random at the maximum. The sample was coated with carbon and tilted at 60° in the SEM for imaging and acquisition of the EBSPs. On each layer, a scan area consisting of a 14 x 14 grid of sectors was characterized. In each of the 196 sectors, a tilt-corrected image was taken at 750x magnification, and 300 uniformly distributed orientation measurements were recorded. After the scan, approximately 7 μm of material was removed and the entire process was repeated until data from five layers was accumulated.

2.1 Reconstruction of the grain boundary network

The initial step in reconstructing the grain boundary network from the recorded data is to combine the SEM images and crystallite orientation measurements to produce high-resolution orientation maps. After correcting spatial distortions in each SEM image, the relative positions of the images in each layer were determined using the algorithm described by Mahadevan and Casasent [5]. Next, the grain boundaries in each sector image were digitized by hand. The relative positions of images were then used to construct large mosaic maps of the grain boundaries and orientation measurements on each layer. To produce orientation maps from these data, every grain in the scan area, defined by contiguous pixels not associated with a grain boundary, was identified and assigned an orientation. Because multiple orientation measurements were made in each grain, minority orientations resulting from errors in the indexing had to be excluded. The remaining majority orientations, which contained some scatter, were averaged to make the final orientation assignment for each grain [6,7].
Once the high-resolution orientation maps from each layer were obtained, it was necessary to align the maps to establish the same reference frame for all layers. Here, we chose the first layer as the global reference frame. The transformation from all subsequent layers to the first layer is given by Ax+t where x is a two dimensional vector which represents the position within a given layer, A is a 2x2 affine transformation matrix, and t is a two dimensional translation vector. To find (A,t) for each layer, we initially find (A,t) that maximizes the area of overlap between positions with the same orientations on adjacent layers. We then use the (A,t) describing the transformation that aligns adjacent layers to calculate the (A,t) that aligns each layer with the initial layer.
After all layers were transformed into the global reference frame, the common grains through all the layers were identified using the following algorithm. The area of overlap between all grain pairs on adjacent layers was determined. The pair of grains that has the largest area of overlap is identified as being two sections of the same grain. The pair with the second largest area of overlap is then assigned in the same way. The process continues until all grains have been assigned or do not overlap any grains that have not been assigned. The success of this algorithm, which we have found to be 99.5% accurate, derives from the fact that the distance between adjacent layers is much smaller than the average grain size. After the grains were identified on all layers, pixels associated with common grain boundaries on adjacent layers were used to create a meshed interfacial surface of triangular elements. The orientations of the crystallites were then reassigned by repeating the process that was used in each layer, but considering all of the orientation data from different section planes of the same grain. Using this process, 2399 distinct grains were identified.

2.2 Grain boundary distribution

After all boundaries were meshed, it was possible to determine the distribution of grain boundaries over all five mesoscopically observable parameters. Using the vertices of the triangles in the mesh, the area and normal vector for each triangle were determined. Next, the normal vector and the orientations of the crystallites bounding each triangle were used to specify all five grain boundary parameters, and the total boundary area associated with each discrete grain boundary can be calculated. The space of grain boundaries is discretized in the following manner. The misorientations, each described by three Eulerian angles (ϕ_1,Φ,ϕ_2), were parameterized by ϕ_1, cos(Φ), and ϕ_2 in the range of zero to $\pi/2$, 1, and $\pi/2$, respectively. The inclinations, described by two spherical angles, θ and ϕ, were parameterized by cos(θ) and ϕ in the range of zero to 1 and 2π, respectively. This domain of grain boundary space was tessellated into cells by dividing the range of each parameter into twelve equal partitions, a resolution of approximately 7.5°, each cell representing a discrete grain boundary type. Within this domain, 36 symmetrically equivalent grain boundaries exist for each characterized grain boundary. Thus, the area of each triangle in the mesh is associated with 36 cells within the domain.

3 Results and Discussion

We have characterized all five mesoscopically observable parameters for approximately 2.5 mm^2 of grain boundary area in a 0.07 mm^3 volume. For every misorientation and inclination pair, there are 36 equivalent pairs in our domain of grain boundary types. Thus, after specifying a misorientation, the value of the distribution at every inclination is taken as the average of all 36 symmetrically equivalent misorientation and inclination pairs. These averaged values for the inclination parameters are then plotted on an inverse pole figure.
The distribution of misorientations alone, averaged over the space of inclinations, exhibited a strong peak at low angle misorientations (17 times random at maximum). Furthermore, large angle misorientations (>15°) were primarily restricted to those described by rotations about <111> due to the sample's axial texture. Fig. 1 shows the distribution of inclination parameters for two fixed misorientations. These plots show strong texture over the inclination parameters. In (a), there appears to be a preference for either pure twist or pure tilt inclinations. There is a strong peak in the distribution at <111>, the pure twist boundary. Furthermore, there appears to be a wide band of relatively high concentrations 90° from <111>, or the pure tilt boundaries, with broad peaks centered on tilt boundaries with [110] planes. In (b), we see that there is a broad minimum at <111> and maxima in the vicinity of tilt boundaries with [211] planes.

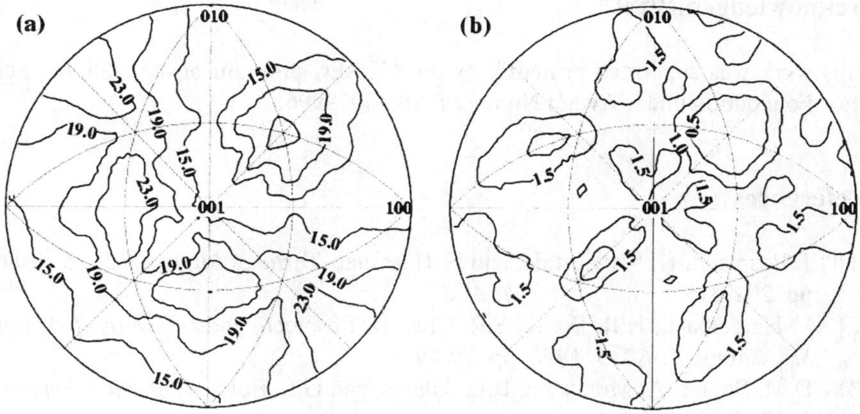

Fig. 1. Distribution of grain boundary inclinations for lattice misorientations corresponding to rotations of (a) 5° about <111> and (b) 60° about <111> ($\Sigma 3$). Contour values are in multiples of a random distribution.

The plots in Fig. 1 represent a small fraction of the entire grain boundary space. However, the apparent trend of preferred inclinations is typical of any of the distributions we have examined at fixed misorientations. While it is well known that grain boundary misorientations are not necessarily distributed randomly over the domain of possibilities, the current results illustrate that the inclinations of the grain boundaries can also exhibit texture. This observation provokes a number of interesting questions regarding the origin of this texture, its relation to the properties of the boundaries, and how it might be controlled. For example, are certain points in the inclination space preferentially occupied because they represent low energy configurations or because these boundaries move more slowly and are eliminated with a lower probability? We intend to address this and other questions in a forthcoming paper that will provide a more complete description of the distribution of grain boundaries over the five mesoscopically observable parameters.

4 Summary

We have developed an experimental technique that allows the geometry and crystallography of large quantities of contiguous crystallites to be characterized. Data extracted from the analysis of 2.5 mm^2 of grain boundary area in a volume of 0.07 mm^3 has enabled us to specify the distribution of grain boundaries over all mesoscopic parameters. We find that strong texture exists in the space of grain boundary inclinations, a domain that has not been previously explored.

Acknowledgement

This work was supported primarily by the MRSEC program of the National Science Foundation under Award Number DMR-0079996.

References

[1] J. Pospiech, K. Sztwiertnia, and F. Haessner, *Textures Microstruct.*, 6, 1986, pp. 201-215.
[2] F. Heidelbach, H.R. Wenk, S.R. Chen, J. Pospiech, and S.I. Wright, *Mater. Sci. Engng. A*, A215, 1996, pp. 39-49.
[3] D.M. Saylor, A. Morawiec, B.L. Adams, and G.S. Rohrer, *Interface Science*, 8, 2000, pp. 131-140.
[4] B.L. Adams, S.I. Wright, and K. Kunze, *Met. Trans.*, 24A, 1993, pp. 819-831.
[5] S. Mahadevan and D. Casasent, *Proceedings SPIE*, submitted 04/01.
[6] M. Humbert, N. Gey, J. Muller, and C. Esling, *J. Appl. Cryst.*, 29, 1996, pp. 662-666.
[7] A. Morawiec, *J. Appl. Cryst.*, 31, 1998, pp. 818-819.

Recrystallization and Grain Growth
Proceedings of the First Joint International Conference
Eds. G. Gottstein and D.A. Molodov, Springer-Verlag, 2001

Mobility of Grain Boundaries and Grain Boundary Systems with Triple Junction: A Direct Comparison

V.G. Sursaeva[1], G. Gottstein[2], and L.S. Shvindlerman[1, 2]

[1] Institute of Solid State Physics, Russian Academy of Sciences, 142432 Chernogolovka, Russia

[2] Institut für Metallkunde und Metallphysik, RWTH-Aachen, D-52056 Aachen, Germany

Keywords: grain boundary, triple junction, migration, mobility.

Abstract. The results of experiments devoted to a direct comparison of the mobility of two types of grain boundary systems – with and without a triple junction – are considered and discussed. It is shown that in certain temperature intervals these structural elements of a polycrystals can drag each other.

1 Introduction

The main result of grain growth is the change of grain microstructure. Since as this change is accomplished by the displacement of the elements of grain microstructure – grain boundaries and triple junctions – the investigation of their motion is extremely important to understand the laws of grain microstructure evolution, its stability, and texture formation in metals and alloys [1]. So far this has been associated with grain boundaries only. For triple junctions it was assumed that their mobility is infinitely large, so they don't affect grain boundary motion, and their role in grain microstructure evolution is reduced to preserve the angles where boundaries meet. Such assessment does not result from a physical analysis, rather ist a conventional tacit assumption. For 2-D (columnar) structures the location of a grain boundary intersection is a line and the discussed assumption leads for uniform grain boundaries to the well-known Von Neumann-Mullins relation [2,3]. The mentioned assumption and relation provides the basis for virtually all theories of grain growth, for the theoretical analysis of experimental data and for a large body of computer simulations of grain growth in 2-D systems [4-7]. To test this assumption by comparing the results of computations with experiments is impracticable since real polycrystals do not conform to the requirements of grain boundary uniformity, thus agreement is obtained in general terms only. To overcome this problem we propose to study model systems, in the present case case connected boundaries in tricrystals, where the limited number of microstructural elements and the given geometry allow to treat the problem on a repro-

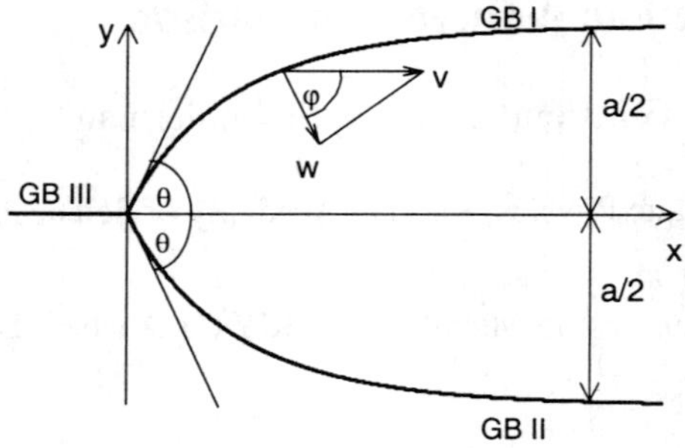

Fig. 1. Geometry of grain boundary system with a triple junction in the course of steady-state motion.

ducible physical basis. The steady-state motion of a grain boundary system with triple junction is only possible in a small set range of geometrical configurations. One of them is shown in Fig. 1 [1,8]. For the uniform grain boundary model, i.e. all grain boundaries possess equal surface tension and mobility irrespective of the misorientation of the adjacent grains and the crystallographic orientation of the boundaries and the mobility of grain boundaries and triple junction are independent of the velocity, the kinetic equation of state can be solved analytically [1,8,9]. The obtained solution describes the shape of a boundary moving in steady-state

$$y(x) = \xi \arccos\left(e^{-\frac{x}{\xi}+C_1} \right) + C_2, \xi = \frac{a}{2\theta}, \qquad C_1 = \frac{1}{2}\ln(\sin\theta)^2, C_2 = \xi\left(\frac{\pi}{2} - \theta\right) \tag{1}$$

The steady state velocity of motion of the system is:

$$V = \frac{2\theta m_b \sigma}{a} \tag{2}$$

where m_b, σ and a are grain boundary mobility, surface tension and the dimension of the system, respectively. On the other hand the steady-state velocity motion of the triple junction can be represented as:

$$V = m_{tj}\sigma(2\cos\theta - 1) \tag{3}$$

where m_{tj} is triple junction mobility.

There is an important relation which links the steady-state value of the angle θ at the vertex of a triple junction to the dimensionless criterion, This criterion Λ reflects the drag influence of the triple junction on the motion of the grain boundary system:

$$\Lambda = \frac{m_{tj}a}{m_b} = \frac{2\theta}{2\cos\theta - 1} \tag{4}$$

It is stressed that Eq. (4) can be equally considered as a criterion which indicates the drag influence of the boundaries on the triple junction motion.

One can see that for a mobile triple junction $\Lambda \gg 1$, the θ tends to $\pi/3$, the equilibrium angle at the triple junction, while the boundary velocity tends to the value of $2\pi m_b \sigma / 3a$. At $\Lambda \ll 1$, the angle tends to zero and the velocity of the system tends to $V = m_{tj}\sigma$, respectively. In other words, Eq. (2) gives us a tool to determine under which kinetic regime – boundary or junction – the system moves.

Such a steady-state motion will take place for the so-called "symmetrical" configuration as well (Fig. 1), where both curved grain boundaries I and II identical and different from the straight boundary III. In this case the Eqs. (3) and (4) read:

$$V = m_{tj}(2\sigma\cos\theta - \sigma_{III}) = m_{tj}\sigma\left(2\cos\theta - \frac{\sigma_{III}}{\sigma}\right) \tag{5}$$

$$\Lambda = \frac{m_{tj}a}{m_b} = \frac{2\theta}{2\cos\theta - \frac{\sigma_{III}}{\sigma}} \tag{6}$$

For the limits ($\Lambda \to 0$ $\Lambda \to \infty$) and we arrive at:

$$\Lambda \to 0:\ V = m_{tj}(2\sigma - \sigma_{III}) \tag{7}$$

$$\Lambda \to \infty:\ V = \frac{2\theta_{eq} m_b \sigma}{a} \tag{8}$$

and

$$\theta_{eq} = \arccos\left(\frac{\sigma_{III}}{2\sigma}\right) \tag{9}$$

Under these conditions the motion of grain boundaries and triple junctions in tricrystals of Zn and Al was investigated. It was shown that at relatively low temperatures the motion of the grain boundary system with triple junction is determined by the triple junction mobility, while at elevated temperatures grain boundary mobility controls the migration of the system. In Figs. 2 and 3 the temperature dependence of θ and Λ for the system in Zn are presented [9]. Comparable results were obtained in experiments on Al tricrystals [10].

Computer experiments by molecular dynamics for the configuration in Fig. 1 indicate that the mobility of the triple junction was reduced when the misorientation of the straight grain boundary represents a low Σ coincidence boundary [11].

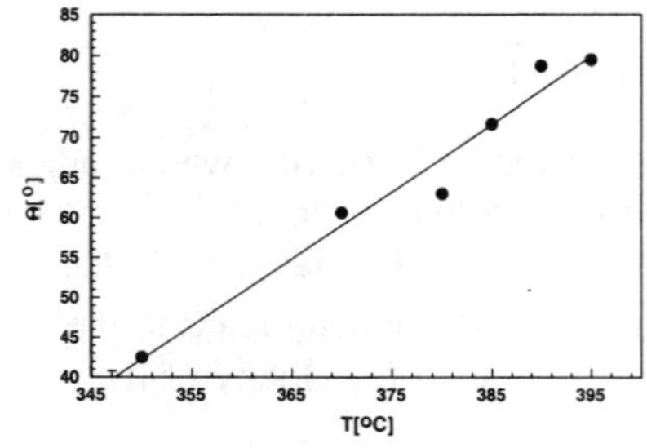

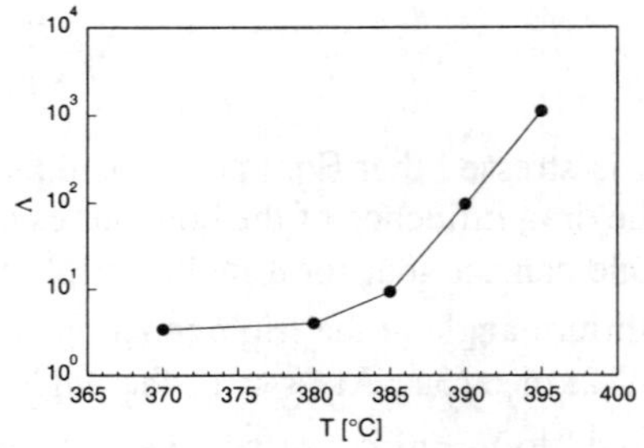

Fig. 2. Temperature dependence of the angle θ in the vertex of a triple junction (configuration Fig. 1) in tricrystals of Zn [10].

Fig. 3. Temperature dependence of the criterion Λ for a symmetrical triple junction in Zn [10].

The prime objective of the current work was a "direct" comparison between the mobility of a grain boundary and a grain boundary system with triple junction. The boundary configuration in Fig. 1 without a triple junction constitutes a grain boundary half-loop (Fig. 4), which has been well investigated theoretically and experimentally [1,12].
The motion of a grain boundary half-loop in Zn of 99.995% purity was studied. Additionally, the motion of practically the same half-loop attached to a straight low-angle boundary (Fig. 1) was investigated. "Practically the same" means that the incorporation of the low-angle grain boundary changes the misorientation of curved grain boundaries only negligibly. Hence, we can conduct a direct comparison between the motion of a grain boundary half-loop and the motion of the same half-loop with a triple junction. In other words, we can determine the mobility of a moving grain boundary system with and without the triple junction.

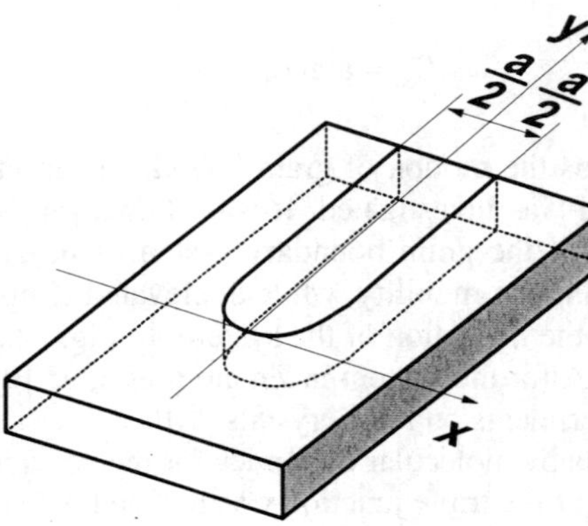

Fig. 4. Geometry of grain boundary half-loop [1].

It was mentioned above that the approach used for an analysis of a uniform grain boundary system (Fig. 1) can also be applied to a symmetrical system, where grain boundaries I and II are the same, but different from the rectilinear grain boundary III. This difference manifests itself in the relation for the dimensionless criterion Λ and in the value of the equilibrium angle θ_{eq} (Eqs. (6) and (9)). An essential issue is the problem how to determine θ_{eq}. One option is to pin the system by notches near the triple junction and to measure θ_{eq} [10]. Alternately, one can use literature data of the grain boundary surface tension in Al. In the present case we used the Read-Shockley equation to estimate the ratio of the surface tension of a low-angle and high-angle grain boundary.

Two grain boundary systems were studied: a $<11\bar{2}0>$ tilt boundary system with 84° misorientation of the curved boundaries (half-loop), the misorientation of the straight boundary was about 3°. The second grain boundary system was a $<10\bar{1}0>$ tilt boundary system with a misorientation of 62° for the half-loop, while the misorientation of the straight boundary was also about 3°. The velocity and mobility of the grain boundary systems for the two types of motion were compared; i.e. in free motion – half-loop – and constrained motion – half-loop with triple junction.

In the following the product of the displacement $\ell(t)$ of the grain boundary vertex and the width of the grain a, i.e. $a \cdot \ell(t)$ will be referred to as reduced displacement.

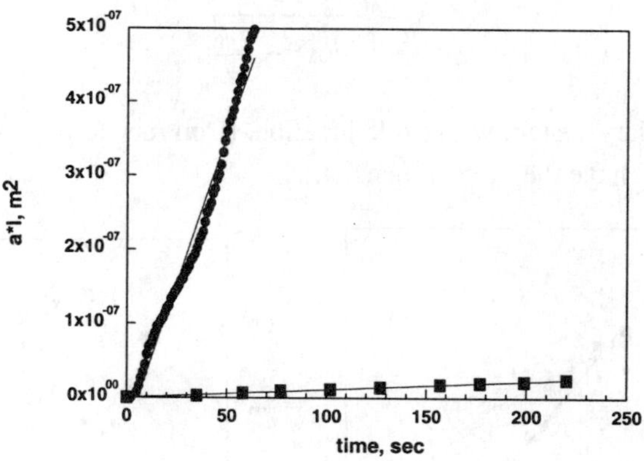

Fig. 5. Time dependence of reduced displacement $a \cdot \ell(t)$ for a grain boundary half-loop (solid circles) and a triple junction configuration (solid squares) at T = 390°C.

In Fig. 5 the dependency of the reduced displacement $a \cdot \ell(t)$ on time for a grain boundary half-loop and a half-loop with triple junction is presented. If half-loops (and half-loops with a triple junction) of different width are studied experimentally, the comparison of the product $a \cdot \ell(t)$ compensates for this variation when determining the major parameter of driving force – the width a. Strictly speaking, the driving force for the half-loop and the half-loop with a triple junction would be $2\sigma/a$ and $2\sigma/a - \sigma_{III}/a$, respectively but because of $\sigma_{III} \ll \sigma$ this difference may be disregarded.

The experiment reveals that the triple junction essentially drags the motion of the grain boundaries. So far the velocity of motion of a grain boundary system without triple junction – i.e. grain boundary half-loop (Fig. 4) and for the same system with triple junction (Fig. 1) were compared and discussed. Now we consider the mobility of grain boundary and triple junction. As the exact value of σ, the grain boundary surface tension, is unknown it is convenient to use the reduced mobility, A_b and A_{tj}, respectively,

$$A_b = \frac{Va}{\pi} = m_b \sigma \tag{10}$$

for a half-loop [1,12]

$$A_b = \frac{Va}{2\theta} = m_b \sigma \tag{11}$$

for a grain boundary system with triple junction

$$A_{tj} = \frac{Va}{2\cos\theta - \frac{\sigma_{III}}{\sigma}} \tag{12}$$

for grain boundary system with triple junction. (Contrary to m_b and m_{tj}, the quantities A_b and A_{tj} have the same dimension).

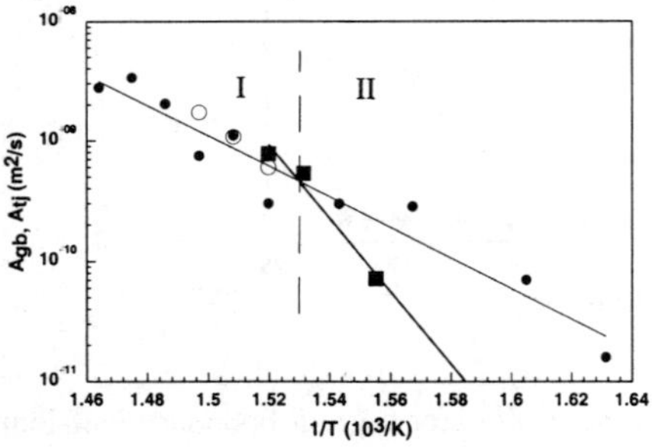

Fig. 6. Temperature dependence of reduced grain boundary mobility A_b (solid circles) and reduced triple junction mobility A_{tj} (solid squares) for the system $<10\bar{1}0>$.

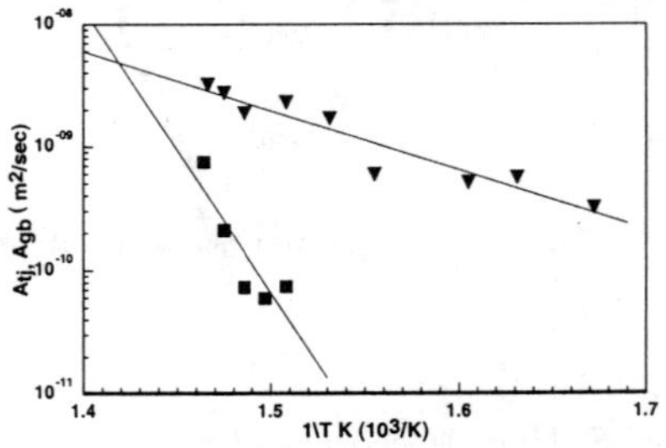

Fig. 7. Temperature dependence of reduced grain boundary mobility A_b (solid triangles) and reduced triple junction mobility A_{tj} (solid squares) for the system $<11\bar{2}0>$.

For the grain boundary systems discussed ($<11\bar{2}0>$ and $<10\bar{1}0>$ tilt boundaries) the value of the ratio σ_{III}/σ was set equal to 0.1.
In Figs. 6 and 7 the temperature dependency of A_b and A_{tj} for system I and II are presented. Apparently the dependency can be subdivided into two regimes. In the regime $A_{tj}<A_b$ the motion of the system is controlled by the triple junction mobility. In the region $A_b<A_{tj}$ the motion of the grain boundary system is determined by the grain boundary mobility.
In Fig. 6 the open circles (left of the point of intersection) represent data for the motion of a grain boundary system with triple junction recalculated according to the equation for boundary kinetics [11]. One can see that the recalculated points are in good agreement with the half-loop mobility (solid circles). There is a substantial difference in the activation enthalpy of grain boundary and triple junction motion, i.e. the activation enthalpy for triple junction motion is much larger than for grain boundary motion. Consequently, at relatively low temperatures the motion of a grain boundary system is controlled by the triple junction mobility while at high temperatures the grain boundary mobility determines the process of migration. The transition temperature depends on the boundary system and can be higher than the melting point. In this special case the motion of the grain boundary system will be governed by the grain boundary mobility in the whole temperature range.
Finally, we would like to point out that the expression“ triple junction drag” is a historical anachronism. Grain boundaries and triple junctions are equally important structural elements in polycrystals. Consequently, ascribing a drag effect exclusively to triple junctions, we accept, by doing so, that grain boundaries are of only concern. Indeed, in region II (Fig. 6) the triple junction mobility is much lower than the grain boundary mobility. In this regime the motion of grain boundary system is entirely controlled by the triple junction mobility and we can state that triple junctions drag grain boundary motion. In region I, however, the triple junction is more mobile than the grain boundary, and, correspondingly, it would will be correct to state that the motion of the system is dragged by grain boundaries.

Acknowledgement

The authors gratefully acknowledge the support of their research by the Deutsche Forschungsgemeinschaft (Grant No.436 Rus 113/539/0) and the Russian Foundation for Fundamental Research (Grant RFFI-DFG 99 02 04017). Special thank of one of the authors (L.S. Shvindlerman) go to the Alexander von Humboldt Foundation for supporting his stay in Aachen through a Humboldt Research Award.

References

[1] G. Gottstein and L.S. Shvindlerman, “Grain Boundary Migration in Metals: Thermodynamics, Kinetics, Applications”, CRC Press, 1999.

[2] J. Von Neumann, in "Metal Interfaces", Americal Society for Testing Materials, Cleveland, 1952, p. 108.
[3] W.W. Mullins, *J. Appl. Phys.*, 27, 1956, p. 900.
[4] K. Lücke, I. Heckelmann, G. Abbruzzese, *Acta Metall. Mater.*, 40, 1992, p. 533.
[5] M.P. Anderson, D.J. Srolovitz, G.S. Grest, P. Sahni, *Acta Metall.*, 32, 1984, p. 783.
[6] V.E. Fradkov, L.S. Shvindlerman, D.G. Udler, *Scripta Metall.*, 19, 1985, p. 1285.
[7] V.E. Fradkov and L.S. Shvindlerman,"Structure and Properties of Interfaces in Metals", Moscow,"Nauka", 1988, p. 213.
[8] A.V. Galina, V.E. Fradkov, L.S. Shvindlerman, *Phys. Met. Metall.*, 63, 1987, pp. 165-168.
[9] U. Czubayko, V.G. Sursaeva, G. Gottstein, L.S. Shvindlerman, *Acta Mater.*, 46, 1998, pp. 5863-5871.
[10] S.G. Protasova, G. Gottstein, D.A. Molodov, V.G. Sursaeva, L.S. Shvindlerman, *Acta Mater.*, submitted.
[11] M. Upmanyu, D.J. Srolovitz, L.S. Shvindlerman, G. Gottstein, *Interface Science*, 7, 1999, pp. 307-319.
[12] V.Yu. Aristov, V.E. Fradkov, L.S. Shvindlerman, *Sov. Phys. Solid State*, 22, 1980, pp. 1055-1059.

Influence of Mechanical Stresses on Twist Grain Boundaries

Myrjam Winning[1], Günter Gottstein[1] and Lasar S. Shvindlerman[1,2]

[1] Institut für Metallkunde und Metallphysik, RWTH Aachen, 52056 Aachen, Germany
[2] Institute of Solid State Physics, Russian Academy of Science, Chernogolovka, Russia

Keywords: twist boundaries, grain boundary motion, X-ray diffraction, external mechanical stress field, aluminium.

Abstract. The common opinion about twist boundaries is that the mobility of twist boundaries is quite small. Moreover the investigation of their motion is impossible on curved twist boundaries under the influence of a capillary force because the pure twist component gets lost in a curved boundary. Recently, it was shown, that planar low and high angle tilt boundaries can couple with a mechanical stress field and that the stress activates the grain boundary motion. From this it seems reasonable to study the behaviour of planar twist boundaries under the influence of a mechanical stress. Therefore the current study is aimed to demonstrate the effect of a mechanical torque on planar twist boundaries.

1 Introduction

Owing to the importance of grain boundaries and grain boundary motion for the microstructure- and texture-development in crystalline materials, there exist a lot of investigations on the motion of tilt grain boundaries, but only few experiments on the motion of twist grain boundaries. One reason for this conspicuous lack of experimental data for the motion of twist grain boundaries is certainly the apparently small mobility which complicates the activation and the measurement of motion. Another reason is the impossibility to study the motion of curved twist boundaries under the influence of a capillary force, because the twist character of a grain boundary gets lost in the curved part. So the only possibility to investigate the grain boundary motion of twist boundaries is to use planar twist boundaries. In this case another driving force is needed for the activation of the grain boundary migration. Most recently it was shown that planar tilt grain boundaries in aluminium interact with a mechanical stress field and that the motion of the grain boundaries can be enforced by a respective stress [1,2]. Therefore the aim of this study is to demonstrate the effect of a mechanical stress on planar <100> twist boundaries in aluminium bicrystals.

2 Experimental Procedure

The experiments were carried out on high purity aluminium bicrystals with planar <100> twist grain boundaries and 7.7 ppm impurity content in a wide range of misorientation angles (11°-33°). To expose the planar boundaries to a mechanical stress field a special specimen holder (Fig. 1) was developed and integrated into a grain boundary tracking device (XICTD [3]) to measure the grain boundary motion in-situ. The specimen holder was designed such as to exert a torque on the sample due to a rotation with a rotation axis perpendicular to the normal of the boundary.

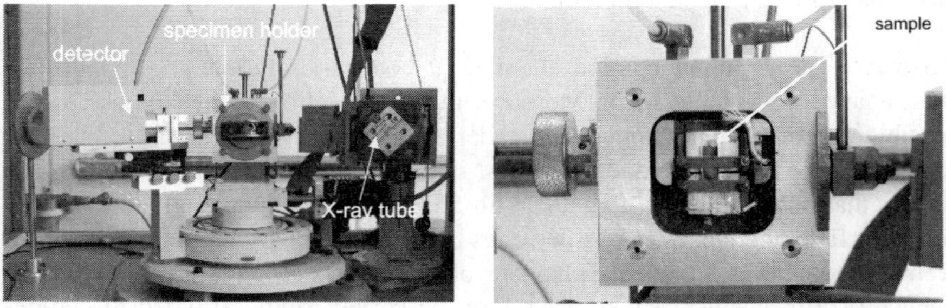

Fig. 1. Specimen holder to apply a torsional stress on the samples with planar twist boundaries.

Fig. 1 is an illustration of the experimental setup to apply a torque to a grain boundary. The sample is mounted on to a hot stage, and the temperature can be controlled during the experiment by a thermo-couple. The lower part of the sample is fixed while the upper part will be rotated, which leads to a torsional elastic strain of the sample. The rotation axis is perpendicular to the grain boundary normal.

3 Results

Owing to the constant shear stress, the investigated boundaries were observed to displace at a constant rate (Fig. 2a). From the displacement-time-diagrams the velocity could be determined for each temperature. To investigate the stress induced grain boundary motion of twist boundaries the velocity of the grain boundary was measured for different temperatures and constant driving forces.
Fig. 2b shows the Arrhenius plot of the mobility of a planar twist boundary. Apparently the stress driven grain boundary motion is a thermally activated process.

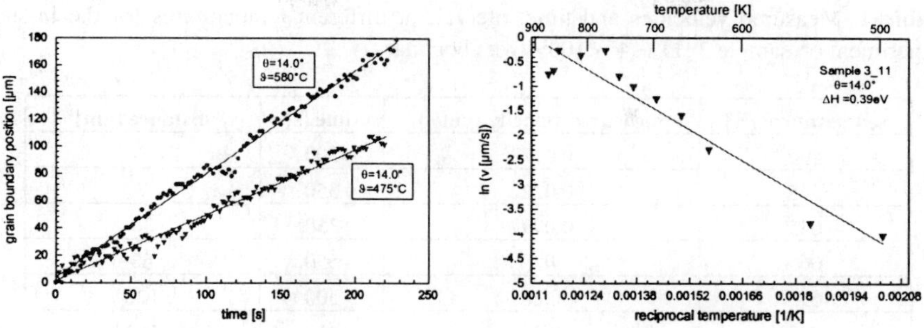

a) b)

Fig. 2. (a) Grain boundary displacement vs. time for a <100> twist grain boundary with θ=14.0° at two different temperatures. (b) Arrhenius plot of the mobility of a planar <100>-twist boundary with θ=14°.

From Eq. 1

$$\ln v = \ln v_0 - \frac{\Delta H}{kT} \tag{1}$$

we can determine the activation enthalpy for grain boundary motion ΔH from the slope of the Arrheniusline. Obviously, it is possible to activate the motion of planar twist grain boundaries. The activation enthalpy for the motion of sample 3_11 (θ=14°) is ΔH=0.39eV.

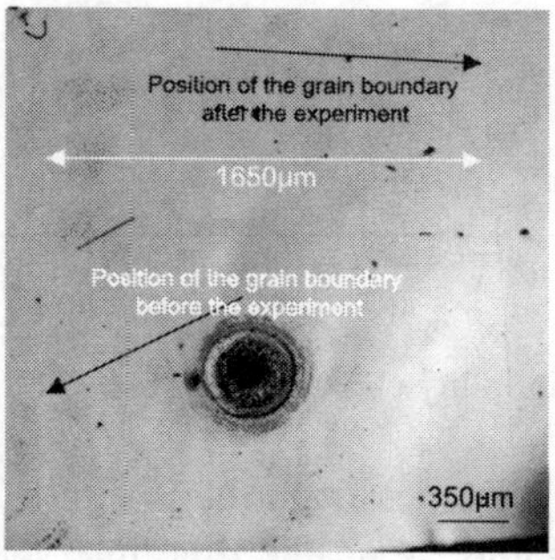

Fig. 3. Macroscopic evidence for the grain boundary motion of the planar <100>-twist grain boundary with θ=14° (sample 3_11).

The optical micrograph in Fig. 3 of the θ=14° <100> twist grain boundary gives macroscopic evidence that grain boundary motion occurred. The distance between the initial and the final position of the boundary, as indicated by respective grooves, is the swept distance during the experiment. The micrograph also proves that the entire boundary moves.

Table 1. Measured velocities and time intervals at different temperatures for the in-situ experiment on sample 3_11 (14° <100> twist boundary).

temperature [K]	measured velocity [μm/s]	time [s]	distance [μm]
490	0.0173	300	7
545	0.0222	350	7
635	0.0997	350	30
665	0.2	550	60
695	0.28	500	126
730	0.36	450	162
750	0.447	350	179
770	0.663	500	299
810	0.68	650	238
850	0.75	350	225
860	0.5	400	150
870	0.46	900	161
			sum: **1584 μm**

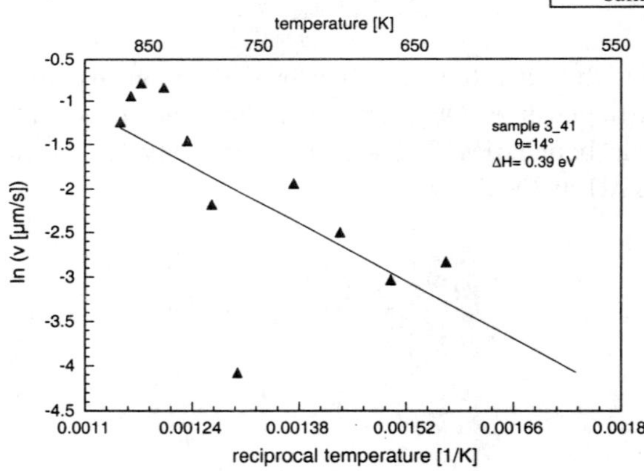

Fig. 4. Arrhenius plot of the mobility of sample 3_41 (θ=14°, <100> twist boundary).

In Table 1 the measured velocities from the in-situ experiments on sample 3_11 are given for each temperature. Also the time intervals for each temperature are given in Table 1. From the velocities and the time intervals the swept distance of the grain boundary for each temperature can be calculated and so the migration distance of the grain boundary during the whole experiment. For the shown experiment on the 14° <100> twist boundary the distance from the in-situ measurement is 1584 μm and in good agreement with the distance which was found on the optical micrograph (1650 μm). In Fig. 4 the Arrhenius plot is given for sample 3_41 (same misorientation angle as sample 3_11). Even here the activation enthalpy is ΔH=0.39eV.

Table 2. Measured velocities and time intervals at different temperatures for an experiment on sample 3_41 (14° <100> twist boundary).

temperature [K]	measured velocity [µm/s]	time [s]	distance [µm]
635	0.059	450	27
665	0.048	900	43
695	0.082	900	74
730	0.144	400	58
770	0.017	300	5
790	0.113	300	34
810	0.23	300	69
830	0.43	300	129
850	0.455	300	137
860	0.39	450	176
870	0.288	300	87
			sum: **809 µm**

In Table 2 the measured velocities and time intervals at different temperatures are given for sample 3_41 (same misorientation angle as sample 3_11). Also for this experiment we found a good agreement between the integral distance, determined from the in-situ measurement, and the distance in the optical micrograph (see Fig. 5). From Fig. 5 we can also see that the entire boundary moves on both sides and that the boundary does not rotate inside the sample.

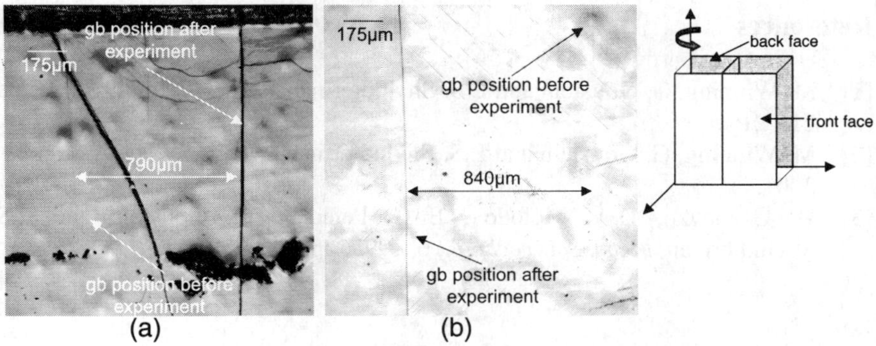

Fig. 5. Macroscopic evidence for motion of the planar <100>-twist grain boundary with θ=14° (sample 3_41). (a) Front face and (b) back face of the sample. (The curved black line in (a) is a surface scratch.)

4 Summary

A method was presented which allows the activation of the motion of planar <100> twist grain boundaries under the influence of a torsional stress. The results prove that planar twist grain boundaries can be driven by a mechanical stress field and that the motion could be investigated in-situ. The motion of twist boundaries under the influence of a mechanical torque was shown to be a thermally activated process. Therefore we can determine the activation enthalpy from the Arrhenius lines. For a <100> twist boundary we found an activation enthalpy of ΔH=0.39eV for different samples with the same misorientation angle (θ=14°).

From the optical micrographs one can see that the entire boundary moves and that the boundary remains planar during the migration. We compared the integral distances of the boundary from the in-situ measurements of the velocities with the distances between the initial and final groove of the boundary in the optical micrographs and found in all cases a good agreement between the distances. From this we conclude, that the measured velocities are really the velocities of the boundary at each temperature.

Acknowledgment

The author would like to express her gratitude to the Deutsche Forschungsgemeinschaft for financial support through contract Wi 1917/1.

References

[1] M. Winning, G. Gottstein and L.S. Shvindlerman, *Acta Mater.*, 49, 2001, pp. 211-219.

[2] M. Winning, G. Gottstein and L.S. Shvindlerman, *Mat. Sci. Eng. A*, in press, 2001.

[3] U. Czubayko, D.A. Molodov, B.-C. Petersen, G. Gottstein and L.S. Shvindlerman, *Meas. Sci. Technol.*, 6, 1995, pp. 947-952.

Steels

Abnormal Grain Growth in Ferritic and Austenitic Stainless Steels due to Second Phase Dissolution

F. Siciliano Jr.[1], J.C. Dutra[2], F.C. Pimenta Jr.[1], and A.F. Padilha[1]

[1] Department of Metallurgical and Materials Engineering, University of São Paulo, Brazil, Av. Prof. Mello de Moraes 2463, 05508-900, São Paulo, SP, Brazil; E-mail: padilha@usp.br

[2] Faculty of Industrial Engineering (FEI), São Bernardo do Campo, SP, Brazil

Keywords: abnormal grain growth, secondary recrystallization, stainless steels.

Abstract. The occurrence of abnormal grain growth in both austenitic and ferritic stainless steels was characterised after several phase dissolution heat treatments. A titanium stabilised austenitic and a superferritic stainless steels were studied. The occurrence of abnormal grain growth has shown to be associated with the dissolution of second-phase particles in both steels. For the austenitic stainless steel, different types of titanium bearing primary precipitates were found and for the ferritic stainless steel, the sigma-phase was present. In the ferritic steel, the temperature range where secondary recrystallization occurs due to sigma-phase dissolution was found to be considerably lower and narrower (1010ºC ± 20ºC) as compared to the range of the austenitic stainless steels (1150°C ± 100°C).

1 Introduction

It is customary to distinguish two types of grain growth: abnormal and normal. During *normal grain growth*, the grains increase their diameter continuously and gradually, displaying a monomodal grain size distribution. During *abnormal grain growth*, also called *discontinuous grain growth* or *secondary recrystallization*, some grains grow exaggeratedly and the mean size of the others remains constant. Its occurrence is usually observed when normal grain growth is impeded.

During the last years, many theories that describe the abnormal grain growth have been proposed [1-5]. They correlate factors such as the presence of precipitates, texture and thickness constraints. More recent theories assume that abnormal grain growth will only occur in the presence of grains with size and mobility (texture) advantage in respect to "normal" grains. The latter characteristic is usually reinforced because abnormal grain growth is often accompanied by texture modification [6].

The occurrence of secondary recrystallization in commercial materials is very often associated with the dissolution of precipitates. The development of stainless steels has resulted in complex steel compositions with substantial amounts of alloying elements. Furthermore, these alloying elements may be in a supersaturated solid solution, in which the precipitation of carbides, nitrides, borides and intermetallic phases occurs in a wide range of temperatures. In order to dissolve these precipitates, a solution annealing is ordinarily specified. However, the onset of abnormal grain growth may occur at the temperature range in which this treatment is carried out. The interaction between the dissolution of these second-phase particles and the occurrence of abnormal grain growth is investigated in this work. The aim of this investigation is to characterise the occurrence of abnormal grain growth in two types of stainless steels: one fully austenitic and one superferritic.

2 Experimental

Two types of stainless steels were selected for the present experiments: one austenitic, A and one ferritic, F. The chemical compositions are displayed in Table 1. The ferritic one was previously heat treated for 94 hours at 850°C to precipitate about 39% of sigma phase in volume. Both steels were fully recrystallized and the grain sizes were: F: 29.5μm, and A: 29.5μm. The solution heat treatments are described as follows: steel A was treated at temperatures of 1090, 1110, 1130, 1150, 1200, 1250, and 1300°C for 5, 10, 15, 20, 25, 60, 125, 625 and 3125 minutes. The specimens annealed for 625 and 3125 minutes were encapsulated in quartz tubes under vacuum. Steel F was treated at 935°C for 4 and 24 hours, 958°C for 22 hours, 990°C for 4 hours, 1014°C for 1and 4 hours and 1040°C for 4 hours as well as several annealing heat treatments at 850 and 900°C for times varying from 1 minute to 167 hours.

Table 1. Chemical compositions of both steels studied: F: ferritic; A: austenitic.

Wt.%	Steel F	Steel A
C	0.01	0.09
Si	0.35	0.46
Mn	0.22	1.70
P	0.014	0.003
S	0.03	0.004
Cr	28.12	14.6
Ni	3.91	15.0
Co	0.02	0.013
Mo	2.44	1.25
Ti	0.01	0.46
Cu	0.04	0.07
B	-	0.0045
Nb	0.31	<0.05
N	0.010	0.010

All the specimens were polished and etched using standard practices to reveal the microstructures. For the steel A, the mean grain size was measured in all specimens according to the standard ASTM-E-112 [7]. At least 20 countings were performed, each with a number of intercepts between 10 and 35. For the speci-

mens that show secondary recrystallization, only the mean "normal" grain size was measured.
The extraction of precipitates in steel A was performed by dissolving the matrix with Berzelius solution [8], followed by filtration.
The precipitates were analysed by X-ray diffraction in a Guinier-Jagodzinski X-ray diffraction technique, using monochromatic radiation of Cuk_{α_1}. The aim of this analysis was to identify the nature of the precipitates.
As a complementary technique, the precipitates were observed by scanning electron microscopy and analysed by EDX. A more detailed description of the procedures is given elsewhere [9,10].

3 Results and Discussion

Steel A (Ti stabilised) shows numerous second-phase particles, such as $(Ti_{0.92}Mo_{0.08})C$, Ti(N,C) and $Ti_2(S,C)$. The former precipitate can be dissolved during annealing and the others are stable at high temperatures. The mean grain size measurements are given in Table 2 for the steel A. The numbers marked in bold indicate the specimens in which secondary recrystallization took place. It can be seen from Table 2 that secondary recrystallization occurred from 1090 to 1200°C. The dissolution of precipitates in steel A falls in the temperature range for the occurrence of abnormal grain growth. The table also shows that the grain sizes are larger for increasing temperatures, for all annealing times. For steel A, the grain sizes for specimens submitted to low temperatures and short times are approximately the same. This strongly suggests that grain growth inhibition is present. Fig. 1 shows the micrograph of one specimen of the austenitic steel A in which abnormal grain growth took place.

Table 2. Mean grain sizes (in μm) of the specimens of steel A submitted to the annealing heat treatments. The bold numbers indicate that abnormal grain growth took place.

Temp.(°C)	Time (min)								
	5	10	15	20	25	60	125	625	3125
1090	59	68	62	62	53	69	***82***	***74***	70
1110	57	***66***	***65***	***66***	***89***	***118***	***61***	64	***83***
1130	72	***77***	***88***	***139***	***116***	***195***	***363***	***79***	***87***
1150	***167***	***217***	***247***	***398***	***378***	***476***	***557***	***96***	***1702***
1200	***102***	82	***258***	-	***359***	***493***	237	936	***960***
1250	364	436	526	586	544	626	1061	539	***798***
1300	389	470	447	455	471	583	582	971	***1316***

Fig. 2 shows the stages of sigma-phase dissolution in steel F. Firstly, the microstructure with full precipitation (a), followed by specimen annealed at 958°C for 1300 minutes where precipitates are still pinning grain boundaries and no grain

Fig. 1. Microstructure of the steel A after annealing at 1090°C for 625 minutes.

growth took place (b). Fig. 2c shows partial dissolution of sigma-phase and abnormal grain growth occurring in specimen annealed at 1014°C for 240 minutes. Finally, Fig. 2d shows the occurrence of normal grain growth, after total dissolution of sigma-phase.

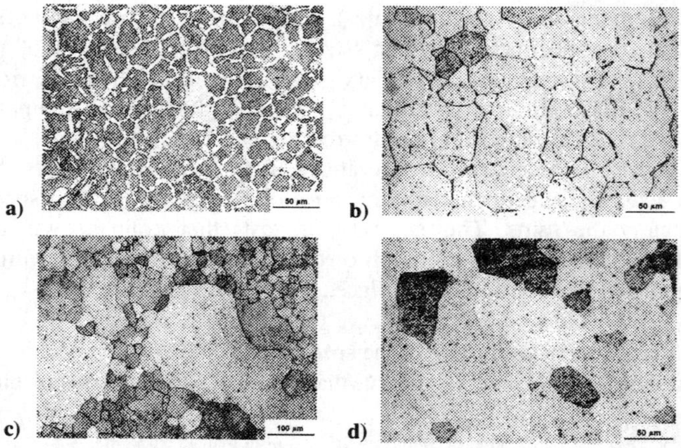

Fig. 2. Steel F microstructures: (a) precipitated at 850°C for 94.25h (condition "P"), (b) P+ solution annealed at 958°C for 21.6h, (c) P+ solution annealed at 1014°C for 4h showing the occurrence of abnormal grain growth, (d) P+ solution annealed at 1040°C for 4h showing normal grain growth.

The partial dissolution of sigma phase was responsible for the secondary grains to appear. Fig. 3 displays the grain growth map for the austenitic steel together with a schematic representation of the process. The points inside the "C" curve exhibited abnormal grain growth. The same diagram is displayed for steel F in Fig. 4. The analysis of both figures shows that, the temperatures for the onset of abnormal grain growth are higher for steel A. The dissolution of precipitates calculated from the model of Aaron and Kotler [11-13], shows that the particles are dissolved in about 125 minutes at 1090°C in steel A. At higher temperatures (e.g. 1300°C)

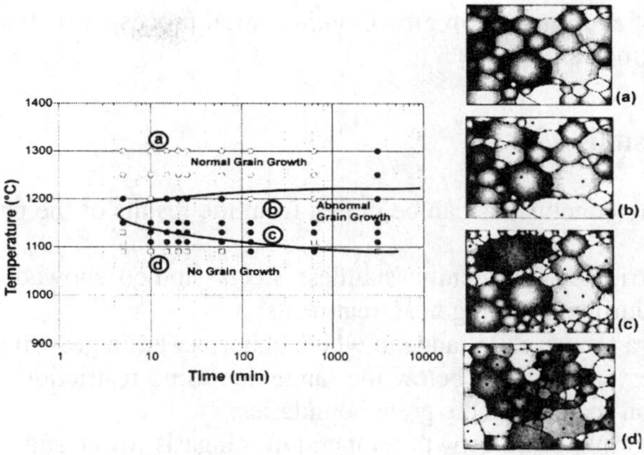

Fig. 3. Grain growth map for steel A. Three types of behaviours can be observed: inhibition of normal grain growth (❑), partial inhibition of normal grain growth (●) no inhibition of the normal grain growth (❍). The schematic representation [15] of the main points in the figure is shown in the right-hand side.

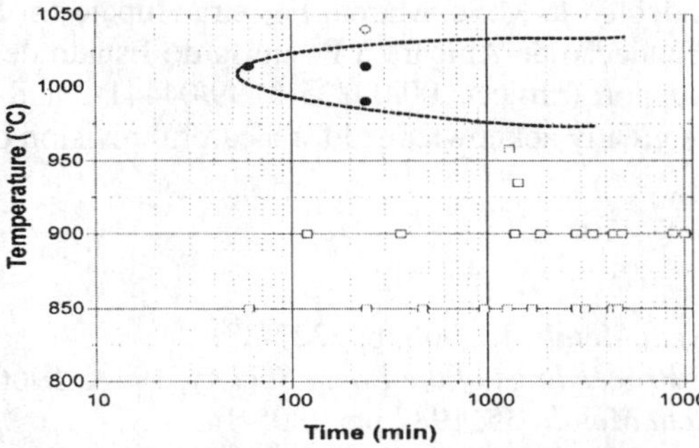

Fig. 4. Grain growth map for steel F. The abnormal grain growth experiments fall inside the area delimited by the curve. Symbols: inhibition of normal grain growth (❑), partial inhibition of normal grain growth (●) no inhibition of the normal grain growth (❍).

the particles dissolve in very short times exerting no restriction to grain growth. It is clear for both steels that abnormal grain growth is caused by partial dissolution of second-phase particles. Steel A, which contains small particles, displayed the highest abnormal grain growth temperature, followed by F which contains massive sigma-phase. The secondary recrystallization occurred at a lower temperature range, in which sigma-phase dissolution takes place. The difference of about 100°C in the abnormal grain growth temperatures for both steels are due to the velocity of second phase dissolution [14]. This fact is supported by two factors: (i) the diffusion in ferrite is about one order of magnitude faster than in austenite and (ii) the sigma-phase dissolves completely in a narrower temperature range as compared to Ti bearing precipitates which will dissolve completely at temperatures higher than 1300°C. Therefore, the present results can be used to prevent the occurrence of abnormal grain growth in industrial processes such as annealing and solution treatments.

4 Conclusions

The following conclusions can be drawn from the results of the present investigation:

- Both ferritic and austenitic stainless steels studied showed abnormal grain growth during annealing heat treatments;
- There is a temperature range in which abnormal grain growth takes place. It is not observed above or below the range due to no restriction (above) and full restriction (below) on the grain boundaries;
- The abnormal grain growth temperature range is lower and narrower for the ferritic steel.

Acknowledgements

The authors are indebted to M.Sc. Márcio Ferreira Hupalo for helpful discussions and to FAPESP-Fundação de Amparo à Pesquisa do Estado de São Paulo, Brazil, for the financial support (projects 00/00835-5, 99/04441-2 and 99/10796-8). Prof. J.D.T. Capocchi is greatly acknowledged for a careful revision of the text.

References

[1] M. Hillert. *Acta Metall.* 13, 965, pp. 227-38.
[2] T. Gladman, *Proceedings of the Royal Society*, 194A, 1966, pp. 298-309.
[3] P.R. Rios, *Acta Metall.* 35, 1987 pp. 2805-14.
[4] Q. Zhengfung et al., *Transactions of the Metals Heat Treatment*. 2, 1985, pp. 9-19 *(in Chinese)*.
[5] P.R. Rios, *Acta Metall.* 45, 1997, pp. 1785-90.
[6] A.F. Padilha, J.C. Dutra, and V. Randle, *Mat. Sci. Techn.*, 15, 1999, pp. 1009-1014.
[7] A.S.T.M, Standard Methods for Determining Average Grain Size (E112), Annual Book of ASTM Standards. Part 11, American Society for Testing and Materials, Philadelphia, Pa. pp. 135-69, 1982.
[8] K.E. Burke, *Metallography*. 8, 1975, pp. 473-88.
[9] A.F. Padilha and G. Schanz, *J. Nucl. Mater.*, 95, 1980, pp. 229-238.
[10] A.F. Padilha, G. Schanz, and K. Anderkom, *J. Nucl. Mater.*, 105, 1982, pp. 77-92.
[11] H.B. Aaron and G.K. Kotler, *Met. Trans.*. 2, 1971, pp. 393-408.
[12] H.B. Aaron : *Met. Sci. Journal*. 2, 1968, pp. 192-3.
[13] M.J. Whelan : *Met. Sci. Journal*. 3, 1969, pp. 95-7.
[14] F. Siciliano Jr. and A.F. Padilha: submitted to *Mat. Sci. Techn.*, 2001.
[15] J.C. Dutra.: Ph.D. Thesis, University of São Paulo, Brazil, 1997, p. 158.

A Tribological Study on Strain-Induced Grain Growth

M. Takashima, M. Kurosawa, and M. Komatsubara

Technical Research Laboratories, Kawasaki Steel Co., Kurashiki 712-8511, Japan

Keywords: texture, temper rolling, grain growth, lubrication, finite element model

Abstract. Effects of tribological conditions of temper rolling on strain-induced grain growth have been investigated in low carbon steel sheets. A sharp Goss texture ({011}<100>) was developed after strain-induced grain growth in the case of the temper rolling with lubricant. On the other hand, without lubricant, the specimen had a random texture. These results show that the tribological conditions of temper rolling have a significant effect on the behavior of strain-induced grain growth. The estimated stored energy of Goss orientation grains increased with the friction of temper rolling. Therefore, the friction of temper rolling is supposed to reduce the driving force of growth of Goss orientation grains.

1 Introduction

As is well known, deformation at the strain of several percents causes strain-induced grain growth. Dunn *et al.* proposed that the driving force of strain-induced grain growth arises from a difference in dislocation density on opposite sides of the grain boundary, showing a grain with lower dislocation density grew into the adjacent grain with higher dislocation density [1,2]. Inoko et al. claimed that edge dislocations and shear bands promote the strain-induced grain growth [3].
These studies imply the possibility of texture control by the strain-induced grain growth. For example, the development of Goss texture ({011}<100>) by temper rolling has been reported by several researchers [4,5]. They investigated the effects of initial grain size and roughness of roller on the texture after strain-induced grain growth. Since Goss texture is one of the most favorable textures for electrical steel sheets used as core laminations of transformers, there is growing technological interest in the texture control by strain-induced grain growth.
The purpose of this study is to investigate the effects of temper rolling conditions on the strain-induced grain growth in low carbon steel sheets. We will show that the tribological conditions of temper rolling have a crucial effect on the texture formation.

2 Experimental Procedure

A steel ingot was hot rolled to 2.2 mm in thickness. The chemical composition of the ingot was as follows; C: 0.02mass%, Si: 0.1mass%, Mn: 0.2mass%. The hot rolled steel sheets were cold rolled to 0.5 mm in thickness and intermediately annealed at 800 for 30s in Ar atmosphere. The temper rolling with and without lubricant was, then, carried out at the rolling speeds varied from 2 m/min to 15m/min. The reduction of temper rolling was 12%. After that, the specimens were finally annealed at 750 for one hour for the strain-induced grain growth.
The texture of finally annealed specimen was studied with pole intensities and (100) pole figure. The grain size of finally annealed specimen was measured with the optical micrograph.

3 Results

The effects of rolling conditions on the grain size of finally annealed specimen are shown in Fig. 1a, and the effects of rolling conditions on the texture are shown in Fig. 1b, and Fig. 2. The increase in lubrication and in rolling speed enlarged crystal grain and reduced (111) pole intensity. A sharp Goss texture was found in the specimen rolled with lubricant. In contrast, the specimen rolled without lubricant had a random texture.

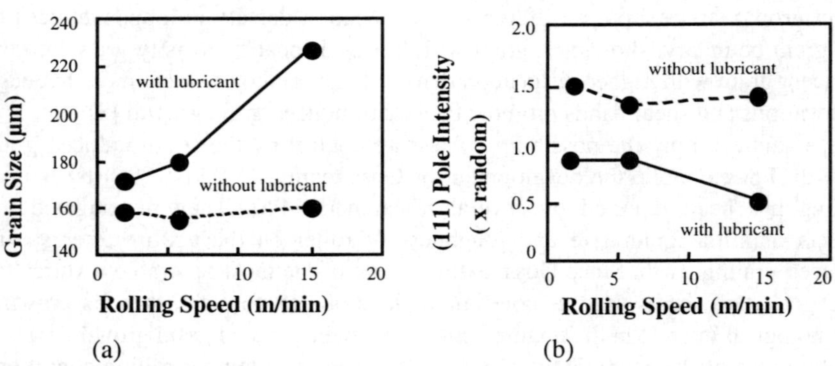

Fig. 1. Effects of rolling conditions on grain size and (111) pole intensity of finally annealed specimen.

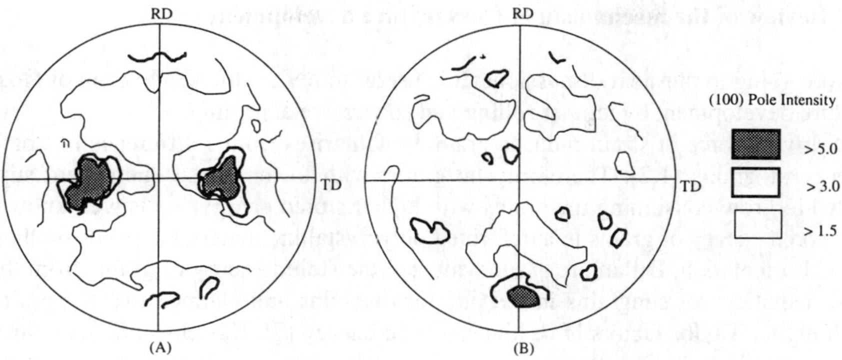

Fig. 2. Effects of lubrication on the (100) pole figure of finally annealed specimen.
(A): The specimen rolled with lubricant at the rolling speed of 15m/min.
(B): The specimen rolled without lubricant at the rolling speed of 15m/min.

4 Discussion

4.1 Temper rolling conditions which enhance Goss texture

Thin film of lubricant existing between the rolled specimen and the roller reduces the friction between their surfaces. The friction is supposed to increase with the decrease of the lubricant film thickness because of the metal-metal contact. Supposing that the viscosity of lubricant is constant during the rolling operation, the thickness of lubricant film h is given by the following equation [6]:

$$h = \frac{3 \cdot \eta \cdot (Vo + V1)}{\sqrt{\frac{t \cdot r}{R}} \cdot Y} \tag{1}$$

Here, is the viscosity of lubricant, Vo is the rolling speed, V1 is the speed of the specimen at the entry, t is the thickness of the specimen at the entry, r is the reduction ratio, R is the diameter of the roller and Y is the deformation resistance of the specimen.
According to Eq. 1, the thickness of lubricant film increases with the rolling speed, which means enhancing the lubrication ability. Therefore, our experimental results obviously show that the increase in lubrication, or the reduction in friction, enhances the Goss texture.

4.2 Review of the mechanism of Goss texture development

Before going to our next discussions, it is useful to review the mechanism of Goss texture development by temper rolling and successive annealing.
The driving force of strain-induced grain growth arises from a difference in stored energy of grains [1,2]. Therefore, the grains with lower stored energy are supposed to grow consuming the grains with higher stored energy. As is well known, the stored energy of grains in a deformed polycrystalline material depends on their crystal orientation. Dillamore *et al.* evaluated the stored energy of grains from the misorientations of subgrains in the interior of grains and claimed that the grains with higher Taylor factors have higher stored energy [7]. Kestens *et al.* examined the mechanism of Goss texture development by temper rolling and successive annealing in low carbon steel sheets. They estimated the stored energy of individual grains using Taylor factors and concluded that the Goss texture is enhanced as Goss orientation grains have low Taylor factors [8].

4.3 Effects of friction of temper rolling on strain-induced grain growth

As described above, tribological conditions of temper rolling has a crucial effect of texture formation. Since the driving force of strain-induced grain growth is a difference in stored energy, we examined the stored energy introduced by temper rolling.
The calculations described later in this section were done in the cases of the friction coefficients 0.14 and 0.30. We will refer to these cases as „the low friction case“ and „ the high friction case“, respectively.
First, in order to assess the straining process of temper rolling macroscopically, the material flow of temper rolled specimen was calculated with a finite element model. This calculation provided the strain rate of the specimen during rolling operation.
Second, the stored energy of temper rolled specimen was estimated with Taylor factors. We assume that the stored energy of grains deformed at a unit strain is proportional to their Taylor factors. Then we introduce , which is supposed to be proportional to the stored energy of grains in temper rolled specimen.

$$\Gamma = \int M d\varepsilon \qquad (2)$$

The integration was done over the temper rolling process. was calculated on the basis of constraint model [9] using the strain rates calculated by the finite element model.
Fig. 3 shows the shear strain rate at the 1/4 thickness layer. The absolute value of shear strain rate in the high friction case is greater than that in the low friction case.

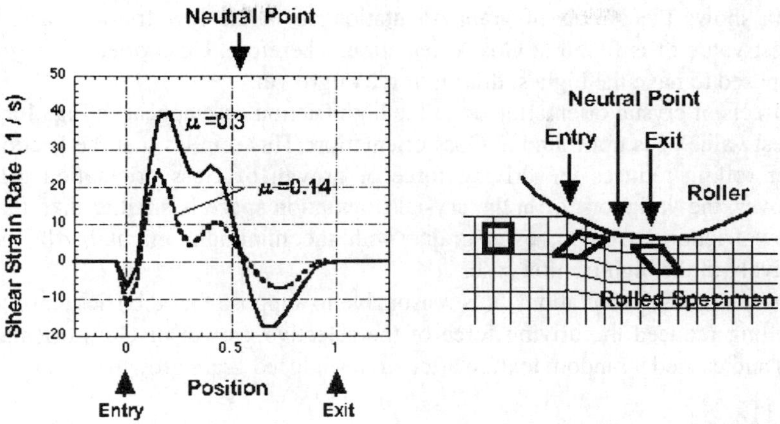

Fig. 3. Shear strain rate at the 1/4 thickness layer.

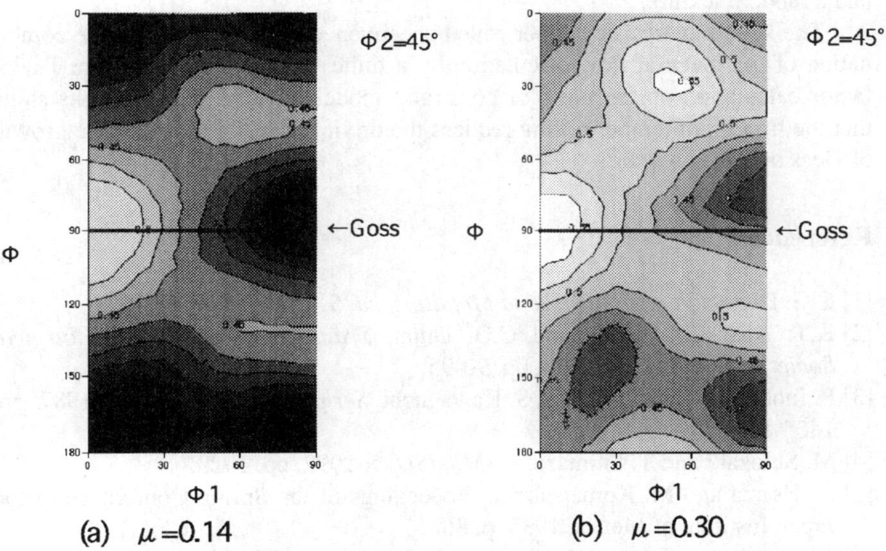

Fig. 4. Effects of crystal orientation and friction coefficient on in the crystal orientation space (Euler Space) of 2=45.

Fig. 4a shows the effects of grain orientation on in the low friction case. The smallest value of is found at Goss orientation. Therefore, Goss orientation grains is supposed to have the highest driving force of growth.
The effects of crystal orientation on in the high friction case appear in Fig. 4b. The smallest value of is not found at Goss orientation. This implies that the friction of temper rolling reduces the driving force of growth of Goss orientation grains. Moreover, the variations of in the crystal orientation space is smaller than that in the low friction case. So, even grains with the minimum might hardly grow selectively in the high friction case.
Considering these calculations, it is reasonable to suppose that the friction of temper rolling reduced the driving force of the selective growth of Goss orientation grains and caused a random texture after strain-induced grain growth.

5 Conclusions

Effects of temper rolling conditions on the strain-induced grain growth have been investigated
1 The specimen rolled with lubricant had a sharp Goss texture after strain-induced grain growth. On the other hand, the specimen rolled without lubrication had a random texture.
2 The stored energy of temper rolled specimen was estimated with the combination of the material flow calculation by a finite element model and the Taylor factor calculation on the basis of constraint model. The calculated results show that the friction of temper rolling reduces the driving force of the selective growth of Goss orientation grains.

References

[1] C.G. Dunn and A.T. Aust, *Acta Metallurgica*, 5, 1957, pp. 368-370.
[2] K.T. Aust, E.F. Koch and C.G. Dunn, *Transactions of the metallurgical Society of AIME*, 215, 1959, pp. 90-95.
[3] F. Inoko, M. Kobayashi and S. Kawaguchi, *Scripta Metallurgica*, 21, 1987, pp. 1405-1410.
[4] M. Shiozaki and T. Shimazu, *CAMP-ISIJ*, 5, 1992, pp. 1923.
[5] A. Honda and M. Komatsubara, Proceedings of the Spring Conference of the Japan Institute of Metals, 1933, p. 86.
[6] A. Azushima, *Tetsuto-Hagane* (Journal of Iron and Steel Institute of Japan), 74, 1978, p. 317.
[7] I.L. Dillamore, P.L. Morris, C.J.E. Smith and W.B. Hutchinson, *Proc. R. Soc. Lond.* A, 329, 1972, pp. 405-420.
[8] L. Kestens, J.J. Jonas, P. Van Houtte and E. Aeroudt, *Metallurgical and Materials Transactions A*, 27A, 1996, pp. 2347-2358.
[9] I.L. Dillamore and H. Katoh, *Met. Sci.*, 8, 1974, p. 21.

Dynamic Observation of Secondary Recrystallization in Cross-Rolled Silicon Steel by Synchrotron X-Ray Topography

Y. Ushigami[1], S. Nakamura[1], T. Kubota[1], and S. Arai[2]

[1] Steel Research Laboratories, Nippon Steel Corporation
20-1 Shintomi, Futtsu, Chiba-ken 293-8511, Japan
[2] Hirohata R&D Laboratories, Nippon Steel Corporation
Fuji-Cho, Hirohata-Ku, Himeji, 671-1188, Japan

Keywords: silicon steel, texture, secondary recrystallization, x-ray topography, coincidence boundary.

Abstract. Dynamics of secondary recrystallization of cross-rolled silicon steel, in which Cube and Goss-type textures evolve, was investigated by in situ observation utilizing synchrotron x-ray topography. Grain growth behavior is strongly affected by grain boundary characteristics and temperature; Cube grain starts to migrate at lower temperature than Goss-type grain. The analysis of primary texture shows that Cube grains and Goss-type grains have a high frequency of Σ7 and Σ9 boundaries to the matrix, respectively. The mechanism can be explained by the assumption that secondary grains grow selectively due to the higher frequency of coincidence boundaries and Σ7 boundary is mobile in the lower temperature range compared with Σ9 boundary.

1 Introduction

Iron crystal has an anisotropy in magnetization and Fe-3%Si alloy which has {100}<001> (Cube) texture is one of the best materials for soft magnetic application because it has two magnetically easy <001>-axes in the sheet plane. Texture control for Cube orientation was studied intensively in the 1950's and one of the proposed methods was cross rolling method; hot rolled sheet is firstly cold rolled parallel to the hot rolling direction and then secondly to the transverse direction for the control of primary texture [1]. In this process, Goss-type ({110}<001> or {110}<110>) texture evolves in addition to Cube texture as a sub-component of secondary recrystallization texture and thus the control of these two components is important from the industrial point of view.

It has been reported that selective growth behavior of secondary recrystallized Goss grains is affected by grain boundary characteristics [2-5] and annealing temperature [6-8]. Coincidence boundaries are supposed to have a special characteristic of lower grain boundary energy, thus suffer lower pinning force from the inhibitor and migrate at lower temperature [3,9].

The authors have found that Cube and Goss-type textures evolve from the different layer of the thickness at the different temperature [10]. The purpose of this paper is to investigate the dynamics of Cube and Goss-type secondary recrystallization to clarify the effect of secondary recrystallization temperature on orientation selectivity for Cube and Goss-type textures.

2 Experimental Procedure

The starting material was the hot band with the following chemical compositions; Si:3.2, C:0.05, Al:0.026, mass%. As the texture of silicon steel is inhomogeneous through the thickness after hot rolling, two specimens were prepared; one was cut from the center layer (specimen-C) and the other from the surface layer (specimen-S) of the hot band to clarify the effect of primary texture on secondary texture. These specimens were then processed as follows; annealing → cross cold rolling (50% reduction parallel to the hot rolling direction followed by 55% reduction perpendicular to the hot rolling direction) → decarburization and primary recrystallization annealing → nitrogen injection treatment (nitrogenized up to 0.02 mass% to form (Al, Si)N precipitate as an inhibitor [11]).
Secondary recrystallization behavior was investigated by the following two annealing conditions.
Experiment-1. These specimens were annealed isothermally at (1)1025°C and (2)1075°C to investigate the effect of secondary recrystallization temperature on secondary recrystallization texture.
Experiment-2. These specimens were annealed in the temperature gradient to evolve some secondary grains in the higher temperature region and then were quenched. Among the secondary grains, Cube and Goss-type grains were selected and grain growth behavior of these secondary grains was observed in situ by the dynamic x-ray topography system at Photon Factory in Japan [12].

3 Results and Discussion

3.1 Structures and orientations of secondary recrystallized grains

In all specimens, coarse grains of 50mm in diameter evolved stably as secondary grains. Orientation distributions of the secondary grains are shown in Fig. 1. Cube and Goss-type ({110}<110>) were the main component of the specimen-C and specimen-S, respectively. Each component evolved selectively at the different temperature. In the case of specimen-C, sharp Cube texture developed at 1025°C. At higher temperature of 1075°C, deviation from the Cube orientation increased. On the other hand in the case of specimen-S, Goss-type secondary grains developed at 1075°C. At lower temperature of 1025°C, deviation from the Goss-type orientation increased and small number of Cube grains were observed.

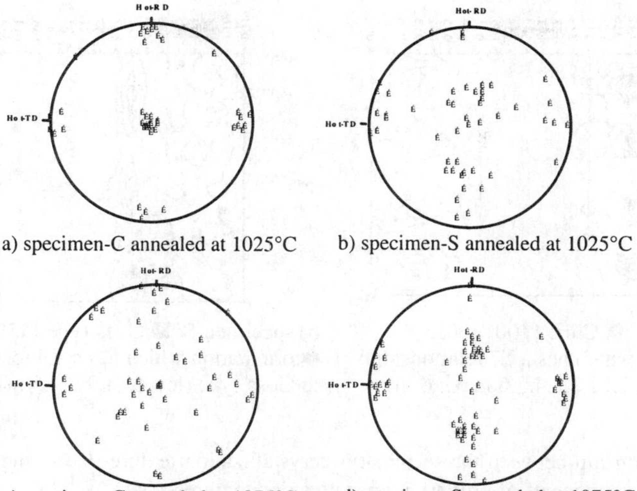

a) specimen-C annealed at 1025°C b) specimen-S annealed at 1025°C

c) specimen-C annealed at 1075°C d) specimen-S annealed at 1075°C

Fig. 1. Orientations of seccondary recrystallized specimens.

3.2 Relationship between primary texture and secondary texture

Fig. 2 shows the three-dimensional orientation distributions of the primary recrystallized specimens by the vector method [13]. The textures of specimen-C and specimen-S were different due to the texture inhomogeneity after hot rolling. The main components of the specimen-C were quadruplet of {211}<124> and {211}<213>, and those of the specimen-S were doublet of {111}<011>. The frequencies of primary recrystallized grains which have coincidence orientation relationships to the secondary grains were calculated by Brandon's criterion [14].

As shown in Fig. 2 and Table 1, Cube orientation has a close correlation to a $\Sigma7$ coincidence relationship to the main component of the primary recrystallized matrix of the specimen-C and Goss-type orientation has a $\Sigma9$ coincidence relationship to the main component of the specimen-S. Therefore, it can be deduced that Cube and Goss-type grains grow selectively due to the high frequency of $\Sigma7$ boundary and $\Sigma9$ boundary to the matrix of specimen-C and specimen-S, respectively.

Table 1. Frequencies of coincidence boundaries.

(a) specimen-C			(b) specimen-S		
Orientation	$\Sigma7$	$\Sigma9$	Orientation	$\Sigma7$	$\Sigma9$
Cube	2.09%	1.65%	Cube	1.14%	1.18%
Goss-type	0.99%	1.07%	Goss-type	1.04%	1.61%
Average	0.99%	1.02%	Average	0.99%	1.02%

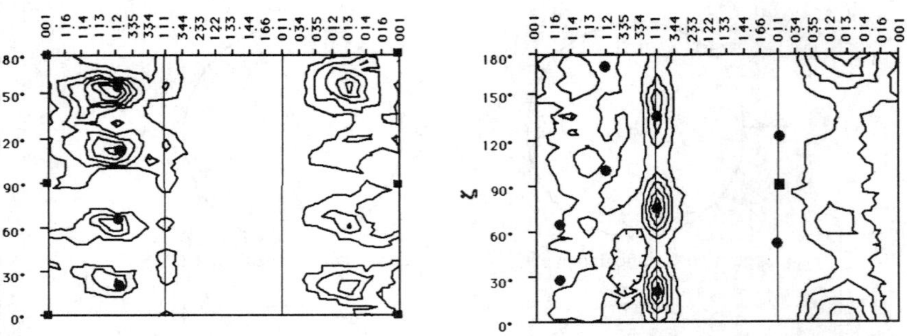

a) specimen-C:■:Cube {100}<001>, ●:orientation which has a Σ7 relationship to Cube. (levels: 1,2,3,4,5,6 max.: 6.1)

b) specimen-S: ■:Goss-type {110}<110>, ●:orientation which has a Σ9 relationship to Goss-type.(levels: 1,2,3,4,5 max.: 5.7)

Fig. 2. Relationship between the secondary recrystallization texture (Cube and Goss-type) and the main components of primary recrystallization textures.

3.3 Secondary recrystallization dynamics

The result of experiment-1 revealed that Cube and Goss-type textures develop due to the different type of coincidence boundary. Then, it was expected that the temperature-dependence of orientation selectivity for Cube and Goss-type is derived from the type of coincidence boundary. To check this point, dynamics of secondary recrystallization was investigated by in situ observation utilizing synchrotron x-ray topography.

Fig. 3a shows the grain growth behavior of Cube secondary grain at 1040°C. Though the migration of secondary grain is locally neither uniform nor stationary due to the retarding force by subboundaries in the secondary grain, it principally depends on its orientation and temperature. Fig. 3b shows the velocities of the two secondary grains; Cube grain of specimen-C and Goss-type grain of specimen-S. It can be seen from this figure that the secondary recrystallization temperature, i.e. the onset temperature of grain growth, is different between the two grains. Cube grain starts to grow at 980°C and Goss-type grain at 1020°C. The relative migration ratio of Cube grain to Goss-type grain, which is a parameter representing selective growth, takes large values in the temperature range from 980°C to 1020°C and decreases as annealing temperature increases. This feature implies that the secondary recrystallization temperature is the crucial parameter of selective growth. The secondary recrystallization temperature is the temperature at which the driving force from the matrix grain exceeds the pinning force derived from the precipitates. As annealing temperature rises, the pinning force deteriorates by dissolution, Ostwaldt ripening and chemical reaction at the surface layer. The pinning force depends on the orientation of the secondary grain through the grain boundary characteristics. Coincidence boundaries are expected to have lower grain boundary energy and suffer lower pinning force from the precipitates, thus trigger the selective growth of secondary grain [3,7,9].

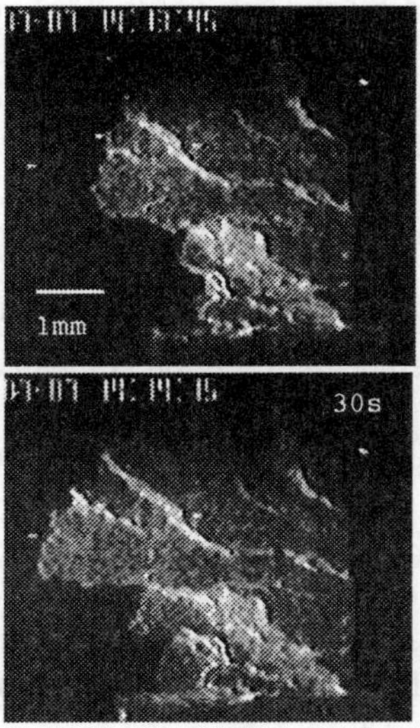

(a)

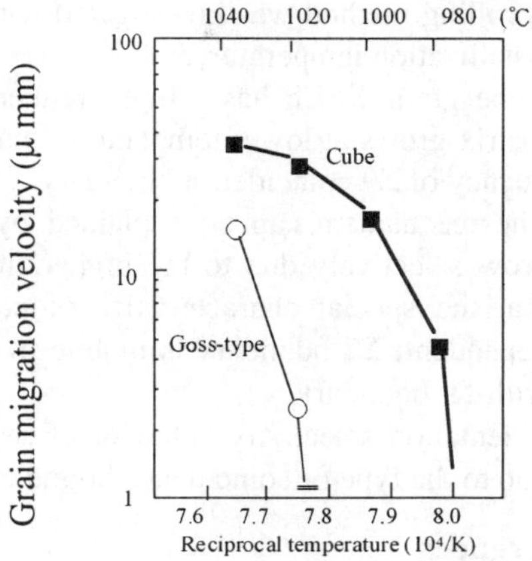

(b)

Fig. 3. Grain growth behavior of Cube grain. (a) Topographic image of secondary grain at 1040°C; (b) effect of temperature on the average grain migration velocity.

These features correspond to the result of experiment-1 and might be attributed to the types of coincidence boundaries. It was reported that the orientation relationships corresponding to low energy grain boundaries (special boundaries) are temperature dependent [15] and special characteristics of coincidence boundaries exist in a finite temperature interval and the transition temperature from a special boundary to a non-special boundary depends on the type of coincidence boundary [16].

Our result can be explained by the assumption that Σ7 boundaries should be mobile in the lower temperature region in comparison with Σ9 boundaries. Similar results were obtained in the case of Goss and deviated-Goss secondary recrystallization of silicon steel in the one stage cold rolling method [6] and in the two stage cold rolling method [8]; in these cases Σ5 boundary was deduced to be mobile in the lower temperature region in comparison with Σ9 boundary. These results coincide with the report on the transition temperatures of coincidence boundaries; in which the transition temperatures of Σ5 boundary and Σ7 boundary are shown to be lower than that of Σ9 boundary [16].

Accordingly, it is deduced that the appropriate secondary recrystallization temperature for the evolution of sharp secondary recrystallization texture changes depending on the types of coincidence boundaries.

4 Conclusions

Dynamics of secondary recrystallization of Cube and Goss-type textures in the cross rolling method was investigated with the special reference to the secondary recrystallization temperature.

(1) Cube grain which has a high frequency of Σ7 coincidence boundary to the matrix grows at lower temperature than Goss-type grain which has a high frequency of Σ9 coincidence boundary.

(2) The mechanism can be explained by the assumption that secondary grains grow selectively due to the higher frequency of coincidence boundaries and that the special characteristics of coincidence boundaries are temperature-dependent; Σ7 boundary is mobile in the lower temperature region compared with Σ9 boundary.

(3) Orientation selectivity behavior of secondary grain is temperature dependent due to the type of coincidence boundary.

References

[1] S. Taguchi and A. Sakakura, U.S. Patent 3287183, 1966.

[2] J. Harase, R. Shimizu and T. Watanabe, Proc.7th RISO Int. Symp. on Met. Mater. Sci., 1986, p. 343.

[3] Y. Ushigami, Y. Suga, N. Takahashi, K. Kawasaki, Y. Chikaura and H. Kii, *J. Mater. Eng,.* 13, 1991, p. 113.

[4] G. Abbruzzese, S. Fortunati and A. Campopiano, *Mater. Sci. Forum,* 94-96 , 1992, p. 405.

[5] B. Hutchinson and H. Homma, Proc. 3rd Int. Conf. of Grain Growth, 1998, p. 387.

[6] Y. Yoshitomi, Y. Ushigami, J. Harase, T. Nakayama, H. Masui and N. Takahashi, *Acta Metall. Mater.*, 42, 1994, p. 2593.

[7] Y. Ushigami, T. Nakayama, Y. Suga and N. Takahashi, *Mater. Sci. Forum* 204-206, 1996, p´. 605, ibid. 157-162, 1994, p. 1081.

[8] K. Murakami, S. Nakamura, M. Ikeda and Y. Ushigami, Proc. 4th Int. Conf. on Recrystallization and Related Phenomena , 1999, p. 981.

[9] Y. Ushigami, T. Kubota and N. Takahashi, *Tex and Microstr.*, 32, 1999, p. 137.

[10] Y. Ushigami, S. Nakamura, T. Kubota and S. Arai, to be published in J.Materials and Prosessing Technology.

[11] Y. Ushigami, F. Kurosawa, H. Masui, Y. Suga and N. Takahashi, *Mat. Sci. Forum*, 204-206, 1996, p. 593.

[12] K. Kawasaki, M. Matsuo, Y. Ushigami and H. Kawata, *Tetsu to Hagane*, 77, 1991, p. 2044.

[13] D. Ruer and R. Baro, *J.Appl.Crystallogr.*, 10, 1977, p. 458.

[14] D.G. Brandon, *Acta Met.*, 14, 1966, p. 1479.

[15] U. Erb and H. Gleiter, *Scripta Met.*, 13, 1979, p. 61.

[16] L.S. Shvindlerman and B.B. Straumal, *Acta Met.*, 33, 1985, p. 1735.

Grain Growth of Iron Oxides during High Temperature Oxidation

Bae-Kyun Kim and J.A. Szpunar

Department of Mining and Metallurgical Engineering, McGill University, 3610 University Street, Montreal, QC Canada H3A 2B2

Keywords: EBSD, high temperature oxidation, steel, texture of oxide.

Abstract. Microstructure of iron oxides is analyzed using electron backscattered diffraction (EBSD) technique. In this study, the orientation imaging microscopy (OIM) allows to follow changes in the oxide structure, grain orientation and grain shape. High temperature oxidation experiments on commercial low carbon steel were performed in the tube furnace up to 950°C. The analysis of microstructure of the cross-sectional area demonstrates that wüstite and magnetite grains have a columnar cell structure parallel to the normal direction to the sheet surface. The process of oxidation and the development of microstructure are described by using data obtained from OIM maps. The different phases of the iron oxide can be distinguished and different characteristics of oxides having different oxidation history are analysed.

1 Introduction

High temperature oxidation of steel has been extensively studied because it affects the quality of the rolled sheets and use of steel in various applications. Practical aspects of steel oxidation are rather well understood because steel-makers have been researching it for many years. During high temperature oxidation the growth of oxide grains is however very difficult to observe. Details of the oxide microstructure cannot be revealed using conventional metallographic methods. Conventional oxidation models for iron describe only the isothermal oxidation with isotropic oxide growth.

In fact, since continuous heating schedule starting at room temperature is applied during the industrial processing operations, different oxidation behaviors are envisaged. Thus, a better understanding of the grain growth at high temperature is necessary for controlling the process of scale formation and removal. For this purpose, the real-time observation of crystal structure by X-ray diffraction, so-called in-situ method is used. Details of the iron oxide microstructures are described using OIM images of the cross-sectional area.

2 Experiments

To determine the corrections necessary for high temperature X-ray diffraction measurements of oxide growth standard oxide powders were measured by X-ray diffraction in situ, in air and in He gas at various elevated temperatures. The measurements were carried out on 99% hematite, 98% magnetite powder (Aldrich) and 99.5% wustite granule (Alfa of Johnson-Matthey). The in-situ measurements of high temperature oxidation were performed using the rotating anode-type Rigaku goniometer with high temperature stage in a two-theta range from 28.8 to 46.8 degrees. The X-ray measurement parameters were 50 kV and 150 mA of CuK_{α} and 0.05 degrees per 1sec in step mode. The heating rate was 20 °C per minute up to 1070 °C.
Specimens for high temperature oxidation experiments were polished with suspension of alumina of up to 0.05 μm particle size. High temperature oxidation was carried out in air in the tube furnace up to 950°C. The surface perpendicular to the normal direction of the specimen with a weak {011}<001> was oxidized. After oxidation, the specimen was rapidly cooled with flowing argon gas in order to prevent the decomposition of wüstite into magnetite and iron. To obtain a good quality Kikuchi pattern, colloidal silica with the size under 0.05 microns was used at the final polishing stage. The scanning electron microscope (SEM) Philips XL-30 FEG with TSL orientation imaging system was used. Transverse direction ,TD, in SEM/OIM geometry is parallel to the direction of oxide grain growth. The SEM operated at 20 kV and the automatic scan was performed over the selected area with a hexagonal grid of 1 micron.

3 Results and Discussion

Before the high temperature oxidation experiments on low carbon steel, non-textured iron oxide powders were examined in-situ by X-ray diffraction. During heating in air, magnetite and wüstite were transformed into hematite at elevated temperature (Fig. 1). For magnetite this transformation starts at 400 °C. In the case of transformation from wüstite, magnetite was formed first at around 400 °C and then oxidized into hematite at 500 °C. When annealed in He gas, both these oxides were stable in the region between room temperature and 1070 °C. Hematite was not transformed into any other phase both during heating in air and in He gas. Diffraction data obtained from the annealing of powders in He gas were used for the calculation of the relative peak intensity ratio of oxide formed on steel specimens.
The result of in-situ X-ray experiments shows the process of oxidation of steel (Fig. 2a). At 400 °C, magnetite peaks are observed in the X-ray diffraction pattern. Hematite starts to form at 600 °C and wüstite at 750 °C. At the temperature of 1070 °C, (006) hematite peak and (400) magnetite peak are observed. Figure 2b shows changes of the ratio of steel oxides diffraction peaks to the standard oxide powders peaks with temperature. Significant changes in the strength of (006)

hematite peak and (400) magnetite peak are recorded; these changes illustrate a strong influence of temperature on oxide texture.

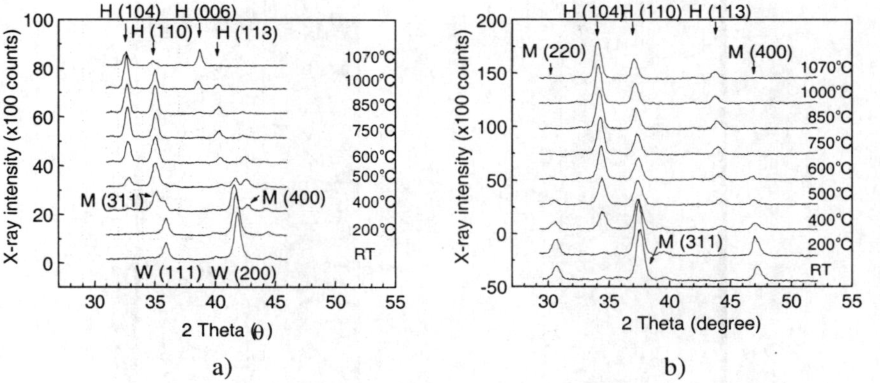

Fig. 1. Phase transformation of wüstite and magnetite powders in air with elevated temperature : (a) wüstite; (b) magnetite.

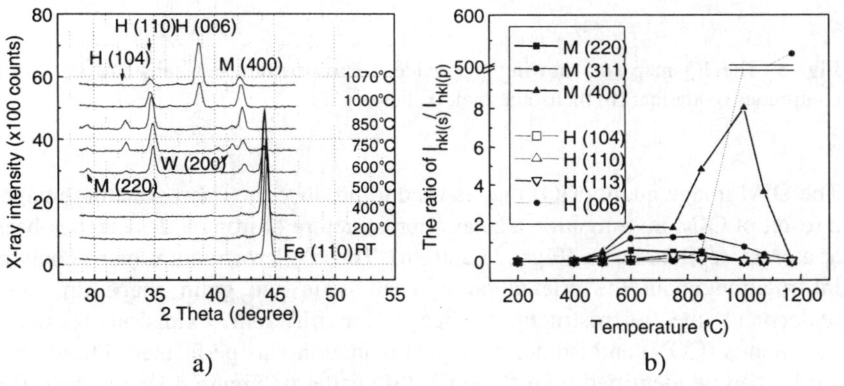

Fig. 2. (a) In-situ X-ray measurement of steel oxidation; (b) Intensity ratio of oxide to powder with elevated temperature.

According to the classical mechanism of iron oxidation at temperature above 570°C [1-3], at the wüstite/magnetite interface wüstite is formed and magnetite is reduced. Magnetite is created and hematite reduced at the hematite/magnetite interface and hematite is nucleated at the hematite/gas interface. The oxidation of steel is controlled by an outward diffusion of iron ions through the oxide layer. Therefore, wüstite formation and growth cannot be detected by in-situ X-ray measurement. The reason that the wüstite peaks at 1070 °C are not observed is a low effective penetration depth of X-ray beam. The relative thickness ratio of the three oxide layers is known to be 95:4:1 for FeO: Fe_3O_4: Fe_2O_3 at 1000 °C [4].

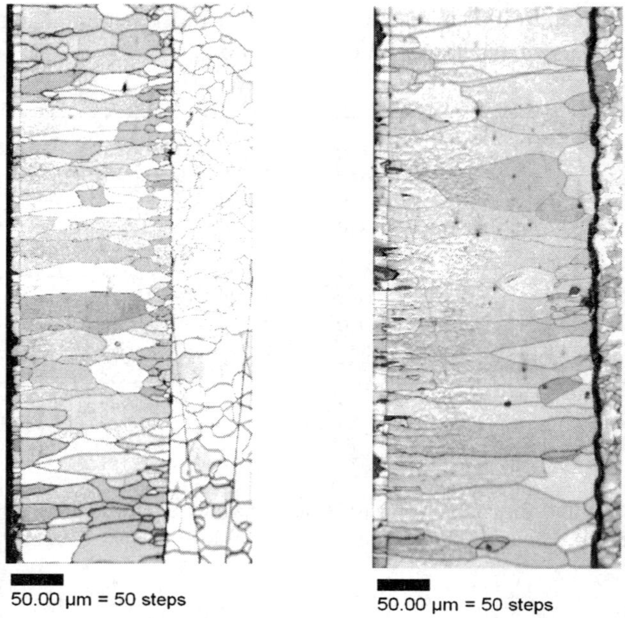

a) b)

Fig. 3. The IQ map for showing the oxide microstructure and substrate steel: (a) the continuous oxidation; (b) the isothermal oxidation.

The OIM image quality (IQ) map is used to produce the microstructure image. As a result of OIM investigation a clear microstructure of iron oxides can be observed at a cross-sectional area (Figs. 3a and 3b). These microstructure maps contain the information about the orientation of each grain and grain shape. In order to understand the microstructural change for different oxidation history, the continuous (CON) and isothermal (ISO) oxidation was performed. Phases of iron oxides can be identified by different EBSD patterns. Figure 4 shows the different Kikuchi patterns obtained from steel, wüstite, and magnetite, respectively. According to the characteristics of the Kikuchi patterns, the left layer in Figs. 4a and 4b is the magnetite layer and the middle layer is wüstite. The right side is steel substrate. In the presented OIM maps, the hematite layer cannot be detected. Hematite is most likely removed during polishing of the specimen, because it is grown like thin blades or fine whiskers [5]. In both of the CON and ISO oxidation, the grains of wüstite have anisotropic, columnar structure. Magnetite has a simple cell structure. In the region of the wüstite/steel interface, there are small, round shape wüstite grains that can represent the nucleation and the growth regions. The main characteristic of these grains is the difference in texture strength; the oxide obtained during the CON oxidation has in the growth region a higher intensity of

(001)-fibre texture and also higher percentage of low angle boundaries (<25 degrees) than the nucleation region [6].

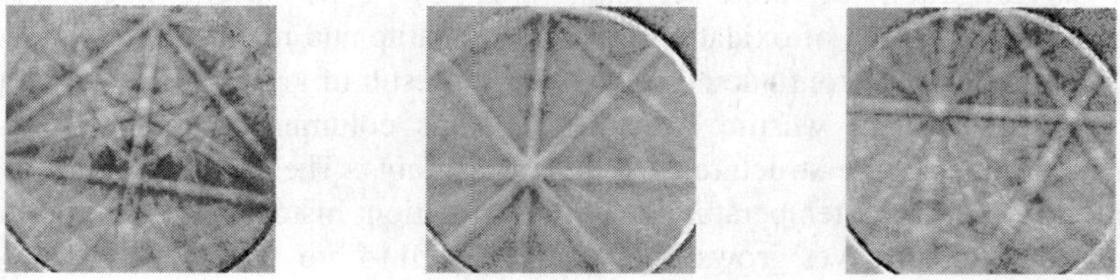

a) b) c)

Fig. 4. Examplary Kikuchi patterns for (a) Fe; (b) wustite; (c) magnetite.

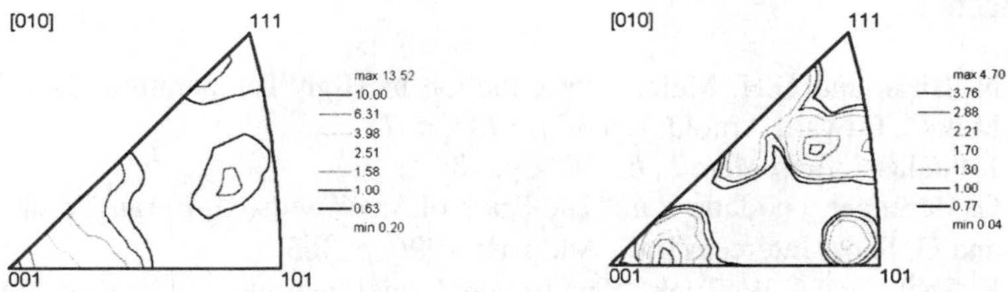

a) b)

Fig. 5. Inverse pole figures for direction parallel to the oxide grain growth direction for wüstite layer (a) the continouns oxidation and (b) the isothermal oxidation.

In Figs. 3a and 3b, the difference between the CON and ISO oxidation is illustrated. The differences in the initial nucleation region are in the oxide morphology and type and strength of texture. In the ISO oxidation, the size of oxide grains in the nucleation region is bigger than that in oxides obtained in the CON oxidation. The thickness of columnar grains is higher that in the case of the CON oxidation. One can explain this observation assuming that in the ISO oxidation the number of nucleation sites is smaller and the growth of nucleus is less constrained by the nuclei that grow in the neighbouring sites. Texture of wüstite in both cases is described by inverse pole figure of the direction parallel to the grain growth direction (Fig. 5). Fibre texture (001) of wüstite is much higher for the CON oxidation than that for the ISO oxidation.

4 Conclusions

The investigation of cross-sectional area by OIM offers microstructural description of details of oxidation process. Hematite and magnetite have a strong {001} texture in the continuous oxidation. As a result of grain growth competition at high temperature wüstite structure develops columnar grain structure and magnetite has simple structure of round type grains. The microstructure of iron oxides depends on temperature and the oxidation history. In the continuous oxidation, wüstite layer grows with a strong (001)-fibre texture in the direction normal to the sheet surface. During the isothermal oxidation, however, wüstite develops a weak texture; this texture is associated with columnar grains that have thickness that is higher than in grains observed during continuous oxidation.

References

[1] N. Birks and G.H. Meier, „Introduction to High Temperature Oxidation of Metal", Edward Arnold, London, 1983, p. 74.
[2] J. Païdassi, *Acta Metall.*, 6, 1958, p. 184.
[3] G. Béranger, Oxidation, in "The Book of Steel", eds. G. Béranger, G. Henry, and G. Sanz, Intercept Ltd., Andover, 1996, p. 363.
[4] K. Sachs and C.W. Tuck, "The Iron and Steel Institute, Publication No. 111", London, 1968, pp. 1-17.
[5] D.A. Voss, E.P. Butler, and T.E. Mitchell, *Metall. Trans. A*, 13A, 1982, p. 929.
[6] B.-K. Kim and J.A. Szpunar, *Scripta Metall.,* submitted.

Microscopic Secondary Recrystallization Process in an Fe-3%Si Alloy

Shigeto Takebayashi[1], Shigeru Suzuki[1], and Yoshiyuki Ushigami[2]

[1]Advanced Technology Research Laboratories, [2]Steel Laboratories,
Nippon Steel Corporation, Shintomi, Futtsu-City, Chiba, 293-8511, Japan

Keywords: Fe-3%Si alloy, secondary recrystallization, EBSP, CSL, Σ9.

Abstract. In order to observe the grain boundary movement during secondary recrystallization in an Fe-3%Si alloy, a pulsive annealing and a following OIM (orientation imaging mapping) observation were alternatively performed for many times. The starting material was an Fe-3%Si alloy which was composed of primary recrystallized grains on one half of a sample sheet and secondary recrystallized Goss grains with {110}<001> orientation on the other half. The crystallographic orientations of the primary grains were studied in view of the CSL relationship with respect to {110}<001> orientation. At the boundary between the primary and the secondary recrystallized grains, the frequency of the near-Σ9 grains was shown to be low, which implies that the Goss grain preferentially consumes the primary grains with Σ9 in the secondary recrystallization process. The orientation evolution process was also studied in detail .

1 Introduction

Secondary recrystallization phenomenon [1] is intensively exploited in Fe-3%Si alloys as the electromagnetic core material, where {110}<001> texture, i.e. Goss texture, is preferred in view of the magnetic property. Since the easy axis of magnetization lies parallel to <001> axis in iron, the Goss texture gives a low watt loss when the field is applied in the RD direction. Therefore, the sharpness of the Goss texture is carefully controlled so that it should lie within 10 degrees or less [2]. The mechanism that gives rise to such a sharp texture has long been investigated and discussed. It has been pointed out that coincidence site lattice (CSL) boundaries, especially Σ9 boundaries, enhance the selective growth of Goss texture due to its lower boundary energy than that of general boundaries [3]. A strong orientation dependence of Goss texture evolution was found to be related to the Σ9 boundaries by Hutchinson [4]. On the other hand, Amiri *et al.* reported that an OIM analysis showed little evidence for the relationship between CSL boundaries and Goss texture evolution [5].
In the present study, OIM was utilized to follow the Goss texture evolution during heat treatment. The orientation change of primary grains in the vicinity of the

secondary grains was analyzed. The orientation distribution was analyzed in view of the deviation angle from that of Σ9 with respect to Goss orientation. In order to investigate the preferential growth of the Σ9 boundaries, also studied was, on the initial sample, the difference in the orientation distribution between the primary grains in touch with and those not in touch with the secondary grain with Goss orientation.

2 Experimental Procedure

The starting material was an Fe-3%Si sheet which was composed of primary recrystallized grains on one half of the sheet and the Goss texture was fully developed on the other half. It was processed under temperature gradient, and the detail of the sample sheet preparation is described elsewhere [7]. The sample was measured by OIM technique in the range of 512x680μm^2 with a step length of 4μm. The samples were analyzed initially, then annealed under hydrogen atmosphere for 2s at 1010°C or 1040°C. The cycle of OIM measurement and annealing was repeated many times for observing the successive orientation change by OIM. The measured orientations were interpreted in view of the deviation angle from that of Σ9 with respect to Goss orientation, which hereafter is referred to as Goss-Σ9 orientation. Prior to analysis, the systematic error due to the sample misalignment was corrected so that the average orientation in the Goss region should take {110}<001> orientation. The possible shift of the observed area was estimated and corrected by statistically correlating the successive two results.

3 Results

3.1 Goss-Σ9 orientation

The measured orientations are interpreted in terms of the deviation angle with respect to Goss-Σ9 orientations which are calculated by rotating {110}<001> orientation matrix by 38.94° around <110> or its symmetrically equivalent axis. Taking also into account the symmetrical equivalence of the {110}<001> orientation, the Goss-Σ9 orientations are sorted into 4 types as shown in Table 1. Observed deviation angle is defined as the minimal angle between the measured orientation and these 4 types of orientations including all of their symmetrical equivalence.

Table 1. Orientations of Σ9 with respect to Goss orientation.

	Orientation index
Type1	{1 1 0}<4 4 7>
Type2	{8 7 7}<7 4 4>
Type3	{4 1 1}<1 4 8>
Type4	{11 5 4}<4 8 1>

3.2 Orientation distributions in and out of the vicinity of Goss boundaries

The present sample has a long Goss boundary going all the way across the specimen. If Goss texture prefers Σ9 boundaries during evolution, the area fraction of Goss-Σ9 orientation is expected to be lowered among the grains in touch with the Goss grain [8]. It was evaluated and compared with that of the grains out of touch with the Goss texture. To evaluate the fraction, 10 areas were observed and then about 5000 primary grains were involved with the investigation. Among those primary grains, about 10% were in touch with the Goss boundaries. Figs. 1a and b show the deviation angle distribution of in and out of touch with the Goss texture, respectively. The solid curves correspond to the Gaussian functions that fit the experimental results. The frequency in Fig. 1a is narrower than that shown in Fig. 1b. According to Brandon's criterion, the allowed deviation is less than 5° [9]. Therefore the fraction at the exact Goss-Σ9 in the present observation is quite low. However, it should be noted that in the small deviation angle region, the frequency is lower in (a) than in (b). To obtain a clear view, the ratio of frequency (a) to (b) is shown in Fig. 2. The ratio increases systematically and linearly in the small deviation angle region less then 10°, where it reaches almost unity. As reported in the literature [8], this depletion in the range within 10° of deviation implies the preferential consumption of the Σ9 grains during secondary recrystallization. And the linear change of the ratio in this region accounts for the dependence of this preferential manner on the off-Σ9 boundaries. The ralationship between CSL and off-CSL boundaries in aluminum was reported by Kokawa et al. [10], who investigated the sliding behavior of the grain boundaries. It should be clarified whether the similar manner holds in the present case. The present discussion is summarized as follows.

(1) The selectivity of Σ9 grains is implied to depend upon the deviation angle from Σ9.
(2) The critical angle of deviation is around 10°, twice as large as Brandon's criterion. And over the critical angle the selectivity is not clear.
(3) The relationship between off-CSL and exact CSL boundaries were investigated in case of aluminum grain boundary sliding [10], where is pointed out that the off-CSL turns into CSL. However, it needs a detailed investigation to clarify whether the similar manner hols in the present case.

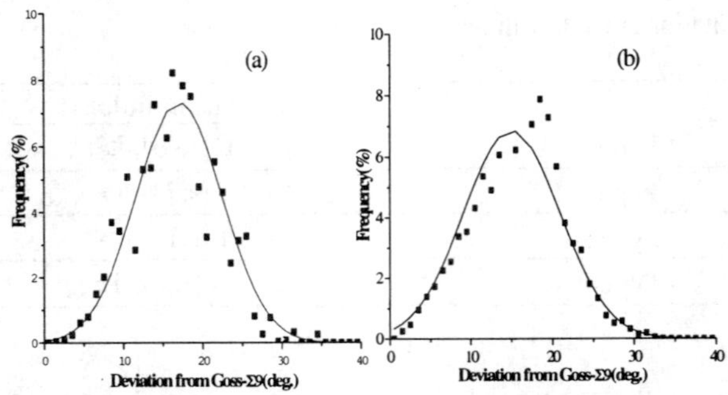

Fig. 1. Distribution of the deviation angle with respect to Goss-Σ9 orientation among the grains in touch (a), and out ou touch (b) with Goss texture. Solid curves are Gaussian fit functions .

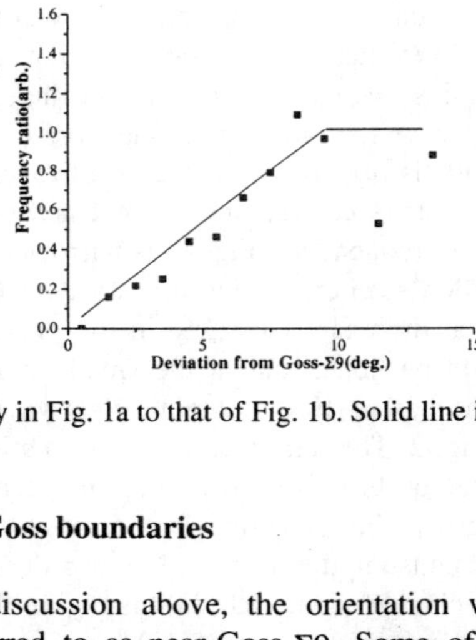

Fig. 2. Ratio of frequency in Fig. 1a to that of Fig. 1b. Solid line is an approximate fit line.

3.3 Local motion of Goss boundaries

As a result of the discussion above, the orientation within 10° from Goss-Σ9 orientation is referred to as near-Goss-Σ9 . Some of the boundary motions depicted during the experiment are shown in Fig. 3 and Fig. 4, which correspond to the case where the annealing temperature is 1010°C and 1040°C, respectively. At 1010°C, the near-Goss-Σ9 grains were initially out of touch with the Goss boundary. As the heat-treatment proceeded, Goss boundary slightly displaced toward the primay recrystallized region homogeneously. At the third cycle of heat treatment, at 6s, a small near-Goss-Σ9 grain, with the orientation 8° away from the type3 Goss-Σ9 shown in Table 1, touched the boundary. It remained and stayed unchanged until 22s, the 11th cycle, which corresponds to Fig. 3a, where the near-Goss-Σ9 touching the boundary is denoted by A. After another heat treatment, the boundary displaced locally as shown in Fig. 3b.

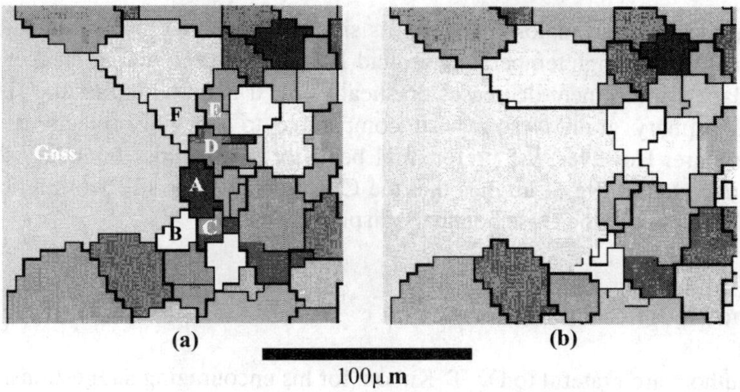

Fig. 3. OIM map of before (a), measured at 22s , and after (b), measured at 24s, the preferential growth. The temperature is 1010°C. Among the grains, A and E are near- Goss-Σ9 grains.

At 1040°C, shown in Fig. 4, the Goss boundary displaced so rapidly that it was difficult to depict the local displacement so clearly as shown in Fig. 3. A near-Goss-Σ9 grain, 7.8° from type1 Goss-Σ9 was observed to be coarsened and impede the Goss boundary displacement and then surrounded almost with Goss texture in 2s, as shown in Fig. 4b. Another coarsened residual grain denoted by B, 25° away from Goss-Σ9, is also observed. Although this observation seems to contradict to the present hypothesis, after additional 6s of heat treatments, the grain A disappears together with other grains while grain B, as shown in Fig. 4b,

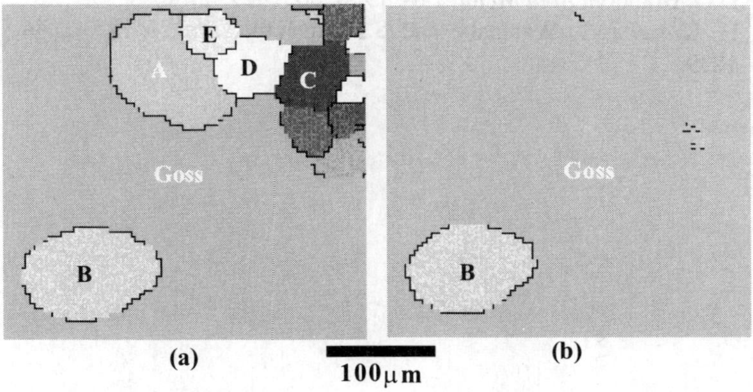

Fig. 4. OIM maps at 1040°C annealing. Cummulative time of annealing is 2s and 8s in (a) and (b), respectively.

still resides without major change in its shape. As reported by one of the present authors (YU), at high temperature around 1040°C, the preferential manner of Goss boundary displacement decreases drastically [6]. It is possible that the size effect of the primary grain is somewhat comparable to the CSL selectivity at high temperature. However, the preferential behavior of the Goss boundary displacement depicted in Fig. 4 implies that the CSL selectivity is still working to give a sounding effect to the Goss boundary displacement.

Acknowledgdement

The authors are grateful to Dr. T. Kubota for his encouraging suggestions. Drs. M. Hashimoto and S. Ito are also greatly acknowledged.

References

[1] J.E. May and D. Turnbull, *Trans. Met. Soc. AIME*, 212, 1958, p. 769.
[2] R. Shimizu and J. Harase, *Acta Metall.,* 37, 1989, p. 1241.
[3] J. Harase and R. Shimizu, *Acta Metall.,* 38, 1990, p. 1395.
[4] B. Hutchinson and H. Homma, ,,Grain Growth in Polycr. Mater. III.'' eds. H. Weiland et al., MMM Soc., 1998, p. 387.
[5] A. Amiri, T. Baudin and R. Penelle.,ibid, p. 567.
[6] Y. Ushigami, K. Murakami and T. Kubota, ibid.,p.491
[7] Y. Ushigami, T. Kubota and N. Takahashi., *ISIJ Int.*, 38, 1998, p. 553.
[8] P. Lin, G. Palumbo, J. Harase and K.T. Aust, *Acta Mater.*, 44, 1996, p. 4677.
[9] D.G. Brandon, *Acta Metall.*, 14, 1966, pp. 1479.
[10] H. Kokawa, T. Watanabe and S. Karashima, *Phil. Mag. A.*, 44, 1981, p. 1239.

Ultra-Fine Grain Structures

Mechanism of Ultrafine Grain Formation during Equal Channel Angular Extrusion in an Aluminum Alloy at Intermediate Temperature

R. Kaibyshev[1], O. Sitdikov[1], I. Mazurina[1], D.R. Lesuer[2]

[1] Institute for Metals Superplasticity Problems RAS, 450001, Khalturina 39, Ufa, Russia
[2] Lawrence Livermore National Laboratory, L-342 P.O. Box 808, Livermore, CA 94551, USA

Keywords: submicrocrystalline structure, aluminum alloys, continuous dynamic recrystallization, equal channel angular extrusion.

Abstract. The mechanism of grain formation during equal channel angular extrusion (ECAE) in a 2219 aluminum alloy has been studied at T=250°C. It was shown that continuous dynamic recrystallization (CDRX) occurred during intense plastic straining and resulted in formation of submicrometer scale grains. An important feature of CDRX is formation of a low energy dislocation structure (LEDS) after first passage, which provides high stability to the evolving subgrain structure. Such a structure transforms into a recrystallized granular structure at a high rate with increasing strain. The mechanism of CDRX is discussed.

1 Introduction

A limited number of studies have been focussed on the evolution of microstructure during intense plastic deformation [1-3]. As a result, no adequate models have been reported in the literature for mechanisms of ultrafine grain formation. Careful examination of the data from studies [1-3] shows that CDRX occurs in pure aluminum. Subgrain structure formation takes place after the first pass in the ECAE process. Additional deformation gradually increases the deformation-induced grain boundary misorientation and results in their conversion into true high angle boundaries. It is known [4] that subgrain structure stabilization is a primary requirement for the onset of CDRX. However, there are no particles providing stabilization of the subgrain structure. In pure aluminum, the sources of stability for boundaries resulting from plastic deformation are not clear. Thus, the aim of the present study is to examine microstructural evolution in an aluminum alloy during intense plastic straining to reveal a mechanism of grain refinement.

2 Experimental Procedure

The 2219 aluminum alloy with the chemical composition Al–6.4%Cu–0.3%Mn–0.18%Cr–0.19%Zr–0.06%Fe was manufactured at the Kaiser Aluminum – Center for Technology by direct chill casting. The alloy was homogenized at 530°C for 6 hours and cooled in air. After this treatment, the alloy had an equiaxed grain structure with an initial grain size of about 120 μm. Then the ingot was machined into rods with a diameter of 20 mm and a length of 100 mm.
The alloy was severely deformed by equal channel extrusion (ECAE) processing [1-3] at T=250°C and a strain rate of 10^{-2} s^{-1} up to true strains 1, 2, 4, 6 and 8. ECAE was done using an isothermal die with a circular internal cross-section. The channel had an L-shaped configuration with an intersection angle equal to 90°. This leads to a strain of e~1 on each passage through the die. After these treatments, the resulting microstructure was examined on 1 mm thick samples which were cut from the centers of the pressed rods with the cutting direction parallel to the extrusion axis. Samples were examined using TEM. For these examinations, samples were cut from the gauge section of extended specimens and thinned to about 0.25 mm. Discs with 3 mm diameter were cut and electropolished to perforation with a Tenupol–3 twinjet polishing unit using a 20% nitric acid solution in methanol at –30°C and 15 V. The thin foils were examined using a JEOL–2000EX TEM with a double-tilt stage at an accelerating potential of 160 kV.

3 Results

The TEM observations showed that the first passage through the die resulted primarily in the formation of a mixed dislocation structure consisting of two structural components. One component of the microstructure was a typical cell structure. The cell boundary thickness was about 100 nm. The other structural component was a well-defined subgrain structure (Fig. 1a) located between the cell bands. The volume fraction of the subgrain structure is slightly higher than that of the cell structure. Subgrains as well as cell structures are essentially equiaxed. The average subgrain size is about 0.7–0.8 μm and the cell size is about 0.4–0.5 μm. With increasing strain up to ~2, dense dislocation walls (DDW) are formed (Fig. 1b). The well-defined LEDS bands alternate with the equiaxed subgrain structure. With increasing strain the volume fraction of DDW areas increased and volume fraction of cell structure tends to reduce. Overall volume fraction of LEDS gradual decreased with strain.
Further deformation up to e=4 resulted in the formation of a mixed network of high-angle and low-angle boundaries. Two distinct microstructural regions were observed. In one of these regions (sub)grains were equiaxed (Fig. 1c) and in the another region (sub)grains were elongated (Fig. 1d). The region containing the

equiaxed (sub)grains was larger than the region containing elongated (sub)grains. The region containing equiaxed (sub)grains is shown in Fig. 1c. High angle boundaries dominated in this structure. The distribution of lattice dislocations is not uniform within both types of crystallites. Individual lattice dislocations are retarded by fine intermetallic particles and, thus, an enhanced lattice dislocation density is observed in secondary phase rich areas. Most high angle boundaries are free from dislocations. Notably, the precipitates of secondary θ−phase as well as the dispersoids were found to be very effective in pinning the sub-boundaries formed and retarding high angle boundary migration. As a result, a large number of particles were observed at (sub)grain boundaries. Increasing deformation up to

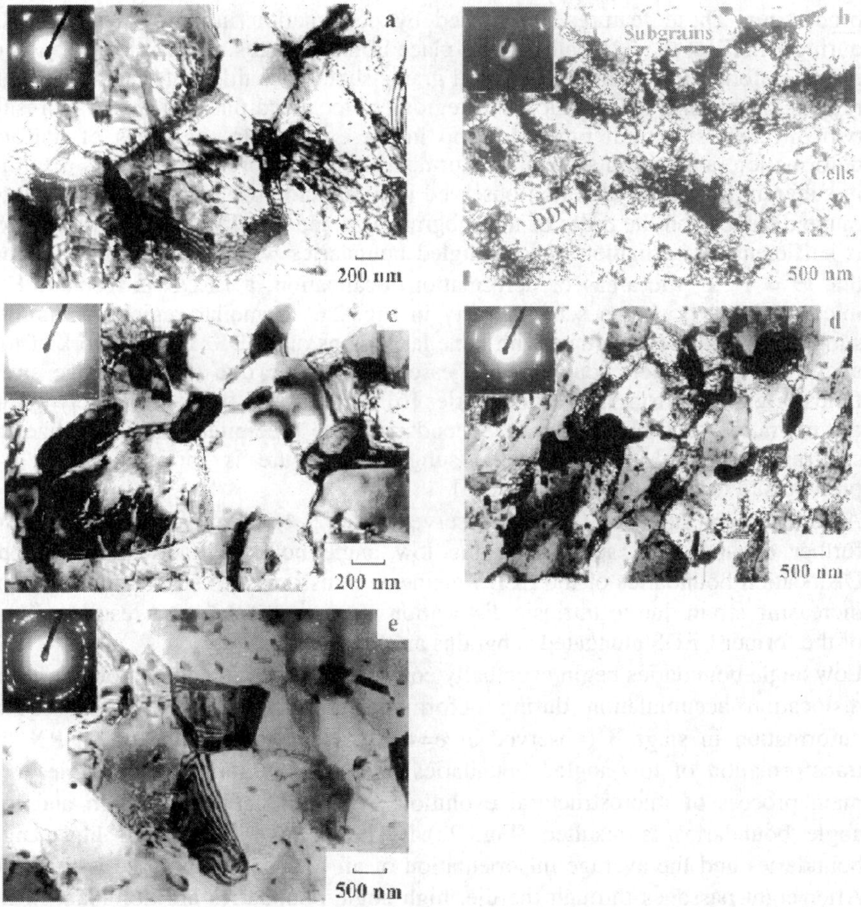

Fig. 1. TEM images showing microstructural change in a 2219 aluminum alloy during ECAE at T=250°C: (a) e=1 (b) e=2; (c) and (d) e=4; (e) e=8.

e=6–8 led to gradual conversion of low angle boundaries into high angle boundaries (Fig. 1e). The shape of grains tended to be essentially equiaxed. In the strain interval e=4–8, the average crystallite size increased by a factor of 1.2 and the lattice dislocation density decreased with increasing strain.

4 Discussion

It is seen that intense plastic straining of the 2219 Al alloy at T=250°C reduced the grain size to a submicrometer level (0.9–1.0 μm) due to CDRX [5]. There are special microstructural features associated with CDRX resulting from ECAE processing. These features are caused by deformation inhomogeneity. CDRX during ECAE of the 2219 alloy takes place in three stages. Intense localization of plastic deformation within the original grains subdivides this material volume into regions of uniform deformation and regions of localized plastic flow. As a result, two different structural types develop in these areas. In the areas of uniform deformation, a subgrain structure is formed. Dislocation rearrangements resulting in subgrain formation are often observed if climb and cross-slip occur over large-range distances on the order of the subgrain size and density of lattice dislocation is sufficient for formation of low angled boundaries with an angle being higher that $\Theta = 1^{\circ}$. In the areas of deformation localization, a LEDS is formed. This microstructure is shown schematically in Fig. 2a. Lamellar regions containing subgrain boundaries alternate with lamellar regions of LEDS. This network of low angle and dislocation boundaries is essentially stable due to long-range stress fields originating from LEDS boundaries [6]. These stress fields effectively hinder the migration of sub-boundaries. Secondary phase precipitates also stabilize the subgrain structure. Thus, a stable subgrain structure is formed, which is a prerequisite for the onset of CDRX [5].

At the second stage of CRDX (observed at e=2–4), the formation of DDWs further enhances the stability of the low angle boundary network (Fig. 2b). Dislocation boundaries of the LEDS elements transform into sub-boundaries with increasing strain due to intrinsic dislocation rearrangements. As a result, in areas of the former LEDS elongated subgrains are formed.

Low angle boundaries begin eventually convert into true high angle boundaries by dislocation accumulation during deformation in stage 2 and at subsequent deformation in stage 3 (observed at e=4–8). At the third stage of CDRX the transformation of low-angled boundaries into true high-angled boundaries is a main process of microstructural evolution. A mixed network of high and low angle boundaries is resulted (Fig. 2c). The volume fraction of high angle boundaries and the average misorientation of all boundaries increase with strain. After eight passages through the die, high angle boundaries are dominant in the microstructure. Strain induced boundary migration results in an equilibrium boundary network configuration as shown in Fig. 2d.

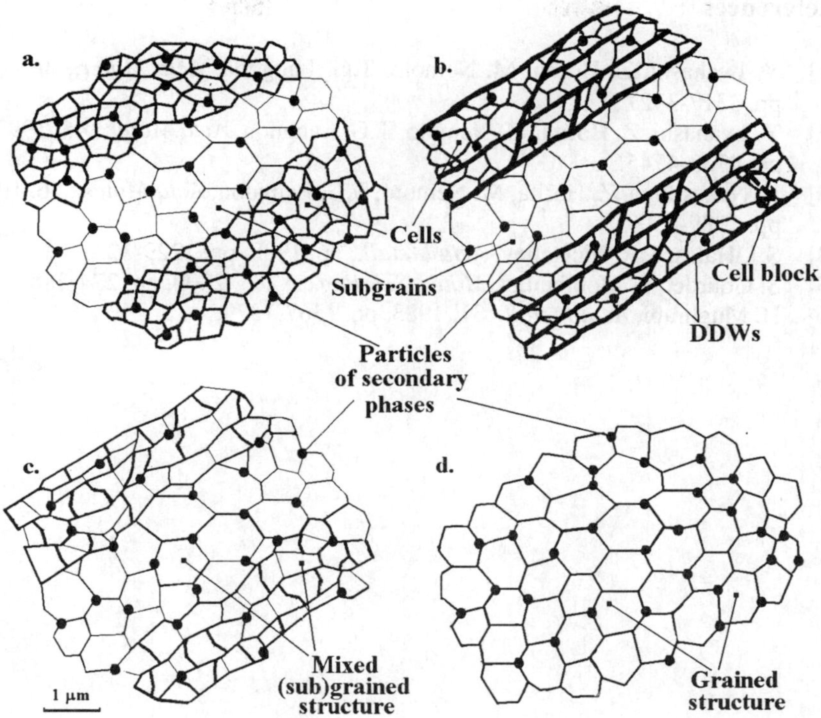

Fig. 2. Schematic illustration of structure evolution during ECAE of the 2219 alloy at T=250°C and strain rate of 10^{-2} s^{-1}: (a) e=1, cell structure alternates with subgrain structure; (b) e=2, banded LEDS alternates with (sub)grain structure; (c) e=4–6, areas of elongated (sub)grains alternate with areas of equiaxed (sub)grains; (d) e=8, equiaxed grain structure.

In summary, it can be concluded that no unique mechanism of grain formation is operative during ECAE at intermediate temperature. New grains form during ECAE by CDRX mechanisms, which are similar to those observed for aluminum alloys deformed by conventional deformation approaches. Severe deformation by ECAE results in significant strain localization and formation of a LEDS concurrently with a subgrain structure. The LEDS stabilizes the subgrain structure. For pure aluminum [1-3], the only source of low-angle boundary stabilization is the LEDS network. For the 2219 alloy, precipitates provide additional sources of low-angle boundary stabilization. CDRX ceases if the subgrain structure is unstable. Thus, strain localization during ECAE promotes the formation of an ultrafine structure.

References

[1] Y. Iwahashi, Z. Horita, M. Nemoto, T.G. Langdon, *Acta Mater.*, 46, 1998, pp. 3317-3327.
[2] Y. Iwahashi, Z. Horita, M. Nemoto, T.G. Langdon, *Acta Mater.*, 45, 1997, pp. 4733-4745.
[3] K. Nakashima, Z. Horita, M. Nemoto, T.G. Langdon, *Acta Mater.*, 46, 1998, pp. 1589-1599.
[4] S.J. Hales, T.R. McNelley, *Acta Metall.*, 36, 1988, pp. 1229-1231.
[5] S. Gourdet, F. Montheillet, *Mater. Sci. Eng.*, A283, 2000, pp. 274-288.
[6] H. Mughrabi, *Acta Metall.*, 31, 1983, pp. 1367-1379.

Nanostructures in Al-Based Alloys After Severe Plastic Deformation

S.V. Dobatkin[1], V.V. Zakharov[2], and R.Z. Valiev[3]

[1] Moscow State Steel and Alloys Institute (technological university) Leninsky prospekt,4, 119991 Moscow, Russia, E-mail: dobatkin@tmo.misa.ac.ru
[2] All-Russia Institute of Light Alloys, Moscow, ul. Gorbunova, 2, 121596 Moscow, Russia
[3] Ufa State Aviation Technical University, ul. K.Marksa, 12, 450000, Ufa, Russia, E-mail: RZValiev@mail.rb.ru

Keywords: severe plastic deformation, nano- and submicrostructure, aging, Al alloys.

Abstract. The structure and microhardness after severe plastic deformation (SPD) in torsion under high pressure of Al-Cu-Mg alloy 2024, binary Al-Sc alloys with contents of Sc-0,14%, 0,4% and 0,55%, and also industrial Al-Mg-Sc alloys 01515, 01535, 01545 and 01570 are investigated. At SPD of quenched alloy 2024 with supersaturated solid solution nanocrystalline structure with the size of a grain 50 nm is obtained due to strain aging. At SPD of binary Al-Sc alloys the only submicrocrystalline structure is formed. Addition more than 3% Mg in Al-Sc alloy results in formation of nanostructure during SPD. The minimal size of a grain - 40 nm is revealed in alloy Al-5,9%Mg-0,3%Sc.

1 Introduction

It is known, that severe plastic deformation (SPD) of aluminum alloys allows to receive rather massive samples with submicrocrystalline structure with the size of a grain more than 100nm [1-3].Thus, there is a task of the further refinement of structure in aluminum alloys up to nanoscale and increases of its thermal stability. In aluminum alloys it is especially important for realization of effect of low-temperature and high-speed superplasticity. Probably, the task can be solved at the expense of phase transformations during SPD [4,5] and additional alloying in particular, with Sc [6-8]. Alloying with Sc of aluminum alloys renders the large influence on their structure and properties.

The purpose of the present work is to study of structure and properties of aluminum alloys, including alloying with Sc, after SPD in torsion under high pressure in condition of a supersaturated solid solution.

2 Experimental Procedure

As a materials of research were taken Al-Cu-Mg alloy 2024 (4,2% Cu; 1,6%Mg; 0,6%Mn), binary Al-Sc and industrial Al-Mg-Sc alloys.An alloy 2024 was studied in two initial states: hot-rolled and quenched (495 ^{0}C). The binary alloys were received by a method of continuous casting from aluminum 99,99% with 0,14; 0,4 and 0,55% Sc. The contents of impurity: 0,02%Fe; 0,01%Si; 0,06%Ti.As industrial Al-Mg-Sc alloys were chosen [8]: 01570 (5,9%Mg; 0,3%Sc; 0,08%Zr), 01545 (5,1%Mg; 0,22%Sc; 0,18%Zr), 01535 (4,1% Mg; 0,16%Sc; 0,13%Zr) and 01515 (3,1% Mg; 0,31%Sc; 0,08%Zr). The alloys, basically, differ by the contents of Mg, though the contents of Sc and Zr also differs a little. An initial state of Al-Mg-Sc alloys: cold-rolled sheets after following heating.
Deformation was performed by torsion under 6 GPa pressure on Bridgman anvil. The samples of 10 mm in diameter and 0,6 mm in thickness had been deformed by torsion to 9 revolutions which corresponds to strain ε=6,4.The structure analyses has been carried out using the transmission electron microscope JEM-100X. Microhardness was defined using the PMT-3 device with 20g and50g load.

3 Results and Discussions

Al-Cu-Mg alloy. The microhardness of 2024 alloy in initial as hot pressed state, as well as that after quenching had approximately the same value of ~ 0.8 GPa (Fig. 1). After quenching with subsequent natural aging the microhardness was 1.2 GPa. Room temperature deformation of initial non-quenched (hot pressed) specimens leads to strong hardening. With increasing strain the microhardness grows and almost attains steady state value 1,9 GPa at $\varepsilon = 6,8$ (5 revolutions).
Room temperature deformation of quenched specimens was accompanied by microhardness increase to 2.2 GPa. Subsequent holding at T = 20 °C for four days resulted in further hardening up to $H_\mu = 2.4$ GPa. Deformation of quenched specimens at the temperature of artificial aging (200 °C) led to lower hardening (after natural aging) $H_\mu = 1.8$ GPa, then deformation at room temperature, $H_\mu = 2.2$ GPa (Fig. 1).
The attempts were made to deform partially aged alloy with initial microhardness of 1.4 GPa after quenching and natural aging for two days. In this case the final hardening after deformation and natural aging is lower ($H_\mu = 1.9$ GPa) then after deformation of as-quenched alloy ($H_\mu = 2.4$ GPa).

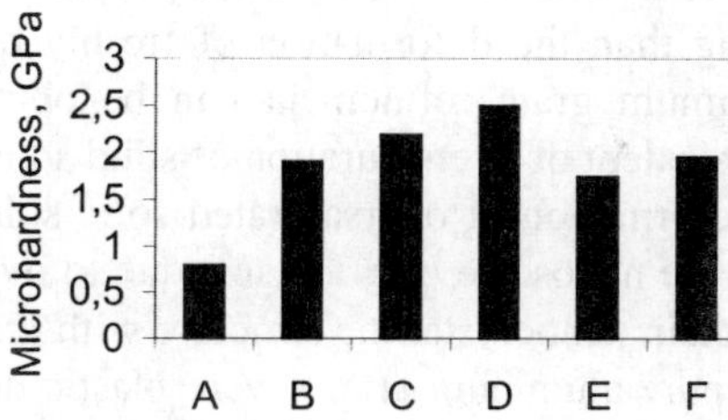

Fig. 1. Microhardness of Al-Cu-Mg-alloy 2024 after different routes of thermal treatment and SPD: A-quenching, B-SPD at 20 ^{0}C of hot pressed state; C- SPD at 20 ^{0}C of quenched state; D-regime C+natural aging; E- SPD at 200 ^{0}C of quenched state; F- SPD at 20 ^{0}C of quenched and pre-aged state

TEM of non-quenched and quenched specimens deformed at room temperature to $\varepsilon = 6{,}8$ (5 revolutions) revealed ultra fine grain structure with predominantly high angle boundaries (Fig. 2). Average grain size obtained from dark field images was ~150 nm for as hot pressed specimen (Fig. 2b) and ~50 nm for as quenched specimen (Fig. 2a).

At the absence of phase transformations a considerable hardening of as-hot pressed Al-Cu-Mg-alloy after severe cold deformation is most likely to occur due to large grain boundary area of submicrocrystaline structure with average grain size of ~ 150 nm (Fig. 2b).

Fig. 2. Structure of Al-Cu-Mg-alloy 2024 after SPD with e=6,8 at 20 ^{0}C (initial state: a) - quenched, b) – hot-pressed).

Deformation of as-quenched alloy at T = 20 °C results in final grain size of 50 nm and higher hardening (Fig. 1, 2a). It is reasonable to assume that the deformation of oversaturated solid solution facilitates precipitation of θ′ and S′-phases even during deformation, that is, in other words, facilitates strain aging. The latter leads to hardening due to both dispersed particles and grain refinement caused by these particles. Strain aging does not come to its completion and subsequent natural aging results in additional hardening (Fig. 1).

Severe deformation of as-quenched and partially aged Al-Cu-Mg-alloy also leads to a lower hardening than the deformation of freshly quenched alloy (Fig. 1). Consequently, maximum grain refinement can be obtained by deformation of alloy with maximum extent of oversaturation of solid solution.

Therefore, severe deformation of oversaturated solid solution of Al-Cu-Mg-alloy allows to overcome the nanoscale of grain size due to overlapping of phase transformations and to obtain nanocrystalline structure with grain size of ~50 μm.

Al-Mg-Sc alloy. In pure aluminum after severe plastic deformation (SPD) in torsion at a room temperature the grain structure is formed with the average size of a grain 660 nm. This structure is usually defined as submicrostructural. High angle misorienation of grains judged by electronograms and double contrast on grain boundaries.

In binary Al-Sc alloys with increase of the Sc contents the size of a grain after SPD monotonously decreases: 570 nm- for Al -0,14%Sc, 370 nm for Al-0,4%Sc and 280 nm for Al-0,55%Sc. But, as it can be seen, the sizes of grains are in submicron range. As the primary particles Al_3 Sc after crystallization in a cast state were absent, and allocation of particles Al_3Sc during deformation we can not define by electronmicroscopic method, it is possible to assume, that the reduction of the size of a grain is connected with increasing of Sc concentration in a solid solution of Al. Though it is impossible to exclude possibility of partial allocation higly disperse particles Al_3Sc during deformation.

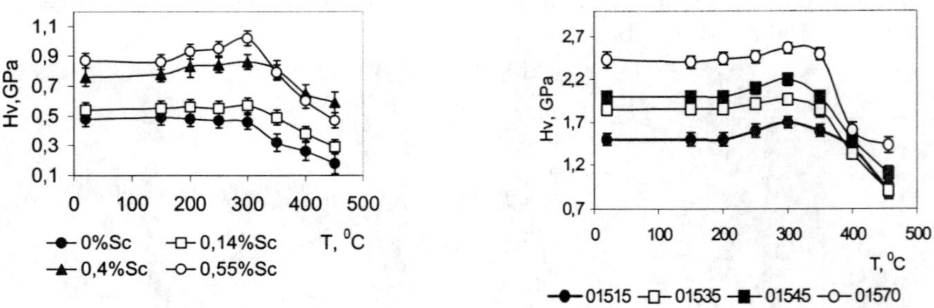

a) b)

Fig. 3. Microhardness dependence on the annealing temperature after SPD: a) Al-Sc alloys, b) Al-Mg-Sc alloys

The microhardness of binary Al-Sc alloys after SPD is much higher, than in a cast state and increases with increase of the Sc contents (Fig. 3). There is an aging process of binary Al- Sc alloys when they are heated, that is shown in increase of microhardness at the expense of allocation disperse particles Al_3Sc (Fig. 3). In an alloy Al- 0,14%Sc the effect of aging is small. The aging more effective in alloys at the higher contents of Sc and also reaches the maximal development at heating on 300 °C. Despite of a fine grain after SPD and aging at heating binary Al-Sc alloys have small durability. For example, in strongest Al-0,55%Sc alloy the

microhardness is hardly more than 1 GPa (Fig. 3). Therefore Al-Mg-Sc alloys after SPD with the various contents of Mg were investigated.
The microhardness of all Al-Mg-Sc alloys after SPD is higher, than in Al-Sc alloys and increases with increase of the contents of Mg. The values of microhardness after SPD are in 1,5 times higher, than in an initial state. The aging at heating passes in the same interval, as in binary Al-Sc alloys: begins at 250 ^{0}C and the maximal effect is observed at 300 ^{0}C. The heating of Al-Mg-Sc alloys in an initial state also has revealed aging, and in the same temperature interval.
The received effect of aging at heating both in an initial state, and in a state after SPD in torsion at a room temperature testifies, first, that at heating of as-cast structure, heating under hot-rolling and heating after cold rolling the decomposition of a solid solution of Sc has passed not completely and the part of Sc has remained in a solid solution, and, secondly, that during SPD at a room temperature there is not or small allocation of particles Al_3Sc, and basically Sc remains in a solid solution of Al.
Electronmicroscopic investigation has revealed nanocrystalline structure in an alloy 01570 after SPD in torsion at T_{def}=20 ^{0}C and ε=6,8 (5 revolutions) with the average size of a grain 50 nm and mainly high angle boundaries. At reduction of the Mg content in alloys the size of a grain is increased: 60 nm in 01545 alloy, 100 nm in 01535 alloy, 150 nm in 01515 alloy.
The heating of Al-Mg-Sc alloys up to 200 ^{0}C after SPD does not result in appreciable change of a grain size and microhardness. For example, for an alloy 01570 grains grow up with 50 up to 90 nm, remaining in nanoscale range.
At heating on 300^{0}C grains in all alloys grow up to 400 … 600 nm. This temperature corresponds to the maximal hardening. Means, the hardening at the expense of allocation of dispersed particles Al_3Sc compensates the softening from grain growth. The heating on 400 ^{0}C causes the further growth of grains up to 800…1000 nm. Hence, it is possible to approve, that nanocrystalline structure (the size of a grain <100 nm) is formed at SPD with T_{def}=20 ^{0}C in Al-Mg-Sc alloys with the content of Mg not less than 3%. For obtaining of the maximal hardening it is necessary, thus, to deform at a room temperature and to heat up on 300 ^{0}C for passage of aging. Increase of the contents of Sc and Mg in the investigated limits promote hardening.

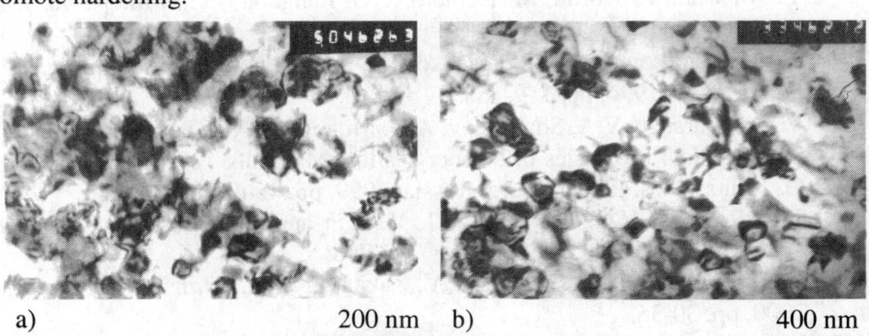

Fig. 4. Structure of Al-Mg-Sc alloy 01570 after SPD at 200 ^{0}C (a) and 400 ^{0}C (b)

To obtain of UFG structure, stable to the subsequent heating under superplasticity, we have tried warm and hot severe deformation of an alloy 01570. After SPD at 200 ^{0}C grain sizes in alloy 01570 is a little increased, but remains in nanoscale frameworks. The microhardness practically does not vary. After deformation at 400^{0}C we have received the grain size about 150 nm, whereas at heating on this temperature of the state deformed at a room temperature with D=50 nm, the size of a grain is increased up to 800 nm. Probably it is possible to explain so small size of a grain at deformation with T=400 ^{0}C only by decomposition of a solid solution during deformation.

4 Conclusions

1) In binary Al-Sc alloys after SPD in torsion at a room temperature the submicrocrystalline structure is formed. The grain size decreases with increase of the Sc content and reaches the minimal value 280 nm at 0,55% Sc.
2) Addition of Mg in binary Al-Sc alloy results in nanocrystalline structure formation after SPD in torsion with the minimal grain size -50 nm (5,9% Mg and 0,3% Sc).
3) The maximal hardening of both Al-Sc and Al-Sc-Mg alloys after SPD in torsion is reached after aging at heating on 300 ^{0}C
   During hot SPD in torsion of Al-Mg-Sc alloy 01570 at T=400 ^{0}C the UFG structure with grain size of 150 nm is formed, apparently at the expense of decomposition of a solid solution.

References

[1] R.Z. Valiev, N.A. Krasilnikov, N.K. Tsenev, *Mater. Sci. Eng.,* A137, 1991, pp. 35- 40.
[2] J. Wang, Z. Horita, M. Furukawa et al., *J. Mater. Res.*, 8, 1993, pp. 2810-2821.
[3] Y. Iwahashi, Z. Horita, M. Nemoto, T.G. Langdon, *Acta Mater.*, 46, 1998, pp. 3317-3331.
[4] A.V. Korznikov, Yu.V. Ivanisenko, D.V. Laptionok et al., *Nanostructured Materials,* 4, 1994, pp. 159-168.
[5] V.V. Sagaradze, V.A. Shabashov, A.G. Mukoseev, N.L. Petcherkina, in "The Structure and Properties of Nanocrystalline Materials", eds. N. Noskova and G. Taluts, RAS, Ekaterinburg, Russia, 1999, pp. 46-63 (in Russian).
[6] T.D. Rostova, V.G. Davydov, V.I. Yelagin and V.V. Zakharov, *Materials Science Forum*, 331-337, 2000, pp. 793-798.
[7] V.G. Davydov, T.D. Rostova, V.V. Zakharov et al., *Mat. Sci. Eng.*, A 280, 2000, pp. 30-36.
[8] Yu.A. Filatov, *J.of Advanced Mat.*, 5, 1995, pp. 386-390.

Dynamically Recrystallized Grain Size of Some Commercial Purity Coppers

V.G. García [1], J.M. Cabrera [1,2], and J.M. Prado [1,2]

[1] Departamento de Ciencia de los Materiales e Ingeniería Metalúrgica, ETSEIB – Universitat Politècnica de Catalunya, Av. Diagonal 647, 08028 – Barcelona, Spain
[2] Centre Tecnològic de Manresa, Av. de les Bases de Manresa 1, 08240 – Manresa, Spain

Keywords : dynamic recrystallization, grain refinement, copper.

Abstract. In commercial purity coppers undergoing hot extrusion operations, the recrystallized grain size can be controlled by an appropriate selection of the residual elements content. Consequently, this study has been focused on deriving the influence of the oxygen content on the dynamic recrystallization conditions, particularly the peak and steady state stresses and the dynamically recrystallized grain diameter after hot compression tests of three different coppers. The oxygen content appears to have a linear relationship with stress, as more hardening is promoted at increasing oxygen contents. On the other hand, finer dynamically recrystallized grain diameters are promoted at increasing Pb, Ni and Ag levels.

1 Introduction

Fire-refined 99.9% coppers made from scrap have become an alternative to Electrolytic Tough Pitch (ETP) coppers because of their more ecological cleanliness. However little direct investigation has been done to clarify the hot flow behavior of these fire-refined that also contain low amounts of oxygen and other residual elements. Much research has been done on alloyed [1] and almost pure copper [2] although the synergetic effect of many residual elements has not been addressed. For instance, it has been reported [3] that increasing oxygen contents in Oxygen Free High Conductivity (OFHC) coppers (O<62 ppm) promote an increase in the dynamic recrystallization (DRX) temperature at the same strain rate. On the other hand, in ETP coppers (175 ppm<O<450 ppm) the behavior is different. Thus in the fire-refined coppers an investigation is needed to insight how much can these low oxygen contents affect the DRX.

2 Experimental Procedure

The relationship between temperature, strain rate and residual chemical content during the DRX of commercial fire-refined copper was studied by hot compression tests on three different alloys (see chemical compositions in Table 1). The first copper selected, copper A, had a low oxygen content (26 ppm) and, relatively low Pb, Ni and Ag amounts (63.51 ppm, 31.70 ppm, 30.84 ppm respectively). The second copper selected, copper B, had a medium oxygen content (46 ppm) and, relatively high contents of Pb, Ni and Ag (169 ppm, 54.27 ppm, 46.49 ppm). Finally the third copper studied, copper C, had the highest oxygen content (62 ppm) but medium contents of Pb, Ni and Ag (133 ppm, 39.97 ppm, 37.71 ppm). The present alloys were supplied as round billets of 231 mm in diameter and 35 mm thick. The supplied billets were examined macrometallografically to assure a homogeneous equiaxial grain and size. Then cylindrical samples were machined, having 15 mm in height and 10mm in diameter. The cylindrical samples were then annealed at 950ºC during 5 minutes for copper A and B, and during 15 minutes for copper C.

Table 1. Chemical composition (ppm) of the three coppers studied and initial grain size d_o (μm) at the beginning of the hot compression test.

ppm	P	Sn	Pb	Ni	Ag	S	Fe	Zn	O	d_o
Cu A	297	86.2	63.5	31.7	30.8	22.0	17.2	15.6	26	637
Cu B	253	120	169	54.3	46.5	9.8	16.3	31.3	46	570
Cu C	153	63.3	133	40.0	37.7	10.1	15.5	13.4	62	530

Before the hot compression tests, samples were previously covered with a solution of boron nitride and acetone in order to minimise friction. The hot compression tests were performed using a software enhanced Instron 4507 electromechanical testing machine to produce constant strain rate. The samples were protected during the tests by a flow of nitrogen gas inside the furnace chamber. The test temperature was monitored by a thermocouple adjacent to the sample. Each copper was tested at eight different temperatures from 600ºC to 950ºC at 50ºC intervals and at six different strain rates, namely $0.3s^{-1}$, $0.1s^{-1}$, $0.03s^{-1}$, $0.01s^{-1}$, $0.003s^{-1}$ and, $0.001s^{-1}$. The strain attained in each test was 0.8. After each compression test the resulting microstructure was retained by quenching in cool water. The compressed samples were then precision cut in half to reveal the axial face, using an Isomet diamond disc machine. The halved samples were then casted so that they could be fitted into an automatic polishing machine according to the contemporary preparation method [4]. The samples were etched using the Cl_3Fe solution before being micrographed using an optical microscope. All micrographs were taken on the middle line of the compressed sample but approximately at four millimetres from the edge that was not in contact with the compressing plates.

The dynamically recrystallized grain diameter was measured over the micrographs using the line intercept or Heyn method.

3 Results and Discussion

In most of the hot flow curves at 600ºC and 650ºC softening took place exclusively by dynamic recovery. Only the curves at 700ºC and 750ºC started to show DRX. Beyond 800ºC DRX was present on all curves, observing the multiple peak type at lower strain rates. Both monotone and multiple peak DRX were capable of refining the initial grain size, contrary to usually stated.

3.1 Effect of the residual elements content on the recrystallized grain size

The recrystallized grain diameter (measured only when a clear steady state was apparent in the flow curves) was finer as the strain rate was increased for a given temperature although the grain diameter tended to stabilise to an asymptotic size. The latter sizes changed when the temperature was varied. As expected, increasing temperatures promoted larger recrystallized grain diameters at the same strain rate. Experimental results of the recrystallized grain sizes versus the strain rate at different temperatures are plotted on Figure 1 for the three coppers studied. It is worth noting that at the lower temperatures similar grain sizes are obtained in the three alloys. Although some scatter is apparent, differences are however noticeable at the higher temperatures (900 and 950ºC), where in global terms, copper A has the largest grain size followed by copper C. The finest sizes are observed in copper B. This order correlates with the amount of the most significant solute elements, i.e. Pb, Ni, and, Ag, but not with the oxygen content. This feature can be explained bearing in mind that the oxygen-copper particles should be dissolved at high temperatures. Consequently at these high temperatures solute drag of Ni and Ag or second phase particles of Pb are probably controlling the grain growth. At lower temperatures, the combined effect of copper-oxygen particles (concurrently precipitating), Pb particles and solute drag are promoting similar grain sizes in the three coppers. Assuming that the size and volume fraction of lead particles remain constant with temperature, and a weak solute drag effect of Ni and Ag, the similar grain sizes in all three coppers at low temperatures can be mainly attributed to the increasing precipitation of oxygen-copper particles.

3.2 Effect of the residual elements on the flow stress.

Regarding the flow stresses, the dependence of the steady state stress on the strain rate for each copper at every temperature is displayed in Fig. 2. As expected, higher stresses are observed at increasing strain rates or decreasing temperatures. It is also evident that no differences are found between the three coppers at the highest temperatures (950°C and 900°C). However below 850°C, copper C shows larger stresses than copper B, while copper A is always the softest. The differences between all coppers increases when the temperature diminishes. Here the oxygen amount correlates well with the latter observation, suggesting a second phase particle hardening as one should expect according to the Cu-O phase diagram. Although some hardening can be also promoted by solute elements, the lack of correlation with the other residual elements suggests that precipitation hardening of copper-oxide particles are playing a definitive role. It must be pointed out that similar behaviours were observed when considering the peak stress values.

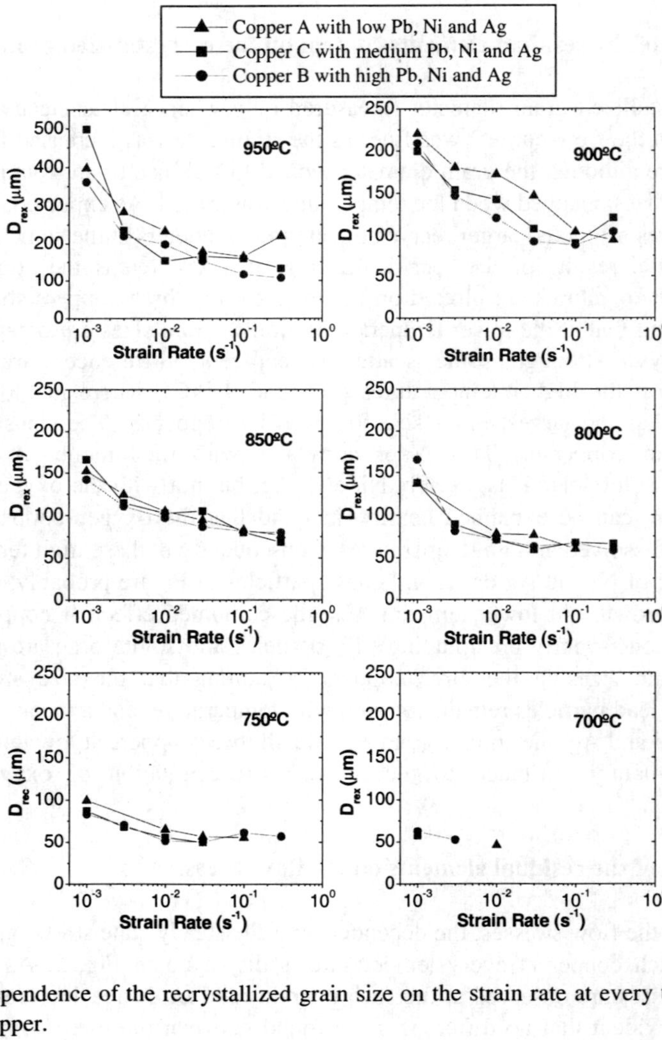

Fig. 1. Dependence of the recrystallized grain size on the strain rate at every temperature for each copper.

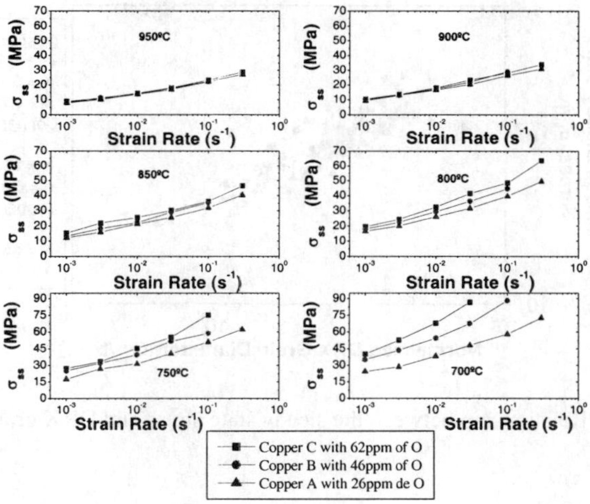

Fig. 2. Dependence of the steady state stress, σ_{SS}, on the strain rate and temperature in the three coppers under study.

3.3 Derby's relationship between σ_{ss} and recrystallized grain size

Derby [5] proposed an equation with physical basis relating the steady state stress, σ_{ss}, and the dynamically recrystallized grain diameter. Such expression was derived as the balance between the formation rate of the deformed substructure and the mean velocity of recrystallising grain boundaries. Accordingly, Derby suggested the following equation:

$$\frac{\sigma_{ss}}{\mu}\left(\frac{D_{rex}}{b}\right)^{n} = K \tag{1}$$

where D_{rex} is the grain size, b is the Burger's vector, σ_{ss} is the steady state stress and μ is the shear modulus. Derby theorised an exponent of n = 2/3 and a K constant between 1 and 10. The mathematical relationship deduced by Derby does not clearly explain which of the three 99.9% coppers promotes a finer grain size under the same hot flow conditions (see Fig. 3). However, the Derby's exponent increases in the same order as the Pb, Ni and, Ag content is increased. The exponent values obtained are similar to others reported in the literature [2,5,6], although slightly different to the theoretical value suggested by Derby. This is consistent with present results bearing in mind that the Derby equation is intended for "free" DRX, that is, in the absence of second phase particles. Finer grain sizes and, consequently larger exponents, are expected in the presence of particles.

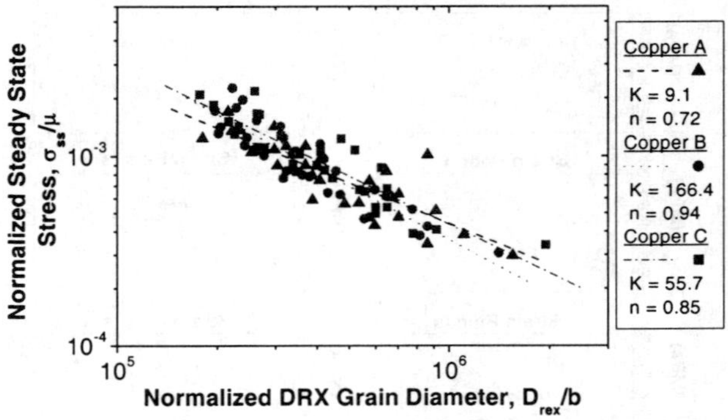

Fig. 3. Derby's Relationship between the steady state stress and DRX grain diameter.

4 Conclusions

The finest grain diameters at high temperature were predominantly obtained with the copper with higher amounts of Pb, Ni and, Ag. In the low temperature range recrystallized grain sizes were independent of the chemical composition. An increment in the oxygen content was directly related to an increase on the steady state and peak stresses. Results were in accordance with the Derby equation, although some deviations in the exponent could be explained by precipitation hardening.

Acknowledgements

The authors are grateful for the co-operation accord between TERTUB and UPC. V.G. G. also wishes to thank the AECI for the financing provided.

References

[1] B. Bayle, Ph. Bocher, J.J. Jonas, F. Montheillet, *Mat.Sci. & Tech.*, 15, 1999, p. 810.
[2] W. Gao, A. Belyakov, H. Miura, T. Sakai, *Mat.Sci. & Eng.*, A265, 1999, p. 237.
[3] N. Ravichandran, Y.V.R.K. Prasad, *Mat.Sci. & Eng.*, A156, 1992, p. 203.
[4] G. F. Vander Voort, *Adv. Mat. & Prs.*, 7, 2000, pp. 38.
[5] B. Derby, *Acta Metall. Mater.*, 39, 1991, p. 955.
[6] L. Blaz, T. Sakai, J.J. Jonas, *Metall. Trans.*, 4, 1973, p. 765.

Characteristics of Ultrafine Ferrite Formed by Strain-Induced Dynamic Transformation in Low Carbon Steels

J.-K. Choi, K.-K. Um, J.-S. Lee, D.-H. Seo, and W.-Y. Choo

Technical Research Laboratories, Pohang Iron & Steel Co., Ltd.,
Pohang P.O.Box 36, Pohang, Kyung-Buk, 790-785 Korea

Keywords: dynamic transformation, ultra fine-grained steel, grain refinement, heavy deformation.

Abstract. Ultrafine ferrite grains are easily obtained by single- or multipass heavy deformation in plain low carbon steels below the Ae3 temperature by utilizing the strain-induced dynamic transformation (SIDT). A dilatometric technique was introduced to measure the fraction of SIDT ferrite during deformation and used for quantitative analysis of the SIDT kinetics. The smaller austenite grain size (AGS) increases SIDT ferrite fraction at a given strain amount, and also makes the final microstructure more uniform because the austenite grain boundary is the most potential nucleation site for SIDT ferrite. The microstructural evolution during static transformation of remaining austenite after SIDT was examined for coarse and fine austenite grains.

1 Introduction

It is now generally accepted that the heavy deformation of low carbon steels results in very fine ferrite microstructure. And it is also understood that SIDT is one of the most effective process for obtaining ultrafine ferrite grains during heavy deformation [1-3]. The dynamically transformed ferrite is extremely fine in size and very stable against coarsening even at the elevated temperature regime. Even though we can obtain the ultrafine-grained ferrite by utilizing SIDT, there need some technical progresses for the application to the actual production because of the limitation in the amount of reduction per pass in the existing mills. In that sense, the maximization of SIDT during successive light rolling passes is practically very important. A fundamental study has been carried out on the effects of basic parameters, such as austenite grain size (AGS) and deformation conditions, on the SIDT in low carbon steel and the post-SIDT microstructural evolution during isothermal holding was also examined.

2 Experimental Procedure

A boron-added low carbon steel specimens (0.15C-0.25Si-1.1Mn-0.01Ti-0.002B) were prepared by vacuum induction melting and hot rolling followed by machining into 10 mm diameter and 15 mm height cylindrical geometry. Hot deformation experiments were performed by using Gleeble 3800MA. The AGS was varied from 20 to 300 μm by changing the austenitizing temperature and the effects of deformation temperature (700 and 650°C), strain rate (1 and $0.1s^{-1}$), and the amount of strain on the SIDT behavior were examined. To investigate the microstructural evolution after SIDT, the specimens were deformed by the true strain of 1.2 at 700°C and held at 650°C up to 320 sec. Some of them were interruptedly quenched after specified holding times and the microstructures were observed.

3 Results

The amount of SIDT ferrite during the deformation could be quantitatively estimated by using dilatometric technique. Fig. 1 shows the relative diametral change of specimen with time after various amount of deformation at 650°C. Since these curves give us the amount of statically transformed ferrite, we can calculate the amount of SIDT ferrite. This helps us to quantitatively investigate the kinetics of dynamic / static transformation of austenite. The necessary strains for obtaining 20% of SIDT ferrite from different AGS were plotted in Fig. 2. It is clear from the figure that more SIDT ferrite is formed in smaller AGS at a given amount of deformation.

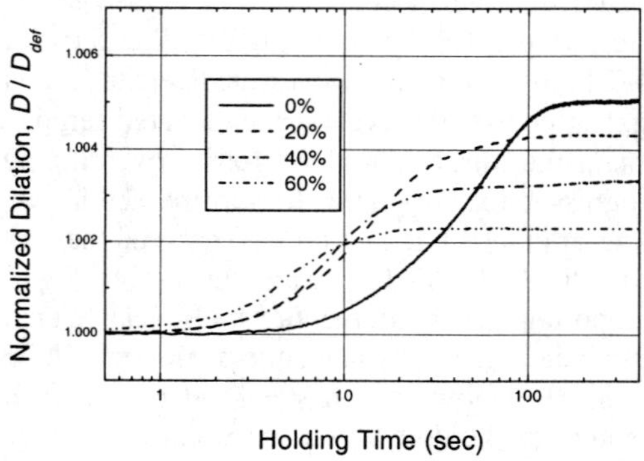

Fig. 1. Relative diametral changes of the specimen with holding time after various amount of deformation at 650°C (strain rate = $1s^{-1}$).

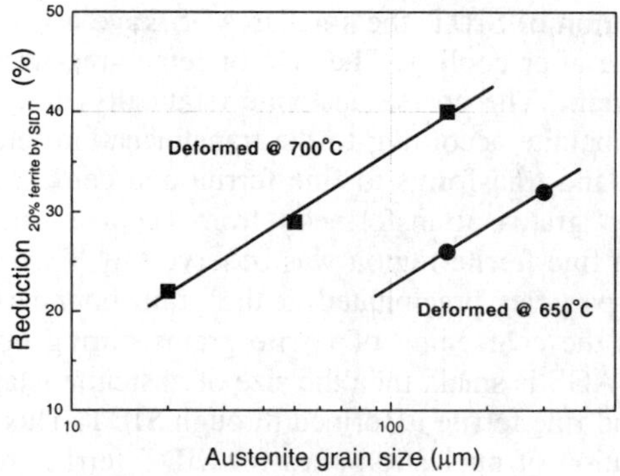

Fig. 2. Effect of austenite grain size on the necessary reduction for 20% SIDT ferrite at different deformation temperature (strain rate = $1s^{-1}$).

Fig. 3 shows very fine SIDT ferrite grains which have mainly nucleated along the prior austenite grain boundaries. In this case, specimen was deformed by the nominal reduction of 60% at 700°C. Even though some SIDT ferrites were nucleated within the prior austenite grains, possibly along the deformation bands, it is noteworthy that the prior austenite grain boundaries are the most potential nucleation sites for SIDT ferrite. The smaller AGS has the more grain boundary area in a given volume and this will result in the larger volume fraction of ferrite on the assumption that the widths of SIDT ferrite agglomerates formed along the prior austenite grain boundaries are the same irrespective of AGS.

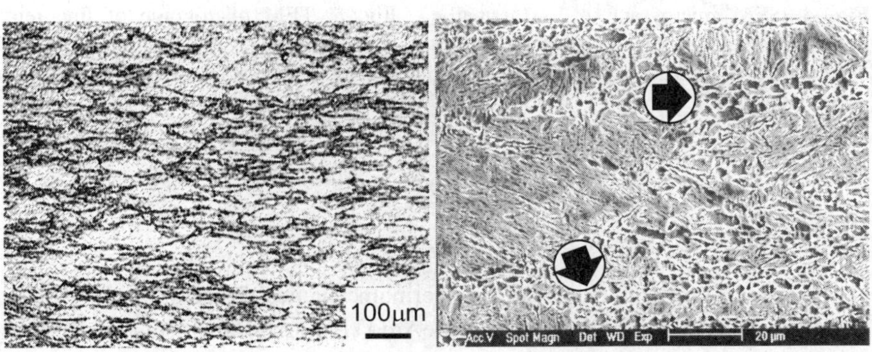

(a) (b)

Fig. 3. (a) Optical and (b) SEM micrographs of SIDT ferrites mainly nucleated at the prior austenite grain boundaries during the deformation by 60% at 700°C. Arrows in (b) indicate SIDT ferrites.

Besides the acceleration of SIDT, the smaller AGS gave a very uniform and fine ferrite microstructure after cooling. The size of remaining austenite islands after SIDT is very important. The coarse austenite (statically) transforms to ferrite + pearlite or ferrite + bainite according to the transformation temperature, whereas the fine austenite island transforms to fine ferrite and cementite particles. Fig. 4 shows the ferrite grains transformed from coarse and fine austenite simultaneously. The fine ferrite region was observed by TEM (Fig. 5). It is seen that fine cementite particles precipitated at the grain boundaries of fine ferrites which can suppress the coarsening of ferrite grains during holding and cooling time. If the original AGS is small, then the size of austenite island is usually small and very uniform and fine ferrite is formed through SIDT. This is probably due to the uniform distribution of nucleation sites for SIDT ferrite in the fine austenite grain microstructure.

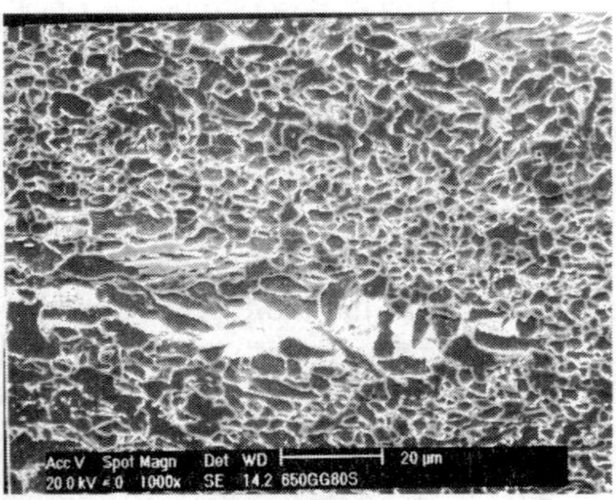

Fig. 4. SEM micrograph of isothermally held specimen for 80 sec at 650°C after deformation at 700°C. Coarse and fine ferrite grains coexist.

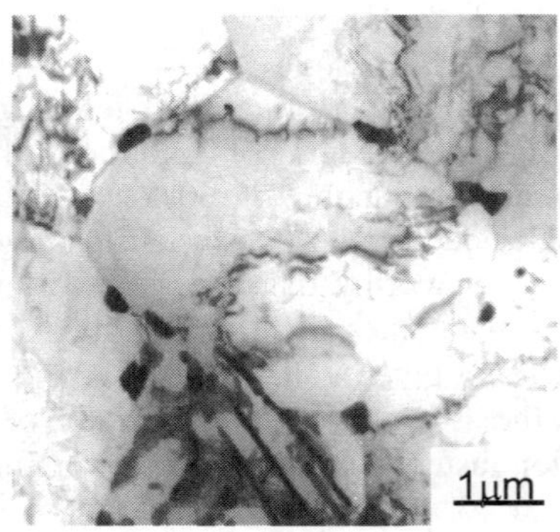

Fig. 5. TEM micrograph of fine ferrite grain region in Fig. 3. Fine cementite particles are precipitated along the ferrite grain boundaries.

References

[1] Priestner, Proc. Int. Conf. On the Thermomechanical Processing of Microalloyed Austenite, eds. A. J. DeArdo *et al.*, AIME, Pittsburgh, 1981, p. 455.

[2] H. Yada, Y. Matsumura, and T. Senuma, Proc. of THERMEC-88, Tokyo, 1988, p. 200.

[3] C.-S. Lee, J.-S. Lee, and W.-Y. Choo, *CAMP-ISIJ*, 13, 2000, p.644.

Grain Growth in Nanocrystalline Ni

B.-K. Kim and J.A. Szpunar

Department of Mining and Metallurgical Engineering, McGill University, 3610 University Street, Montreal, QC, Canada H3A 2B2

Keywords: EBSD, nanocrystalline Ni, annealing, grain growth, texture.

Abstract. The effect of annealing temperature and time on the grain growth of nanocrystalline Ni was investigated. Crystallographic texture measurements and orientation imaging microscopy were used to study the annealing of nanocrystalline Ni. In annealing of nanocrystalline Ni with a strong (001)-fibre texture, at temperature from 250°C to 300°C, the growth of (111) texture component is observed. The same result is obtained during annealing from 10 to 30 minutes at 300°C. The initial (001)-fibre texture is however dominant at each annealing conditions. It is found that grain growth of nanocrystalline Ni is accompanied by the increase of Σ3 boundaries with annealing temperature and time.

1 Introduction

Nanocrystalline materials are polycrystalline materials with the grain size of the order of nanometers (typically up to 50 nm in diameter) [1,2]. The scientific and technical interests are focused on mechanical, chemical and physical properties that are different than those of conventional polycrystalline solids. The reason is an increase in the volume fraction of grain boundaries and triple junctions.

For various applications of these materials a better understanding of annealing processes that affect the stability of structure is important. The reason is that nanocrystalline materials are thermodynamically unstable [2]. Their special properties depend on the grain size.

Several papers have reported studies on the grain growth in nanocrystalline materials during annealing [1,3]. The mechanism of the grain growth is complex and there is no agreement, at present, as to what is the main driving force for the grain growth. In order to understand and control the unique properties of nanocrystalline materials, the systematic study of structural transformation during grain growth is required. At present, the grain boundary structure in nanocrystalline materials and a role it plays during the annealing is not well understood. Therefore, it is also important to analyze the grain boundary characteristics during annealing because the grain boundaries play an important role in understanding the grain growth and texture transformation [5].

2 Experiments

Nanocrystalline Nickel (Nano-Ni) with a strong (001)-fibre texture was prepared using the electrodeoposition [1,3,4]. The average grain size was 15-25 nm as evaluated from the Scherrer formula. In order to investigate structure changes in annealing, X-ray diffraction patterns were collected in situ during annealing. The heating rate was 10°C per minute and the residence time for in-situ X-ray measurement was 16 minutes at each temperature. The temperature was changed stepwise, in steps of 50°C, up to 550°C.

In another experiment the Nano-Ni was annealed in a reflective furnace with Ar and quenched in water. The heating rate was 120°C per minute. The annealing conditions used are summarized in Table 1. The texture of annealed Nano-Ni was measured using the Siemens D-500 texture goniometer with a Mo target. The inverse pole figures of normal direction were calculated from orientation distribution functions (ODFs), which were obtained using the arbitrary defined cell (ADC) method [6]. For this purpose, the texture software package, HTexTool v.3.0, was used [7]. The microstructure of annealed samples was investigated using image quality (IQ) and inverse pole Figure (IPF) map obtained for the direction normal to the specimen surface from electron backscattering patterns in a Philips XL-30 FEG using TSL orientation imaging microscopy (OIM). The detail of description and definition of IQ can be found elsewhere [8]. The accelerating voltage was 20KV with about 130-150 pA of beam current.

Table 1. Annealing conditions.

Temp.	1 min.	10 min.	30 min.	20 hrs.
250°C	-	O	-	-
300°C	O	O	O	-
400°C	-	O	-	O

3 Results and Discussion

3.1 In-situ measurement of X-ray diffraction and texture transformation

The results of in-situ X-ray diffraction measurement show a change of intensities of (111) and (002) peak at the temperature range between 250 and 300°C (Fig. 1a). Peak breadth of each peak and the integrated intensity ratio of (111) to (002) peaks were changed at this temperature region (Fig. 1b). The (111) crystallographic diffraction peak is grown during this transformation. These results indicate that nanocrystalline structure becomes unstable and transforms into microcrystalline state at the temperature range between 250 and 300°C. Texture evolution can be considered as an indicator of microstructural changes in the specimens during annealing. The results of texture measurement after annealing are summarized in Table 2. The pole intensity of (001)-fibre and (111)-fiber texture was obtained from their inverse pole figures of the normal direction and is

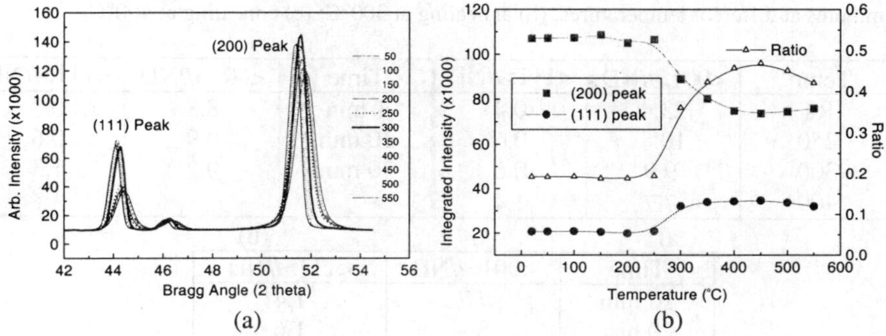

Fig. 1. The peak intensity changes illustrating the competition between (111) and (002) peak as function of temperature: (a) the result of in-situ X-ray measurement; (b) the variation of the integrated intensity of (111) and (002) peak and the ratio of (111) to (002) ratio integrated intensity with temperatureelevated temperature.

presented as a function of annealing temperature and time at 300 and 400°C. No other texture component was observed. In as-prepared nanocrystalline Ni, there is only a strong (001) fibre texture with the intensity 23.4 times random. As the annealing temperature is increased or the annealing time at 300°C is increased, the (111)-fibre texture component grows but never exceeds the pole intensity of (001)-fibre texture. In annealing at 400°C, an increase of (001)-fibre texture is higher than that of (111)-fibre texture. After 20 hours at 400°C, the main component, (001)-fibre texture is still stronger. These results, however, are quite different from those reported in the papers on annealing of nanocrystalline materials with a major (001) and a minor (111)-fibre texture components, such as nanocrystalline Ni and Ni-Fe alloy [9,10]. In these cases, the pole intensity of (001)-fibre texture rapidly decreases and (111)-fibre texture develops and becomes stable in annealing. It indicates that the stable texture component after annealing may depend on the initial texture of a specimen.

3.2 Microstructure development and grain boundary characteristics

The microstructure of nanocrystalline Ni specimens was analyzed using the OIM to determine the grain size and the area fraction of (001)-fibre and (111)-fibre texture components respectively. Also the grain boundary character distribution was investigated, namely the grain misorientation distribution and the frequency of coincidence site lattice (CSL) boundaries. The scanning area was 1.5x3 μm^2 with 20 nm steps for 250 and 300°C. For 30 minutes at 300°C and 10 minutes and 20 hours at 400°C, the area of 8x16 μm^2 with a 40 nm step size was scanned. In order to reduce the experimental error, the interpretation of data was done only

Table 2. Pole intensity in the normal direction inverse pole figure: (a) annealing for 10 minutes at different temperatures; (b) annealing at 300°C; (c) annealing at 400°C.

Temp.	<001>//ND	<111>//ND
RT	23.4	0.4
250	10.8	0.6
300	9.9	0.6
400	7.7	1.4

(a)

Time	<001>//ND	<111>//ND
1 min	8.8	0.8
10 min	9.9	0.6
30 min	9.2	1.3

(b)

Time	<001>//ND	<111>//ND
10 min	7.7	1.4
20 hrs.	8.3	1.6

(c)

for data points that have the confidence index (CI) higher than 0.09. The grain tolerance angle with 15° was normaly used.

As a result of the annealing treatment, the average grain size and the area fraction of grains having (001)-fibre and (111)-fibre texture components is changed as shown in Fig. 2. In Fig. 2a and b, the results obtained at temperatures 250 and 300°C demonstrate that nanostructure has not been changed significantly at these temperatures. The average grain size of (001)-fibre and (111)-fibre texture component is 70 and 50 nm respectively. The area fraction of grains having these texture components is increased slightly during annealing at these temperatures.

In annealing at 400°C for 10 minutes, nano grains were grown up to a few hundreds nanometers. The average diameter of grains with (111)-fibre texture component is 270 nm and is higher than those with (001)-fibre texture component. The area fraction of grains having (001)-fibre texture is largely increased in annealing at this temperature. These results are similar as the results obtained in annealing for longer time at 300°C (Fig. 2c, d). In annealing with time at 400°C, at longer annealing times the grains of (001)-fibre texture component increase in size from 230 to 400 nm (Fig. 2e, f) while the grains of (111)-fibre texture component remain relatively unchanged. The OIM results illustrate that the major (001)-fibre texture component grows in annealing before transformation into microcrystalline state. The grains having (111)-fibre texture grow in the next stage of annealing. This indicates that the transformation of nanocrystalline Ni into microcrystalline solid is accompanied with the development of (111)-fibre texture. In order to understand why the grains with (111)-fibre texture grow preferentially, the characteristics of grain boundary character distribution (GBCD) were investigated. As a result, one can observe that the area fraction of grain boundaries with misorientation angle between 55 and 60° increased up to 41% with annealing temperature. The GBCD changes as well with time and the same tendency is observed at 300°C. The grain boundaries having the misorientation angles

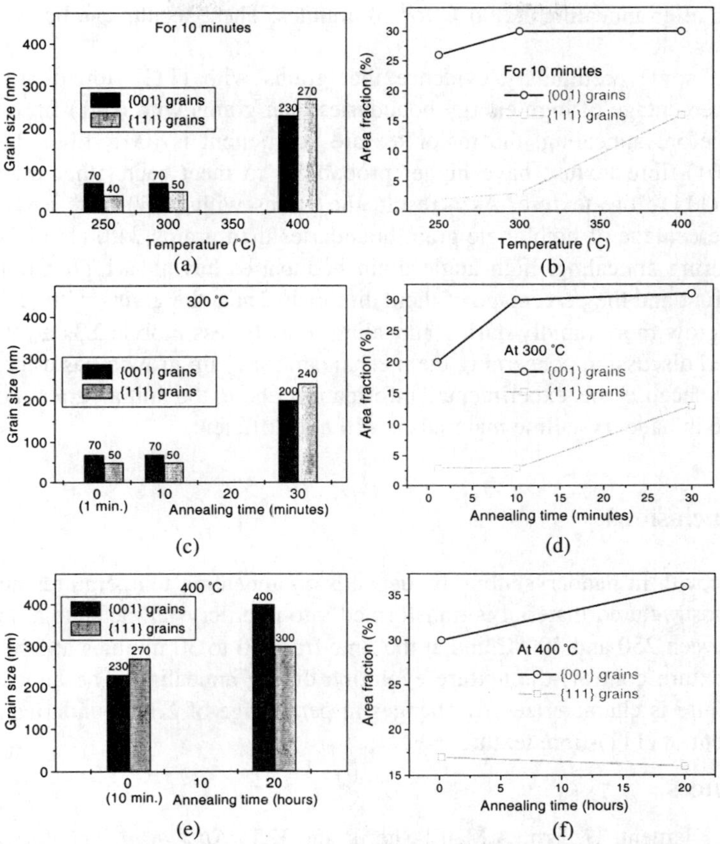

Fig. 2. Evolution of average grain size of {001} and {111} grains during annealing and area fraction of each component in each annealing condition: (a) and (b) annealing at 250, 300, and 400°C; (c) and (d) annealing with time at 300°C; (e) and (f) annealing with time at 400 °C

around 60° are known as Σ3 boundaries or (Σ3s) when the Brandon criterion [11] is applied. The Σ3s are known to have low mobility and low grain boundary energy. Thus, the Σ3s are not destroyed during the processes of grain growth and new twin boundaries may be also created during annealing. Since the increase in the area fraction of Σ3s is associated with the increase of size of grains with (111)-fibre texture, one can assume that the (100) grains are pinned more strongly during the growth than (111)-fibre texture grains.

In the case of Ni-20% Fe specimen having a major (001)-fibre and a minor (111)-fibre component, the area of Σ3s boundaries is rapidly increased up to 48% of all CSL boundaries. In annealing at 370°C for 30 minutes, a major (111)-fibre texture is 3.6 times random and a minor (001)-fibre texture 1.6 times random. The area fraction of Σ3s is 1.8 times higher than that in a strongly textured specimen obtained after annealing at 300°C for 30 minutes. These results can be interpreted as follows.

There is some preliminary evidence that grains with (111)-fibre texture have higher percentage of high energy boundaries than grains with (001)-fibre texture. Since, before annealing, the major texture component is (001)-fibre the grains with (001)-fibre texture have higher probability to meet each other than grains having (111)-fibre texture. As a result, the grains with (001)-fibre texture have lower percentage of high angle grain boundaries than grains with (111)-fibre texture. During annealing, high angle grain boundaries having a high mobility are moved first and the percentage of them diminish. Thus, the grains with (111)-fibre texture grow more rapidly during annealing. Finally less mobile Σ3s are left. The presented discussion concerning the mechanism for grain growth has to be further verified because the experimental information about the initial grain boundary structure in nanocrystalline materials is still not sufficient.

4 Conclusions

Grain growth in nanocrystalline Ni depends on annealing temperatures and time. The nanostructured material is transformed into microcrystalline state at temperature between 250 and 300°C and at the time from 10 to 30 minutes at 300°C. The initial texture controls the texture evolution during annealing. The annealed microstructure is characterized by changes in percentage of Σ3s boundaries and development of (111)-fibre texture.

References

[1] U. Klement, U. Erb, A.M. El-Sherik and K.T. Aust, *Mat. Sci. Eng.*, A203, 1995, p. 177.

[2] H. Gleiter, *Progress in Mater. Sci.*, 33, 1989, p. 223.

[3] U. Erb, A.M. El-Sherik, G. Palumbo and K.T. Aust, *Nanostructural Materials*, 2, 1993, p. 383.

[4] K. Boylan, D. Ostrander, U. Erb, G. Palumbo and K.T. Aust, *Scripta Metall.*, 25, 1991, p. 2711.

[5] T. Watanabe, in "Grain Boundary Engineering", eds. U. Erb and G. Palumbo, K.T. Aust Symp.,CIM, 1993, p. 57.

[6] K. Pawlik, J. Pospiech and K. Lücke, *Texture Microstruct.*, 14-18, 1991, p. 25.

[7] L. Houlong 2000, HTex Tools, ver. 3.0.

[8] D.P. Field, *Ultramicroscopy*, 67, 1997, p. 1.

[9] N. Rajimohan, Ph. D. Thesis, 2000, McGill University, Montreal.

[10] F. Czerwinski, H. Li, M. Megret and J.A. Szpunar, *Scripta Metall.*, 37, 1997, p. 1967.

[11] D.G. Brandon, *Acta Metall.*, 14, 1966, p. 1479.

Grain Growth in Nanocrystalline Ni-20%Fe Alloy

Y.B. Park[1], J. Park[1], S.S. Kim[1], and T.H. Yim[2]

[1] Department of Materials Science and Metallurgical Engineering, Sunchon National University, Sunchon 540-742, Korea
[2] Materials Process Technology R&D Center, Korea Institute of Industrial Technology, Chonan 330-820, Korea

Keywords: electrodeposition, grain growth, nanocrystalline permalloy, textures.

Abstract. Grain growth was investigated in a nanocrystalline Ni-20wt-%Fe alloy fabricated by an electrodeposition method, and its effect on the texture evolution was examined. The as-deposited specimens revealed a strong {100} and a weak {111} fibre-type textures. Grain growth in the material occurred on annealing beyond 370°C and resulted in the texture change that the {111} fibre strongly developed at the expense of the {100} fibre. It was observed using orientation imaging microscopy (OIM) that the {111} grains are much coarser than other oriented grains in the early stages of grain growth. The relationship between the texture evolution and the microstructural change accompanied by grain growth is interpreted and discussed in terms of the orientation dependency of grain growth.

1 Introduction

Nanocrystalline materials consisting of nanometer-sized crystallites contain a large number of interfaces, and thus, a large volume fraction of the atoms are associated with the intercrystalline region [1]. Due to this morphological characteristic, in general, nanocrystalline materials are far away from thermodynamic equilibrium and as a result reveal quite different properties in comparison with conventional coarse-grained materials [2]. For example, in ferromagnetic materials, the soft magnetic properties can be significantly modified by reducing its grain size to nanometer scale [3-5]. In fact, a particular attention has been paid to the synthesis of nanocrystalline materials for better soft magnetic properties. As an attempt for these materials, extensive studies [6-7] on structure-property correlations in nanocrystalline Ni-Fe alloys have been carried out in recent years.

In the present work, a nanocrystalline Ni-20%Fe alloy was fabricated using electrodeposition. Ni-Fe alloys exhibit different physical properties with different chemical compositions, among which the Ni-20%Fe alloy, referred to as permalloy, has widely been used owing to an outstanding combination of soft magnetic properties. Such nanocrystalline electrodeposits are in a high non-equilibrium (metastable) state, i.e. a high energy state due to a large number of interfaces [1] resulting in grain growth at relatively lower temperatures. The present study was aimed at investigating grain growth that takes place during the annealing of the nanocrystalline Ni-20%Fe alloy electrodeposits. In particular, early stages of grain growth were examined with the aid of orientation imaging microscopy (OIM) techniques and discussed in terms of the evolution of textures.

2 Experimental Procedure

A nanocrystalline Ni-20wt-%Fe alloy was fabricated by an electrodeposition method in a modified Watts-type bath containing nickel chloride, iron sulfate, boric acid and saccharine. Deposition was carried out at a temperature of 45°C and at a current density of 50 mA/cm^2. The final thickness of the as-deposited specimens was approximately 50 μm. The grain size of the electrodeposits was calculated to be about 10 nm using XRD peak broadening.
Thermal behavior of the nanocrystalline Ni-20%Fe alloy electrodeposits was examined using a DSC, where the sample heated from room temperature up to 450°C at a heating rate of 10°C/min. On the basis of the DSC data, the specimens were annealed at temperatures ranging from 250°C up to 400°C for various heating periods in an infrared heating furnace in an argon atmosphere.
X-ray pole figures were measured using Co K_α to determine the development of macrotextures. For the analysis of microtextures, the samples were observed in a field emission gun scanning electron microscope (FEGSEM) with an automated orientation imaging microscopy (OIM) system. OIM scans were carried out in a hexagonal grid of 20 nm step size.

3 Results

Fig. 1 shows a typical DSC curve obtained during heating of the nanocrystalline Ni-20%Fe alloy fabricated in the current work. The exothermal processes are indicated by the appearance of two heat release peaks. A small amount of heat release that takes place at about 330°C might be related to a structural relaxation of the interfacial component. The major heat release peak starts at 373.1°C, i.e. the onset temperature and reaches the peak maximum at 391.4°C. This exothermal process is attributed to the occurrence of grain growth in the sample. Grain growth occurs in polycrystalline materials to decrease the interfacial energy and hence the total energy of the system. Since nanocrystalline materials are in a high energy state due to a large volume fraction of the interfacial component, the driving force for grain growth is so large that the grain boundaries can readily move far below temperatures at which grain growth is expected to occur in microcrystalline materials. It is estimated that the heat energy released during grain growth in a

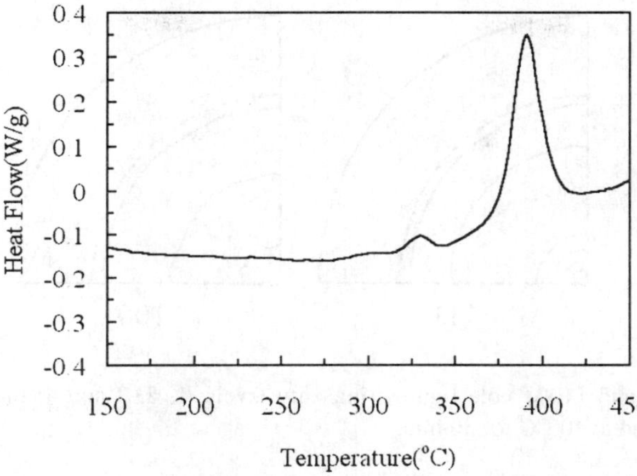

Fig. 1. DSC curve of the nanocrystalline Ni-20%Fe alloy on heating at 10°C/min.

5 nm-sized material is about 2-3 orders of magnitude larger than that in a 5 µm-sized material [8].

The occurrence of grain growth is accompanied not merely by the change in dimension but also by a structural rearrangement, i.e. a change of crystallographic textures. Fig. 2 depicts the {111} and {100} pole figures in the as-deposited specimen, the texture of which is characterized by a strong <100> fibre and a weak <111> fibre. This textural state was completely reversed on annealing the sample at 400°C for 5 min. As shown in Fig. 3, the development of the <111> fibre texture became predominant while the intensity of the <100> fibre texture decreased to a low level. Therefore, it is obvious that during annealing, <111>//ND grains grow faster than <100>//ND grains and other oriented grains.

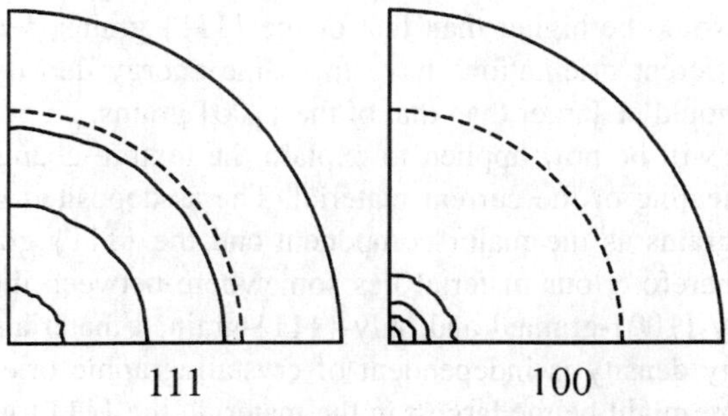

Fig. 2. {111} and {100} pole figures (intensity levels: 1, 2, 3 and 4; max = 4.43) in the as-deposited specimen.

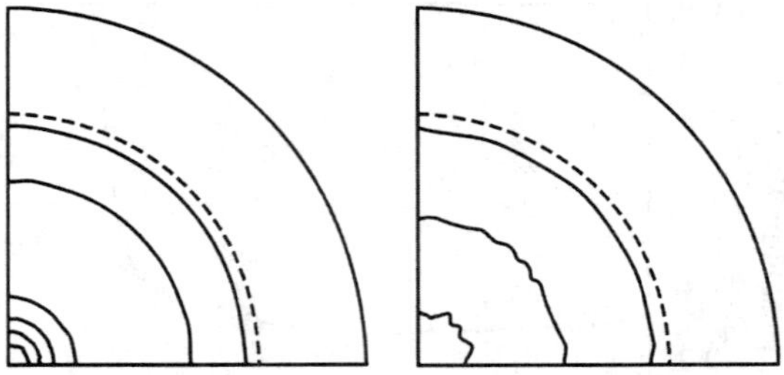

Fig. 3. {111} and {100} pole figures (intensity levels: 1, 2, 3 and 4; max = 4.94) in the sample annealed at 400°C for 30 min.

4 Discussion

Recently, Li, Czerwinski and Szpunar [9] calculated the orientation dependency of surface energy by using atomistic interactions by Lennard-Jones (LJ) in electrodeposited Ni-Fe alloys. According to their calculation, the surface energy is the lowest for the {111} plane and increases in the order of {100}, {110}, {311}, {210}, etc. It follows that the {111} texture is expected to strongly develop when the electrodeposition is carried out under a condition close to equilibrium. As the deposition condition becomes farther from equilibrium, crystallographic planes having higher surface energies would increase rather than {111}. Suppose two nanocrystalline electrodeposits consisting of {100} and {111} grains, respectively. It is apparent that the free energy of the {100}-grained material is higher than that of the {111}-grained material. There can exist two possibilities that the energy difference between the two materials results in microstructural difference: first, in case that the two materials have the same grain size, energy per unit volume of the {100} grains would be higher than that of the {111} grains; second, in case that grains with different orientations have the same energy density, the size of the {111} grains would be larger than that of the {100} grains.

This approach will be now applied to explain the texture change that took place during the annealing of the current material. The as-deposited specimen consists of the {100} grains as the major component and the {111} grains as the minor component. Therefore, our material lies somewhere between the two extremities above, i.e. only-{100}-grained and only-{111}-grained materials. Under assumption that energy density is independent of crystallographic orientation, since the {111} grain size might be the largest in the material, the {111} grains would grow faster due to the size advantage to dominate the annealing texture. On the other hand, if grain size is the same regardless of the orientations of grains, the

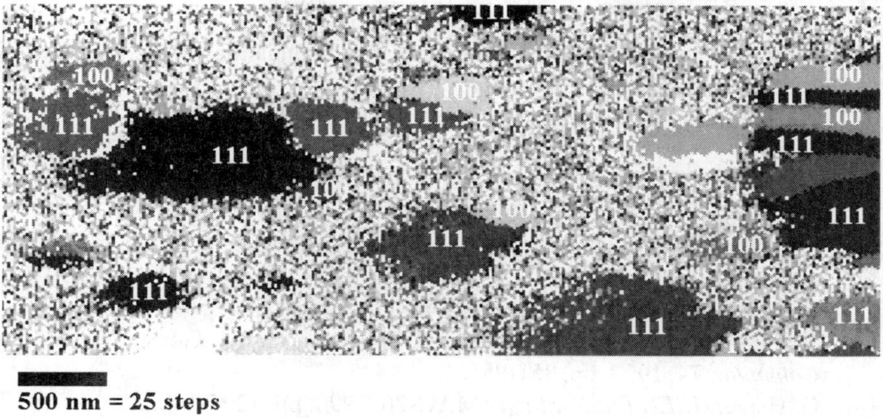

Fig. 4. OIM map showing that the {111} grains are much coarser than the {100} and other oriented grains in the sample annealed at 380°C for 1 min.

{111} grains having the lowest energy density may grow into the different oriented grains to decrease the total energy of the system.
In reality, the OIM map obtained in the sample annealed at 380°C for 1 min shows that the {111} grains are fast grown to be much coarser than the {100} and other oriented grains as in Fig. 4. It is apparent that the {111} grains abnormally grow in the early stages of grain growth and dominate the final texture as depicted in Fig. 3. From OIM observation of the as-deposited specimens, there is no evidence that the size of {111} grains is larger than that of other oriented grains. Therefore, the abnormal growth of the {111} grains would not be attributed to the size advantage but to their lower energy density as discussed above.

5 Summary

Grain growth that takes place during annealing of nanocrystalline Ni-20wt-%Fe alloy electrodeposits was investigated and discussed in terms of its possible effect on the texture evolution. The texture of the as-deposited specimen was characterized by a strong <100> and a weak <111> fibres. Abnormal grain growth of {111} in the early stages of grain growth occurred on annealing beyond 370°C and resulted in the texture change that the <111> fibre strongly developed at the expense of the <100> fibre. This texture evolution was interpreted by means of the orientation dependency of grain growth.

Acknowledgement

This work has been supported by the Research Foundation of the Engineering College, Sunchon National University.

References

[1] G. Palumbo, S.J. Thorpe, and K. Aust, *Scripta Metall. Mater.*, 24, 1990, pp.1347-1350.
[2] C. Suryanarayana, *Int. Mater. Rev.*, 40, 1995, pp. 41-64.
[3] S. Ramasamy, J. Jiang, H. Gleiter, R. Birringer, and U. Gonser, *Solid State Commun.*, 74, 1990, pp. 851-855.
[4] G. Herzer, *IEEE Trans. Magn.*, MAG-26, 1990, pp. 1397-1402.
[5] M.J. Aus, B. Szpunar, A.M. El-Sherik, U. Erb, G. Palumbo, and A.T. Aust, *Scripta Metall. Mater.*, 27, 1992, pp. 1639-1643.
[6] F. Czerwinski, H. Li, M. Mergret, J.A. Szpunar, D.G. Clark, and U. Urb, *Scripta Mater.*, 37, 1997, pp. 967-1972.
[7] F. Czerwinsky, J.A. Szpunar, and U. Erb, *J. Mater. Sci.*, 11, 2000, pp. 243-251.
[8] K. Lu, *Scripta Metall. Mater.*, 25, 1991, pp. 2047-2052.
[9] H. Li, F. Czerwinski, and J.A. Szpunar, *Nanostructured. Mater.*, 9, 1997, pp. 673-676.

Recrystallization and Grain Growth
Proceedings of the First Joint International Conference
Eds. G. Gottstein and D.A. Molodov, Springer-Verlag, 2001

Annealing Behaviour of Ultra Fine-Grained Iron with Dispersed Particles

A. Belyakov*, Y. Sakai, T. Hara, Y. Kimura, and K. Tsuzaki

Frontier Research Center for Structural Materials, National Research Institute for Metals, 1-2-1 Sengen, Tsukuba, Ibaraki 305-0047, Japan
*On leave from the Institute for Metals Superplasticity Problems, Ufa, Russia

Keywords: steel; dispersed particles; mechanical milling; static recrystallization; normal grain growth.

Abstract. Annealing behaviour of an ultra fine-grained iron with iron oxide particles was studied in a temperature range from 600°C to 900°C. Isochronal annealing for about 3.6 ks does not lead to considerable changes in microstructure at temperatures below 800°C, where the specimens maintain very fine grain sizes. In such conditions, the long term annealing behaviour can be mainly characterised by a normal grain growth accompanied by recovery. The annealing mechanisms operating in very fine-grained matrix are discussed in some detail.

1 Introduction

The development of fine-grained microstructures in metallic materials is one of the most interesting trends in materials science. The mechanical properties of metals and alloys are believed to be significantly improved by decreasing the grain size to a sub-micron scale [1, 2]. Several methods have been proposed to produce materials with sub-micron grains: rapid solidification, vapour condensation, severe plastic deformation, etc. [1-3].
Recently, mechanical milling followed by consolidating warm rolling has been applied to develop high strength steels with ultra fine-grained structures, which contain dispersed oxide particles [4, 5]. Both the high internal distortions evolved under the consolidation process and the dispersed particles homogeneously distributed throughout the fine-grained matrix may result in a specific annealing behaviour taking place in such materials. Most of researches have been focused mainly on the investigations of a thermal stability for the ultra fine-grained structures, i.e. the effect of temperature on the grain size, resistance to discontinuous grain growth, etc. [1, 2]. Ohtaguchi et al. have shown [6] that the presence of fine oxide particles is quite effective for retarding the grain growth. Such ultra fine grain structures are thermally stable in the ferrite range. However,

further experimental studies are necessary for a fundamental understanding of the annealing behaviour during the heating of submicrocrystalline structures.
The aim of the present work is to study the annealing behaviour of an iron – iron oxide alloy with ultra fine-grained microstructure, which was developed by mechanical milling followed by warm rolling.

2 Experimental Procedure

Three kinds of iron – iron oxide powders with different amounts of oxygen, i.e. 0.2, 0.6 and 1.5 mass% (Table 1), were used. The powders were mechanically milled by a planetary ball mill in an argon atmosphere for about 100 hours using a stainless steel pot (SUS403) and steel balls (SUJ2). After mechanical milling, the powders were canned in a steel pipe with an inner diameter of 30 mm and then multiple bar rolled to a total nominal reduction in area of 86% at 700°C for consolidation. The main structural parameters evolved after consolidation, i.e. grain (D) and oxide particle size (d), are also shown in the Table 1.
The consolidated bars were cut into 10 mm sample pieces and then annealed in argon atmosphere at temperatures (T) from 600 to 900°C. The structural analysis was performed on the sections parallel to the rolling direction using a JEM-2010F transmission electron microscope operating at 200 kV. The average grain size was measured by the linear intercept method in two directions, i.e. the long intercept and the short intercept. The restoration kinetics was studied by means of a fractional softening (X) calculated from the equation: $X = (Hv_\varepsilon - Hv_t) / (Hv_\varepsilon - Hv_o)$. Hv_ε is the hardness for the as-processed ultra fine-grained material, and Hv_t and Hv_o are those for partially and fully annealed samples, respectively.

Table 1. Chemical composition of iron - oxide powders (mass%), and sizes of grains (*D*) and oxide particles (d) evolved after consolidation.

	O	N	C	Si	Mn	P	S	*D*, μm	*d*, nm
Fe-0.2%O	0.18	0.003	0.002	0.010	0.150	0.010	0.010	0.7	9
Fe-0.6%O	0.58	0.005	0.010	0.009	0.110	0.009	0.007	0.35	10
Fe-1.5%O	1.51	0.008	0.030	0.005	0.003	0.005	0.001	0.2	12

3 Results and Discussion

3.1 Effect of the annealing temperature

Typical microstructures evolved in samples after isochronal annealing for about 3.6 ks at various temperatures are shown in Fig. 1. This figure also presents the initial microstructure developed in the 1.5%O sample for reference. It is clearly seen that the final microstructure evolved in the samples after annealing depends strongly on both the oxygen content and the heating temperature. Assuming that most of the oxygen is the Fe_3O_4-type oxide, the volume fractions (f) of the oxide particles are about 1, 3 and 8% in samples containing 0.2, 0.6 and 1.5 mass% oxygen, respectively [5]. Such a variation in the volume fraction of dispersed particles produces different pinning effects. The maximum retarding force results in a cessation of rapid grain growth when $D \approx 4\,d / 3\,f$ [7]. Such critical grain sizes are calculated as 1.2, 0.45 and 0.2 μm, for the specimens with 0.2, 0.6 and 1.5%O, respectively. These critical D are roughly similar to the initial grain sizes developed after consolidation. Therefore, the dispersed oxides are effective for retarding the rapid coarsening during processing.

Fig. 1. Typical structures developed in ultra fine-grained iron – iron oxide alloys after annealing. (a) as processed 1.5%O sample, (b) 0.2%O sample annealed at 800°C for 3.6 ks, (c) 1.5%O sample annealed at 800°C for 3.6 ks and (d) 1.5%O sample annealed at 900°C for 3.6 ks.

Fig. 2 presents the effect of the annealing temperature on the ultra fine-grained microstructures in studied alloys quantitatively. Two temperature regions with a separating range at 800 ~ 900°C can be clearly distinguished. Annealing at T ≥ 900°C results in a rapid grain coarsening, and is accompanied by a softening. Referring to [6], we can conclude that such an effect of the annealing temperature on the present steels is closely related to the phase transformation, i.e. the $\alpha \rightarrow \gamma \rightarrow \alpha$ transformation breaks the initial ultra fine-grained ferritic structures completely.

In contrast, an annealing at T ≤ 800°C does not lead to considerable changes in the microstructures. The average grain size increases about 1.5 times, and the hardness decreases slightly, compared to the as-processed state, with increasing the annealing temperature from 600 to 800°C. It is interesting to note that a ratio of the initial grain size to the annealed one (that is about 0.7 in Fig. 2) seems to be independent of the oxygen content. In other words, the relative annealing kinetics is not affected strongly by the variation in the oxide volume fraction.

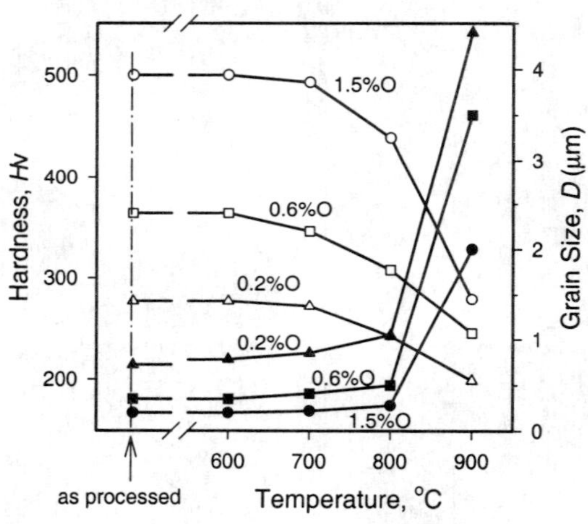

Fig. 2. Effect of annealing temperature on the hardness (open symbols) and the grain size (closed symbols) for the fine-grained iron – iron oxide alloys. The annealing time was 3.6 ks.

3.2 Restoration mechanisms operating during annealing

The results suggest that the ultra fine-grained microstructures developed in the samples are remarkably stable against dramatic grain coarsening during heating in the ferrite range. Nano-scaled dispersed oxide particles that homogeneously distributed through the sample can be effective barriers for retarding the rapid grain growth. Moreover, materials with a large fraction of strain-induced high angle grain boundaries are believed to be resistant to discontinuous recrystallization regardless of apparently high stored energy, and only slow normal grain growth can operate in such structures [1]. Let us now consider the structural restoration mechanisms operating during annealing in more detail, using the 1.5%O steel as an example.

Experimental studies on primary recrystallization kinetics are usually based on the correlation with the Avrami model by plotting ln(1/(1-X)) against annealing time (t) in double logarithmic scale [8]. Fig. 3 shows the Avrami plot for the fractional softening and the relationship between the average grain size and the annealing time at 800°C. The slope of about 0.5 for the Avrami plot in Fig. 3 is quite small, compared to those from other studies on primary recrystallization [8]. This suggests that any other restoration processes including recovery may be responsible for the annealing behaviour in this study. A short incubation period also counts in favour of the recovery, which can play an important role during annealing.

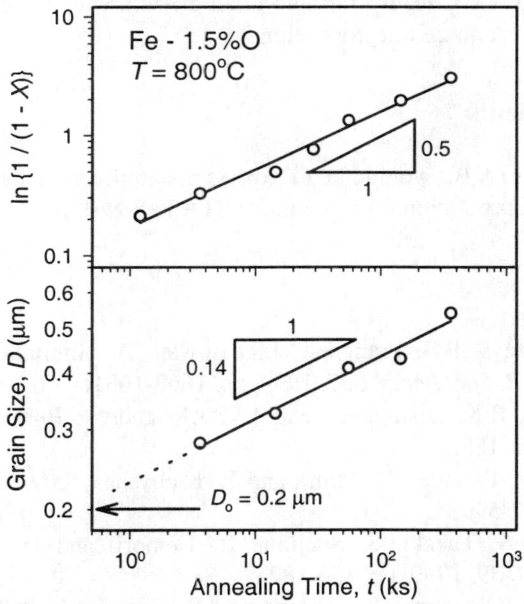

Fig. 3. Avrami plot for fractional softening (X) and grain size changes during annealing of the fine-grained iron containing 1.5%O at 800°C.

Normal grain growth in metallic materials can be described by a power law equation: $D = K t^{1/n}$ [8], where the grain growth exponent, n, is usually in the range from 2 to 10. The (1/n) = 0.1 ~ 0.3 was obtained for a pure iron [9] and for a ferrite containing fine cementite particles [10]; however, the grain sizes were 1~2 order of magnitude coarser than in the present study. The data presented in Fig. 3 indicate that the normal grain growth accompanied by recovery may be considered as the main structural mechanism operating in the ultra fine-grained steels during annealing. The controlling factors for the normal grain growth in particle containing steels (i.e. particle coarsening, dragging the particles with grain boundaries, etc.), are an important topic and should be studied in more detail.

4 Conclusions

The annealing behaviour of the ultra fine-grained iron – iron oxide alloys produced by mechanical milling followed by consolidating working was studied at temperatures from 600 to 900°C. The main results can be summarized as follows:

1. The fine-grained microstructure is essentially stable against a rapid coarsening during annealing at temperatures below 800°C.
2. The long term annealing behaviour at $T \leq 800$°C can be characterized mainly by a normal grain growth accompanied by recovery.
3. Annealing at 900°C results in rapid grain growth, which is closely related to the $\alpha \rightarrow \gamma \rightarrow \alpha$ phase transformation.

Acknowledgements

One of the authors (A.B.) would like to express his gratitude to the Japan Science and Technology Corporation for providing a STA Fellowship.

References

[1] F.J. Humphreys, P.B. Prangnell, J.R. Bowen, A. Gholinia and C. Harris, *Phil. Trans. R. Soc. Lond.*, 357, 1999, pp. 1663-1681.

[2] R.Z. Valiev, R.K. Islamgaliev and I.V. Alexandrov, *Progr. Mat. Sci.,* 45, 2000, pp. 103-189.

[3] A. Belyakov, T. Sakai, H. Miura and R. Kaibyshev, *ISIJ International*, 39, 1999, pp. 592-599.

[4] Y. Kimura, S. Takaki, S. Suejima, R. Uemori, and H. Tamehiro, *ISIJ International*, 39, 1999, pp. 176-182.

[5] Y. Sakai, M. Ohtaguchi, Y. Kimura and K. Tsuzaki, in "Ultrafine Grained Materials", eds. R.S. Mishra et al., TMS, 2000, pp. 361-370.

[6] M. Ohtaguchi, K. Tsuzaki and K. Nagai, in "Recrystallization and Related Phenomena (ReX'99)", eds. T. Sakai and H.G. Suzuki, The Japan Institute of Metals, 1999, pp. 495-500.

[7] C.S. Smith, *Transactions A.I.M.E.*, 175, 1948, pp. 15-51.

[8] F.J. Humphreys and M. Hatherly, "Recrystallization and Related Annealing Phenomena", Pergamon Press, Oxford, 1996.

[9] G. Wiener, *Transactions A.S.M.*, 44, 1952, pp.1169-1185.

[10] P. Hellman and M. Hillert, *Scandinavian Journal of Metallurgy*, 4, 1975, pp. 211-219.

Recrystallization and Grain Growth
Proceedings of the First Joint International Conference
Eds. G. Gottstein and D.A. Molodov, Springer-Verlag, 2001

Nano- and Submicrocrystalline Low-Carbon Steels Obtained by Severe Plastic Deformation

S.V. Dobatkin[1], P.D. Odessky[2], N.A. Krasilnikov[3], G.I. Raab[3], and V.N. Konenkova[1]

[1] Moscow State Steel and Alloys Institute (technological university), Leninsky prospekt 4, 119991 Moscow, Russia
[2] Institute of Building Constructions, Moscow, 2-Institutskaya ul. 6, 109451 Moscow, Russia
[3] Ufa State Aviation Technical University, ul. K.Marksa, 12, 450000 Ufa, Russia

Keywords: low-carbon steel, severe deformation, torsion, ECA pressing, nanostructure.

Abstract. The opportunity of ultra-fine grained structure formation in low-carbon steels during cold and warm severe plastic deformation (SPD) in torsion under pressure 6GPa and equal channel angular (ECA) pressing is investigated. At cold SPD of 0,09CMnSi and 0,1CMnVTi steels the nanoscale structure is formed: oriented like-cell fragments and grains. The deformation of initial quenched steels with nonequilibrium structure results in the greater refinement of structure, than deformation of initial hot-rolled state. At warm ECA pressing of 0,21CMnSiV and 0,25CMnSi steels the mixed structure is formed: dynamically recovered and submicrocrystalline with the grain sizes 300…500 nm. Such partially submicrocrystalline structure causes a high-strength state: YS > 1000 MPa, EL = 10 %.

1 Introduction

Ultrafine-grained (UFG) materials obtained by SPD at lower temperature call for special interest because of their unusual properties [1,2]. The investigations were carried out, basically, on pure metals and plastic alloys [3,4]. Industrial steels are investigated a little. Certainly, it is impossible to use widely methods of severe plastic deformation in an industry now. But on our sight, it is necessary to study limiting structural states of industrial materials and properties, appropriate to them.

The purpose is to study of structure and microhardness of low-carbon 0,09CMnSi; 0,1CMnVTi ; 0,21CMnSiV and 0,25CMnSi steels after SPD and heating.

2 Experimental Procedure

As materials of research were chosen low carbon steels having rather low YS and used, basically, as elements of designs (Table 1).

Table 1. Chemical composition of low-carbon steels studied

Steel	Alloying element, wt.%						
	•	Si	Mn	V	Ti	P	S
0,09CMnSi	0.12	0.65	0,85	-	-	<0.030	<0.035
0,1CMnVTi	0,10	-	1,12	0,08	0,07	-	-
0,21CMnSV	0,21	0,75	0,85	0,17	-	<0,030	<0,025
0,25CMnSi	0,25	0,77	1,4	-	-	0,015	0,031

0,09CMnSi and 0,1CMnVTi steels were used in both hot-rolled initial state with ferritic-pearlitic structure and in quenched initial state with bainitic structure. 0,21CMnSiV and 0,25CMnSi steels were heated at 950 °C (30 min.) and cooled with the furnace before deformation.

Deformation was performed by torsion under 6 GPa pressure on Bridgman anvil. The samples of 10 mm in diameter and 0,6mm in thickness had been deformed by torsion up to 5 revolutions which corresponds to logarithmic strain e = 6,8.

ECA pressing has been carried out using the samples of 20 mm in diameter and 120 mm in length at the temperature 550 °C with an angle of intersection of channels 90° for 2 and 4 passes with 180° turn around the axis of pressing after each pass.

The structure analyses has been carried out using the transmission electron microscope JEM-100X. DRON-3 diffractometer was used for X-ray analyses. Microhardness was defined using the PMT-3 device with 50g load.

3 Results and Discussion

TEM study of quenched deformed 0,1CMnVTi steel revealed an oriented fragmented cell-like structure with equiaxed grains of average size of ~65 nm inside it. With deformation temperature increasing up to 500 °C the nature of this structure remained unchanged (Fig. 1a). However, grains grew to average size of ~85 nm and the volume fraction of grains increased. The mean grain size was approximately equal to the transverse dimension of an oriented fragment. Earlier, the emergence of nanostructure was reported to occur due to deformation bands, twins and stacking faults during room temperature SPD [5]. In the present case of a BCC polycrystals the deformation bands most probably formed as oriented fragments.The average grain size inside the oriented fragments after SPD was larger in hot rolled steel than in quenched steel, and was equal to ~95 nm at T_{def} = 20 °C and to ~120 nm at T_{def} = 500°C. Individual coarse grains (200 nm and more) were also observed inside three-four oriented fragments. These grains were evidently formed by as-coalescence mechanism. The fraction of such grains increased with temperature. 0,09CMnSi steel after SPD in both initial states exhibited a microstructure similar to that in 0,1CMnVTi steel with, however, smaller grain size: 40 nm and 60 nm for quenched and hot rolled initial states, respectively.

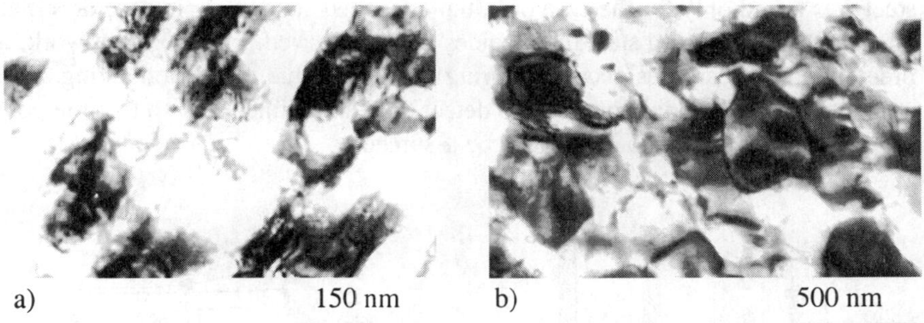

Fig. 1. Structure of 0,1CMnVTi steel after SPD by torsion and heating: a) SPD, t=500 °C; b) SPD, t=20 °C and heating to 500 °C (30 min.)

Thus, the SPD of low carbon steels in torsion at T = 20 – 500 °C with strains of e = 6,8 did not lead to a completely nanocrystalline structure.

X-ray analysis revealed a considerable difference in reflexes for 0,1CMnVTi steel in the initial and deformed states. In deformed steel, reflexes were less intensive and wider. The variation of physical broadening of (110) and (220) reflexes, as well as of the width of (110), (200), (211) and (220) reflexes in initial and deformed states was similar (Fig. 2). Physical broadening and width of reflexes in initial quenched state were larger than in initial hot-rolled state due to higher dislocation density from phase transformation. Further SPD led to larger β and B.

Lattice parameters computed by extrapolation technique from 4 reflexes were the same for all treatments (Table 2). The microstrain and block sizes were calculated using the approximation technique (Table 2). The variation of microstrain was similar to that of the physical broadening and the width of reflexes. For the deformed state it was found that in initially quenched state the blocks were slightly smaller (64 nm) than in initially hot rolled state (85 nm), although the fairly large experimental error should be taken into consideration. Notable is a good agreement between the results of X-ray analysis and TEM studies where the sizes were measured as 65 and 95 nm, respectively.

SPD in torsion within the studied temperature range with formation of partially nanocrystalline structure led to considerable hardening (Fig. 3). In both steels after SPD the microhardness increased in 2.5 – 3 times in case of hot rolled state and in 1.5 – 2 times in case of quenched state, as compared to the initial state. For a given steel in a given condition the values of microhardness were almost constant within the entire temperature range. Also within this temperature range, the microhardness of both steels after SPD was higher in a case of quenching initial state than in a case of hot rolling. The increased microhardness of the initially quenched steel after deformation can be attributed to higher dislocation density due to phase transformation, smaller size of fragment -cells and grains, as shown by TEM results, and, perhaps, to higher carbon content in solid solution. The dissolution of carbides under SPD was noted in numerous studies [6-8]. In [9] the dissolution of carbides during SPD of 0,2C-2Mn-B.

steel was found only for the case of initial quenched state while in case of initially quenched and tempered state the carbides were preserved. In the present work, the precipitation of carbides occurred during quenching due to self-tempering. Small amounts of carbides did not allow to detect them experimentally so that the possibility for their dissolution could only be assumed.

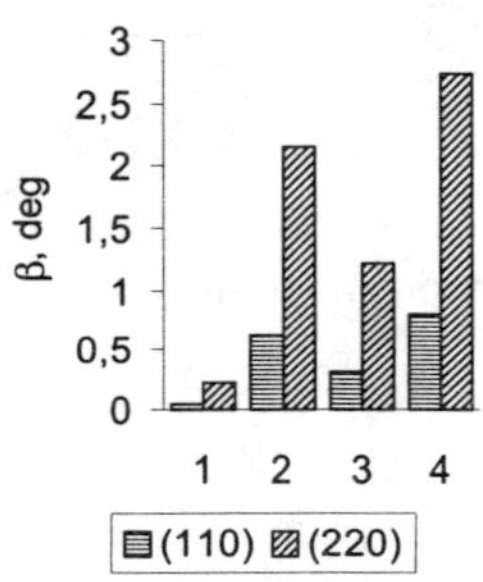

Fig. 2 (left). Physical broadening of X-ray (110) and (220) reflexes. States: 1- hot-pressed, 2– quenched, 3- after SPD (initial state: hot-pressed), 4 - after SPD (initial state: quenched).

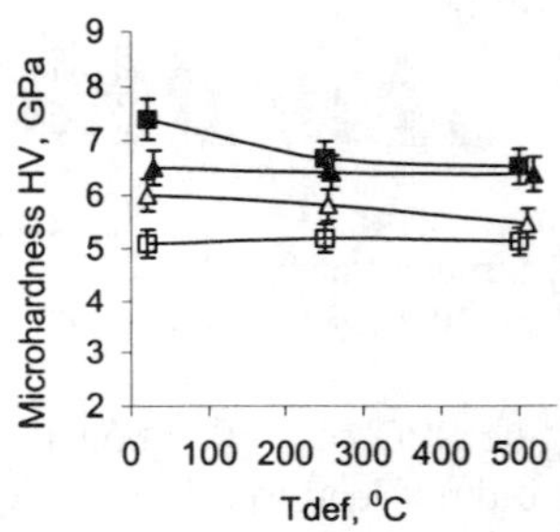

Fig. 3 (right). Microhardness dependence at deformation temperature of low-carbon steels after SPD by torsion. Initial states: for 0,09CMnSi : ▲ - quenched, △ - hot-pressed; for 0,1CMnVTi : ■ - quenched, □ - hot-pressed

It should be noted in general that no advantages of V and Ti additions were revealed in the present work, because the reheating at 950 °C for quenching did not allow to dissolve carbonitrides. For this reason, there was a little difference in microhardness between steels 0,09CMnSi and 0,1CMnVTi (Fig. 3). When both steels deformed under SPD in torsion at 20 °C were reheated to the temperature 500 °C the microhardness decreased in more than 2 times. This contrasted to very weak variations of microhardness during SPD with the temperature up to 500 °C. The reheating at 500 °C resulted in a submicrocrystalline structure with 200–500 nm grains (Fig. 1b). Some retained oriented fragments are also observed.

Table 2. Parameters of a lattice and of a thin crystalline structure for 0,1CMnVTi steel

State of steel	Lattice parameter, A	Parameters of a thin crystalline structure	
		Size of blocks D, A	Microstrains ε, %
Initial state: hot-pressed	2.8673 ± 0.0002	> 1500	< 0.04
After SPD (initial state: hot-pressed)	2.8670 ± 0.0005	850 ± 350	0.35 ± 0.04
Initial state: quenched	2.8673 ± 0.0002	> 1500	0.23 ± 0.02
After SPD (initial state: quenched)	2.8676 ± 0.0004	640 ± 280	0.45 ± 0.05

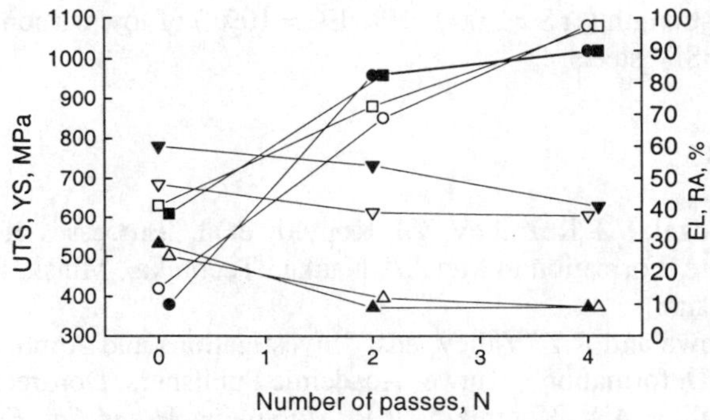

Fig. 4. Mechanical properties of 0,21CMnSiV and 0,25CMnSi steels after SPD by ECAP at 550 °C with N=4. Filled symbols - 0,25CmnSi, hollowed symbols - 0,21CmnSiV: ● - YS, ■ - UTS, ▲ - EL, ▼ - RA.

Consequently, SPD at 20 – 500 °C allows to obtain a partially nanocrystalline microstructure. The initial as quenched state is more favorable for the refinement of microstructure than the initial as hot rolled state. The dimensions of specimens deformed in torsion do not allow for standard mechanical tests. The mechanical properties were determined after warm ECA pressing of low carbon steels 0,21CMnSiV and 0,25CMnSi. The temperature of 550 °C for warm deformation was selected with account for the capability of the deformation tools. Also, as indicated by the studies of SPD in torsion, the microstructure and microhardness of low-carbon steels did not vary much at 20 – 500 °C. After deformation with N = 4 that corresponds to the true strain of e = 2,1, both steels exhibited a mixed microstructure comprised of dynamically recovered and submicrocrystalline portions which differed in misorientation (medium or high) of the boundaries of the structure elements. The sizes of these two structure elements were found to be 300 – 500 nm for both steels.

This partially submicrocrystalline structure is responsible for a considerable increase in strength with preservation of satisfactory ductility. The values of yield stress exceeded 1000 MPa and were close to the values of tensile strength (Fig. 4).

4 Conclusions

1. SPD of 0,09CMnSi and 0,1CMnVTi steels in torsion under high pressure at room temperature results in the emerging of nanoscale structure comprised of oriented cell-like fragments and fine grains. The volume fraction of grains increases with deformation temperature increasing to 500 °C.
2. SPD in torsion of the initially quenched 0,09CMnSi and 0,1CMnVTi steels leads to a more dispersed nanoscale structure than that of initially hot rolled steels.
3. Partially submicrocrystalline structure after warm ECA pressing is favorable for high strength (YS >1 000 MPa, EL = 10%) of low carbon 0,25CMnSi and 0,21CMnSiV steels.

References

[1] V.M. Segal, V.I. Reznikov, V.I. Kopylov et al. "Processes of Plastic Structure Formation in Metals", Nauka i Technika , Minsk, Belorus, 1994 (in Russian).

[2] T.C. Lowe and R.Z. Valiev, eds. "Investigations and Applications of Severe Plastic Deformation", Kluwer Academic Publishers, Dordrecht, 2000.

[3] R.Z. Valiev, A.V. Korznikov, R.R. Mulyukov, *Mater. Sci. Eng.*, A168, 1993, pp. 141-148.

[4] Y. Iwahashi, Z. Horita, M. Nemoto et al., *Acta Mater.*, 46, 1998, pp. 3317-3331.

[5] A.M. Patselov, V.P. Pilyugin, E.G. Tchernyshov et al., in "The Structure and Properties of Nanocrystalline Materials",eds. N.Noskova and G. Taluts, RAS, Ekaterinburg, Russia, 1999, pp. 37-44 (in Russian).

[6] V.N. Grignev, V.G. Gavrilyuk, *Metallofizika*, 4, 1982, pp.135-152 (in Russian).

[7] A.V. Korznikov, Yu.V. Ivanisenko, D.V. Laptionok et al., *Nanostructured Materials*, 4, 1994, pp. 159-168.

[8] V.V.Sagaradze, V.A.Shabashov, A.G.Mukoseev, N.L.Petcherkina, in "The Structure and Properties of Nanocrystalline Materials",eds.N. Noskova and G. Taluts, RAS, Ekaterinburg, Russia, 1999, pp. 46-63 (in Russian).

[9] M.V. Degtyarev, T.I. Tchashuchina, L.M. Voronova et al., *FMM*, 77, 1994, pp. 141-146 (in Russian).

The Recrystallization of a Highly Deformed Al-Fe-Mn Alloy

H. Jazaeri and F.J. Humphreys

Manchester Material Science Centre, University of Manchester and UMIST,
Grosvenor St., Manchester, M1 7HS, UK

Keywords: continuous recrystallization, sub-micron grains, large strain.

Abstract. The deformation and annealing behaviour of an Al-Fe-Mn alloy cold rolled to a strain of 3.9 has been investigated using high resolution EBSD. After strains of up to 2.6 (93% reduction) the material undergoes discontinuous recrystallization on subsequent annealing. However, as the strain increases to 3.9 (98% reduction), the recrystallization becomes continuous, producing a sub-micron grained microstructure which remains stable during low temperature annealing. It is shown that the inhomogeneous deformation associated with the large second-phase particles in the Al-Fe-Mn alloy, is an important factor in producing stable fine-grain microstructures during large-strain deformation.

1 Introduction

Recent research (e.g. [1-4]) has shown that large strain rolling deformation of some aluminium alloys may produce microstructures which are stable against normal (discontinuous) recrystallization. In such cases a large fraction of the boundaries are of high angle (HAGB), and on subsequent annealing at low temperatures, minor local boundary rearrangements result in microstructures which, apart from their scale, are similar to "recrystallized" material. As these microstructural changes occur uniformly throughout the material, with no apparent "nucleation" of the recrystallization, such a process can be phenomenologically classified as one of "continuous recrystallization". The microstructural stability of such highly deformed alloys can be explained in terms of the large fraction (0.7-0.75) of high angle grain boundaries in such microstructures [5-7].
The aims of the present project are to investigate the nature of the transition between discontinuous and continuous recrystallization and the effects of strain and second-phase particles on this transition.

2 Experimental Procedure

A 30 mm hot rolled plate of Al-1.45Fe-0.43Mn (AA8006) was supplied by Alcan International. The material was found to be recrystallized, with a grain size of

~12 μm. The alloy contained large (2-10 μm) iron-rich intermetallic constituent particles and smaller (~100 nm) $MnAl_6$ dispersoids.
The material was cold rolled to strains of up to 3.9 (reduction of 98%). After deformation, samples were annealed for one hour at temperatures 150-400°C and water quenched. The texture and microstructure of both deformed and annealed samples were studied by using high resolution EBSD in a Philips XL30 FEG-SEM. High and low angle boundaries were defined as having misorientations greater than 15° and 2°, respectively and are shown in black and white respectively on the maps. All maps were of the ND-RD plane and ND is vertical in all figures. X-ray texture analysis was carried out on samples prepared from the mid-section of the material, parallel to the rolling plane.

3 Results and Discussion

3.1 Microstructural development during high strain deformation

Examples of EBSD maps of the deformed material are shown in Fig. 1, for samples deformed to low (1.2) and high (3.9) strains. With increasing reduction the grains become elongated in the rolling direction, the distance between HAGBs parallel to the rolling plane is decreased and the proportion of boundaries which are high angle increases, as shown in Fig. 2. Fig. 1b shows the microstructure after a strain of 3.9, where a sub-micron grained structure (0.7 μm) with more than 70% HAGBs has been formed. The grains are elongated in the rolling direction with a mean aspect ratio ~2.45. The HAGB spacing parallel and perpendicular to the rolling direction was 0.42 μm and 1.03 μm, respectively.

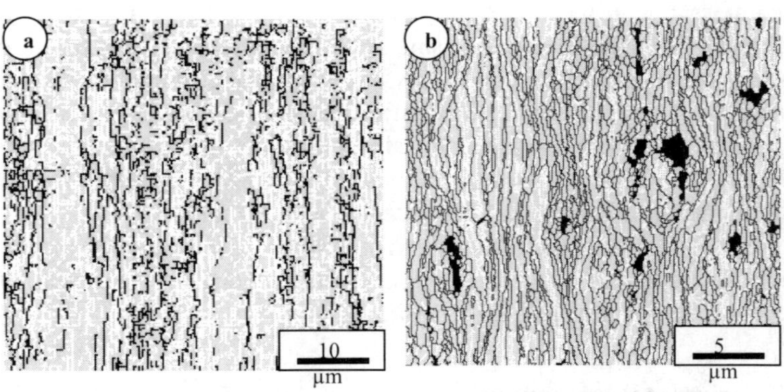

Fig. 1. EBSD maps of cold rolled AA8006 at strains of (a) 1.2 and (b) 3.9.

As may be seen in Fig. 1b, the deformed microstructure was significantly affected by large second phase particles breaking up the planarity of the rolling structure [8]. The textures of the rolled alloys, summarised in Fig. 3, contain the expected rolling texture components, dominated by the β-fibre, with S being strongest. With increasing strain, the Goss and Cube components decrease and the rolling components increase up to a strain of ~2.5, after which there is some decrease.

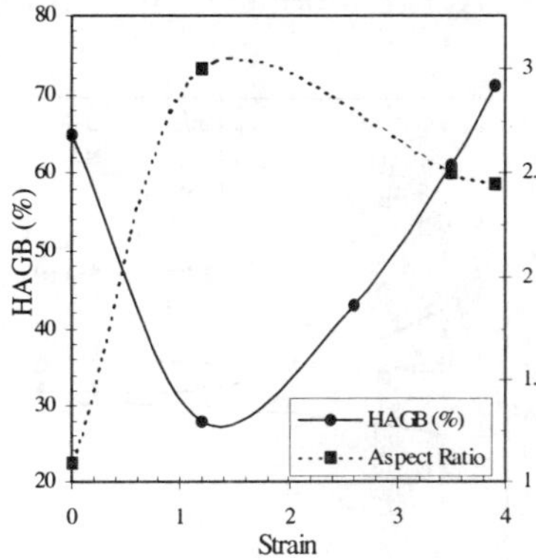

Fig. 2. The change of % HAGB and grain aspect ratio with strain.

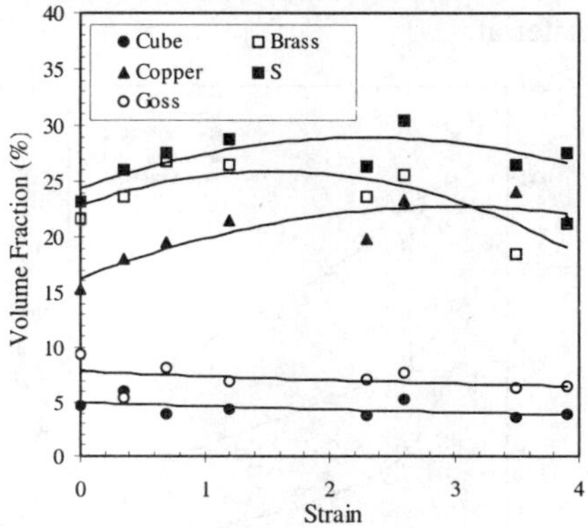

Fig. 3. The effect of strain on texture.

3.2 Annealing behaviour

Lower strain deformation: On annealing the alloy deformed to lower strains (e.g. 1.2), recrystallization started at ~225°C. In a partially recrystallized sample annealed at 250°C (Fig. 4a), the distinction between unrecrystallized and recrystallized regions of the microstructure is clear, demonstrating that the sample

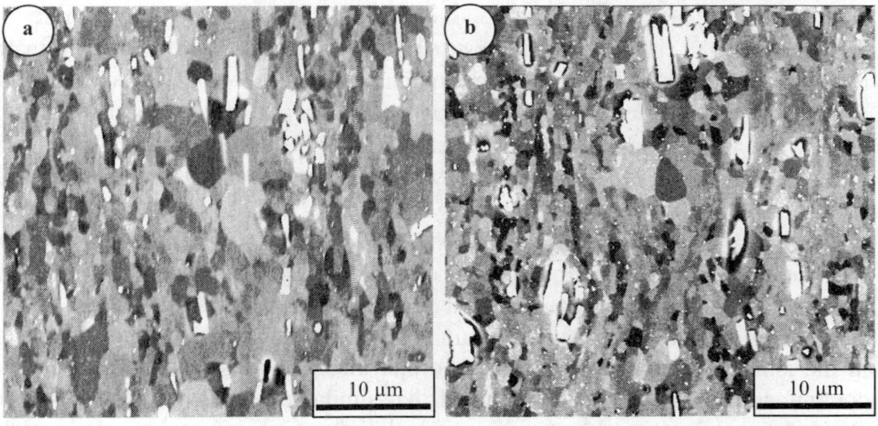

Fig. 4. Microstructures of partially recrystallized AA8006 annealed at 250°C after deformation in strains (a) 1.2 and (b) 2.6.

is undergoing **discontinuous** recrystallization. The recrystallization process is completed at 350°C. At strains of up to 2.6 (reduction of 93%), similar discontinuous recrystallization occurs as may be seen Fig. 4b, although the recrystallization is completed at a lower temperature. The effect of strain on the recrystallization kinetics and on the final grain size is shown in Fig. 5. It is notable that a very small grain size (~5μm) can be formed by discontinuous recrystallization in this material.

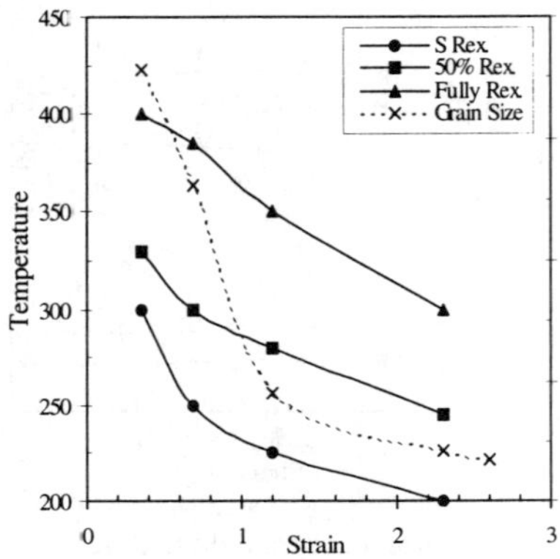

Fig. 5. Recrystallization kinetics of the deformed AA8006 up to a strain 2.3.

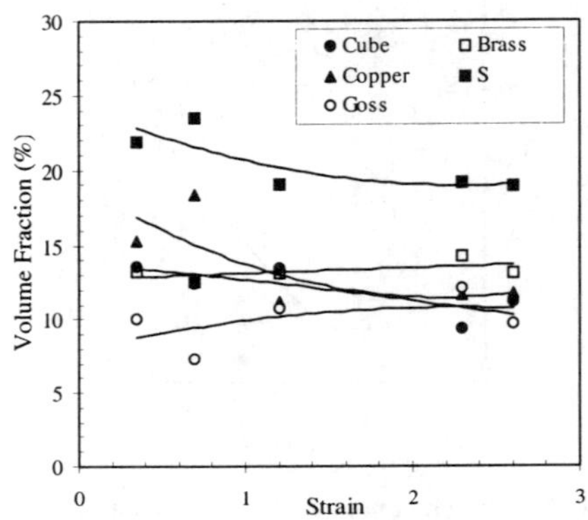

Fig. 6. The effect of strain on texture after recrystallization.

The volume fractions of texture components in the fully recrystallized samples are shown in Fig. 6. The rolling texture components remain strong after recrystallization, although they are weaker than after deformation. However, there is a significant increase in the cube texture component on recrystallization.

High strain deformation: Fig. 7 shows the microstructure of the highly deformed material (ϵ=3.9) after annealing at 200° and 250°C. During annealing at these low temperatures, the fine-grained microstructure formed during deformation remains relatively stable. Minor boundary migration leads to the grains becoming more equiaxed and a little coarser. At this stage the grain size is 1.6 μm and the grain aspect ratio is 1.8. Comparison with the annealing behaviour of material deformed to a moderate strain of 2.3 (Fig. 5) shows that this material would have begun to recrystallize at 200°C and be 50% recrystallized at 250°C. These results confirm earlier work [2] on a similar alloy (AA8014), in which it was shown that a uniform coarsening of the microstructure occurred during low temperature annealing when the microstructure contained at least 70% HAGBs. The present results clearly demonstrate that when the strain is increased sufficiently, there is a change in the annealing behaviour from the normal discontinuous recrystallization to a continuous process. It should be noted that similar effects have been found for alloys deformed to even larger strains by techniques such as equal channel angular extrusion [9,10].

The large rolling strains tend to break-up the microstructures and lead to a large amount of high angle boundary. However, the microstructures of the heavily deformed material are not homogeneous even at the largest strains. Fig. 1 shows that smaller and more equiaxed grains are formed near the large second-phase particles, with more elongated lamellar grains in regions remote from the particles, and this heterogeneity remains after annealing as is seen in Fig. 8a. The large particles clearly play an important role in developing a fine-grain microstructure during the deformation and low temperature annealing.

On annealing at higher temperatures, the microstructure of the highly strained material becomes coarser and more equiaxed, and at 300°C (Fig. 8b), the grain size is 1.9 μm and the aspect ratio has decreased to 1.2. At temperatures above 350°C the microstructure becomes relatively unstable, and discontinuous grain growth occurs, resulting in a heterogeneous microstructure of mean grain size ~5μm.

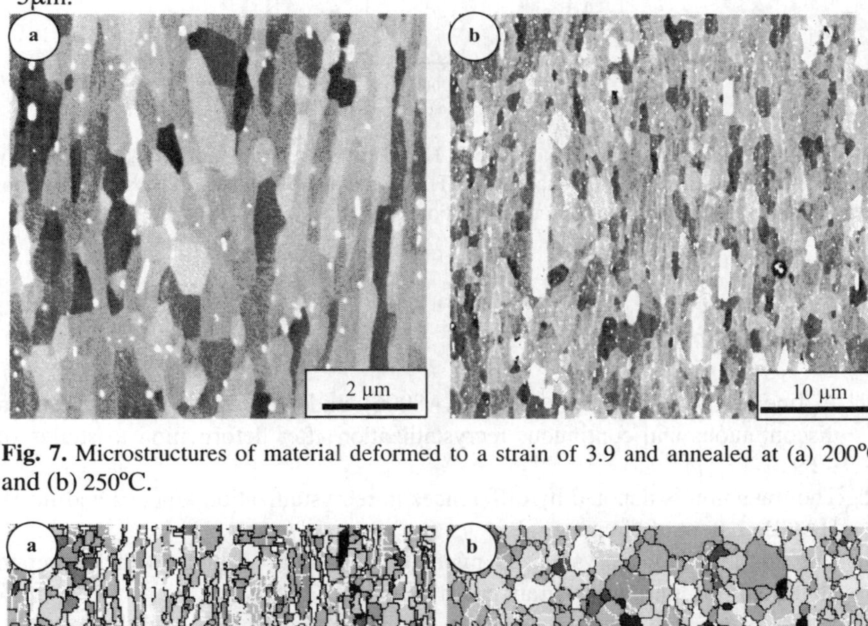

Fig. 7. Microstructures of material deformed to a strain of 3.9 and annealed at (a) 200°C and (b) 250°C.

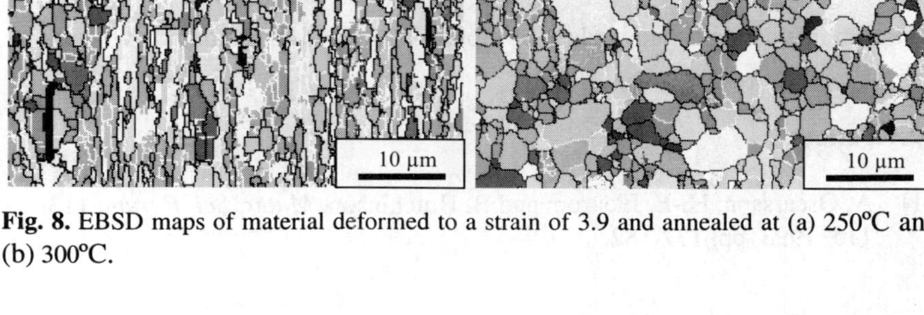

Fig. 8. EBSD maps of material deformed to a strain of 3.9 and annealed at (a) 250°C and (b) 300°C.

The difference in the annealing behaviour of continuously and discontinuously recrystallizing samples is also shown by the change in %HAGB during annealing (Fig. 9). During discontinuous recrystallization, there is a large increase in %HAGB but for samples undergoing continuous recrystallization there is little change in the HAGB content, even at the higher annealing temperatures where grain growth is occurring.

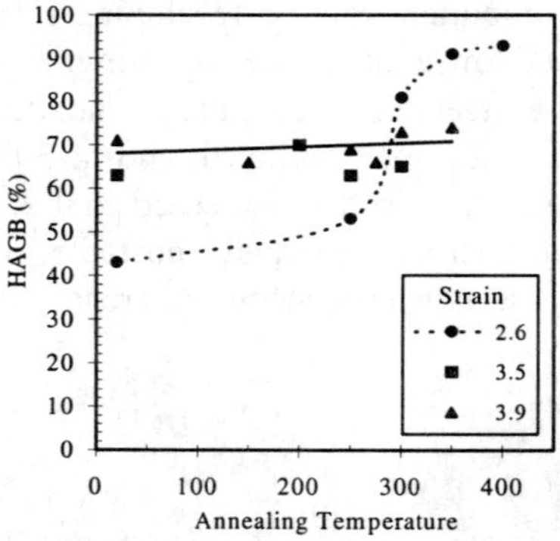

Fig. 9. The effect of strain on the.change in HAGB content during annealing. The heavily deformed materials show little change in %HAGB compared with the low strain material which undergoes discontinuous recrystallization.

4 Conclusions

1. A fine-grained Al-Fe-Mn alloy (AA8006) undergoes a transition between discontinuous and continuous recrystallization after deformation to strains of ~2.9.
2. The transition is denoted by differences in recrystallization kinetics and in the HAGB content of the material.
3. The continuously recrystallized microstructures are unstable with respect to abnormal (discontinuous) grain growth above 350°C.

Acknowledgement

The authors wish to acknowledge the support of Alcan Int. Banbury for the work.

References

[1] A. Oscarsson, H.-E. Ekstrom and B. Hutchinson, *Mater. Sci. Forum*, 113-115, 1993, pp. 177-182.

[2] F.J. Humphreys, P.B. Prangnell, J.R. Bowen, A. Gholinia, and C. Harris, *Phil. Trans. R. Soc., London*, A357, 1999, pp. 1663-1681.
[3] O. Engler and M.-Y. Huh, *Mater. Sci. Eng.*, A271, 1999, pp. 371-381.
[4] R.K. Davies, V. Randle and G.J. Marshall, *Acta Mater.*, 46,1998, pp. 6021–6032.
[5] F.J. Humphreys, *Acta Mater.*, 45,1997, pp. 4231-4240.
[6] F.J. Humphreys, *Acta Mater.*, 45,1997, pp. 5031-5039.
[7] A. Oscarsson, B. Hutchinson, B. Nicol, P. Bate and H.-E. Ekstrom, *Mater. Sci. Forum*, 157-162, 1994, pp. 1271-1276.
[8] F.J. Humphreys and M. Hatherly, Recrystallization and Related Annealing Phenomena, Pergamon Press, 1995.
[9] A. Gholinia, P.B. Prangnell and M.V. Markushev, *Acta Mater.*, 48, 2000, pp. 1115-1130.
[10] Z. Horita, M. Furukawa, M. Nemoto, A.J. Barnes and T.G. Langdon, *Acta Mater.*, 48, 2000, pp. 3633-3640.

Dimensional Effect at the Normal Grain Growth in Nanostructured Aluminium

S. Protasova and V. Sursaeva

Institute of Solid State Physics, Russian Academy of Sciences
Chernogolovka, Moscow district, 142432, Russia

Keywords: nanostructured aluminium films, grain growth, grain boundary, triple junction.

Abstract. Microstructure evolution and grain growth in nanostructured aluminium films has been examined. The films were produced by vacuum evaporation on NaCl (100) substrate. Time dependences of mean grain area are presented. It has been revealed that in the films the normal grain growth takes place. The obtained data on the normal grain growth were compared with those for 2-D aluminium strips and foils. It was concluded that in the temperature range studied the grain growth kinetics is determined by triple junctions dragging force.

1 Introduction

Properties of polycristalline materials are strongly influenced by their structure, namely by mean grain area, grain size distribution and the distribution of grain orientations [1,2]. In turn these latter are determined by grain nucleation and grain growth. Grain growth proceeds by generation and migration of grain boundaries and triple junctions. Grain boundaries and triple junctions are not alike in their properties, especially in their kinetics. To control the microstructure during grain growth it is necessary to understand what determines the motion kinetics of grain boundaries and triple junctions. It is especially of importance in nanostructured materials, where grain boundaries and triple junctions make up a considerable part of the structure. In many studies concerning grain boundary migration and grain growth it was assumed that triple junctions do not have any influence on the motion of the boundaries, and their role in this process was reduced to establish the thermodynamic equilibrium angles at the junction during boundary motion (e.g., see [1]).

Recently we carried out investigations on triple junction motion in Al tricrystals. The geometry of the tricrystals is shown in Fig. 1. The effect of triple junction on grain boundary motion is determined by the dimensionless criterion Λ [3]:

$$\Lambda = \frac{m_{TJ} a}{m_{GB}} = \frac{2\theta}{2\cos\theta - 1}, \tag{1}$$

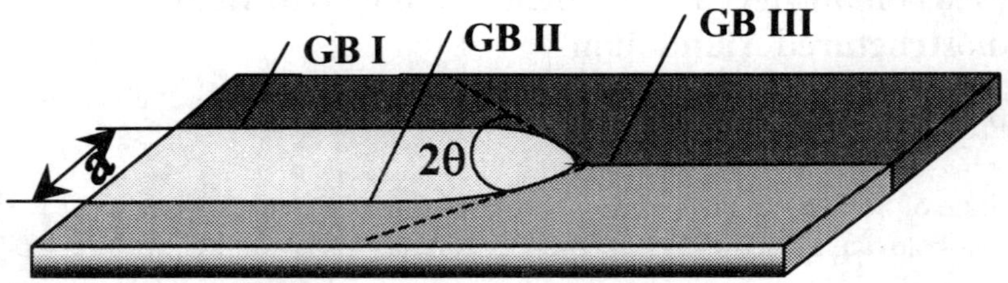

Fig. 1. Scheme of the grain boundary system with a triple junction. GB I, GB II, GB III – tilt grain boundaries with different misorientations; a – width of the middle grain; 2θ – value of the vertex angle.

where m_{TJ} and m_{GB} are the mobilities of triple junction and grain boundaries correspondingly.

One can distinguish two limiting situations:

- Λ tends to infinity. In this case the angle θ tends to the thermodynamically equilibrium value. $\theta=60^{\circ}$. The motion of grain boundary system does not depend on the triple junction mobility and is governed by the mobility of grain boundaries, i.e. there is grain boundary kinetics.
- Λ tends to zero. For this situation the angle θ tends to zero. In other words the motion of the system is governed by triple junction mobility, i.e. there is triple junction kinetics.

There is temperature at which a transition from junction kinetics to grain boundary kinetics is observed. This temperature will be called transition temperature ($T_{transition}$).

It can be seen from (1) that Λ also depends on grain size a: the smaller the grain size the smaller Λ value, and grain growth will be determined by triple junction kinetics. The presented paper concerns the influence of triple junctions on the normal grain growth in nanostructured aluminium films.

2 Experimental

Aluminium 99.999% nanostructured films were produced by vacuum evaporation on NaCl (100) substrate. The film thickness was about 800 Å. In order to obtain non-textured nanocrystalline structure the substrate temperature was maintained at 100°C. The mean grain size R was 200–400 Å. The films were annealed in vacuum furnace at 200–400°C for 1, 4, 8 and 10 minutes; the temperature was maintained with an accuracy of 1°. The annealed films were investigated by transmission electron microscopy (TEM). The dark field TEM technique was used to determine the grain size R and grain area S. In order to obtain reliable data the measurements were made for more than 500 grains.

3 Results

The term normal grain growth is used to describe the steady – state grain growth process with power growth law of the mean grain size, with monomodal and time self – similar grain size distribution and some other features. This kind of grain growth is observed both for 2-D and 3-D systems [1,2].
Results of grain growth study in aluminium films are presented in Figs. 2–5 and Table 1. Fig. 2 shows the time dependence of the mean grain area for annealing at T = 200°C. Relative grain number vs. relative grain size $R/\bar{R}$ for different annealing time is presented on Fig. 3. Fig. 4 shows the grain area distribution vs. relative grain area $S/\bar{S}$ for different annealing time. This dependence corresponds to the equation $f(\xi) = \exp(-\xi)$, where $\xi = S/\bar{S}$ – is the relative grain area. The cumulative frequency is determined as $-\ln(1-\int_0^{\xi} f(\xi)d\xi$ [1]. Fig. 5 is the grain size distributions in aluminium films for various annealing time.
All the results confirm that the process that we observe in aluminium films can be described as normal grain growth:

- Time dependence of mean grain area observed in experiment is linear (Fig. 2) that corresponds to parabolic law.
- Distribution of the relative grain size is self – similar (Fig. 3).
- The grain area distribution is linear (Fig. 4).
- Monomodal grain size distribution is observed (Fig. 5).

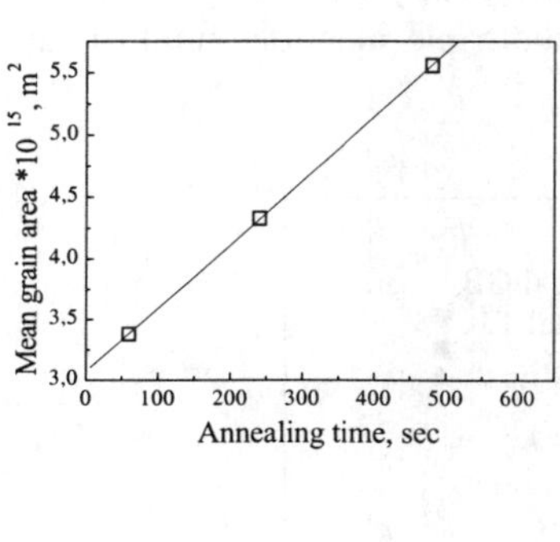

Fig. 2. Time dependence of mean grain area increment at T = 200°C.

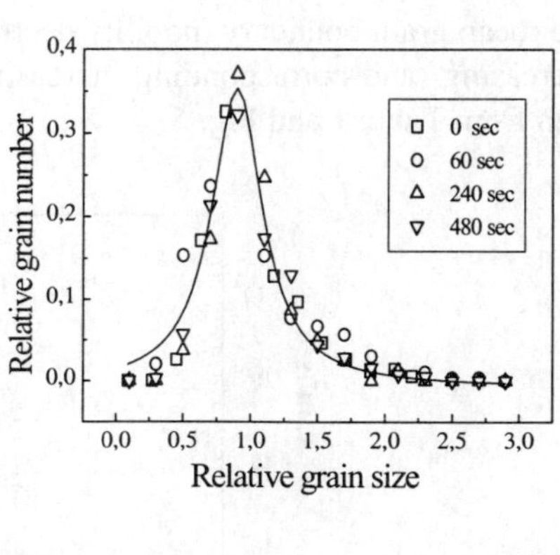

Fig. 3. The grains distribution vs. relative size at T = 200°C.

As it was mentioned above the most of the theoretical studies concerning grain growth supposed that at the stage of normal grain growth the triple junction have no influence on the grain boundary motion. However we received a great amount of experimental data on the triple junctions motion [4,6] and on the 2-D normal

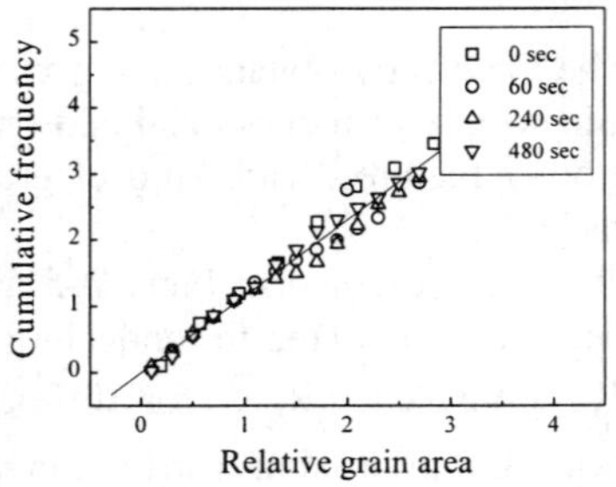

Fig. 4. Grains area distribution (cumulative frequency).

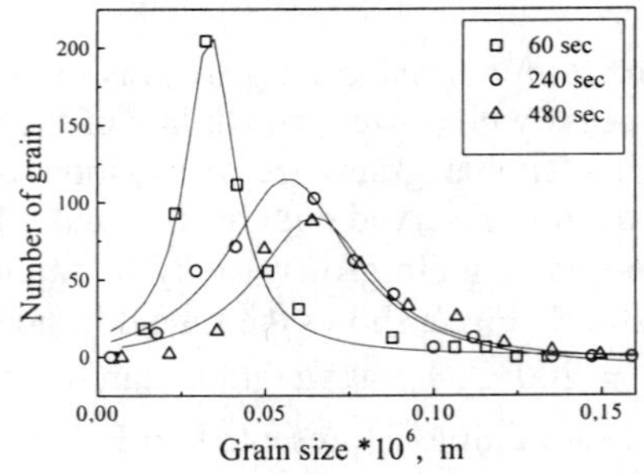

Fig. 5. Monomodal grain size distribution at T = 200°C.

grain growth in polycrystals [4,7]. Results for individual grain boundaries motion were taken from [5]. These data demonstrate that there is a dragging force of triple junctions that influence grain boundary motion. The dimensionless criterion $\lambda = m_{TJ}.a/m_{GB}$ depends on grain size, a, and reflects the drag influence of the triple junction on the migration of the system.

It interesting to compare the obtained data on grain growth in aluminium films with those for 2-D structure in aluminium with different grain size. In order to do it we extrapolated mean grain boundary mobility $\overline{A}$ and mean grain growth velocity $d\overline{S}/dt$ in films for T = 500°C. The values of mean grain growth velocity and mean grain boundary mobility decreases sufficiently with the grain size decreasing (and corresponding increasing of the triple junctions number) as it is seen from Table 1 and Fig. 6.

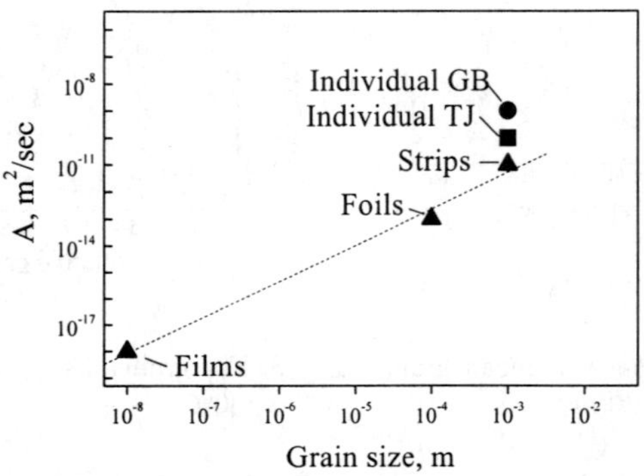

Fig. 6. Mean grain boundary mobility A for the different grain size in aluminium (films, foils, strips) in comparison to the mobility for the individual grain boundaries and triple junctions (see Table 1).

Table 1. Kinetic parameters of grain growth with different grain size (in aluminium films, foils and strips) and grain boundary and triple junction motion at T = 500°C

		Mean grain boundary mobility $\bar{A}$, m^2/s	Mean grain growth velocity dS/dt, m^2/s	Grain size, m
Films		$7.18\cdot10^{-18}$	$1.5\cdot10^{-17}$	10^{-8}
2-D foils		$5.4\cdot10^{-13}$ [7]	$9.2\cdot10^{-13}$ [7]	10^{-4}
2-D strips		$3.0\cdot10^{-11}$ [4]	$7.5\cdot10^{-11}$ [4]	10^{-3}
Individual grain boundaries		$10^{-9}-10^{-7}$ [5]		10^{-3}
Individual triple junctions	TJ kinetic	$10^{-10}-10^{-7}$ [4,6,9]		10^{-3}
	GB kinetic	$10^{-9}-10^{-8}$ [4,6,9]		10^{-3}

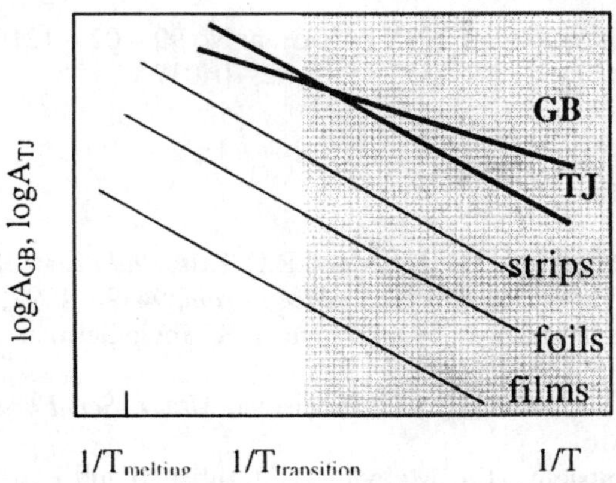

Fig. 7. Sketch of the temperature dependence of the grain boundary mobility A_{GB} and triple junction mobility A_{TJ} for different grain size in aluminium films, foils and strips.

As it is known [6,8] the motion kinetics of the system of grain boundaries with a triple junction depends on its temperature. For the most of the individual triple junctions in aluminium $T_{transition}$ is higher then 500°C. The data obtained for individual triple junctions demonstrate that at $T<T_{transition}$ the most of such systems moves by junction kinetics, thus the triple junctions drag the motion of grain boundary system. Also there are triple junctions that remain immobile at this temperature [9]. It can be supposed that all the results both for aluminium films,

strips and foils were obtained at temperature lower than $T_{transition}$. The temperature $T_{transition}$ is unequal for the structures with different grain size a. For a very small grain size a, e.g. nanostructured material, the $T_{transition}$ is comparatively high (Fig. 7). Fig. 7 shows schematically the temperature dependences of the triple junction mobility, grain boundary mobility and mean grain boundary mobility for aluminium strips, foils and films.
At a given temperature T, as it is seen from Table 1, Figs. 6 and 7, the mean grain boundary velocity is different for different grain size (strips, foils, films). The value of mean grain boundary mobility at a given temperature is distinctly lower for fine grained (films) than for coarse grained (strips) microstructures, and hence the high thermal stability of nanostructured materials is observed [10].

4 Conclusions

1. The normal grain growth was observed in 3-D aluminium films.
2. Triple junctions have a dragging force on the normal grain growth in aluminium foils.

Acknowledgment

This work is supported by INTAS Grant No 99 – 02 – 1216, INTAS Reference No YSC 01 – 4271, RFBR – DFG 99 – 02 – 04017

References

[1] V.E. Fradkov, D.G. Udler and R.E. Kris, *Phil Mag.*, 58, 1998, pp. 277–283.
[2] C.V. Thompson, *Mater. Science Forum*, 94–96, 1992, pp. 245–256.
[3] A.V. Galina, V.E. Fradkov and L.S. Shvindlerman, *Phys. Met. Metall.*, 63, 1987, p. 165.
[4] V.G. Sursaeva and S.G. Protasova, *Mater. Sci. Forum,* 294–296, 1999, pp. 513-516.
[5] G. Gottstein, D.A. Molodov, U. Czubayko and L.S. Shvindlerman, *J. Phys. IV*, C3, 5, 1995, pp. 89-106.
[6] S.G. Protasova, G. Gottstein, D.A. Molodov, V.G. Sursaeva, and L.S. Shvindlerman, *Acta Materialia*, 2001, in print.
[7] V. Sursaeva, U. Czubayko and A. Tuflin, *Texture and Microstructure*, 32, 1999, pp. 187–195.
[8] V.G. Sursaeva, U. Czubyko, G. Gottstein and L.S. Shvindlerman, *Mater. Sci. Forum*, 294–296, 1998, pp. 517–520.
[9] S.G. Protasova and V.G. Sursaeva, *Defect & Diff. Forum*, 2001, submitted.
[10] G. Gottstein and L.S. Shvindlerman, „Grain Boundary Migration in Metals: Thermodinamic, Kinetics, Applications“, CRC Press, 1999.

Recrystallization and Grain Growth
Proceedings of the First Joint International Conference
Eds. G. Gottstein and D.A. Molodov, Springer-Verlag, 2001

Formation of Micro-, Submicro- and Nanocrystalline Structures in Nickel-Base Alloys

V. A. Valitov, O. A. Kaibyshev, Sh. Kh. Mukhtarov, N. R. Gajnutdinova

Institute for Metals Superplasticity Problems, 39 Khalturina Street, Ufa 450001, Russia

Keywords: grain boundary, submicrocrystalline, nanocrystalline, superalloy.

Abstract. Regularities of structure formation in nickel base alloys subjected to various types of strengthening in a wide temperature-strain rate interval of deformation have been studied systematically. The character of operating mechanisms of plastic deformation and recrystallization and their interaction with processes of second phase dissolution and precipitation have been analyzed. It has been established that a microduplex type microcrystalline structure is resulted from transformation of a subgranular structure to a granular type and concurrent transformation of partially coherent interphase boundaries to non-coherent boundaries. It has been shown that during severe plastic deformation the process of structure formation is a step by step character.

1 Introduction

There is presently substantial research in materials with submicro- (SMC) and nanocrystalline (NC) structures due to their unique physical and mechanical properties [1, 2]. Grain refinement up to a submicron size promotes essential expansion of the temperature – strain rate range of superplasticity and decreasing the flow stress [3, 4]. However, the features of MC, SMC and NC structures formation in complex nickel base alloys are not yet clear. The goal of the present paper is to establish processing regimes for generating MC, SMC and NC states in nickel base alloys with different chemical composition and constituent phases.

2 Experimental Procedure

The chemical compositions of the nickel base alloys that were studied are shown in Table 1. The materials selected for this study were considered to be typical examples of precipitation hardened nickel base alloys with a range of strengthening phases: Inconel 718 -γ''+δ, and EP962 - γ'. PDS alloy - dispersion-strengthened with phases γ'+Y_2O_3. Samples of 718 and EP962 alloys were machined from 200 mm diameter billets. The powder PDS alloy was produced by mechanical alloying and was in the form of a 22 mm diameter rod.

Table 1. Compositions of the alloys that were studied.

	Concentrations (wt. %)										
Alloy	Cr	Co	W	Mo	Fe	Al	Ti	Nb	C	Ni	Other
718	19	-	-	3.1	18	0.5	1.0	5.1	0.05	Base	
EP962	13.1	10.7	2.8	4.6	0.6	3.2	2.6	3.4	0.1	Base	0.01 B
PDS	10.1	-	7.5	-	-	6.8	-	-	0.1	Base	1.2 Y_2O_3

Compression test was conducted over a wide range of temperatures (600-1100°C) and strain rates ($1.33\cdot10^{-4}$-$10^{-2}s^{-1}$). Cylindrical samples, 10 mm in diameter and 15 mm in height, were compressed in a SCHENK-PSA testing machine. Larger samples (40-80 mm in diameter and 10-12 mm tall), with SMC structures, were generated on 100 and 630 ton presses using processing conditions derived from the results of the smaller samples. Transmission electron microscope (TEM) investigations were performed using a JEM-2000EX.
For attainment of high strains required for generating NC structures the method based on shear deformation under high pressure on the Bridgeman anvil was used. The dimensions of an initial sample were 4 mm in diameter and 0.7 mm thick. The true strain value e for points of a sample located on the radius r is determined as shear strain [5]:

$$e = 2\pi rN/\sqrt{3}\,L \tag{1}$$

where N – the number of anvil rotations, L – sample thickness, mm.

3 Results

3.1 Initial state

The microstructure of hot-deformed precipitation-hardened alloys 718 and EP962 in the initial state is completely recrystallized coarse-grained with a mean grain size of 40-80 μm. The TEM studies have shown that within the alloy 718 grains there are coherent disk-type γ''-precipitates, which are uniformly distributed. The diameter of γ''-phase disks is about 60 nm, and their thickness is 20 nm. The γ''-phase precipitates are also observed on grain boundaries. In the EP962, coherent precipitates of γ'-phase, 0.1 μm in size, are observed within the grains. Grain boundaries precipitates of γ'-phase are partly coherent and larger by 2-3 times.
The powder dispersion-strengthened nickel base alloy, PDS, was produced by mechanical alloying. This generated an SMC structure with an initial grain size of 0.8 μm. The SMC microstructure is characterized by high density of dislocations both within the grains and along grain boundaries. Dispersion particles of yttrium phase, 0.025μm in size, are uniformly distributed within grains.

3.2 MC, SMC and NC formation

In alloys 718 and EP962, the microduplex type MC structures were formed using high strain thermo-mechanical processing. During processing intense dynamic recrystallization was accompanied by significant changes in morphology of second phase precipitates and their distribution in the matrix. Analysis of the evolution of the microstructure during hot deformation allowed establishment of the bases of transformation of the coarse-grained structure to the MC structure [6]. Moreover, some features of the constituent phase influence were revealed.

The scheme of microcrystalline structure formation for the alloy 718 is shown in Fig. 1a. At the initial stage of deformation at 0.76-0.7T_{ml} one can observe almost complete dissolution of dispersed γ''-phase particles. Development of intragranular dislocation slip increases the density of dislocations and leads to formation of dislocation walls rested against δ-phase plates. Occurrence of local overstresses at the γ/δ interface leads to formation of transverse dislocation boundaries inside of particles. With increasing strain value one can observe the formation of subgranular structure within the matrix and the increase of misorientation between subgrains to the high angle one first of all in the vicinity of initial high angle grain boundaries. Intensification of diffusion processes and grain boundary sliding at interphase and intergranular boundaries leads to fragmentation of plates and chages orientation ratio between δ-phase precipitates and matrix subgrains. The increase in the strain value up to 50-75% leads to transformation of subgrains to grains with high angle boundaries and transformation of δ-phase fragments to globular type particle-grains with arbitrary orientation in respect to matrix grains.

In the EP962 alloy the character of microstructure transformations during hot is slightly different. In the initial stage of hot deformation the intragranular dislocation slip leads to formation of subgranular structure with a subgrain size comparable with an intraparticle distance. Further increase in the strain value results in transformation of partially coherent γ/γ' interphase boundaries to non-coherent common type boundaries and transformation of γ-phase subgrains to grains with high angle boundaries. The grains of γ- and γ'-phases become comparable in size, while interphase and intergranular boundaries have mainly banded contrast intrinsic to high angle grain boundaries.

High strain thermo-mechanical processing was used for formation of the SMC and NC structures in the alloys under study. This process includes isothermal deformation with intermediate annealing and a gradual step by step reduction in the temperature of processing. Although there are significant differences in the chemical compositions and constituent phases of the alloys studied the formation of the SMC and NC structures possess certain similarities.

For formation of the SMC and NC structures in the alloys under study the strain thermo-mechanical processing was conducted which includes isothermal deformation with intermediate annealing and a gradual step by step decrease in the temperature of processing. In spite of significant distinctions in chemical composition and constituent phases of the alloys under study the character of the SMC and NC structure formation in these alloys is analogous in many ways.

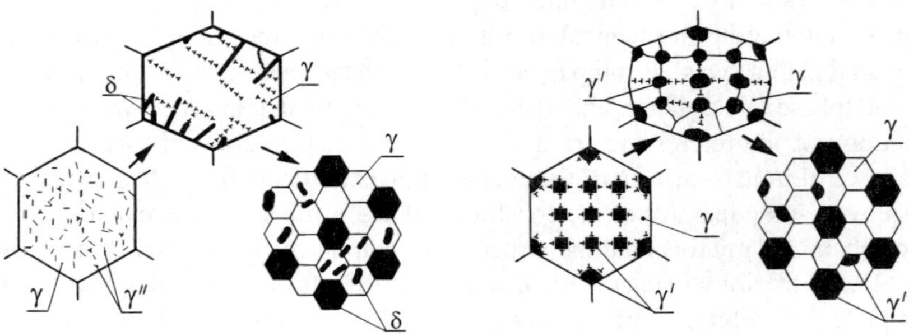

Fig. 1. Formation of MC structure in alloys: a – alloy 718; b - EP962.

High-strain thermo-mechanical processing of alloy 718 and EP962 in the temperature regimes 0.6-0.78T_{ml} generated refinement of the MC structure to a grain size of 0.5-0.3 μm. In both alloy 718 and EP962, some coarse second phase particles (up to 1 μm) were still present. This suggests localization of the deformation in the softer matrix, which was probably due to insufficient accumulated strain. The noteworthy feature of the SMC structure of the alloys under study is the non-equilibrium nature of the structure as described by the high dislocation density both within the grains and along the grain boundaries. Grain boundary contrast typical of high-angle grain boundaries was not observed. Post-deformation annealing contributes to formation of the SMC structure with equilibrium grain boundaries [4]; however, elastic stresses have not been completely removed.

In order to achieve additional grain refinement, thermo-mechanical processing of the PDS alloy was performed at temperatures of 0.71-0.67T_{ml}. This processing refined the initial γ and γ′-phase grains to a size of ~0.5 μm.

Additional deformation at much lower temperatures of 0.51-0.64T_{ml} generated NC structure in precipitation hardened alloys with grain sizes of 0.02-0.05 μm. TEM analysis of images of alloys in the NC state revealed that the microstructures are similar to those observed in other NC materials and characterized by a high level of stresses that testifies noticeable azimuth blurring of reflexes.

The most significant microstructure refinement towards the sizes close to NC ones has been obtained as a result of severe plastic deformation at room temperature using the combined scheme (upsetting with torsion on the Bridgeman anvil).

The results obtained during microstructure and microhardness (Fig. 2, 3) investigations after severe plastic deformation made it possible to establish that the formation of the NC structure during severe plastic deformation occurs in several stages that is in conformity with the results of investigations obtained on other materials [7]. At the same time there are some peculiar features that are probably due to the nature of this alloy and structural parameters. At the early stage of deformation a cellular structure is formed.

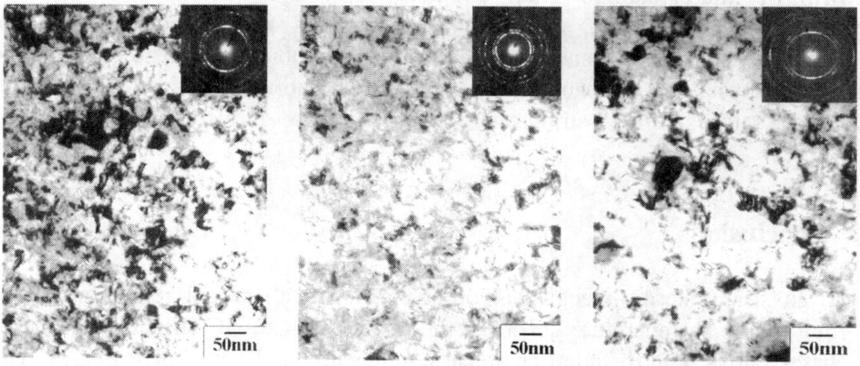

Fig. 2. Microstructure of alloys after severe plastic deformation: a – alloy 718; b-EP962; c-PDS.

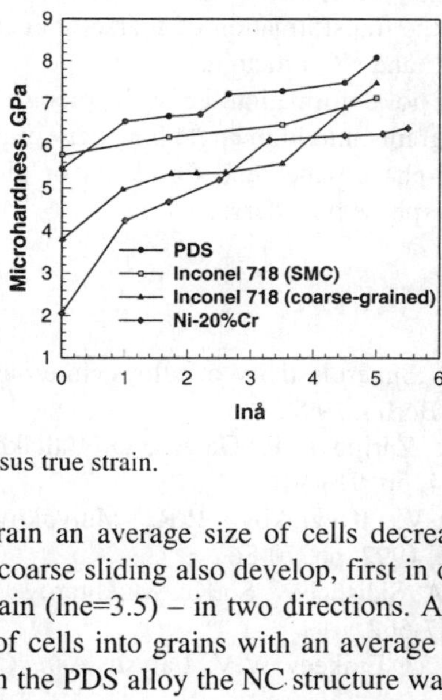

Fig. 3. Microhardness versus true strain.

With increasing the strain an average size of cells decreases for all alloys. For alloy 718 the bands of coarse sliding also develop, first in one direction (lne=2.5), and with increasing strain (lne=3.5) – in two directions. At strain e=4.5-5.5 there occurs transformation of cells into grains with an average size of 20-50 nm (Fig. 2a). At the same time in the PDS alloy the NC structure was observed at an earlier stage (lne=3.5) than in the alloy 718 (lne=4.5). The analogous character of microstructure and microhardness changes was revealed in the Ni-20%Cr alloy, which is the base of nickel alloys. However, the microhardness level of the Ni-20%Cr alloy was lower than that in complex nickel base alloys.

These results show that in nickel base alloys with different chemical compositions and constituent phases the microstructure can be refined to a scale of several ten nanometers by high strain thermo-mechanical processing. During that dynamic recrystallization was accompanied by dissolution, precipitation and coagulation of second phases. The structure refinement up to the SMC and NC scale leads to a growth of intercrystalline grain boundary area and qualitative changes in microstructure. The SMC and NC microstructures are characterized by the significant elastic stresses and non-equilibrium grain boundaries. The later and the large intercrystalline grain boundary area are, probably, main factors providing low temperature and high strain rate SP in nickel base alloys with SMC and NC structures.

4 Conclusions

1. High strain thermo-mechanical processing with a gradual reduction in the deformation temperature generates the MC, SMC and NC structures in nickel bases alloys with different chemical composition and constituent phases. The NC structure can be formed also during severe plastic deformation at room temperature.
2. During severe plastic deformation the mutual influence of dynamic recrystallization and precipitation of second phases is observed. These processes promote the transformation of coarse-grained structure in to the MC and further the SMC and NC structures.
3. In alloy 718 during hot deformation the MC structure is evolved as a result of transformation subgrains into high angle boundary grains, accompanied by the fragmentation of δ-phase plates and formation of globular δ-phase particles with high angle interphase boundaries.

References

[1] O.A. Kaibyshev, „Superplasticity of alloys, Intermetallides and Ceramics“, Springer-Verlag, Berlin, 1992.
[2] G. Salishchev, R. Zaripova, R. Galeev, O. Valiakhmetov, *Nanostructured Materials*, 6, 1994, pp. 913-916.
[3] R.Z. Valiev, A.V. Korznikov, R.R. Mulyukov, *Phizika metallov i metallovedenie,* 6, 1992, pp. 70-86.
[4] V.A. Valitov, G.A. Salishchev, Sh.Kh. Mukhtarov, *Mater. Sci. Forum*, 243-245, 1997, pp. 557-562.
[5] F.Z. Utyashev, F.U. Enikeev, V.V. Latysh, *Ann. Chim. Fr.,* 21, 1996, pp. 379-389.
[6] O.A. Kaibyshev, V.A. Valitov, G.A. Salishchev, *Phisika metallov i metallovedenie,* 75, 1993, pp. 110-117.
[7] Yu.V. Ivanisenko, A.V. Korznikov, I.M. Safarov, R.Z. Valiev, *Nanostructured Materials,* 6, 1995, pp. 433-436.

The Formation of Uniform Fine-Grain Microsructures in Two-phase Ti-6Al-4V Titanium Alloy during Hot Deformation

S.V. Zherebtsov[1], G.A. Salishchev[1], R.M. Galeyev[1], O.R. Valiakhmetov[1], M.F.X. Gigliotti[2], B.P. Bewlay[2], C.U. Hardwicke[2] and R.S. Gilmore[2]

[1] Institute for Metals Superplasticity Problems, Russian Academy of Sciences Khalturina str. 39, Ufa 450001, Russia

[2] GE Corporate Research and Development, Schenectady, NY. USA 12301

Keywords: uniform fine-grain structure, titanium alloy, continous recrystallization, discontinous recrystallization, globularization.

Abstract. The effects of the initial microstructure and thermomechanical conditions of deformation on the formation of fine-grain microstructure in alloy Ti-6Al-4V were examined. Continuous dynamic recrystallization (DR) takes place over a wide temperature range within the ($\alpha+\beta$)-region. The conversion of α/β phase-boundaries from a semi-coherent state into an incoherent one controls the transformation of lamellar microstructure into a globular one. Production of a uniform fine-grain microstructure with minimal microtexture is possible using the correct combination of ($\alpha+\beta$)-deformation conditions.

1 Introduction

The $\alpha+\beta$ titanium alloys with uniform fine-grain microstructure have high technological and service properties. This increases both the efficiency of their hot working and the reliability and service life of articles out of them. Moreover, the uniform fine-grain microstructure (a grain size of 10-20 μm) in billets is of a special importance for ultrasonic testing. Ultrasonic inspectability is a function of billet microstructure: billets containing acoustic entities with sizes comparable to the wavelength of the ultrasound exhibit high noise [1]. This noise may hide or mask critical flaws in material that may limit applications of the titanium alloy. Traditional processing for formation of a fine-grain microstructure in $\alpha+\beta$ titanium alloys comprises deformation in both the β-region and the ($\alpha+\beta$)-region. High rates of recovery at deformation in β-region retard recrystallization and formation of a uniform recrystallized structure with the finest β-grain size. Note that formation of microstructure in β-region occurs due to discontinuous and/or continuous recrystallization. The later is not favourable since in this case there is no significant misorientation between β-grains. During following ($\alpha+\beta$)-deformation this fact can lead to creation of areas of microstructure with microtexture, non-uniform distribution of phase grains and their different shapes. This microstructure, even though it may possess small grain size, provides considerable noise levels during ultrasonic inspection, which of course is undesirable. The aim of the present work is to investigate the influence of initial microstructure and temperature–strain rate conditions of ($\alpha+\beta$)-deformation on formation of uniform fine-grain microstructure in the Ti-6Al-4V.

2 Experimental Procedure

The forged rod of 230 mm diameter out of Ti64 (wt%: 6.5Al; 5.1V; 0.1Fe; 0.03Si; 0.02C; 0.01N; balance Ti) was used in the as-received condition. The beta transus temperature of this material was 995°C. Section of the Ti64 rod was forged at the temperatures of β-phase field to yield the microstructure with 250 μm β-grain size. Samples cut from this section were heat treated in the β-region to produce the microstructures with different β-grain size and thickness of α-plates. Samples with a 800 and 2000 μm grain sizes and approximately similar thickness of α-plates were obtained via a heat treatment of 1050°C for 1 hr and 1200°C for 2 hrs, respectively. Samples with β-grain size of 300 μm and a mean thickness of α-plates ~1.5 and ~0.6 μm were obtained through a heat treatment, respectively, of 1010°C/5 min followed air cooling (condition A) and 1010°C/5 min + water quenching + annealing at 650°C for 0.5 hr (condition B).

Compression tests were conducted on Ti64 samples to determine the effect of microstructure and strain rate-temperature conditions on flow behavior and structure evolution during deformation in $\alpha+\beta$ phase field. Cylinders, 10 mm diameter 15 mm height, machined from the heat treated samples were upset at range of temperature 650-970°C and strain rate 10^{-4}...10^{-1} s^{-1}.

The isothermal processing to achieve the uniform fine-grain microstructure was conducted due to multi-step deformation of section taken from the as-received Ti64 rod. Multiple deformation series included a sequence of compression and drawing operations with multiple rotations of the billet. Besides, for comparison the conventional fine-grain billet was produced from the section of Ti64 rod due to traditional technique of processing at the same temperatures of heating.

3 Results

Investigation of the Ti64 rod microstructure in the as-received condition shows the difference in grain size and the presence of significantly large areas with almost similar crystallographic orientation of grains that is revealed due to their similar etching (Fig. 1). The mean grain size in various areas ranges from 650 to 1000 μm, and the size of areas of low misorientated grains is about 3-8 mm.

The σ-ε curve for two conditions (A and B) of Ti64, obtained at temperatures 970, 900 and 650°C, are shown in Fig. 2. The σ-ε curves are characterized by the presence of flow stress maximum and subsequent softening. With decreasing temperature of deformation the flow stress increases significantly. The initial condition of Ti64 slightly influences on flow stress. Flow stress in the alloy with thinner α-plates (Condition B) is a bit lower than in Condition A and differences increase with decreasing temperature of deformation.

Fig. 1. Macrostructure of Ti64 rod in as-received condition (the arrow shows area of low misorientated β-grains).

The measurements of the strain rate sensitivity of the flow stress (m) were conducted by means of strain rate change tests at height reduction 10 and 50%. The 'm' value does not depend on the alloy condition and its value rises with increasing strain. The maximum m values were observed at temperatures 970 and 900°C. For example, in Condition A at 900°C and the strain rate 10^{-4} s^{-1}, m value increased from m(10%) = 0.12 to m(50%) = 0.33. This testifies that Ti64 with a lamellar structure in definite strain-strain rate regimes of deformation displays features of superplasticity (SP).

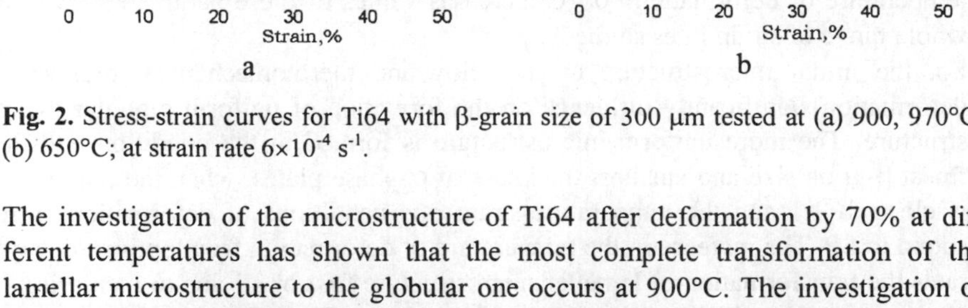

Fig. 2. Stress-strain curves for Ti64 with β-grain size of 300 μm tested at (a) 900, 970°C; (b) 650°C; at strain rate 6×10^{-4} s^{-1}.

The investigation of the microstructure of Ti64 after deformation by 70% at different temperatures has shown that the most complete transformation of the lamellar microstructure to the globular one occurs at 900°C. The investigation of microstructure evolution at this temperature of deformation has shown that in the initial stage of deformation transverse subgrains are formed in plates of phases. With increasing strain value the misorientation of subgrains in α-plates increases that results in formation of grooves on interphase surface and fission of plates. (Fig. 3a). Furthermore, the particles forms are spheroidized that leads to formation of structure consisting of globular particles of α and β phases. The process of formation of transverse intergranular boundaries, which in the process of deformation transforms the low angle misorientation to the high angle one is similar to DR in terms of its continuous mechanism. Evidently, in the process of deformation the interphase boundaries change their structure and become non-coherent that also contributes to creation of grooves and division of plates. The transformation of the semi-coherent interphase boundary to the non-coherent one can be observed during deformation at 650°C, at which β-phase remains after cooling. The transformation of the interphase boundary is seen from the change in their contrast (Fig. 3b).

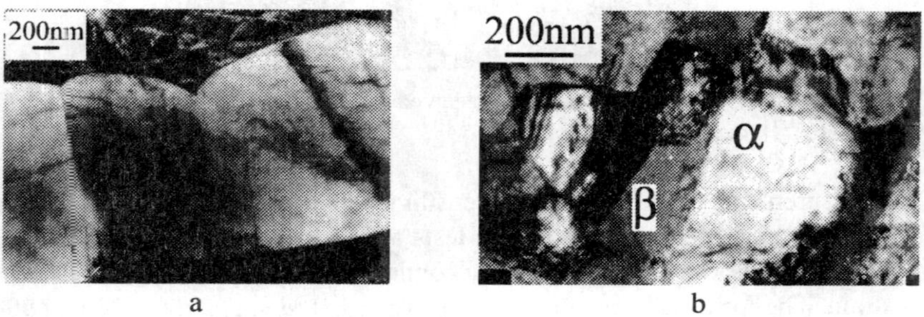

Fig. 3. (a) TEM image of α -plate after deformation of Ti64 at 900°C, ε=30%, 6×10^{-4} s^{-1}; (b) structure of Ti64 after deformation at 650°C, ε=50%, 6×10^{-4} s^{-1}.

Fig. 4a shows the influence of the strain rate on parameters of microstructure forming in the alloy with a β-grain size of 300 μm at ε = 70% and deformation temperature 900°C. With decreasing strain rate to 10^{-4} s^{-1} a value of thickness/length ratio of α-plates (aspect ratio B/L) rises and volume fraction of equiaxed α-зерен (V_α) increases. Values of these parameters in condition B are higher than in condition A. The β-grain size exerts a significant influence on parameters B/L and V_α (Fig. 4b). The decrease in the β-grain size from 2000 to 800 μm at the temperature of deformation 900°C increases values of these parameters over the whole range of strain rates studied.

So, the initial microstructure of the alloy and thermomechanical regimes of deformation significantly influence on the formation of uniform globular microstructure. The more uniform microstructure is formed in the condition with the finest β-grain size and smallest thickness of α-phase plates, when the phase ratio is close to 1:1 and the temperature-strain rate conditions of deformation correspond to SP. The increase in the temperature of deformation from optimal deteriorates the transformation of lamellar microstructure first of all due to increasing β-phase amount in which recovery occurs intensively. The influence of the decrease in the deformation temperature is twofold. On the one hand, the amount of alloying elements in β-phase is increased that facilitates the accumulation of dislocation, loss of coherence of interphase boundaries because of an essential change of lattice parameters of phases and transformation of plates. On the other hand, diffusion processes slow down that retards globulization.

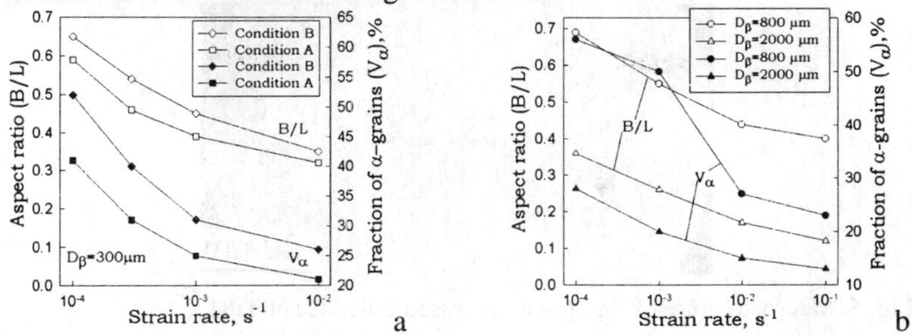

Fig. 4. Strain-rate dependence of aspect ratio (thickness/length of α-plates) and volume fraction of equiaxed α-grains in Ti64 tested at 900°C, ε=70%.

The favourable influence of the decrease in the β-grain size and the thickness of plates is connected with increasing deformation uniformity. This leads to more uniform occurrence of continuous DR in phase plates and transformation of initial semi-coherent interphase boundaries to non-coherent ones. The later fact is determining for transformation of the lamellar structure to the globular one [2]. Fission of plates into separate fragments and their following spheroidization intensify the development of grain boundary sliding (GBS). As a result, the increased strain rate sensitivity of flow stress is observed. It was shown [3] that uniform globular microstructure is formed under deformation conditions close to temperature-strain rate conditions of SP. The operation of GBS leads to uniform redistribution of phase grains and provides texture-free condition of material.

In case of increasing β-grain size and, consequently, increasing deformation non-uniformity, the formation of globular grains in areas of deformation concentration will retard transformation of other areas of microstructure. Under conditions of isothermal deformation, i.e. low strain rates, such areas show less resistance to deformation than areas with non-globular microstructure. For removing this effect it is necessary to change conditions of deformation to suppress localized flow within the material interior with globular microstructure by decreasing temperature or increasing strain rate. In this case deformation characteristics of globularized and non-globularized areas become more similar. Moreover, deformation characteristics of phase also become similar especially as temperature of deformation decreases. As a result, non-globularized areas or soft phase are subjected to hardening and the process of structure transformation within the sample interior occurs more completely. Thus, for formation of uniform microstructure in a billet the versions of processing taking into account the mentioned features of transformation can be proposed. One of such versions is isothermal deformation in temperature-strain rate regimes of SP. For providing uniform deformation within the whole billet, a multilateral deformation of the billet via its multiple rotation in respect to deformation load can be used. In case of such scheme of loading, because of the non-uniform strain rate distribution within the billet interior the non-globularized areas are subjected to deformation at each rotation and, as a result, the uniform microstructure is formed.

Fig. 5. Macrostructure of fine-grain processed billet out of Ti64.

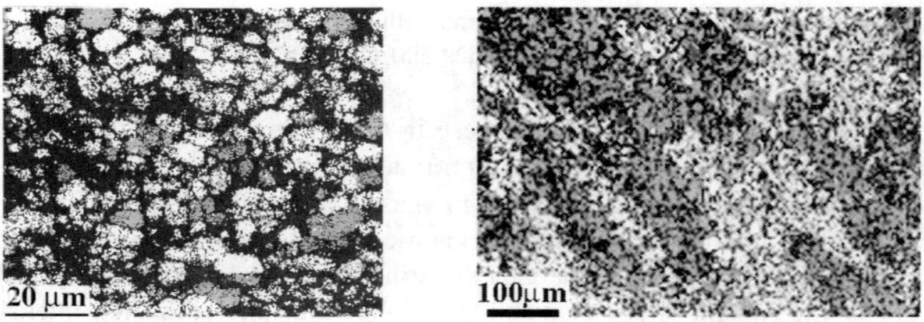

a b

Fig. 6. EBSD images of (a) uniform fine-grain Ti64 billet and (b) conventional forged billet

Fig. 5 shows the macrostructure of the billet after isothermal multiple step deformation in (α+β)-field at 900-950°C. The macrostructure is uniform and has dull surface. The microstructure of the billet characterizes the electron backscatter diffraction (EBSD) pattern (Fig. 6a). The EBSD image, in which the [0001] pole inclination was represented in colors, soundly shows crystallographic microtexture of a scanned microstructure. Colors close to one another represented inclinations that are similar. The microstructure of fine-grain billet is characterized by high grain size uniformity and contains α-particles, which are not in colonies. The orientation of these α-particles is random. The conventional billet microstructure contains α-particles arranged in large colonies of similar crystallographic orientation (Fig. 6b).

References

[1] B. Ginty, P. Hallam, C. Hammond, G. Jackson and C. Robb, in "Titanium'80, Science and Technology", eds. H. Kimura and O. Izumi, TMS/AIME, Warrandale PA, 1980, pp. 2095-2103.

[2] M.I. Mazurski and G.A. Salishchev, *Phys. Stat. Sol. (b)*, 188, 1995, pp. 653-658.

[3] G.A. Salishchev and R.M. Galeev, in "Proc. th Int. Conf. Recrystallization '90", ed. T. Chandra, MM&MS, 1990, pp. 603-608.

Advanced Experimental Methods

Recrystallization and Grain Growth
Proceedings of the First Joint International Conference
Eds. G. Gottstein and D.A. Molodov, Springer-Verlag, 2001

Application of Laser-Ultrasonics (LUS) to Studies of Recrystallisation in Metals

E. Lindh-Ulmgren, D. Artymowicz and B. Hutchinson

Swedish Institute for Metals Research, Drottning Kristinas väg 48, S-11428 Stockholm, Sweden

Keywords: recrystallisation, laser, ultrasonic waves, kinetics.

Abstract. The paper describes how measurements of ultrasonic wave propagation can be used to obtain information about microstructures and accordingly the state of recrystallisation in metals. The method also offers the possibility of making dynamic observations during annealing. Examples of application of the technique to recrystallisation of copper and stainless steels are presented.

1 Introduction

Although recrystallisation is one of the most important processes occurring in metals from both scientific and practical viewpoints, the study of it is frequently hampered by the difficulty of dynamic observations. Generally, it is necessary to section a material and prepare it for the microscope before examining its condition of recrystallisation. This difficulty may be one of reason why so much dispute has characterised the field. Furthermore, the lack of a non-intrusive technique capable of detecting the condition of recrystallisation has so far prevented on-line control of microstructure and properties for quality control in the metallurgical industry. In fact, a number of non-destructive methods do exist to quantify recrystallisation but they all suffer from limitations that restrict their general application, Table 1.

Table 1. Available techniques for dynamic studies of recrystallisation

Technique	Characteristics and limitations
Calorimetry e.g. [1]	Only suitable for very pure metals, limited range of heating rates, gives no microstructural information
Hot stage optical or scanning microscopy	Surface observations only – may not represent bulk behaviour, limited heating rates
Hot stage TEM e.g. [2]	Specimen thickness necessitates HVEM, allows detailed observation but poor statistics, limited heating rates
Hot stage x-ray Diffraction	Measures thin surface layer, limited heating rates
Hot stage neutron Diffraction e.g. [3]	Allows bulk observations, requires nuclear reactor with special instrumentation, limited heating rates
Synchrotron x-ray Diffraction [4]	Allows observations in bulk, statistical information (?), requires synchrotron and special instrumentation

It is evident that all of these techniques have limitations that prevent them from being used in many laboratory applications and certainly none of them is suitable to use in an industrial environment. The present paper describes an alternative and relatively new method, which offers much more opportunity for applications, based on measurements of ultrasonic waves.

2 Principles of the Laser-Ultrasonic Technique (LUS)

Laser-ultrasonics have been developed in recent years for non-destructive testing and characterisation of materials [5]. A short pulse (~20 ns) of infrared laser light is shot at the surface of the material where it generates an ultrasonic signal due to ablation and/or thermoelastic effects. The arrival of the waves is detected at some other position using a sensitive interferometer based also on laser principles. A schematic view of such an experimental set-up is given in Fig. 1. Many different types of ultrasonic waves are generated simultaneously by the laser pulse, e.g. Rayleigh surface waves, compression (P) , shear (S) and guided (Lamb) waves, as well as a vast range of frequencies. By choice of geometry and interferometer it is possible to select the waves most suitable for the purpose of structure analysis.
Two principally different types of measurements are possible. Velocity of the waves is governed by the bulk elastic properties of the material and hence by crystallographic texture. When recrystallisation occurs, the accompanying change in texture is revealed through a change in the wave velocity. Secondly, the attenuation of ultrasonic waves is a strong function of grain size in a metal so that changes in grain structure are revealed by the decay of the signal. Examples of these effects in copper and steel have been presented previously, e.g. [6,7].

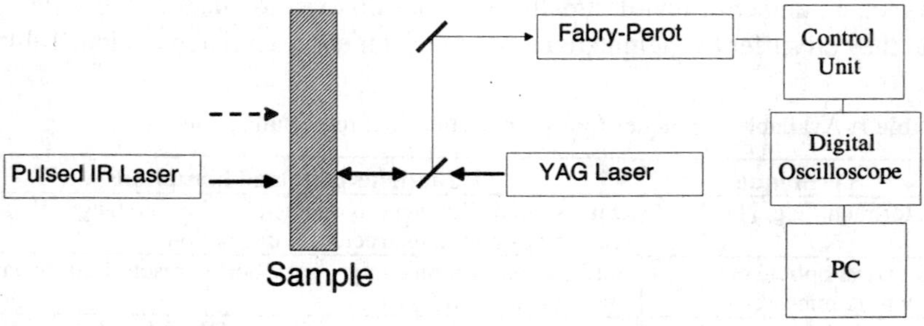

Fig. 1. Schematic diagram showing how LUS measurements can be made.

3 Measurements and Results

An 11 mm thick plate of hot rolled copper was cold rolled to 3.3 mm (70% reduction) and part of this was then recrystallised by annealing for 10 minutes at 600°C. Measurements were made with the two laser beams aligned on opposite sides of the samples (epicentric condition) using a Fabry-Pérot interferometer having a central frequency of 18 MHz. Spectra for the two material conditions are compared in Fig. 2. As can be seen, there are a large number of signals from which information can potentially be gained and there are evident differences depending on the condition of the copper. The main signals in the present set-up are the arrivals of the compression wave as it passes 1,3 5,... times through the plate with reflections at both the free surfaces. The velocity of the compression wave travelling normal to the plate can be determined with high accuracy from successive arrival times. For the cold rolled condition, this velocity was 4710 m/s whereas it changed to 4845 m/s after recrystallisation. The velocity can therefore be used as a fingerprint of the microstructural state although rigorous analyses based on ODF data are also possible [8]. Another type of measurement concerns the attenuation of the ultrasonic signal as the wave passes through the material. The decay of the successive compression wave arrivals is much faster in the annealed than in the cold rolled plate. The fine starting grain size of the hot rolled plate combined with the effect of the cold rolling results in a rather homogeneous material with relatively little scattering of the waves. However, recrystallisation produces a coarser grain size which increases the attenuation level. This parameter can, therefore, be another means of following the microstructural transformation as the material recrystallises. Yet another parameter is the strength of the shear wave (S1) which is sensitive to the anisotropy.

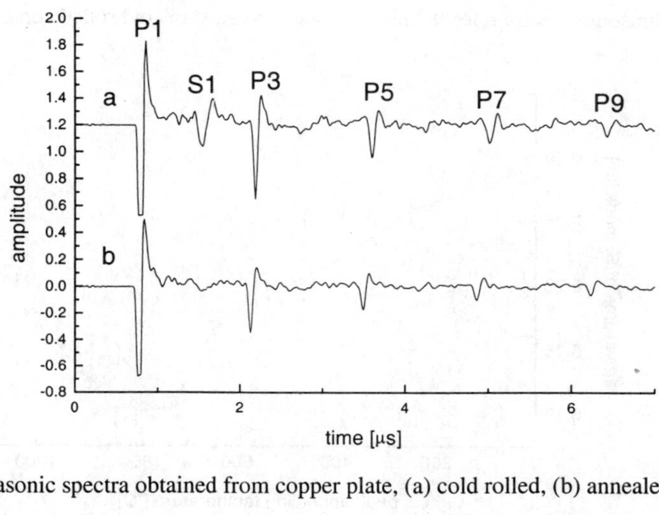

Fig. 2. Ultrasonic spectra obtained from copper plate, (a) cold rolled, (b) annealed.

Further experiments have been carried out on 0.3 mm thick cold rolled 301 grade stainless steel sheet. In this case, the pulsed IR laser beam was translated sidewards 30 mm to examine the propagation of Lamb waves. Strips of the steel 30 mm wide were annealed by direct electric resistance heating at approximately 80°C/s to various temperatures and then immediately cooled before making measurements at room temperature. In this case, the velocity of the S_0 Lamb wave was selected as shown in the spectrum in Fig. 3. The variation of wave velocity is plotted in Fig. 4 as a function of the peak temperature during the anneal. Microstructural observations were made on electropolished samples in the SEM using back-scattered electron channelling contrast and examples of these for four different temperatures are shown in Fig. 5.

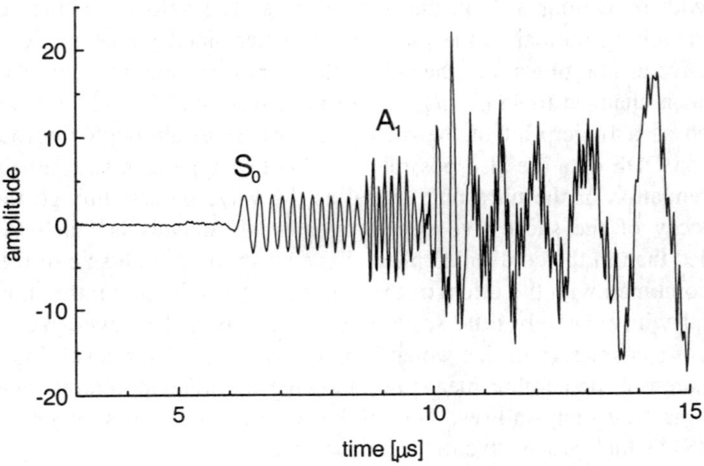

Fig. 3. Ultrasonic spectrum for 0.3 mm 301 stainless steel in cold rolled condition.

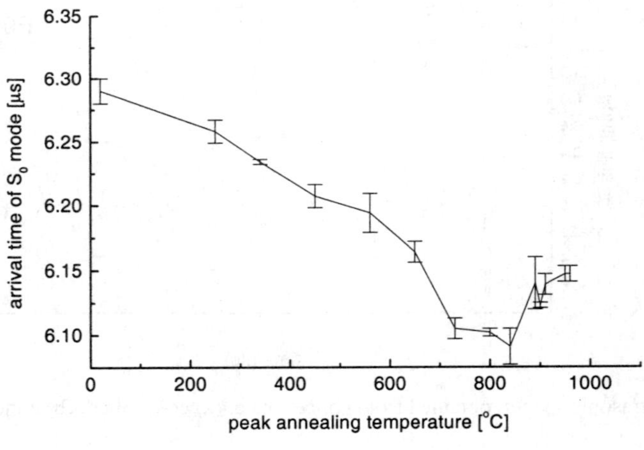

Fig. 4. Variation of Lamb wave velocity with peak temperature for annealed 301 stainless steel.

Fig. 5. Microstructures (SEM-BSE) of 301 stainless steel after annealing to peak temperatures of (a) RT, (b) 730°C, (c) 900°C and (d) 950°C.

The 301 stainless steel used here is, in fact, a rather complex material. The cold rolling treatment resulted in a large content of deformation-induced α-martensite. This was evident from the strongly ferromagnetic condition of the as-received sheet. During annealing to progressively higher temperatures, this martensite transformed thermally back to austenite and the steel was virtually non-magnetic at 730°C indicating that all the martensite had been eliminated by this stage. However, as shown by the microstructure (Fig. 5b) the austenite was still in a heavily deformed condition. On further heating to between 900° and 950°C recrystallisation of the austenite was evident in the SEM studies (Fig. 5c and d).

The changes in crystal structure and microstructure described above are clearly identifiable in the ultrasonic wave velocity data shown in Fig. 4. The velocity is a maximum in the cold rolled state due to the presence of the deformation-induced martensite. As this disappears, the density increases and there is probably an associated change in the elastic properties, together resulting in a decrease of the wave velocity. After annealing between about 730° and 840°C the velocity is essentially constant as little further change in microstructure takes place in this range. An increase then takes place after higher temperatures of annealing which is associated with recrystallisation as a result of the accompanying textural change. Since the crystal structure of austenite is highly anisotropic, some of this anisotropy is inherited in the polycrystalline steel whenever this contains a crystallographic texture. On recrystallisation the texture changes and so do the elastic properties (including wave velocity) in any given direction.

It should be remarked that the present observations have all been carried out at room temperature to simplify the analysis at this preliminary stage of the work. Dynamic *in-situ* experiments are planned for the near future. Previous work [6] has shown that temperature presents no hindrance to LUS measurements although changes in wave velocity arising from thermal expansion and variation in elasiticity need to be taken into account

4 Summary and Conclusions

Measurements of ultrasonic wave propagation are capable of revealing changes in structure associated with phase transformations and recrystallisation. This has been demonstrated here for two different materials. By using laser techniques for both generation and detection of the ultrasonic signal, the method can be made non-intrusive and therefore suitable for real-time investigations of microstructures in metals. This allows the possibility for direct measurements at elevated temperatures and can provide a technique for in-line process control of microstructures, textures and mechanical properties in the metallurgical industry.

Acknowledgements

This work has been carried out as part of the general research programme of the Swedish Institute for Metals Research with financial support from IRECO. The authors thank Lena Ryde, Stefan Sundin, Chris Scruby and Brian Moss for their advice and assistance.

References

[1] M.B. Bever, D.L. Holt and A.L. Titchener, *Prog. In Mat. Sci.*, 17, 1972.
[2] R.K. Ray, W.B. Hutchinson and F.M.C. Besag, *Acta Metall.*, 23, 1975, p. 831.
[3] N. Hansen, T. Leffers and J.K. Kjems, *Acta Metall.*, 29, 1981, p. 1523.
[4] D. Juul-Jensen and H.F. Poulsen, in „Proc. 21st Risö Symposium on Materials Science“, 2000, p. 103.
[5] C.B. Scruby and L.E. Drain, „Laser Ultrasonic Techniques and Applications“, Adam Hilger, 1990.
[6] C.B. Scruby, C.E. Bull, R.M.K. Young and F.J. Humphreys, „AEA Technology Report TSD 0909“, 1996.
[7] S. Sundin and D. Artymowicz, submitted to *Metall. Transactions*, 2001.
[8] D. Artymowicz, S. Sundin, B. Hutchinson, A. Astill and M. Nogues, „*Swedish Institute of Metals* Report IM-2000-063“, 2000.

Monitoring Austenitic Grain Growth Kinetics by Laser-Ultrasonics

S. Kruger, A. Moreau, and G. Lamouche

Industrial Materials Institute, National Research Council of Canada, 75, Boul. de Mortagne, Boucherville, Québec, J4B6Y4, Canada.

Keywords: grain growth kinetics, austenite, laser-ultrasonics.

Abstract. Laser-ultrasonics is a non-destructive and non-contact technique that can be used to evaluate the kinetics of recrystallization, grain growth and phase transformation at high temperatures and in real time. In this paper, the technique is applied to monitor phase transformations and austenitic grain growth during fast thermal cycles that simulate those received by the heat affected zone during welding. The ultrasonic attenuation observed during the thermal cycle allowed the characterization of the ferrite-to-austenite phase transformation, the onset of grain growth, the kinetics and extent of grain growth and the austenite-to-ferrite transformation.

1 Introduction

Prior austenite grain size is directly related to the fracture toughness of the heat affected zone of a welding joint. The coarse-grained region is generally the location of initiation of brittle fracture [1]. Precipitates like nitrides of Ti, Al, Nb or V are usually used to control the grain growth of the austenite, but the characterization of this control requires a substantial amount of metallographic work. Previous work [2,3] demonstrated the ability of laser-ultrasonics to determine the austenite grain size of various steel grades. Laser-ultrasonics is a non-contact technique where ultrasonic waves are generated by a pulsed laser and detected by laser interferometry [4]. The technique can be applied to materials at high temperatures and can monitor the evolution of the microstructure in real time. In this paper, laser-ultrasonics is used to follow the microstructure of steel samples during very rapid heat treatments that simulate those received by the heat affected zone of welding joints. The advantage of laser-ultrasonics for this application is that a complete characterization of the grain size and phase transformation kinetics can be measured during a single thermal cycle in a single experiment. In addition, the technique can provide grain size estimates when the prior austenite grain size cannot be evaluated easily by metallographic methods.

2 Experimental Procedure

Laser-ultrasonic measurements were performed on low carbon (C-Mn-Nb) steel samples with dimensions of 2 x 11 x 70 mm^3 and with flat and parallel faces. The heat treatment was performed in a Gleeble 3500 thermo-mechanical simulator equipped with windows transparent to laser radiation wavelengths. The Gleeble allowed the fast heating rate needed to simulate the heat-affected zone of a welding joint. The cooling part of the heat treatment was helped with a continuous helium quench. Rosenthal cycles with peak temperature of 1250 °C, 1300 °C and 1350 °C and cooling time from 800 °C to 500 °C of 30 seconds were programmed into the Gleeble. Fig. 1 shows the programmed cycle for peak temperature of 1350 °C. The difference between the temperature measured by a thermocouple welded on the sample and the programmed temperature was smaller than 4 °C.

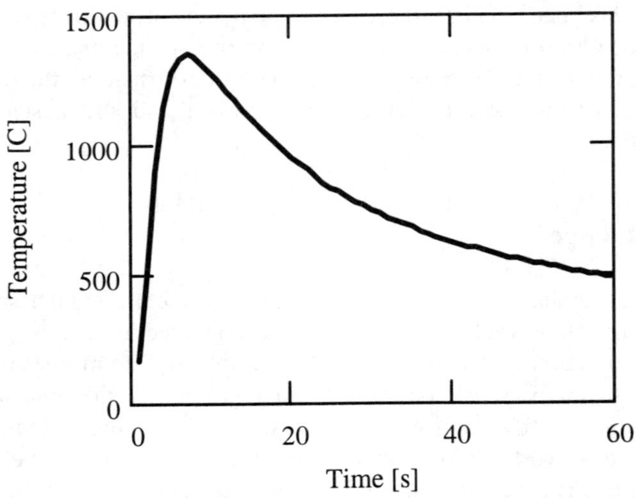

Fig. 1. Applied thermal cycle for peak temperature of 1350 °C.

The phase transformations (ferrite to austenite and vice versa) temperatures were determined with a dilatometer attached to the sample. In laser-ultrasonics, ultrasound is generated and detected with lasers. The generation laser was an excimer laser operating at 248 nm and delivering 400 mJ pulses with duration of about 12 ns. The ultra-violet pulse was focused onto a region of about 2 mm in diameter. The optical power density was high enough to ablate a very small portion of the sample surface. This induced a deformation that propagated in the material: the ultrasonic pulse. Ultrasonic displacements were detected from the other face of the material with a pulsed Nd:YAG laser operating at a 1.06 μm wavelength, and having a pulse energy of 130 mJ and a pulse of duration 60 μs.

The infrared pulse was focused to a region of about 2 mm in diameter. Due to the Doppler effect, the surface motion modulated the reflected light in frequency. This modulation was analyzed with a Fabry-Pérot interferometer operating in the reflection mode [5]. The signal was digitized at a sampling rate of 500 MS/s and transferred in real time to the computer. For the fastest part of the cycle (first 30 seconds), the signals were acquired at a rate of up to ten per second. The ultrasound attenuation was evaluated from the ratio of the Fourier spectra of the first two echoes. For the current study, the ultrasonic frequency of 15 MHz was chosen to take advantage of the previous work performed at that frequency [2,3]. The propagation of ultrasound did not occur in a simple planar or far-field regime due to the sizes of the generation and detection spots. A diffraction correction was applied to the attenuation measurement. At 15 MHz and with our setup, this correction amounts to a small value of 1 dB, as measured on a sample of the same thickness (2 mm) that has very low ultrasonic attenuation. The samples were metallographically evaluated after the cycle to measure the final prior austenitic grain size. The ASTM standard comparison method was used to measure the grain size.

3 Results

3.1 Relationship between ultrasonic attenuation and austenite grain size

Previous work at IMI/NRC [2,3] demonstrated the effectiveness of a simple calibration curve to relate ultrasonic attenuation measured at 15 MHz to the prior austenitic grain size. Fig. 2 shows this calibration curve. The reason for choosing 15 MHz is that it is high enough so that grain scattering dominates the total attenuation (for lower frequencies, the temperature dependent absorption can dominate) and low enough to allow attenuation measurements on large grains samples having high attenuation and a low signal-to-noise ratio. This simple approach showed to be very useful for thin samples (1 or 2 mm). For thicker samples, a more complex approach using the entire attenuation spectrum with adequate temperature correction provides similar results. The attenuation measurements used for the calibration curve were either obtained at 1100 °C or extrapolated to 1100°C with a correction of 2.4×10^{-3} dB/mm°C. The grain size is the equivalent area diameter, that is the diameter of a circle having an area equal to the mean area of the measured grains. The exponential fit to the experimental data was found to be

$$GS = 13.6\alpha^{1.65} \quad (1)$$

Where GS is the grain size in microns and α is the attenuation expressed in dB/mm.

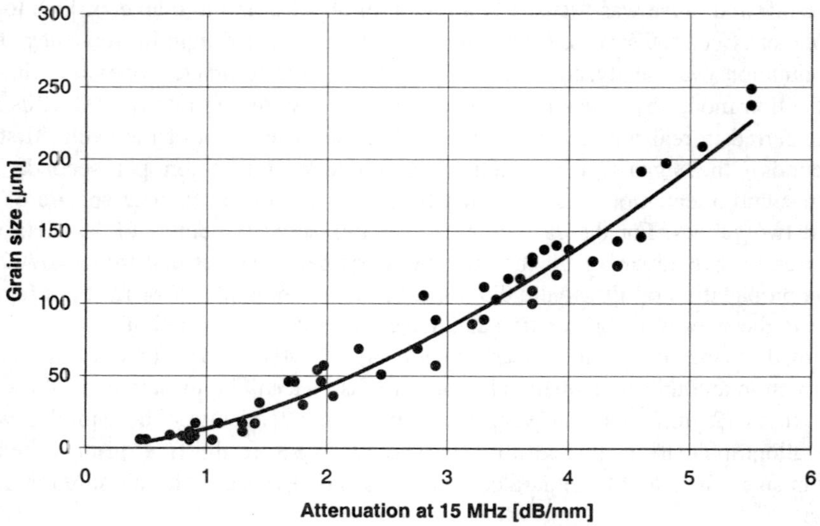

Fig. 2. Grain size versus attenuation at 15 MHz obtained for different steel grades (A36, 1008, 1020, 1035 and 1074). The solid line is a power law fit to the experimental data.

3.2 Monitoring the microstructure of heat affected zones

Figure 3 shows a typical evolution of the attenuation during the thermal cycle. During heating, the attenuation remains almost constant near 0.2 dB/mm up to 850°C where there is a ferrite to austenite phase transformation. Above 900°C, the attenuation increases to 1 dB/mm at 1100 °C and then increases rapidly up to about 3.5 dB/mm. During cooling the attenuation decreases to 2.5 dB/mm at about 900°C and remains almost constant to 570°C where it drops abruptly due to the austenite to ferrite phase transformation. The decrease of the attenuation during cooling between 1200 °C and 900 °C is attributed to the effect of temperature on attenuation, since little or no microstructural transformation should occur. The origin of this temperature dependent attenuation can be a thermally activated absorption mechanism and/or a change of the elastic constants anisotropy of the grains. For the 1350 °C peak cycle, the attenuation continues to increase after the peak temperature, when cooling down to about 1250 °C (figure not shown), indicating a grain growth during cooling in this temperature range. This increase in attenuation during heating, discounting the temperature effect, should be associated with austenite grain growth, according to the calibration curve of Fig. 2. The final austenite grain size estimated with the calibration curve for an attenuation of 2.7 dB/mm at 1100 °C is about 70 microns. No significant austenitic grain growth is believed to occur for temperatures below 1100°C during cooling and, since this is the calibration temperature, there is no need for temperature corrections. The metallographic prior austenite grain size measured was about 85 microns. For cycles having peak temperature of 1300 °C and 1350 °C, the measured metallographic grain sizes were 140 μm and 280 μm, respectively, while the ultrasonic estimates of grain sizes were 110 μm and 230 μm, respectively. The discrepancies between the metallographic and ultrasonic grain sizes are within the range of the measurement accuracy of the metallographic technique (± 25%). During the austenite to ferrite transformation, grain refinement causes an abrupt drop in the attenuation near 575 °C. Nevertheless, the attenuation remains higher than before the heat treatment, indicating a coarsening of the initial microstructure.

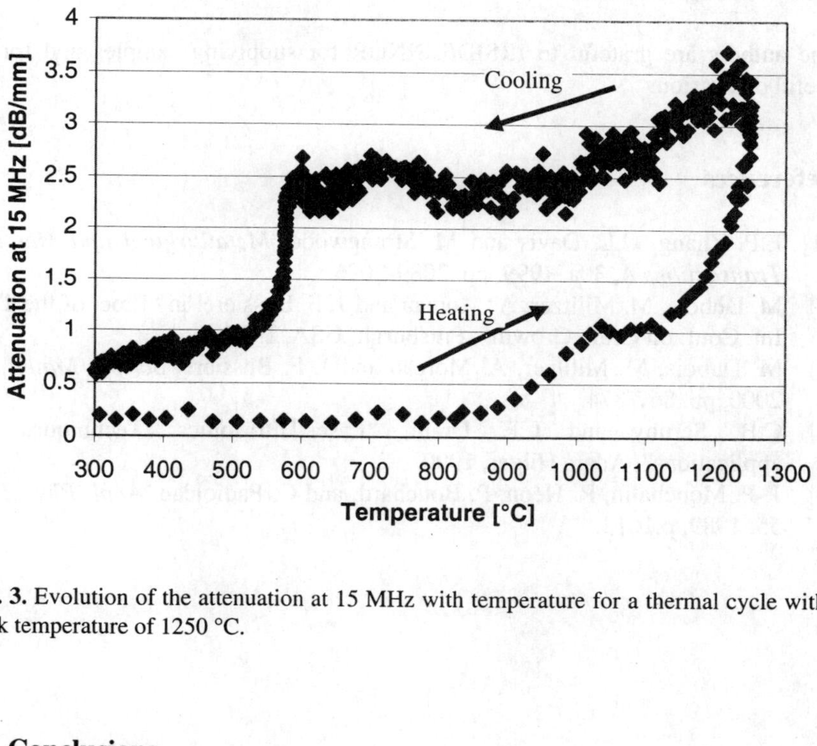

Fig. 3. Evolution of the attenuation at 15 MHz with temperature for a thermal cycle with a peak temperature of 1250 °C.

4 Conclusions

The usefulness of the laser-ultrasonic technique to follow the evolution of the microstructure was demonstrated for fast thermal cycles simulating the conditions of the heat-affected zone during welding. The evolution of the 15 MHz ultrasound attenuation coefficient allowed the observation of the ferrite-to-austenite transition, austenite grain growth, the austenite-to-ferrite transition, and the resultant coarsening of the microstructure. This was achieved using a previously published calibration relating the prior austenite grain size to the 15 MHz attenuation measured at 1100°C. Three thermal cycles with peak temperatures of 1250°C, 1300°C, and 1350°C were applied to low carbon (C-Mn-Nb) steel samples and the measured prior austenite grain sizes agreed with the values obtained by metallography within the metallographic measurement accuracy. In contrast with lengthy metallographic techniques, laser-ultrasonics is a fast and in-situ tool that allows real-time characterization of grain size evolution. Since it is non-destructive and non-contact, it also has a high potential for the online characterization of the evolution of the microstructure in-situ during welding as well as for any thermo-mechanical processes.

Acknowledgdement

The authors are grateful to IRSID/USINOR for supplying samples and for the useful discussion.

References

[1] L.P. Zhang, C.L. Davis and M. Strangwood, *Metallurgical and Materials Transactions A*, 30a, 1999, pp. 2089-2096.

[2] M. Dubois, M. Militzer, A. Moreau and J. F. Bussiere, in "Proc. of the Third Int. Conf. on Grain Growth", Pittsburgh, USA, 1998.

[3] M. Dubois, M. Militzer, A. Moreau and J. F. Bussiere, *Scripta Mater.*, 42, 2000, pp. 867-874.

[4] C.B. Scruby and L.E. Drain, "Laser-Ultrasonics: Techniques and applications", Adam Hilger, 1990.

[5] J.-P. Monchalin, R. Héon, P. Bouchard, and C. Padioleau, *Appl. Phys. Lett.*, 55, 1989, p.1612.

Growth Kinetics of Individual Cube Grains as Studied by the 3D X-Ray Diffraction Microscope.

E.M. Lauridsen[1], S. Schmidt[1], L. Margulies[1,2], H.F. Poulsen[1], and D. Juul Jensen[1]

[1] Materials Reaseach Department, Risø National Laboratory, DK-4000, Roskilde, Denmark

[2] European Synchrotron Radiation Facility, BP 220, F-38043, Grenoble Cedex, France

Keywords: *in-situ* studies, synchrotron radiation, growth kinetics, cube grains.

Abstract. A new method is presented for *in-situ* measurements of growth kinetics in polycrystalline materials. The method provides information on the recrystallization kinetics of the individual grains and allows for orientation dependencies of the recrystallization kinetics to be revealed. The method is implemented at the 3D X-ray Diffraction microscope, recently installed at the European Synchrotron Radiation Facility. True bulk measurements are ensured by the use of focused high energy x-rays (40-100 keV) with a penetration depth of up to 4 cm in aluminium. The potential of the method is illustrated by results for 90% cold rolled commercially pure aluminium. Focus is on the recrystallization kinetics of cube oriented grains.

1 Introduction

For fcc metals of medium to high stacking fault energy, experimental observations show that during recrystallization the volume fraction of grains with the cube orientation increases significantly, and in some cases completely dominate the final recrystallization texture [1]. A vast number of experimental techniques have been used with the aim of understanding this behavior, including neutron diffraction, electron microscopy (SEM & TEM), optical microscopy and standard x-ray diffraction. In general these techniques only provide information on average growth rates. Furthermore, all techniques except neutron diffraction probe the surface. As such they can not be used for *in-situ* bulk characterization.
By the development of the 3-Dimensional X-Ray Diffraction (3DXRD) microscope [2,3] at the European Synchrotron Radiation Facility (ESRF), Grenoble, a new tool for studies of recrystallization kinetics has become available. The 3DXRD microscope utilizes focused high energy x-rays (E > 40keV) and enable

in-situ measurements of the growth kinetics of the individual grains embedded within the bulk of the material.
The present paper gives a description of the method and its potential is illustrated by an investigation of the growth kinetics of "near" cube grains. The material used is commercially pure Aluminium (AA1050) cold rolled 90% and annealed at 270°C. An earlier study of this material done at the 3DXRD microscope showed a significant spread in the growth rates of the individual grains [4]. Large variations were found not only between the different grains, but also for the individual grains themselves. In this first study no attempt was made to correlate these variations with grain orientations.

2 Experimental Procedure

The experimental setup is sketched in Fig. 1. High energy x-rays (E = 50keV) are focused by a bent Laue crystal while the illuminated volume is defined by slits positioned in front of the sample. In order to obtain a broad homogenous beam profile the Laue crystal is slightly defocused and the sample is moved away from the focal point as illustrated in Fig. 1.

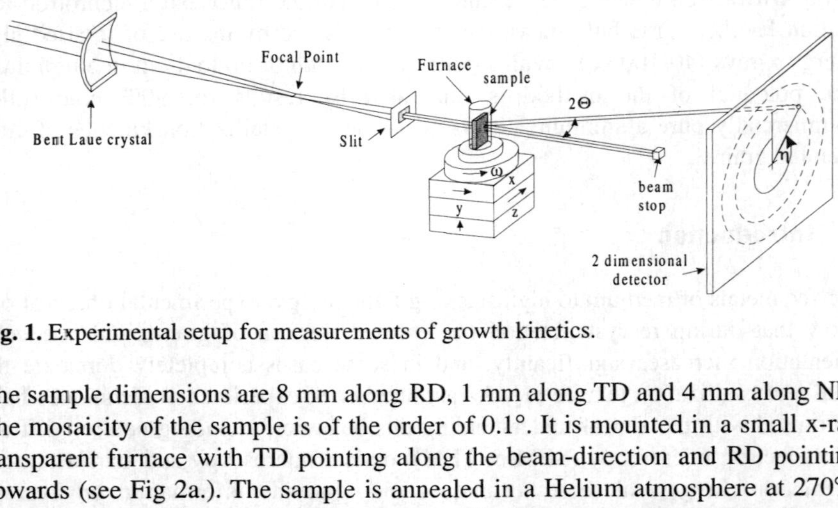

Fig. 1. Experimental setup for measurements of growth kinetics.

The sample dimensions are 8 mm along RD, 1 mm along TD and 4 mm along ND. The mosaicity of the sample is of the order of 0.1°. It is mounted in a small x-ray transparent furnace with TD pointing along the beam-direction and RD pointing upwards (see Fig 2a.). The sample is annealed in a Helium atmosphere at 270°C for 180 minutes after which it is approximately 95% recrystallized. The distance from the sample to the detector is chosen such that the {111}, {200}, {220} and {311} Debye-Scherrer rings are visible on the detector.
Before start of annealing the intensity profiles along the Debye-Scherrer rings are smooth with broad peaks reflecting the texture of the deformed microstructure. But as recrystallization occurs sharp peaks appear on top of the smooth texture background and continue to grow in intensity while the texture background vanishes.

Data is taken at 5 consecutive ω settings spanning the range from ω = -0.5° to 0.5° with an oscillation range of Δω = 0.25°. (For a definition of instrumental angles see Fig. 1). At each ω setting images are acquired with exposure times of 1s, 2s, and 5s. The use of multiple exposure times allows both small slowly growing grains and big fast growing grains to be monitored without saturating the detector. Using a gauge volume of 100 μm x 100 μm x 1 mm, images are taken repeatedly during the annealing period using the 5 ω settings and the 3 exposure times as described above, leading to a time resolution of four minutes. Every 5'th loop the gauge volume is increased to 140 μm x 140 μm x 1 mm by opening the slits.
When analyzing the data it is important to the make sure that the whole grain volume is included in the analysis. This is taken care of by three requirements that a grain has to fulfill in order to be valid. First, the diffraction peak has to lie completely within the oscillation range, i.e. no tails of the peak must be present in images at the neighboring ω settings. This explains the need for consecutive ω settings. Second, the grain has to lie completely inside the gauge volume, i.e. the intensity of the diffraction peak should not rise when the gauge volume is increased. Thirdly, there must be no peak overlap, i.e. the peak should be well separated from neighboring peaks along the η direction as well as along the ω direction to ensure measurement of true integrated intensity. For grains fulfilling these requirements, the background corrected integrated intensities are directly proportional to the grain volume (for details see Ref. [4]).

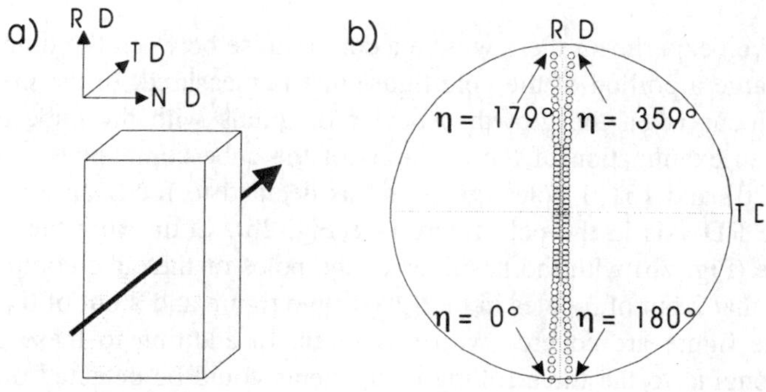

Fig. 2. a) Sample orientation with respect to the incident x-ray beam. b) Corresponding trace measured in the pole figure.

3 Orientation Analysis

The 3DXRD method reveals not only the growth kinetics of the individual grains, but also provides information on the crystal orientation of the grains.
By a stereographic projection of the scattering vectors onto the (RD,TD)-plane of the sample, there is a one-to-one correlation between the coordinates of a

diffraction peak $(\omega,\eta,2\theta)$ and the (α,β)-coordinate in the corresponding pole figure. This is illustrated in Fig. 2b, where the trace of the Debye-Scherrer ring is shown in pole figure coordinates for the setup used. By rotating the sample around the ω-axis full pole figures can be generated [3].

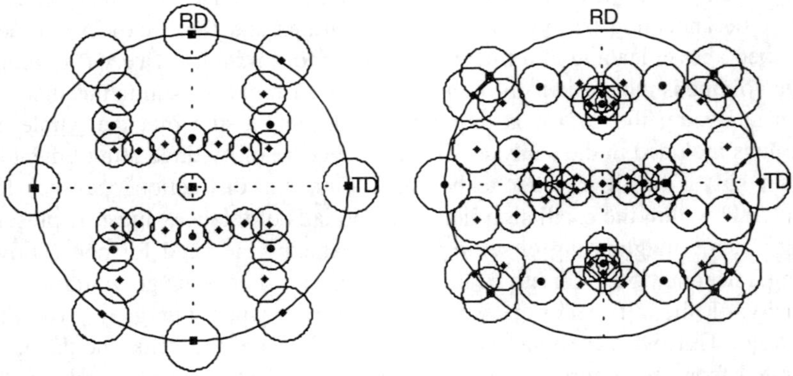

Fig. 3. Illustration of the influence of multiplicity on the overlap of regions surrounding the ideal standard texture components. The standard components shown are: ■ Cube, ♦ Brass, • Copper and ∗ S. The lines surrounding the standard components corresponds to a 10° deviation. a) {200} pole figure and b) {220} pole figure.

For any given experiment there will be a compromise between the time resolution and how large a portion of the pole figure that is measured. In the present study the main focus is on the growth behavior of grains with the cube orientation. Based on an examination of the positions of the cube component in the {111}, {200}, {220} and {311} pole figures it was decided to focus on a small region around the RD-axis in the pole figure (see Fig. 2b). Comparing the trace in the pole figure (Fig. 2b) with the positions of the poles of the cube component (Fig. 3), we see that 2 out of 3 poles in the {200} pole figure and 2 out of 6 poles in the {220} pole figure are covered with this setup. In addition to these grains also grains belonging to the three rolling components could be detected on all of the four measured Debye-Scherrer rings.

While the growth kinetics of a grain is fully described by the intensity changes of a single diffraction peak, determination of the full orientation of the grain requires at least two reflections. The strategy for the present experiment is, however, to measure growth rates with a reasonable good time resolution and then to use the orientation information available from just one diffraction peak to identify grains with "near" cube orientation. This is done by locating the areas in the pole figure corresponding to a 10° deviation from the ideal cube orientation (see Fig. 3). For each diffraction peak on the detector, the corresponding pole figure coordinates are calculated and if they fall within 10° from the ideal cube component it is classified as "near" cube orientation.

This simple procedure for orientation analysis implies some limitations. This is illustrated in Fig. 3 where the main components of the recrystallization texture are plotted in a {200} and {220} pole figure, respectively. The lines surrounding the ideal components correspond to a 10° deviation. For the {200} pole figure, having the lowest multiplicity, the degree of overlap between the different texture components are negligible. For pole figures with higher multiplicity, like the {220} pole figure, the degree of overlap becomes much more severe and determination of the orientation becomes much less unambiguous.

It is in principle possible to overcome this problem by extending the experiment. The development of a new indexing algorithm for polycrystalline materials [5] has opened the possibility of determining the full orientation of each grain within a polycrystalline material. In order for the algorithm to work one would need to scan a broad ω range to ensure that a number of diffraction peaks from each grain are measured. Assuming that grains do not disappear, such a scan could for instance be done at the end of the annealing. This does however imply a further complication of the data analysis.

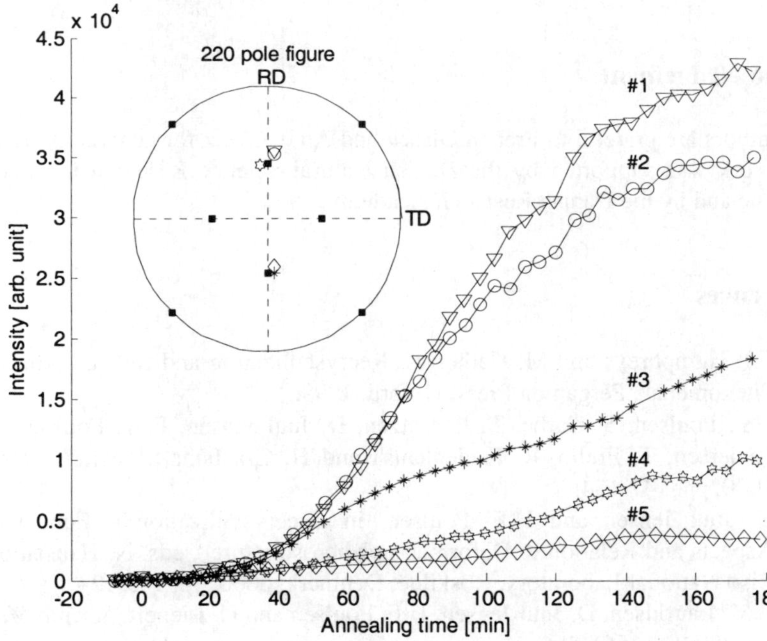

Fig. 4. Growth curves for the 5 grains with an orientation less than 10° degrees away from the ideal Cube orientation. It should be noted that grain 1 and 2 are also within 10° of the Cu and Br texture components. Insert: Pole figure coordinates of the 5 grains. (The ideal cube component is marked with ■).

4 Results

In total 50 grains are found to be valid, and out of these 5 is found to have an orientation that deviates by less than 10° from the ideal cube orientation. Here it should be noted that grain #1 and #2 is also within 10° from the S and Cu components. All 5 "near" cube grains are found on the {220} Debye Scherrer ring. The intensities versus annealing time of the 5 grains are shown in Fig. 4. The insert shows the corresponding positions in a {220} pole figure.
This is the first time the recrystallization kinetics of individual cube grains could be followed in-situ. The 5 "near" cube grains show very different recrystallization kinetics, and variations in the growth rate of the individual grains are visible as well. This shows that average growth rates do not give the complete picture and stresses the importance of understanding the influence of the in-homogeneity of the deformed microstructure on the growth kinetics of the individual grains. A need is hereby identified for recrystallization models that can incorporate distributions of growth rates.

Acknowledgement

The authors are grateful to Preben Olesen and Andy Goetz for general asstistance. This work was supported by the Danish Natural Science Research Council via Dansync and by the Danish Research Academy.

References

[1] F.J. Humphreys and M. Hatherly, „Recrystallization and Related Annealing Phenomena“, Pergamon Press, Oxford, 1999.
[2] H.F. Poulsen, S. Garbe, T. Lorentzen, D. Juul Jensen, F.W. Poulsen, N.H. Andersen, T. Frello, R. Feidenhans'l and H. Graafsma, *J. Synch. Rad.*, 4, 2000, pp. 147-154.
[3] D. Juul Jensen and H.F. Poulsen, in „Recrystallization – Fundamental Aspects and Relation to Deformation Microstructure“, eds. N. Hansen et al., Risø National Laboratory, Roskilde, Denmark, 2000, pp. 103-124.
[4] E.M. Lauridsen, D. Juul Jensen, H.F. Poulsen and U. Lienert, *Scripta Mater.*, 43, 2000, pp. 561-566.
[5] E.M. Lauridsen, S. Schmidt, R.M. Suter and H.F. Poulsen, submitted for publication.

Recrystallization and Grain Growth
Proceedings of the First Joint International Conference
Eds. G. Gottstein and D.A. Molodov, Springer-Verlag, 2001

Stereology and 3D Quantitative Characterization of Microstructure for Grain Growth Studies

G. Liu[1], C. Wang[1], S. Liu[1], X. Qin[1], H. Yu[1,2] and X. Song[1,3]

[1] School of Materials Science and Engineering, University of Science and Technology Beijing, Beijing 100083, China. Email: G.Liu@ustb.edu.cn
[2] Department of Applied Physics, Beijing Institute of Technology, Beijing 100081, China
[3] Technische Universität Darmstadt, Fachbereich 21 Materialwissenschaften, physikalische Metallkunde, Petersenstr. 23, 64287 Darmstadt, Germany

Keywords: grain growth, microstructure characterization, stereology, image modeling.

Abstract. In studies of grain growth in a single-phase material, it is essential to be able to quantitatively characterize grain size, grain size distribution, grain topology and their evolution during the grain growth process. It is rather easy to do such quantification for two-dimensional grain growth studies, however, systematic quantitative characterization of microstructure evolution progress in three-dimensional space is extremely difficult. In this paper, availability of possible stereological methods, especially modern ones along with the relationship among different stereological microstructure parameters and their usefulness in grain growth studies are discussed.

1 Introduction

In studies of grain growth in a single-phase material, either application-oriented or fundamental studies, it is essential to be able to characterize grain size, grain size distribution, grain topology and their evolution during the grain growth process at different quantification levels [1-9]. However, systematic quantitative characterization of microstructure evolution progress in three-dimensional (3D) space is extremely difficult. Almost all the experimental observations of real material grains are made on their two-dimensional (2D) cross-sections instead of directly on the structure itself in 3D space. Providing quantitative geometric information in 3D inaccessible by any other methodology, stereology [10-15] has thus become a powerful and essential tool for three-dimensional grain growth (3DGG). In this paper, availability and limitations of possible stereological methods, especially modern ones along with the relationship among different stereological microstructure parameters are reviewed and discussed.

2 Geometric Features and Properties of 3D Microstructures

The three-dimensional microstructure, X_3, in any material can be expressed geometrically in set theory as

$$X_3 = Y_3 \cup Y_2 \cup Y_1 \cup Y_0 \quad (1)$$

where the subscripts denote the dimensionality of the set X_3 and of the subsets (or called feature classes in stereology) Y_i (i = 0,1,2,3) embedded in X_3. Even in the simplest case such as one-phase polycrystalline microstructures, X_3 always contain grains having a range of sizes, shape and orientations. Feature classes that may exist in one-phase X_3 and their geometric properties are summarized in Table 1 based on References [11,12,14,16-18].

2.1 Grain sizes

In studies of three-dimensional grain growth studies, the most important microstructural descriptors are mean grain size, grain size distribution, and grain topology parameters.
The term "grain size" in polycrystalline materials is not rigorously defined. ASTM grain size number (G) is widely used in engineering applications, but it is not very useful for an accurate understanding of the 3DGG process. The power law kinetics with values of the growth exponent ≤ 0.5 has been experimentally observed commonly in normal grain growth processes in real materials [5,6]. The "average grain size" d used in this law is actually the mean grain intercept and is a function of the grain boundary area per unit volume S_V, which can be evaluated with the aid of standard stereology methods as shown in Table 1. It should be pointed out that the distribution of grain intercept values does not provide much useful information on three-dimensional grain size distribution in a 3DGG process.
On the other hand, sometimes the average grain section area is also used as a descriptor of "mean grain size". As shown in Table 1, this parameter is exactly related to the edge length per unit volume L_V other than the grain boundary area per unit volume S_V. Similar to the distribution of grain intercepts, the distribution of grain section areas does not provide much useful information on three-dimensional grain size distribution in a 3DGG process either. Grain volume, sphere-equivalent grain diameter, and tangent grain diameters are better descriptors of three-dimensional grain size and their distributions. Unfortunately, they are not accessible from simple two-dimensional measurements.

2.2 Grain topology

A complete theory of grain growth must take into account both the topological space-filling requirements and the kinetics of local boundary migration [6]. The techniques required for determination of grain topology parameters, such as the number of faces F, edges E, and corners C of individual 3D grains and their corresponding means $\overline{F}, \overline{E}, \overline{C}$, are summarized in Table 2.

Table 1. Feature classes that may exist in one-phase space-filling microstructures and their geometric properties

Feature Class	Geometric Properties	
	I. Accessible using classical stereology	II. More difficult to be evaluated
Grains (Y_3)	Grain intercept (l) and its distribution; Mean grain intercept ($\bar{l}$); Number of grains per unit length ($N_L = 1/\bar{l}$); The widely-used parameter "average grain diameter": $d = \bar{l} = 1/P_l = 1/N_l = 2/S_V$	Grain volume (v); Number-weighted grain volume distribution; Volume-weighted grain volume distribution; Sphere-equivalent diameter ($D_{eq} = 6v^{-1/3}/\pi$), etc.
	Grain section area (a) and its distribution; Mean section area ($\bar{a}$); Number of Grains per unit area ($N_A = 1/\bar{a}$)	
	ASTM grain number ($G = -3.2877+6.6439\log_{10}N_L$, or $G = -2.9542+3.3219\log_{10}N_A$, where N_L is in mm^{-1} and N_A in mm^{-2} according to ASTM Standard 112)	Number of grains per unit volume ($N_V = 1/\bar{v}$)
	Volume fraction of grains with given crystalline orientations (V_V)	
GB Surfaces (Y_2)	Grain boundary area per unit volume ($S_V = 2N_L = 2\ P_L = 2/\bar{l}$); Number of GB intersects per unit testing length (P_L); Integral mean curvature of grain faces per unit volume (M_V)	Grain boundary area of a grain (S_G); Number of faces per unit volume (F_V); And that of a grain (F_G)
GB Triple Edges (Y_1)	Edge length per unit volume ($L_V = 2P_A = 4/\bar{a}$); Number of edges appearing per unit testing area (P_A)	Length of an edge; Mean edge length of a grain and that averaged over all grains
		Number of edges per unit volume (E_V); that of a grain (E_G); that of a grain face (E_F)
Grain Corners (Y_0)		Number of corners per unit volume (C_V) and that of a grain (C_G)

Table 2. Characteristic parameters measured by different stereological methods

Stereological Method	Possible parameters able to be evaluated			
	Mean size parameters	Volume distribution parameters	Mean topological parameters	Topological distribution parameters
(1) Serial sectioning method[a]	$\overline{V}_N$, $\overline{V}_V$, $\overline{V}_V^{\,n}$, $\overline{V}_N^{\,n}$	$CV_N(v)$, $CV_V(v)$, $CS_N(v)$, $CS_V(v)$, $CE_N(v)$, $CE_V(v)$	C_V, E_V, F_V, N_V, $\overline{C}, \overline{E}, \overline{F}, \overline{n}$	$CV_N(F)$ $CS_N(F)$ $CE_N(F)$ E, F, C, n
(2) Disector method	$\overline{V}_N$		C_V, E_V, F_V, N_V, $\overline{C}, \overline{E}, \overline{F}, \overline{n}$	
(3) Point-sampled intercepts	$\overline{V}_V$			
(4) Point-sampled areas	$\overline{V}_V^{\,2}$			
(5) Selector method	$\overline{V}_N$		C_V, E_V, F_V, N_V, $\overline{C}, \overline{E}, \overline{F}, \overline{n}$	
(2)+(3)		$CV_N(v)$		
(5)+(3)		$CV_N(v)$		
(3)+(4)		$CV_V(v)$		
(2)+(3)+(4)		$CS_N(v)$		
(5)+(3)+(4)		$CS_N(v)$		

The volume, number of faces, edges, and corners of each grain can be determined individually by using the serial sectioning analysis or grain-separation technique.
The symbols used in this table are described as follows:
The volume v and the number of faces F, edges E, and corners C of individual 3D grains and their corresponding means such as $\overline{Vs}$ (number-weighted) and $\overline{Vv}$ (volume-weighted), and $\overline{F}, \overline{E}, \overline{C}$;
The number of grains per specimen volume N_V; the number of grain faces F_V, edges E_V, and corners C_V per volume;
The coefficient of variation $CV_V(v)$, the kurtosis $CE_V(v)$, and the skewness $CS_V(v)$ of the volume-weighted grain volume distribution;
The coefficient of variation $CV_N(v)$, the kurtosis $CE_N(v)$, and the skewness $CS_N(v)$ of the number-weighted grain volume distribution; $CV_N(v) = [(\overline{Vv} / \overline{Vs}) - 1]^{1/2}$.

The coefficient of variation CVN(F), the kurtosis CEN(F), and the skewness CSN(F) of the number-weighted grain face number distribution; and so on. The number of sides of grain faces n, and its number-weighted mean $\bar{n}$. Since the grain growth process has been found topology-dependent, the experimental determination or the theoretical modeling of grain size-topology relations and grain topology evolution in three-dimensional space is very necessary.

2.3 Grain size distribution and grain topology distribution

The grain size and/or topology distributions in three dimensions are much more difficult to be determined experimentally. When the mathematical form of three-dimensional grain size distribution (3D GSD) varies during the grain growth process, the conventional unfolding calculation of 3D GSD from 1D or 2D GSD [10-12,14] will give useless results. Table 2 provides some experimental methods to determine the 3D GSDs, in which the number-weighted grain volume distribution is preferred in grain growth studies. It shows that serial sectioning technique may still be necessary if the whole distribution frequency curve and characteristics are needed. Fortunately, our recent work was just demonstrated that analysis of about 300 uniform equiaxed grains with this method is adequate to obtain reliable results of the grain size and topology distributions [19]. However, more grains should be analyzed for a reliable coefficient of variation (CV) of 3D GSD [19].

For a log-normal grain volume distribution, the relationship $CV_N(v) = CV_V(\dot{v})$ holds. Unfortunately, both the theoretical and computer simulation studies show that the grain size distribution in normal grain growth process cannot be well approximated by a log-normal distribution [1,8,9].

Serial sectioning is very tedious and time-consuming. Thus, the disector method for space-filling grains proposed by Liu et al. [18], which requires only two dependent sections of the 3D structure, is suggested. The average grain volume is determined as the reciprocal value of unbiased number of grains per unit volume. The average topological parameters, such as the number of faces per grain, can also be obtained by the disector. On the other hand, the unbiased volume-weighted average grain volume can be obtained by the point-sampled intercept method, which needs only independent sections. A reliable coefficient of variation (CV) of 3D GSD can also be obtained by combination of disector and point-sampled intercepts [18].

3 Concluding Remarks

The evaluation of grain boundary surface and of triple grain edge length per unit volume can be made in unbiased ways, so that is the most suggested practice.

In order to obtain the unbiased number of grains contained in unit volume, the disector method designed for space-filling polycrystal material is suggested.

In order to obtain reliable coefficients of variation (CV) of three-dimensional grain size distribution, a combination of disector and point-sampled intercepts methods is suggested. The distributions of grain intercepts or sectional areas may be very misleading to the three-dimensional grain size distribution.
Almost all the existing three-dimensional grain growth theories are based on grain radius other than grain volume. Is it possible to rewrite 3DGG theories in grain volumes? If yes, their experimental verification may become much easier.

Acknowledgement

The authors are grateful to the National Natural Science Foundation of China for financial support of the projects No. 59872003 and No. 59771070.

References

[1] M. Hillert, *Acta mater.*, 13, 1965, p. 227; *Mater. Sci. Forum*, 204-206, 1996, p. 3.
[2] F.N. Rhines, K.R. Craig, *Metall. Trans.*, 5A, 1974, p. 413.
[3] G. Abbruzzese, K. Lücke, *Mater. Sci. Forum*, 204-206, 1996, p. 55.
[4] G. Liu, H. Yu, X. Song, and X. Qin, *Materials and Design*, 22, 2001, p. 33.
[5] H.V. Atkinson, *Acta mater.*, 36, 1988, pp. 469-491.
[6] F.J. Humphrey, M. Hatherly, "Recrystallization and Related Annealing Phenomena", Pergamon Press, Oxford, 1995, pp. 281-325.
[7] M.P. Anderson, G.S. Grest, D.J. Srolovitz, *Phil. Mag.*, B59, 1989, p. 293.
[8] F. Wakai, N. Enomoto, and H. Ogawa, *Acta mater.*, 48, 1999, p. 1297.
[9] X. Song, G. Liu, and N. Gu, *Script Mater.*, 43, 2000, p. 355.
[10] S.A. Saltykov, "Stereometric Metallography", 1st Edn, 1948; 2nd Edn, 1958 (in Russian, Moscow); 3rd Edn, 1970 (in Russian, Moscow) and 1974 (in German, Leipzig).
[11] R.T. DeHoff, F.N. Rhines, "Quantitative Microscopy", McGraw-Hill, New York, 1968.
[12] E.E. Underwood, in "Stereology and Quantitative Metallography", ASTM STP 504, American Society for Testing and Materials, 1972, pp. 3-38.
[13] J. H. Steels, Jr., Ibid, pp. 39-58.
[14] Y. Yu, G. Liu, "Stereology–Principle and Applications of Quantitative Analysis of Structures", Metallurgical Industry Press, Beijing, 1989.
[15] C.V. Howard, M.G. Reed, "Unbiased Stereology–Three-Dimensional Measurement in Microscopy", BIOS Scientific Publishers, Oxford, 1998.
[16] R.T. DeHoff, *Image Analysis and Stereology*, 19, 2000, p. 1.
[17] G. Liu, *J. Microscopy*, Oxford (UK), 171, 1993, p. 57.
[18] G. Liu, H. Yu, and W. Li, *Acta Stereologica*, 13, 1994, p. 281.
[19] G. Liu, H. Yu, *Image Analysis and Stereology*, 19, 2000, p. 91.

Recrystallization and Grain Growth
Proceedings of the First Joint International Conference
Eds. G. Gottstein and D.A. Molodov, Springer-Verlag, 2001

Substructural Development in Cold Rolled Al-0.13Mg

P.J. Hurley and F.J. Humphreys

Manchester Materials Science Centre, UMIST and The University of Manchester, Grosvenor St, Manchester, United Kingdom, M1 7HS

Keywords: substructure, FEG-SEM, angular resolution, shear band, cell block.

Abstract. Microstructural development in a high purity Al-0.13wt% Mg alloy cold rolled to reductions of 20% and 50% has been analysed using high resolution EBSD. Use of data averaging techniques improves the relative angular resolution to <0.5°, allowing more detailed characterisation of the deformed structures than previously possible. Important features of the microstructure are discussed, including the development, after 20% reduction, of a laminar cell structure of approximately alternating orientation across entire grains, and at higher reductions, the onset of microshear bands with a common orientation intersecting this laminar structure.

1 Introduction

Until recently, the main technique available to researchers for quantitative analysis of deformation substructures in metals was transmission electron microscopy (TEM), and this has been sucessfully used for the detailed characterisation of microstructures in aluminium (e.g. [1]). However, automated high resolution electron backscatter diffraction (EBSD) with a FEGSEM has the spatial resolution necessary to characterise cell/subgrain structures >0.1 μm and has the additional advantages that complete orientation data are obtained and large areas of the surfaces of bulk samples may be examined. As discussed by Huang and Jensen [2], the main disadvantage of EBSD compared to TEM is the angular resolution of ~1° which is a problem in many deformed metals where cell misorientations are typically in the range 0.5-3° [3]. However, it has recently been shown that the relative angular resolution for EBSD can be reduced to <0.5° by orientation averaging and filtering [4], and the current paper discusses the application of this technique to the characterisation of deformed aluminium.

2 Experimental Procedure

High purity Al-0.13wt% Mg was DC cast as a 15kg ingot. A 40 mm section cut from the cast was homogenised at 550°C for 24 hours and cold rolled 75%. The resulting material was annealed at 300°C for 2 hours leading to recrystallisation with a final average grain size of 82 μm. Individual sections of the plate were subsequently cold rolled to 20% and 50% reductions. Some of the material rolled to 50% was given a recovery anneal at 200°C for 1 hour. Metallographic specimens were subsequently prepared from the ND-TD plane by standard metallographic methods, followed by electropolishing. High resolution SEM was performed using a Philips XL30 and a Camscan 2040S, both equipped with a Field Emission Gun (FEG). Both systems were fitted with an HKL Channel4 EBSD acquisition system. Orientation mapping was carried out using pixel step sizes of 0.1-0.15 μm. Data analysis was performed using VMAP, an in-house software package. TEM, on material cut from the ND-TD plane, was also carried out on the deformed materials using a Philips CM200 operating at 200KV.

3 Results and Discussion

3.1 Improving the angular resolution

Fig. 1a is a transmission electron micrograph of the Al-0.13Mg alloy rolled at room temperature to 50% reduction. The microstructure of dislocation cells which are relatively free of dislocations is typical of aluminium deformed to both low and high levels of applied strain [1].
EBSD mapping was performed on a small region of the sample, and the pattern quality map of Fig. 1b, reveals a cell structure similar to that revealed by TEM. EBSD patterns from regions of lattice distortion or from cell boundaries are of lower contrast than those from more perfect regions because they contain either diffuse or overlapping diffraction patterns, and they appear dark in pattern quality maps. Fig. 1a and b show that the substructure is dominated by the cell walls and that, within the cells, the orientation of the matrix is essentially constant. The accuracy of a single orientation measurement from EBSD increases with the number of averaged points of that orientation, and recently, a novel approach to orientation averaging involving the use of an edge-preserving data processing filter, referred to as a modified Kuwahara filter, was introduced, which enables the angular resolution to be improved to <0.5°, providing that a suitably small step size relative to the scale of the cells is used during EBSD mapping [4]. By applying this procedure to the EBSD data for the map in Fig. 1b, the amount of orientation noise is reduced significantly. For example, the raw data from the map in Fig. 1b was used to reconstruct a cell structure which included all boundaries with a misorientation >0.5° (Fig. 1c), and it is clear that the amount of noise in the reconstructed map is too high to allow any useful quantitative information to be obtained. However, it can be seen that after two passes of the modified Kuwahara filter (Fig. 1d), the noise in the reconstructed map has been essentially removed, leaving an orientation map with all boundaries >0.5° revealed.

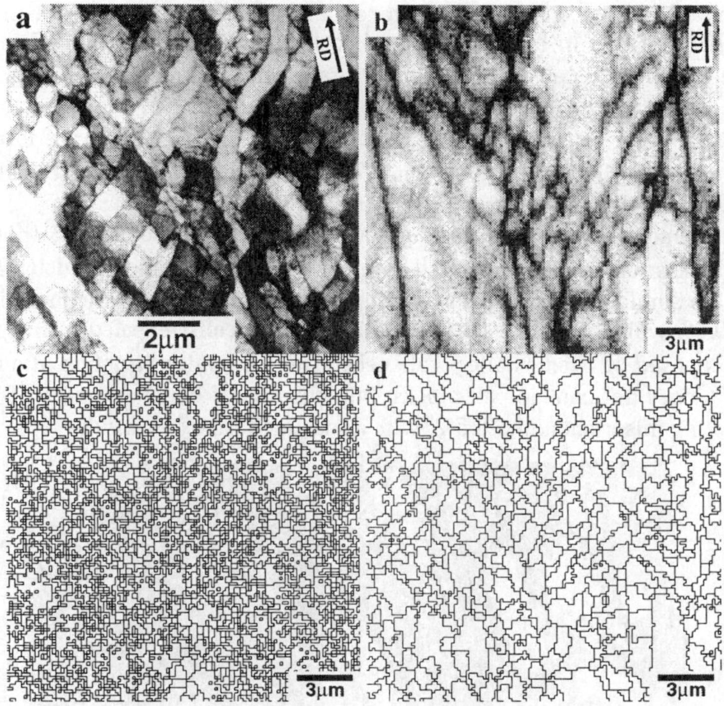

Fig. 1. Al-0.13Mg rolled to 50% (a) TEM; (b) EBSD (band contrast); (c) reconstructed using raw data; (d) reconstructed after 2 passes of the modified Kuwahara filter.

3.2 Applying EBSD to the study of substructural development

After 20% reduction, TEM revealed that the substructure consisted of bands of elongated dislocation cells bounded by dense dislocation walls, often termed "geometrically necessary boundaries" (GNBs) [1], along their lengths (Fig. 2a). Thinner, more randomly oriented dislocation walls termed "incidental dislocation boundaries" (IDBs) [1], generally separated the ends of these long cells, although occasionally a dense dislocation wall was observed to intersect a series of the elongated cells, such as the boundary marked with arrows in Fig. 2a. High resolution EBSD with the improvement in angular resolution discussed, was used to study the deformed state. Fig. 2b shows a small region in relative Euler contrast [4], in which each pixel is assigned a shade of grey depending on its orientation relative to a reference pixel. The power of EBSD is now revealed since it can immediately be seen that the adjacent elongated cells appear to alternate in orientation. Along the cells, the orientation also alternates but the orientation changes are smaller. This effect is more obvious when this map is plotted in colour. It is usually found that GNBs have larger misorientation than IDBs and delineate regions, known as cell blocks, which contain cells separated by lower misorientation (IDBs) [1,2]. In Fig. 2c and d, all cell boundaries >0.5° have been plotted (grey) to show the cells in this region. In addition, boundaries >1° (Fig. 2c) have been plotted to determine whether cell blocks are distinguishable. Although groups of cells with boundaries >1° are found (Fig. 2c), they are typically no wider than the dimensions of a single cell, confirming the contrast in the EBSD map shown in Fig. 2b and the TEM observations of Fig. 2a.

When boundaries >2° are superimposed on the 0.5° cell boundaries (Fig. 2d), no cell block structure is revealed. The magnitude of cells measured by TEM and EBSD is similar, suggesting that most cell boundaries have lattice rotations of at least 0.5°, which rules out the possibility that the cells revealed in Fig. 2c and d are composed of even smaller scale cells. Thus, the microstructure at a reduction of 20% does not contain cell blocks, but, consists of lamellar arrays of GNBs of alternating misorientation ~0.5-2^0 with a spacing of one cell width.

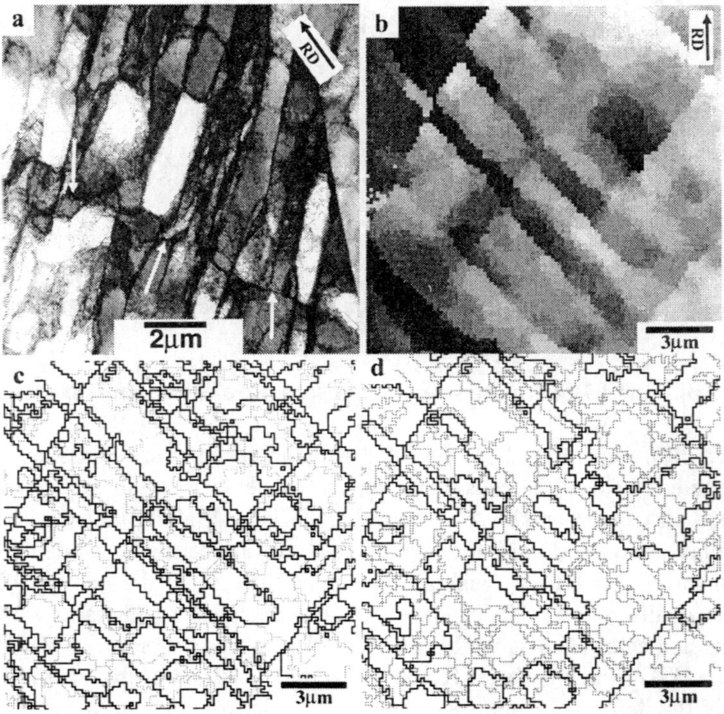

Fig. 2. Substructure after 20% rolling (a) TEM; (b) EBSD (relative Euler angle map); (c) reconstructed boundaries with 0.5° boundaries (grey) and 1° boundaries (black); (d) reconstructed boundaries with 0.5° boundaries (grey) and 2° boundaries (black).

A similar investigation on the sample rolled to 50% reduction is presented in Fig. 1a. This shows that the cells have become more equiaxed in nature and their size has decreased. Microshear bands have begun to form in the substructure, visible as a cluster of fine cells intersecting and offsetting the elongated cells formed earlier in the process (Fig. 1a). These features, formed at approximately 20° to the rolling plane, are easily distinguishable in the EBSD relative Euler contrast map of Fig. 3a as the dark bands intersecting, and breaking up, the elongated cells, which still appear to have maintained an approximately alternating orientation. The cells making up the microshear bands have a common orientation. It may be that the slight undulations in orientation parallel to the elongated bands observed after 20% rolling represent the early stages of microshear band formation. With further rolling, these orientation undulations increase in magnitude and gradually develop into regions of localised shear composing fine-scale cell structures with a common orientation. During their development, the localised shears associated with these bands, offset the elongated cells which they intersect, breaking these down into the more equiaxed cells observed after 50% rolling. To test for the existence of cell blocks in this sample, boundaries of misorientations >1°, 2° and 3° have been superimposed on the cell structure with boundaries >0.5° in Figs. 3b,c,d. It is seen

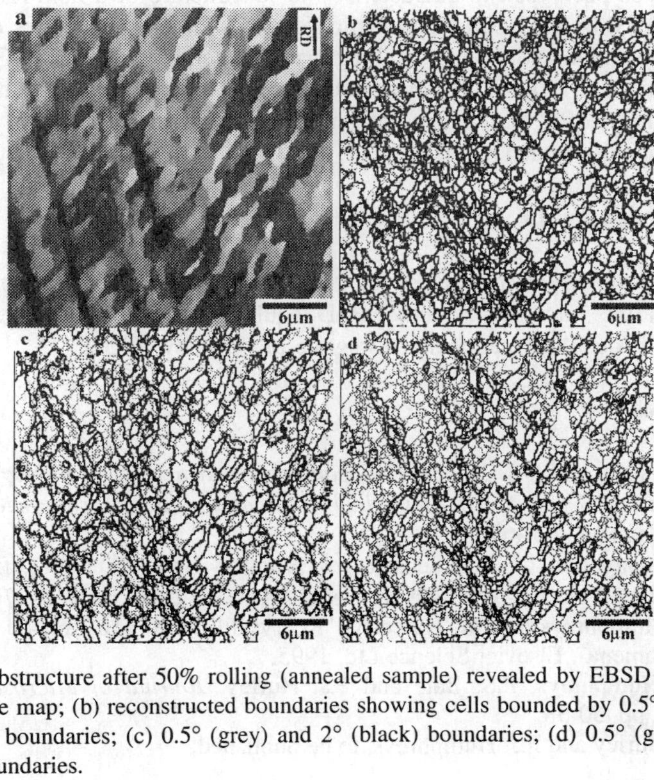

Fig. 3. Substructure after 50% rolling (annealed sample) revealed by EBSD (a) relative Euler angle map; (b) reconstructed boundaries showing cells bounded by 0.5° (grey) and 1° (black) boundaries; (c) 0.5° (grey) and 2° (black) boundaries; (d) 0.5° (grey) and 3° (black) boundaries.

that cell blocks are particularly prominent when boundaries >2° are plotted (Fig. 3c), but less so when boundaries >3° are plotted, indicating the formation of cell blocks containing IDBs of misorientation 0.5-1° separated by GNBs of misorientation ~2-3°.

These results show that cell blocks are only formed at larger strains, which appears to conflict with the general consensus that cell blocks first appear at low strains [3] and become thinner and more lamellar at larger strains.

In addition to the analysis of boundary types and distribution discussed above, the EBSD results provide detailed statistics on grain and subgrain size and misorientation as a function of grain orientation, strain and misorientation, and these will be presented elsewhere [5].

4 Conclusions

1) Application of orientation averaging to EBSD maps, using an edge-preserving filter, improves the relative angular resolution to approximately 0.4° in metals where the true orientation is constant within the substructure (i.e cell forming metals such as aluminium).

2) EBSD has been used to characterise the substructure of a cold-rolled single-phase Al-0.13Mg alloy. After a reduction of 20%, lamellar structures of alternating misorientation, consisting of long parallel cells bounded by GNBs with misorientations of ~0.5-2°. By 50% reduction, the microstructure comprised cell blocks delineated by boundaries with misorientations in the order of 1-3°, containing cells misoriented by ~0.5°.

Acknowledgements

The authors would like to express their gratitude to Mr Ian Brough for his assistance with the microscopes and EBSD equipment. The financial support of EPSRC and Alcan International is also gratefully acknowledged.

References

[1] N. Hansen and D.J. Jensen, *Philosophical Transactions of the Royal Society of London Series A - Mathematical Physical and Engineering Sciences*, 357, 1999, pp. 1447-1469.

[2] X. Huang and D. Juul Jensen, in "Electron Backscattered Diffraction in Materials Science", eds. Schwartz et al., Kluwer, 2000, pp. 265-276.

[3] F.J. Humphreys and M. Hatherly, "Recrystallisation and Related Annealing Phenomena", Elsevier Science Ltd, 1995.

[4] F.J. Humphreys, P.S. Bate and P.J. Hurley, *Journal of Microscopy*, 201, 2001, pp. 50-58.

[5] P.J. Hurley and F.J. Humphreys, to be published.

Characterization of Abnormal Contrast Observed in OIM Images Measured on a Rough Surface

Shigeto Takebayashi[1], Shigeru Suzuki[1] and Yoshiyuki Ushigami[2]

[1] Advanced Technology Research Laboratories,
[2] Steel Laboratories, Nippon Steel Corporation, Shintomi, Futtsu-City, Chiba, 293-8511, Japan

Keywords: EBSP, OIM, Kikuchi pattern, mis-assignment, Goss orientation

Abstract. EBSP (Electron back-scattering pattern) is often used for determining the crystallographic orientation of a small area. Also, OIM (Orientation Image Microscopy) has been developed to observe a microstructure using image mapping of the orientation based on EBSP. However, an abnormal contrast is sometimes observed in an OIM image. This may be attributed to mis-indexing of the orientation, which result from poor contrast of EBSP images. The poor contrast may be related to insufficient sample conditions such as roughness and thin oxide layer. Such is often encountered in observing {110}<001> orientation in a poorly conditioned Fe-3%Si sample. In the present study, the occurrence of the pattern mis-indexing in the assignment of {110}<001> orientation in a roughly etched sample was investigated in detail. It is found that the Kikuchi pattern is frequently mis-assigned as the pattern rotated by about 30°, 60°, or 90° around its (111) pole.

1 Introduction

EBSP has been exploited as a powerful tool in determining the crystallographic orientation distribution of the materials [1-3], where a set of Kikuchi lines are captured and each of the lines are assigned and evaluated at each point of measurement. However, without proper care of the conditions, troublesome phenomenon sometimes occurs in assigning the indices to the captured set of Kikuchi lines that leads to a noisy result [4]. Many authors have intensively studied the conditions that bring about the noisy results and how they can be avoided or corrected [5,6]. Although the proper calibration of the system is one of the most important conditions necessary, the surface condition of the specimens is also playing an important role. Furthermore, some patterns of the set of Kikuchi lines are subject to mis-assignment more often than other patterns. In the present study, possible mis-assignment of {110}<001> orientation, which is hereafter referred to as Goss orientation, was investigated in a Fe-3%Si alloy. In order to enhance the mis-assignment, a roughly etched sample was used.

2 Experimental Procedure

The starting material was an Fe-3%Si sheet which was composed of primary recrystallized grains on one half of the sheet and the Goss texture was fully developed on the other half. It was processed under temperature gradient, and the detail of the sample sheet preparation is described elsewhere [7]. In order to enhance the mis-assignment frequency, the sample sheet was insufficiently etched to end up with a rough surface. Then the manual EBSP measurement was carried out at some points in the secondary recrystallized region to count the frequency of mis-assignment. Finally, the scanning measurement was also carried out and analyzed. A conventional SEM with a W-filament equipped with an ordinary EBSP system was used in the measurement.

3 Results

3.1 Kikuchi pattern on a rough surface

Even on a rough surface, there are some points that give a correct assignment. Fig.1a shows a Kikuchi pattern observed at such a point in the secondary recrystallized region. Attention should be paid to that it does not give the pattern normal to the sample surface since this pattern is a projection of the crystal orientation to the screen coordinate system, which is determined by the sample tilt angle and the geometrical relationship between the incident electron beam and the screen location. In a correct assignment, the pattern was indexed as shown in Fig. 1b. The orientation given by this pattern assignment is automatically transformed into the sample coordinate system, which consists of rolling, transverse and normal direction, to give a correct orientation, Goss orientation in the present case. In Fig. 1 is shown a (111) pole in the upper right region of the screen and some characteristic poles and triangles in the other region. At other points where mis-assignment occurs, the pattern was indexed as shown in Fig.2a-c. In these mis-assigned patterns, (111) pole is located at nearly the same position, in the upper right region of the screen. A close look at these figures identifies that the anticlockwise rotation of the correct pattern in Fig. 1b around its (111) pole by 30°, 60° and 90° leads to the mis-assigned patterns shown in Fig. 2a-c, respectively.

The (100) pole figure of the orientation given by these assignments is shown in Fig. 3, where the pole points are on a certain circle. The relationship among these poles reflects the rotational relationship among the Kikuchi patterns assigned in the screen coordinate system. The center coordinate x and the radius r of the circle are, in the RD TD coordinate system shown in the figure, obtained as Eqs. (1) and (2), respectively.

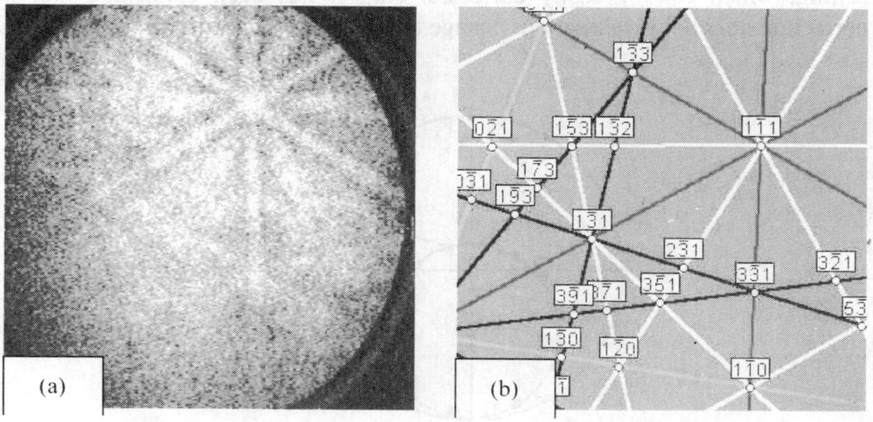

Fig. 1. Kikuchi pattern (a) and its assignment (b) that gives a correct Goss orientation measured at a point in the secondary recrystallized region.

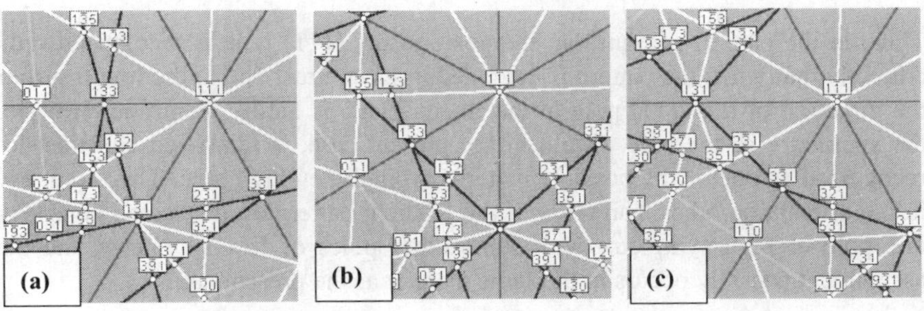

Fig. 2. The patterns of mis-assignment of Goss orientation. (a) through (c) correspond to the anticlockwise rotation of the correct pattern in Fig. 1b by 30°, 60° and 90° respectively. The center of rotation is (111) pole shown in the upper right region.

$$x = \left(\frac{1 - \tan^2 22.5^\circ}{2}, 0 \right) \tag{1}$$

$$r = \frac{1 + \tan^2 22.5^\circ}{2} \tag{2}$$

In Fig. 3, the poles of the Goss orientation show some deviation from the ideal positions. This is due to the small amount of mis-alignment that arouse in sample setting, which can be easily corrected. And it has been confirmed that the correction does not give a serious change to the present analysis.

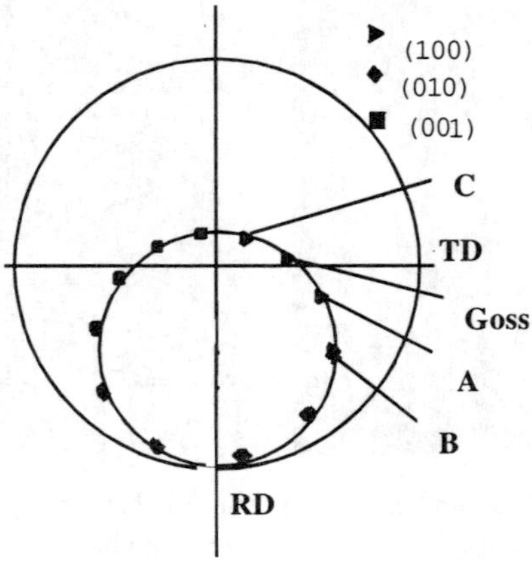

Fig. 3. (100) pole figure of the orientations indexed as shown in Figs. 1 and 2. Points A, B, and C represents the poles of the pattern (a-c), respectively, in Fig. 2.

Thus, in the pattern recognition, the position of (111) pole is recognized rather correctly, but when the pattern is degraded in its contrast due to the rough surface, the direction of the (111) pole in the screen plane is hardly identified. However, the symmetry of the pole, 6-fold in the present case, is identified and then it is recognized as one of the possible patterns around the (111) pole. Therefore, even in the orientation other than Goss, if the Kikuchi pattern on the screen consists of one pole of high symmetry with some other poles of low symmetry, the mis-assignment possibly occurs in the same manner as the present case.

3.2 OIM maps of a rough surface.

The OIM map of a rough surface reflects the above-mentioned mis-assignment and results in a noisy one. Fig. 4 shows an orientation map in the normal direction and its (100) pole figure obtained of a rough surface in the secondary recrystallized region. The map is so noisy that it is not applicable to an ordinary crystallographic orientation analysis. However, in view of the pattern mis-assignment analysis, it provides us with a sample to apply the analysis described in the previous section.

In the pole figure shown in Fig. 4b, it is clearly seen that the poles sit around the circumference of the same circle as seen in Fig. 3. Therefore, the noisy grains shown in the OIM map are derived from the mis-assignment of the Kikuchi patterns corresponding to Goss orientation as analyzed in the previous section.

Thus, it is clear that the observed area is simply a region with Goss orientation rather than is made up of the fine noisy grains as shown in Fig. 4a. In the present case, where it is secured that the area of OIM observation consists of a simple Goss texture, the noisy grains derived from the Kikuchi pattern mis-assignment can intentionally be deleted and converted into the orientation corresponding to the Goss texture. It is accomplished by converting the pattern (a), (b) and (c) in Fig. 2 into the pattern (b) in Fig. 1 or the poles A, B and C into Goss in Fig. 3. Thus converted OIM map and its (100) pole figure are shown below in Fig. 5.

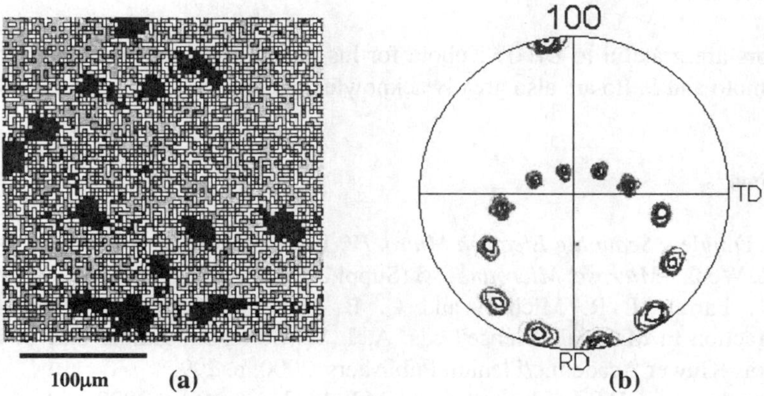

Fig. 4. OIM map (a) and its (100) pole figure (b) measured on a rough surface.

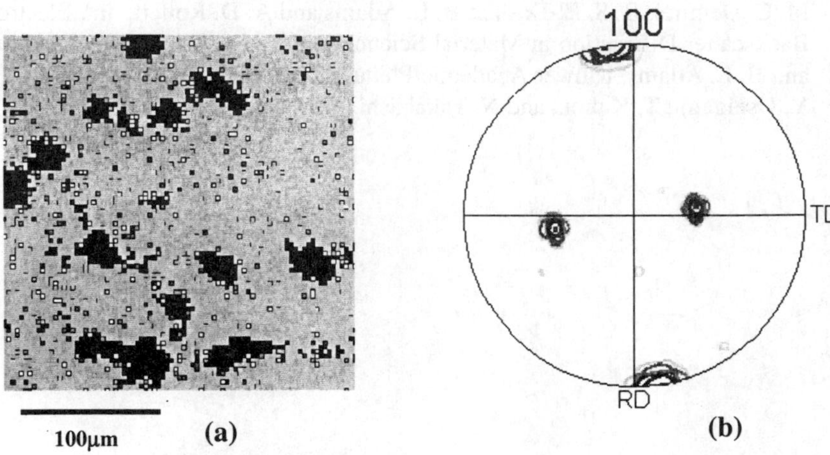

Fig. 5. The converted OIM map and its pole figure. The noisy grains are deleted and converted into Goss orientations.

Most of the noisy small grains are deleted and converted into the orientations concentrated around the Goss orientation as can be clearly seen in the figures. As described in the previous section, the Goss texture deviates somewhat from the ideal orientation due to mis-alignment of the specimens. However, it can be corrected easily, the method of which is somewhat beyond the scope of the present study. The present method of mis-assignment conversion seems to work well in case where the roughness-derived mis-assignment should be handled in analyzing the crystallographic orientations of the materials.

Acknowledgdement

The authors are grateful to Dr. T. Kubota for his encouraging suggestions. Drs. M. Hashimoto and S. Ito are also greatly acknowledged.

References

[1] D. J. Dingley, *Scanning Electron Micro. IV,* 1981, p. 273.
[2] H. R. Wenk., *Microsc. Microanal.,* 5 (Suppl 2: Proceeding), p. 228.
[3] J. K. Farrer, J. R. Michael and C. B. Carter, „Electron Backscatter Diffraction in Material Science" eds. A. J. Schwartz, M. Kumar and B. L. Adams Kluwer Academic/Plenum Publishers, 2000, p. 299.
[4] S. Suzuki, in „ OIM Academy ",TexSEM Labs. Japan office, 2000, p. 1.
[5] L. Ryde, „Channel Users Meeting Abstract Volume'' HKL Technology, 1998, p. 44.
[6] M. C. Demirel, B. S. El-Dasher, B. L. Adams and A. D. Rollett, in „Electron Backscatter Diffraction in Material Science", eds. A. J. Schwartz, M. Kumar and B. L. Adams, Kluwer Academic/Plenum Publishers, 2000, p. 65.
[7] Y. Ushigami, T. Kubota and N. Takahashi., *ISIJ Int.*, 38, 1998, p. 553.

Recrystallization and Grain Growth
Proceedings of the First Joint International Conference
Eds. G. Gottstein and D.A. Molodov, Springer-Verlag, 2001

Recrystallized Volume Fraction Analysis

J. Tarasiuk[1], P. Gerber[2], and B. Bacroix[2]

[1] Wydzial Fizyki i Techniki Jadrowej, Akademia Gorniczo-Hutnicza,
30-059 Krakow, Al. Mickiewicza 30, Poland

[2] LPMTM, CNRS, Université Paris XIII, Institut Galilée,
Av. J. B. Clément, 93430 Villetaneuse, France

Keywords: annealing, recrystallization, electron diffraction.

Abstract. Two methods, based on the analysis of differences of EBSD images in recrystallized and non-recrystallized parts of the material, are proposed. In opposite to some other methods, the proposed ones are non-destructive and one of them provides rather good error estimation. It is tested on several samples of various copper alloys presenting with different recrystallized fractions (obtained by annealing after cold rolling). The so-obtained recrystallized fractions are then compared with these determined from micro-hardness measurements. The limits of the present approach are then discussed and some other potential applications are also briefly presented.

1 Introduction

Recrystallization is a frequent technological process. Very often it follows the plastic deformation. Many material characteristics are changed during recrystallization as eg.: hardness, electrical resistance or magnetic permeability. Usually material properties depend on the ratio between recrystallized and non-recrystallized volume part of material, therefore estimation of recrystallized volume fraction is important tool for material properties analysis.

The simplest way to determine recrystallized volume fraction is to analyze microhardness plots [1,2]. This method is based on the observed fact that metals become softer during recrystallization. The main problem of this method is the material heterogeneity. The correct relation between recrystallized volume and microhardness is also defficult to find. Microhardness is measured in small spot. Usually a result is not representative for the whole sample. A typical procedure is to measure many spots and take into account an average value of microhardness. A great development of EBSD technique in last years gives new methods of recrystallized volume determination. The best one is the method based on image recognition [3,4]. A set of electron microscopy pictures of material cross section is analyzed.

Grain growth can be recognized on them. Calculating the area of growing grains one can estimate the recrystallized volume. This method is very accurate but very laborious and time-consuming.

2 Fast Method for Xv Estimation

Typical EBSD data is a set of points located in the square or hexagonal net. For each point several material characteristics are measured. For fast Xv estimation one needs only crystal lattice orientation in each point. Orientation of crystal lattice is changing form point to point. Scale of these changes depends on different parameters but the most important one is the degree of recrystallization. Material after deformation is very inhomogeneous even in small scale. Crystal lattice is strongly deformed and neighboring points can have slightly different orientation.

Fig. 1. EBSD view of deformed (a) and recrystallized (b) sample. One degree border lines are marked.

Typical deformed material is presented in Fig.1a, while the same material after recrystallization is shown in Fig.1b. Neighboring points having misorientations greater than 1° are shaded. Fig. 2 presents distributions of crystal lattice

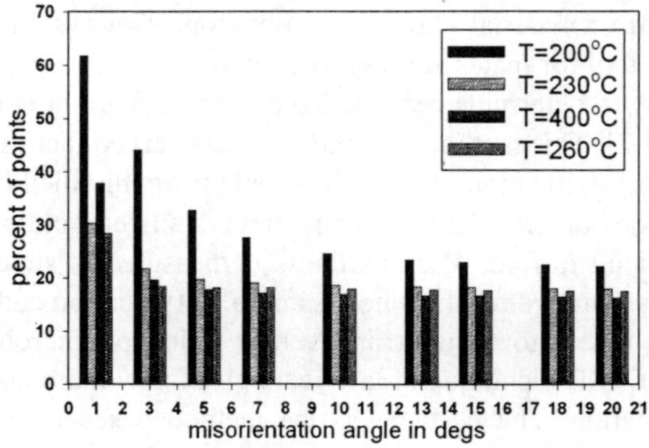

Fig. 2. Distribution of crystal lattice misorientation in various annealing temperatures.

misorientations between neighboring points for the same material in different recrystallization stages. The misorientation in degrees was marked on the horizontal axis. Vertical axis corresponds to number of points having neighbors with a given or higher misorientation angle. It means that value 34% at 3° should be interpreted that 34% of all measured points in the sample have neighbors with misorientation not lower than 3°. We can observe that for misorientation angle higher than 5° differences between samples recrystallized in various temperatures are negligible. On the other hand, for lower misorientation angle, especially for 1°, percentage of point having neighbors at this misorientation is clearly different and can be use for the estimation of recrystallized volume. The concept of Xv estimation is very simple. We should count number of points having 1° (or higher) misoriented neighbors in deformed material (N_D) and in completely recrystallized material (N_R). For any intermediate state we count number of points (n) fulfilling the described condition. Recrystallized volume fraction can be estimated as:

$$X_v = \frac{n - N_D}{N_R - N_D} \tag{1}$$

3 Precise Method for Xv Estimation

Much more precise method can be defined basing on another parameter obtained from EBSD measurements i.e., on quality index (q) [5-7]. Factor q is usually interpreted as a value linked to dislocation density. Lower values of q factor correspond to higher dislocation density and vice versa. It's obvious that for deformed state more measured points should have lower q values and relatively small number of point should have high q factor values. For completely recrystallized material q factor distribution should be inverted. Changes of quality index factor before and after recrystallization for the same sample are shown in

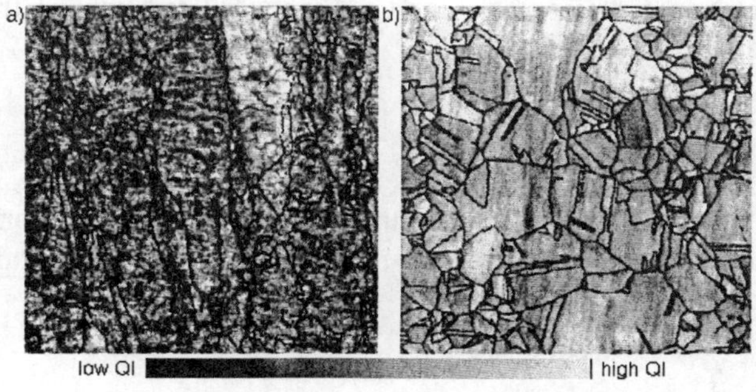

Fig. 3. EBSD view of deformed (a) and recrystallized (b) sample. Quality index is presented on gray scale.

Fig.3. A typical q factor distribution for deformed f(q) and partially recrystallized g(q) material is presented in Fig.4, while g(q) distribution was set against fully recrystallized one h(q) in Fig.5. Distributions of q factors are normalized to 100%. It means that area under distribution curves concerns all measured points and percent of points having quality index between a and b can be calculated as:

$$X_{ab} = \int_a^b f(q) \quad dq \tag{2}$$

The shaded area on Fig.4 presents points that were recrystallized. All points above f(q) and below g(q) correspond to points not existing in deformed state but appearing in a partially recrystallized state. It means that shaded area can be interpreted as minimal value of recrystallized volume fraction, what can be expressed as:

$$X_{min} = \int_{q:g(q)>f(q)} \left[g(q) - f(q) \right] \quad dq \tag{3}$$

Similar interpretation can be applied for distributions presented in Fig.5. However, in this case the shaded area corresponds to points, which are not recrystallized.

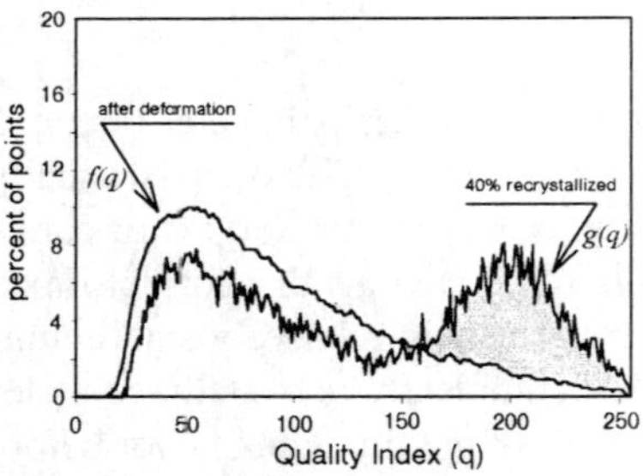

Fig. 4. Recrystallized part (see text).

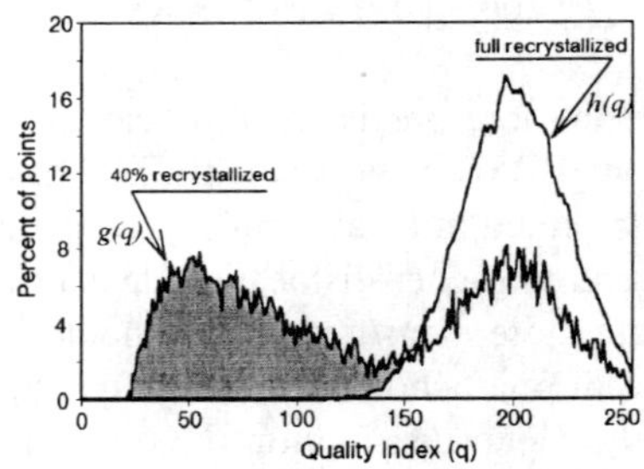

Fig. 5. Not recrystallized part (see text).

This interpretation gives as upper limit for recrystallized volume fraction, which can be formally expressed as:

$$X_{max} = 1 - \int_{q:g(q)>h(q)} \left[g(q) - h(q) \right] \quad dq \tag{4}$$

Thus, it can be said that if EBSD measurements were done for deformed, partially and fully recrystallized state then the range estimation of recrystallized volume fraction is possible:

$$X_V \in \langle X_{min} \ , \ X_{max} \rangle \tag{5}$$

The range of uncertainty is determined by lower and upper limit for X_V:

$$\Delta X = X_{max} - X_{min} \tag{6}$$

If one needs only one number to characterize the recrystallized volume fraction the possible solution is to assume that the uncertain ΔX is at least recrystallized in the same degree as whole sample; this means that global estimation X_v will be:

$$X_V = \left(X_{min} + X_{min} \cdot \Delta X\right) \tag{7}$$

4 Example of Use

Both methods were tested on the same samples set. The copper sample was cold rolled till 70% reduction. Next samples were annealed during 15 minutes in various temperatures. The samples were mechanically and electrolitically polished. EBSD measurements were performed on a Cambridge S360 scanning electron microscope with the automatic OIM software from TSL. For each temperature the recrystallized volume fraction was estimated using both described methods. The obtained results were compared with recrystallized volume fraction obtained from micro-hardness measurements made on the same samples. Comparison chart is presented in Fig. 6.

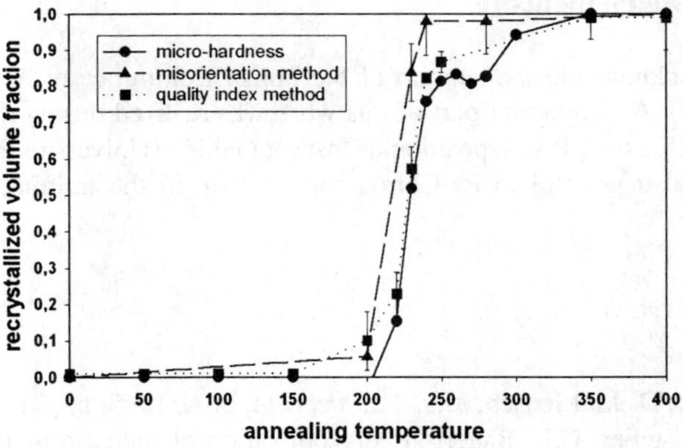

Fig. 6. X_v estimateby three different methods.

It is seen that methods based on quality index and micro-hardness are in good agreement. The new method seems to be more sensitive. The method recognizes beginning of recrystallization a little earlier than micro-hardness method. This can be understood easily: in EBSD, a fraction of 10% is much higher than the precision limit of the method whereas for low values of this fraction, the micro-hardness measurements have good chance to include only non recrystallized material. The method based on points misorientation gives X_v estimation a little shifted on the temperature scale. Possible explanation is that before or during annealing some recovering processes undergo. Recovering process decrease number of very low angle misorientations which can be "seen" by the method as earlier recrystallization.

5 Conclusions

Both presented methods give good approximation of recrystallized volume fraction. Method based on points misorientation can be used as fast test. This method can be easily applied for various EBSD measurements and gives first X_v approximation without any additional measurements. The second presented method is very accurate and can be use instead of micro-hardness measurements. Obviously this method is more complicated than micro-hardness, therefore it is pointless to use it only for X_v determination; if, however, the EBSD measurements are performed for other purposes, it seems to be reasonable to use presented method rather than micro-hardness measurements.
It is also worth to note that proposed methods are non-destructive, which may be very important in thin layer experiments.

6 Acknowledgements

J. Tarasiuk acknowledges a support of the Polish Committee of Scientific Foundation (KBN). An important part of this work was realized during the stay of one of authors (J.T.) on a PAST position at Institut Galilée (Université Paris 13). The authors are also grateful to R. Chiron for his help in the training in SEM and EBSD.

References

[1] E. Woldt, D. Juul Jensen, *Met. mat. trans. A*, 26A, 1995, p. 1717.
[2] M.S. Masteller, C.L. Bauer, "Recrystallization of metallic materials", ed. F. Haessner, Dr.Riederer Verlag, Stuttgart, 1978, p. 263.
[3] M.P. Black, R.L. Higginson, *Scripta metall. mater.*, 41, 1999, p. 125.
[4] H. Réglé, A. Miroux, Y. Raulet, L. Chapuis, in Proc. 21st International Symposium on Materials Science : Recrystallization – Fundamental Aspects and Relations to Deformation Microstructure, eds. N. Hansen et al., Risø National Laboratory, Roskilde, Denmark, 2000, p. 525.
[5] A. Miroux, PhD thesis, Université Paris13, 1999.
[6] K. Kunze, S.I. Wright, B.L. Adams, and D.J. Dingley, *Textures and Microstructures*, 20, 1993, p. 41.
[7] Th. Baudin, Y. Chastel, R. Penelle, *Rev. Mét. Sci. et Génie des Mat.*, 5, 1998, p. 611.

Author Index

Keyword Index